Carnivoran Evolution

Members of the mammalian clade Carnivora have invaded nearly every continent and ocean, evolving into bamboo-eating pandas, clam-eating walruses, and, of course, flesh-eating sabre-toothed cats. With this ecological, morphological, and taxonomic diversity and a fossil record spanning over sixty million years, Carnivora has proven to be a model clade for addressing questions of broad evolutionary significance. This volume brings together top international scientists with contributions that focus on current advances in our understanding of carnivoran relationships, ecomorphology, and macroevolutionary patterns. Topics range from the palaeoecology of the earliest fossil carnivorans to the influences of competition and constraint on diversity and biogeographic distributions. Several studies address ecomorphological convergences among carnivorans and other mammals with morphometric and Finite Element analyses, while others consider how new molecular and palaeontological data have changed our understanding of carnivoran phylogeny. Combined, these studies also illustrate the diverse suite of approaches and questions in evolutionary biology and palaeontology.

Anjali Goswami is a lecturer in palaeobiology in the Department of Genetics, Evolution and Environment and the Department of Earth Sciences at University College London. Her research focuses on large-scale patterns of evolution and development, integrating data from embryos to fossils to understand influences on morphological evolution in mammals.

Anthony ('Tony') Friscia is a faculty member at the University of California–Los Angeles, where he has won a number of teaching awards for his work teaching evolution and human anatomy to non-science majors, and where he also works to shape their science general education curriculum. His research covers small carnivores, both extant and extinct, and he is particularly interested in questions about ecomorphology and community structure.

Cambridge Studies in Morphology and Molecules: New Paradigms in Evolutionary Biology

This new Cambridge series addresses the interface between morphological and molecular studies in living and extinct organisms. Areas of coverage include evolutionary development, systematic biology, evolutionary patterns and diversity, molecular systematics, evolutionary genetics, rates of evolution, new approaches in vertebrate palaeontology, invertebrate palaeontology, palaeobotany, and studies of evolutionary functional morphology. The series invites proposals demonstrating innovative evolutionary approaches to the study of extant and extinct organisms that include some aspect of both morphological and molecular information. In recent years the conflict between "molecules vs. morphology" has given way to more open consideration of both sources of data from each side making this series especially timely.

Carnivoran Evolution: New Views on Phylogeny, Form, and Function
Edited by Anjali Goswami and Anthony Friscia

Carnivoran Evolution

New Views on Phylogeny, Form, and Function

EDITED BY

Anjali Goswami and
Anthony Friscia

CAMBRIDGE UNIVERSITY PRESS
Cambridge, New York, Melbourne, Madrid, Cape Town, Singapore,
São Paulo, Delhi, Dubai, Tokyo

Cambridge University Press
The Edinburgh Building, Cambridge CB2 8RU, UK

Published in the United Kingdom by Cambridge University Press, Cambridge, UK

www.cambridge.org
Information on this title: www.cambridge.org/9780521515290

First published 2010

Printed in the United Kingdom at the University Press, Cambridge

A catalogue record for this publication is available from the British Library

Library of Congress Cataloging in Publication data

Carnivoran evolution : new views on phylogeny, form, and function / edited by Anjali Goswami,
Anthony Friscia.
    p. cm. – (Cambridge studies in morphology and molecules)
   Includes bibliographical references and index.
   ISBN 978-0-521-51529-0 (Hardback) – ISBN 978-0-521-73586-5 (pbk.)
   1. Carnivora–Evolution. 2. Carnivora–Morphology. 3. Carnivora–Phylogeny.
I. Goswami, Anjali. II. Friscia, Anthony. III. Title. IV. Series.
   QE882.C15C365 2010
   599.7'138–dc22

2010019261

ISBN 978-0-521-51529-0 Hardback
ISBN 978-0-521-73586-5 Paperback

Contents

Contributors   *page* vii
Preface   ix
Acknowledgements   xii

1   Introduction to Carnivora   I
    ANJALI GOSWAMI

2   Phylogeny of the Carnivora and Carnivoramorpha,
    and the use of the fossil record to enhance understanding
    of evolutionary transformations   25
    JOHN J. FLYNN, JOHN A. FINARELLI,
    AND MICHELLE SPAULDING

3   Phylogeny of the Viverridae and 'Viverrid-like' feliforms   64
    GERALDINE VERON

4   Molecular and morphological evidence for Ailuridae
    and a review of its genera   92
    MICHAEL MORLO AND STÉPHANE PEIGNÉ

5   The influence of character correlations on phylogenetic
    analyses: a case study of the carnivoran cranium   141
    ANJALI GOSWAMI AND P. DAVID POLLY

6   What's the difference? A multiphasic allometric analysis
    of fossil and living lions   165
    MATTHEW H. BENOIT

7   Evolution in Carnivora: identifying a morphological bias   189
    JILL A. HOLLIDAY

8   The biogeography of carnivore ecomorphology   225
    LARS WERDELIN AND GINA D. WESLEY-HUNT

9   Comparative ecomorphology and biogeography of Herpestidae
    and Viverridae (Carnivora) in Africa and Asia   246
    GINA D. WESLEY-HUNT, REIHANEH DEHGHANI,
    AND LARS WERDELIN

10 Ecomorphological analysis of carnivore guilds
in the Eocene through Miocene of Laurasia   269
MICHAEL MORLO, GREGG F. GUNNELL, AND DORIS NAGEL

11 Ecomorphology of North American Eocene carnivores:
evidence for competition between Carnivorans and Creodonts   311
ANTHONY R. FRISCIA AND BLAIRE VAN VALKENBURGH

12 Morphometric analysis of cranial morphology in pinnipeds
(Mammalia, Carnivora): convergence, ecology, ontogeny,
and dimorphism   342
KATRINA E. JONES AND ANJALI GOSWAMI

13 Tiptoeing through the trophics: geographic variation in
carnivoran locomotor ecomorphology in relation to environment   374
P. DAVID POLLY

14 Interpreting sabretooth cat (Carnivora; Felidae; Machairodontinae)
postcranial morphology in light of scaling patterns in felids   411
MARGARET E. LEWIS AND MICHAEL R. LAGUE

15 Cranial mechanics of mammalian carnivores: recent
advances using a finite element approach   466
STEPHEN WROE

Index   486

The colour plates are situated between 274 and 275

Contributors

Matthew H. Benoit, Joint Sciences Department, The Claremont Colleges, Claremont, California, USA

Reihaneh Dehghani, Department of Zoology, Stockholm University, Stockholm, Sweden

John A. Finarelli, Michigan Society of Fellows and Department of Geological Sciences, University of Michigan, Ann Arbor, Michigan, USA

John J. Flynn, Division of Paleontology, American Museum of Natural History, New York, New York, USA

Anthony R. Friscia, Departments of Undergraduate Education Initiatives and Physiological Science, University of California–Los Angeles, California, USA

Anjali Goswami, Department of Genetics, Evolution, and Environment and Department of Earth Sciences, University College London, London, UK

Gregg F. Gunnell, Museum of Paleontology, University of Michigan, Ann Arbor, Michigan, USA

Jill A. Holliday, Department of Biological Science, Florida State University, Tallahassee, Florida, USA

Katrina E. Jones, Department of Earth Sciences, University of Cambridge, Cambridge, UK

Michael R. Lague, NAMS-Biology, The Richard Stockton College of New Jersey, Pomona, New Jersey, USA

Margaret E. Lewis, NAMS-Biology, The Richard Stockton College of New Jersey, Pomona, New Jersey, USA

Michael Morlo, Forschungsinstitut Senckenberg, Frankfurt am Main, Germany

Doris Nagel, Department of Palaeontology, Universität Wien, Wien, Austria

Stéphane Peigné, Centre de Recherche sur la Paléobiodiversité et Paléoenvironnements, Départment Histoire de la Terre, Muséum National d'Histoire Naturelle, Paris, France

P. David Polly, Department of Geological Sciences, Indiana University, Bloomington, Indiana, USA

Michelle Spaulding, Department of Earth and Environmental Sciences, Columbia University, New York, New York, USA

Blaire Van Valkenburgh, Department of Ecology and Evolutionary Biology, University of California–Los Angeles, California, USA

Geraldine Veron, Département Systématique et Evolution, Muséum National d'Histoire Naturelle, Paris, France

Lars Werdelin, Department of Palaeozoology, Swedish Museum of Natural History, Stockholm, Sweden

Gina D. Wesley-Hunt, Biology Department, Montgomery College, Rockville, Maryland, USA

Stephen Wroe, Evolution and Ecology Research Centre, School of Biological, Earth and Environmental Sciences, University of New South Wales, Sydney, Australia

Preface

With its high taxonomic, morphological, and ecological diversity and excellent fossil record, the placental mammal order Carnivora has proven to be a model group for addressing questions of large evolutionary significance. Recent work has resulted in a well-resolved phylogeny of extant taxa, as well as for many extinct clades, allowing for rigorous analysis of a wide range of evolutionary questions. Although the order is named after its meat-eating members, the dietary breadth of living carnivorans (members of the order Carnivora) extends from frugivorous to insectivorous taxa, durophagous taxa, as well as the hypercarnivorous taxa that are usually associated with the group. Carnivoran locomotor diversity is also remarkable among mammals, with fully aquatic, semi-aquatic, arboreal, terrestrial, and fossorial taxa. Recent studies have shown that this diversity extends to their early fossil representatives. Multiple ecological and morphological convergences of carnivorans and distantly related clades, including the extinct creodonts and extant and extinct carnivorous marsupials, also strengthen the utility of carnivorans for comparative ecomorphological and biomechanical studies. This volume focuses not only on the current advances in our understanding of mammalian carnivoran evolution, but especially on how carnivorans are being used as a model clade for testing new methodologies and addressing fundamental issues in palaeontology, which can ultimately be applied to clades with poorer fossil records.

The subtitle of this volume – 'New Views on Phylogeny, Form, and Function' – while being pleasantly alliterative, highlights some of the most exciting fields of study in evolutionary biology and palaeontology to which carnivorans have lent themselves. In recent years, mammalian carnivorans have been the focus of extensive phylogenetic analyses, both molecular and morphological, and incorporating both extant and fossil taxa, which have resolved many long-standing issues in carnivoran relationships. Flynn *et al.* (Chapter 2) provide an overview of the state of overall carnivoran phylogeny, erect a new clade, and demonstrate some of the patterns that can be studied using a phylogenetic framework. Contributions by Veron (Chapter 3) and Morlo and Peigné (Chapter 4) look closely at some of the more interesting evolutionary problems within Carnivora. Veron tackles the viverrids, a diverse Old World group that has undergone extensive revision in recent years. Morlo and Peigné look closely at the evolution of the red panda and its relatives in the Ailuridae. Although represented by only one taxon today, the fossil history of this family reveals more diversity in taxonomy and form.

All three of the phylogenetic studies use the latest in techniques, including combining molecular and morphological data, and especially data from fossil specimens, in their analyses. Goswami and Polly (Chapter 5) use carnivorans to investigate a potential issue in phylogenetic methodologies – the correlated evolution of characters. Using simulations and empirical data, they test whether character correlations influence discrete character states over evolutionary time scales, and assess methods to identify these correlations in existing analyses. Benoit (Chapter 6) addresses another issue in phylogeny and taxonomy – how to identify species. Using lions as an example, Benoit uses sophisticated analyses of allometric trajectories to assess the strength of characters used to justify the American lion as a separate species.

With this solid taxonomic and phylogenetic framework and their excellent fossil record, carnivorans easily lend themselves to interesting studies of macroevolutionary patterns and more fundamental issues of influences on diversity. Holliday (Chapter 7) uses a phylogenetic framework to assess biases in morphological evolution, testing whether hypercarnivory tends to limit further morphologic change and testing hypotheses of a macroevolutionary ratchet that leads hypercarnivores into an evolutionary 'dead end'. Both Werdelin and Wesley-Hunt (Chapter 8) and Wesley-Hunt et al. (Chapter 9) address the biogeographic distribution of carnivoran ecomorphologies. The broader study of Werdelin and Wesley-Hunt looks at morphological disparity across the entire order and compares disparity among families and among continents. Their companion piece, Wesley-Hunt et al., focuses on just two of the more widely distributed families, civets and mongooses, to investigate how they divide up community ecospace in different regions.

Ecologies within communities can be traced across time as well as space, and both Morlo et al. (Chapter 10) and Friscia and Van Valkenburgh (Chapter 11) use this approach to look at the earliest carnivorans. In the Paleogene, carnivorans shared the meat-eating niche with the extinct creodonts, and how this temporal overlap affected their respective ecologies, as well as the history of their diversity, has long been a topic of research and debate. The broader, global study of Morlo et al. addresses how ecospace and guild structure vary temporally and spatially across the history of carnivorous mammals, as taxonomic membership varies. Friscia and Van Valkenburgh test the more specific question of whether creodonts were actively replaced by carnivorans, as the former decrease in diversity during the Eocene of North America.

While most studies of carnivorans focus exclusively on the terrestrial clades, the pinnipeds represent one of the most extraordinary transitions in mammal evolution. The study of Jones and Goswami (Chapter 12) bridges form and function, by using pinnipeds as a case study for investigating reproductive and ecological influences on cranial morphology, as well as identifying types of convergence across the three extant clades of pinnipeds.

Polly (Chapter 13) returns to the terrestrial realm, using sophisticated analyses of biomes and limb morphology to assess the relationship between locomotory styles and the environment in North American carnivorans and exploring its potential as a tool to reconstruct past environments. Lewis and Lague (Chapter 14) follow on the postcranial theme, comparing limb morphology in machairodontid sabretooth felids to modern felids to assess whether they employed similar locomotory and hunting styles, or were as distinct in the postcranium as they were in cranial and dental morphology.

Lastly, Wroe (Chapter 15) provides an overview of the latest in 3D imaging and finite-element techniques and presents several comparisons of skull mechanics in placental and mammalian carnivores. Finite element analysis has proven to be a unique and fascinating tool for reconstructing the mechanical capabilities of extinct morphologies, and Wroe's chapter details how these methods reveal surprising differences between superficially similar carnivores.

Few vertebrate groups can claim such a diversity of topics that can be rigorously tested with fossil and extant taxa as carnivorans can. Advances in development, genetics, phylogenetics, morphometrics, finite element analysis, and 3D imaging have all been extensively applied to carnivorans, keeping this clade at the forefront of research in evolutionary biology and palaeontology. We hope that this volume will serve not only as an overview of recent advances in carnivoran evolution, but also as a methodological guide for studying large-scale patterns in the fossil record and for addressing fundamental questions in evolutionary biology with morphological and palaeontological data.

Acknowledgements

As with most things in palaeontology, this book started with beer. At the American Society of Mammalogists' meeting in Amherst in June 2006, we were at the lobster bake, talking over our favourite beverage, and discussing how much we loved the taxa we studied – carnivorans. Through the course of our conversation we realised that there had never been a symposium on mammalian carnivores at the Society of Vertebrate Paleontology (SVP) meetings, and that the classic Gittleman volumes on carnivore behaviour, ecology, and evolution were already over 10 years old and quite out of date with respect to phylogeny and quantitative, macroevolutionary analyses (also with quite a different focus than a dedicated volume on evolution). Sure, the carnivore researchers tended to cluster together at the meetings, but we hadn't had a formal meeting of minds. A few emails, and a proposal to the SVP Program Committee a few months later, and our symposium (with the same name as this volume) was a reality at the 2007 SVP meeting in Austin, TX. So the first people we have to thank are Jason Head and the rest of the Program Committee of that meeting for allowing us to have that gathering which ultimately led to this volume. Many of the talks that were part of that symposium made it into this volume as chapters, and we thank all the participants of that day. It was especially gratifying for us all to gather over a meal afterwards and discuss what we had just shared with each other. Carnivorans are often called a 'charismatic' group, and the same can be said of the people who study them.

The creation of this volume took significantly longer than the year to get the symposium together, and we thank the contributors for their patience throughout this process, and for their continued faith that two, relatively young, researchers could pull this together. We believe that our contributors represent the full range of carnivoran workers, from the well-known heavyweights in the field to the recent graduates who are bringing new approaches, methods, and enthusiasm to the study of carnivoran evolution, making this quite a unique volume. Of course, all of the contributions were subject to rigorous scrutiny, and we have many people to thank for reviewing the contributed manuscripts. Jill Holliday, Stéphane Peigné, Geraldine Veron, Xiaoming Wang, Marcelo Sánchez-Villagra, John Finarelli, Alistair McGowan, Margaret Lewis, K. Elizabeth Townsend, Blaire Van Valkenburgh, Graham Slater, Lars Werdelin,

Annalisa Berta, Stephen Wroe, Olaf Bininda-Emonds, Rob Asher, Eleanor Weston, Norberto Giannini, Matt Benoit, P. David Polly, John Damuth, Vera Weisbecker, Laura Porro, and Alistair Crosby all provided an immense service in improving the quality of each chapter and the volume as a whole and we thank them all for the time and care they put into their reviews. John Finarelli in particular was an endlessly reliable reviewer and source for ideas, opinions on issues, and last-minute fact checking. We also of course have to thank our mentors, John Flynn and Blaire Van Valkenburgh, respectively, who have inspired us not only to pursue research in carnivorans, but to take a deeper view of evolution patterns and processes. A large part of the great advances in the study of carnivoran evolution discussed in this book are a reflection of their decades of work on this topic, and certainly we have them to thank directly for any of our own success.

Lastly, we would like to thank Rob Asher and Russell L. Ciochon for inviting us to submit our proposal for a volume in the series *Morphology and Molecules* at Cambridge University Press, to Dominic Lewis for guiding us through the proposal process and the early stages of preparation, and to Rachel Eley, our Assistant Editor, who answered all of our many questions as we came to the end of this project. It has been a long and occasionally trying experience, but we are very proud of this volume and grateful to all of the people who helped us get to this point.

Anjali Goswami and Tony Friscia

1

Introduction to Carnivora

ANJALI GOSWAMI

Why Carnivora?

The placental mammal order Carnivora encompasses many charismatic taxa, from dogs and cats to bears, otters, hyaenas, and seals. Perhaps more than any other mammalian clade, carnivorans are a source of fascination for humans, partially due to our intimate observation of the domesticated species that reside in many of our own homes. Beyond our quirky cats and loyal dogs, however, carnivorans have long and often been the subject of a variety of studies and documentaries of natural history concerning behaviour, ecology, and evolution, and for many good reasons. With over 260 living species, Carnivora is one of the most species-rich clades of mammals. It should be noted that the term 'carnivoran' is a phylogenetic classification, in contrast to 'carnivore', an ecological classification describing any meat-eater.

Evolutionarily, Carnivora is divided into two major branches (Flynn *et al.*, this volume, Chapter 2, Figure 2.2): Feliformia (including cats, linsangs, civets, mongooses, fossas, falanoucs, and hyaenas; Figure 1.1) and Caniformia (encompassing dogs, bears, seals, sea lions, walruses, the red panda, raccoons, skunks, weasels, badgers, otters, and wolverines; Figure 1.2) (Wozencraft, 2005; Myers *et al.*, 2008). As that list suggests, this taxonomic diversity is well matched by their ecological breadth. While the name Carnivora usually conjures up images of tigers and wolves, carnivorans range in diet from pure carnivores to species that specialise on fruit, leaves, and insects, as well as the full spectrum of mixed diets; carnivorans are represented by omnivorous bears, frugivorous raccoons, and even insectivorous hyaenas. Even better for students of evolution, many carnivoran families have given rise to multiple different ecomorphs. This ecological diversity is possibly best exemplified by the species-poor but ecologically diverse bears, which have evolved folivorous, frugivorous, omnivorous, insectivorous, and hypercarnivorous forms (Wozencraft, 2005). In fact, as discussed by Holliday (this volume, Chapter 7), the hypercarnivorous forms

Carnivoran Evolution: New Views on Phylogeny, Form, and Function, ed. A. Goswami and A. Friscia. Published by Cambridge University Press. © Cambridge University Press 2010.

Figure 1.1 Feliformia. A, Felidae; *Panthera leo*, lion; B, Felidae: *Smilodon fatalis*, sabre-toothed cat; C, Viverridae: *Arctictis binturong*, binturong; D, Hyaenidae: *Crocuta crocuta*, spotted hyaena; E, Herpestidae: *Mungos mungo*, banded mongoose; F, Eupleridae: *Cryptoprocta ferox*, fossa. Photo credits: A, D, A. Goswami; B, P. Goswami; C, Klaas Lingbeek-van Kranen, iStockphoto®; E, N. Smit, iStockphoto®; F, J. Weston, iStockphoto®.

Figure 1.2 Caniformia. A, Mustelidae: *Lontra canadensis*, northern river otter;
B, Procyonidae: *Nasua narica*, coati; C, Ailuridae: *Ailurus fulgens*, red panda;
D, Mephitidae: *Mephitis mephitis*, striped skunk; E, Odobenidae: *Odobenus rosmarus*,
walrus; F, Otariidae: *Zalophus californianus*, California sea lion; G, Ursidae: *Ursus
arctos*, brown bear; H, Canidae: *Vulpes vulpes*, red fox. Photo credits: A, F, H,
FreeDigitalPhotos.net; B, G. Brzezinski, iStockphoto®; C, S. Peigné; D, J. Coleman,
iStockphoto®; E, T. Shieh, iStockphoto®; G, K. Livingston, iStockphoto®.

that we usually think of as representing Carnivora may well be the least successful members of the clade.

Carnivoran diversity does not end with diet, as carnivorans display a broad range in styles of locomotion, including cursorial, arboreal, fossorial, and aquatic species. Carnivorans inhabit all of the world's oceans and five of the continents, with only Australia and Antarctica lacking native terrestrial carnivorans, prior to introduction by humans. However, aquatic members of the clade have colonised those regions as well. The semi-aquatic to fully aquatic species, including otters, walruses, sea lions, and seals, have evolved systems to extract molluscs from their shells, filter krill, and mate on sea ice (Myers *et al.*, 2008). The deepest diving carnivoran, the northern elephant seal, can reach depths of over a kilometre, while its distant relative, the cheetah, can cross that distance on land in less than a minute. Arboreal forms are no less specialised, with prehensile tails evolving multiple times in carnivoran evolution, including in living kinkajous and binturongs, as well as possibly in some fossil forms (Flynn *et al.*, this volume, Chapter 2).

This last point highlights one of the primary reasons that research into carnivoran evolution is such an exciting field of scientific research: in addition to their remarkable living diversity, carnivorans have an excellent fossil record, spanning almost the whole of the Cenozoic (Flynn and Wesley-Hunt, 2005). We know of nearly three times as many extinct carnivoran genera as extant genera (approximately 355 and 129, respectively; McKenna and Bell, 1997). The precise origins of Carnivora are poorly understood, but one possibility is that they evolved from a *Cimolestes*-like ancestor, a late Cretaceous–early Paleocene insectivorous mammal. The earliest known stem carnivorans, or the first carnivoramorphans, as defined by Wyss and Flynn (1993), are from the earliest Paleocene (65–61 Mya) of North America (Fox and Youzwyshyn, 1994). These stem carnivorans are very different from the forms seen today, but they share with living carnivorans a characteristic dental modification called carnassials. Carnassials are the blade-like upper fourth premolar and lower first molar, which shear against each other for enhanced meat-slicing ability. While some of the frugivorous and folivorous carnivorans have subsequently modified their carnassials, it is the key character uniting crown group and stem carnivorans in Carnivoramorpha (Wyss and Flynn, 1993; Flynn and Wesley-Hunt, 2005; Flynn *et al.*, this volume, Chapter 2).

The relationship of Carnivora to other placental mammals

The recent proliferation of molecular phylogenetics has vastly changed our understanding of carnivoran relationships, both to other mammals and to each other. Recent studies divide placental mammals into four superorders.

Carnivora falls within the superorder Laurasiatheria, which also includes the orders Perissodactyla (horses, tapirs, and rhinoceroses), Cetartiodactyla (whales and even-toed ungulates), Chiroptera (bats), Soricomorpha (shrews and moles), and Pholidota (pangolins). The other placental mammal superorders are Euarchontaglires (primates, rodents, rabbits, tree shrews, and colugos), Afrotheria (elephants, sea cows, hyraxes, aardvarks, tenrecs, and sengis), and Xenarthra (sloths, armadillos, and anteaters). Together, Laurasiatheria and Euarchontaglires form the clade Boreoeutheria, reflecting their hypothesised northern hemisphere origin (Murphy *et al.*, 2001, 2007). Among the most surprising results of these analyses is the possibility that pangolins, scaly anteater-like mammals, are the closest living relatives to Carnivora (Murphy *et al.*, 2001).

Introduction to the major carnivoran clades and their fossil record

Stem carnivorans

The earliest fossil representatives of the living families of Carnivora appeared in the late Eocene. However, as noted above, there are many earlier fossils with the diagnostic carnassial teeth that represent the stem leading to the living families. There are two major groups of stem carnivorans: Viverravidae (not to be confused with civets in the family Viverridae) and Miacoidea. It was previously thought that feliforms evolved from viverravids, and caniforms from miacoids. However, many new well-preserved fossils of Paleocene (65–55 Mya) and Eocene (55–34 Mya) carnivorans have resolved much of the early history of the group (Wesley-Hunt and Flynn, 2005).

Viverravids (Figure 1.3) are probably the most basal group of Carnivoramorpha and were small- to medium-sized terrestrial animals that incorporated insects as a large part of their diet (Flynn *et al.*, this volume, Chapter 2). Miacoidea is a group of terrestrial and arboreal early carnivoramorphan species that appear to represent a series of intermediate forms between the more basal viverravids and the true (=crown clade) carnivorans. New fossils support a single origin of the living carnivoran families from 'Miacoidea', which suggests that the living families may have separated almost 15 million years later than previously thought, although the precise interrelationships are still contentious (Wesley-Hunt and Flynn, 2005; Polly *et al.*, 2006; Flynn *et al.*, this volume).

By the late Paleocene (61–55 Mya), viverravids and miacoids are known from Asia and North America, spreading to Europe by the early Eocene (55–49 Mya). Both Viverravidae and 'Miacoidea' were extinct by the late Eocene (37–34 Mya). Also in the late Eocene (37–34 Mya), the first representatives of several

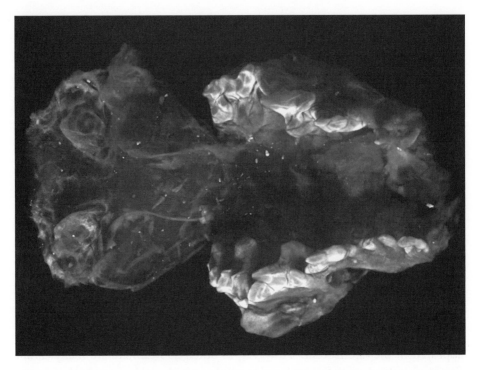

Figure 1.3 Viverravidae. Ventral view of a computerised microtomography rendering of the cranium of *Viverravus acutus* (UM 67326) from the Early Eocene of the Bighorn Basin, Wyoming (Polly *et al.*, 2006). Rendering by G. R. Davis, Queen Mary, University of London, using Drishti Volume Exploration and Presentation Tool (A. Limaye, Australia National University).

crown group carnivoran families (Canidae, Mustelidae, Ursidae, Amphicyonidae, and Nimravidae) appear on the northern continents, discussed in more detail below; however, modern feliform families do not appear until the Oligocene (34–24 Mya). Carnivorans do not invade the southern continents (Africa and South America) until the Miocene (24–5 Mya). While all caniform families have a global distribution, feliforms, except for Nimravidae and Felidae, are largely restricted to the Old World throughout their history (Flynn and Wesley-Hunt, 2005).

Feliformia (Figure 1.1)

Feliforms are often thought of as less diverse than caniform carnivorans, although there is little support for this view in terms of modern taxonomic diversity; there are 56 extant feliform genera and 73 extant caniform genera (Myers *et al.*, 2008). However, when extinct genera are included, caniforms far outnumber feliforms, with 244 extinct caniform genera to 76 extinct feliform genera, if nimravids are included with feliforms (McKenna and Bell, 1997). This difference in taxonomic diversity is often coupled with the idea that feliforms are also ecologically and morphologically less diverse, perhaps driven by the observation that domestic cat breeds have a more limited range of variation than domestic dog breeds (Wayne, 1986). However, while feliforms lack the ecological and morphological breadth represented by some caniforms, particularly pinnipeds, there is much unappreciated diversity in feliforms.

Felidae

The most speciose feliform clade is, perhaps surprisingly, Felidae, with 41 extinct and extant genera (McKenna and Bell, 1997; Myers *et al.*, 2008). Felids are generally hypercarnivorous, with some of their distinguishing features including a short, blunt rostrum, retractable claws, well developed carnassials, and reduction of the postcarnassial dentition. The earliest records of felids are from the Oligocene of Eurasia, but in the Miocene, felids expand their range to include every continent other than the isolated Australia, Antarctica, and South America (which they quickly invaded following the formation of the isthmus of Panama in the late Pliocene) (Marshall *et al.*, 1982; McKenna and Bell, 1997; Flynn and Wesley-Hunt, 2005).

Extant felids (Figure 1.1a) are perhaps some of the rarest and most captivating of animals, being generally solitary, stalking predators with exquisite camouflage. Extinct felids are comparably fascinating, including some of the most popular fossils, machairodontine sabre-toothed cats (Figure 1.1b). However, felid diversity is often dismissed with the observation that lions are essentially scaled-up house cats (Wayne, 1986; Sears *et al.*, 2007). While there is certainly some truth to this generalisation, Benoit (this volume, Chapter 6) and Lewis and Lague (this volume, Chapter 14) demonstrate that felid allometry is not as straightforward as previously thought.

Viverridae

After Felidae, the most taxonomically diverse feliforms are the much-revised Viverridae, with 28 recognised genera (McKenna and Bell, 1997), even after removal of taxa now incorporated in the families Nandiniidae (West African palm civet), Prionodontidae (Asian linsangs), Herpestidae (mongooses), and Eupleridae (Malagasy carnivorans), as discussed by Veron (this volume, Chapter 3). As its long history as a wastebasket taxon suggests, Viverridae is a group of relatively generalised, medium-sized carnivorans restricted to the Old World. Civets have well-developed carnassials and long, pointed snouts, and one of their most distinguishing characters is the presence of a perineal gland.

Most are arboreal and nocturnal, feeding on a variety of small vertebrates and invertebrates, but there are some interesting specialisations within this clade. Many of the palm civets (Paraxodurinae) are primarily frugivorous and highly arboreal, and, as noted above, one member of this clade, *Arcticis binturong* (Figure 1.1c), has evolved a prehensile tail (Myers *et al.*, 2008). Viverrids have a lengthy fossil record, first appearing in Eurasia in the Oligocene before spreading into Africa in the Miocene (McKenna and Bell, 1997; Flynn and Wesley-Hunt, 2005).

Hyaenidae

Hyaenidae is the next most speciose feliform clade, with 20 extinct genera representing a far greater taxonomic diversity than the 4 extant genera, all of which are now confined to Africa and South to Southwest Asia (McKenna and Bell, 1997; Myers *et al.*, 2008). The first hyaenids appear in the early Miocene of Europe and Africa, quickly moving to Asia by the middle Miocene, and briefly invading North America in the late Pliocene (McKenna and Bell, 1997; Flynn and Wesley-Hunt, 2005). While many of the extinct hyaenids were bone-crackers, similar to the better-known modern species (Figure 1.1d), some converge on canid morphologies, possibly occupying a similar niche to that of modern dogs in the Miocene and Pliocene of Eurasia and Africa (Werdelin, 1996b; Van Valkenburgh, 2007). The only hyaenid to make it to North America, *Chasmaporthetes*, was one of these 'hunting hyaenas', with a more canid-like stance and dentition well adapted for cursoriality and pursuit predation (Berta, 1981).

One of the most unusual living feliforms is a hyaenid, *Proteles cristata*, the aardwolf. In contrast to the massive molars observed in most hyaenids, the aardwolf has drastically reduced their postcanine dentition to a variable number of peg-like premolars and molars. Aardwolves eat termites almost exclusively, a specialisation that is reflected in its reduced dentition, broad tongue, sticky saliva, and small body size (Wozencraft, 2005). There is disagreement on the divergence date of aardwolfs from the other modern hyaena species, with estimates ranging from the middle to late Miocene (Werdelin and Solounias, 1991; Koepfli *et al.*, 2006), but it certainly represents an extreme shift in ecology and morphology from its hypercarnivorous ancestors.

Herpestidae

Herpestidae, a clade of relatively small and primarily African feliforms, has 14 extant and only a single extinct genus. Most herpestids are carnivorous, feeding on a variety of small vertebrates and insects, although they are often associated with the ability of some species to kill snakes. The social mongooses

(Figure 1.1e), several closely related genera of herpestids, are well known for having evolved complex social systems, most famously *Suricata suricatta*, the meerkat (Flynn *et al.*, 2005; Myers *et al.*, 2008), although many other herpestids are solitary. Some species, including meerkats, are semi-fossorial, while others are semi-aquatic, such as *Atilix paludinosis*, the marsh mongoose. For the most part, herpestids are terrestrial and relatively generalised, although agile, carnivores (Myers *et al.*, 2008).

With their similarly long, pointed snouts, herpestids were originally considered a subclade of Viverridae. In fact, herpestids are most closely related to the Malagasy carnivorans and to hyaenids (Veron, Chapter 3; Flynn *et al.*, Chapter 2). Herpestids first appear in the early Miocene of Europe and Africa, moving into Asia by the late Miocene (McKenna and Bell, 1997; Flynn and Wesley-Hunt, 2005).

Eupleridae

The Malagasy carnivorans, Eupleridae, include several genera that were originally included in Herpestidae, commonly described as Malagasy mongooses, as well as three taxa that were included in Viverridae (Myers *et al.*, 2008). The cat-like *Cryptoprocta ferox* (Figure 1.1f) and the vermivorous and insectivorous *Eupleres goudotii* are some of the unusual forms that have evolved during this clade's long isolation on Madagascar, and their divergence from Herpestidae has been estimated to around 18–24 million years ago (Yoder *et al.*, 2003).

Nandiniidae

The most basal extant feliform clade is also the smallest, Nandiniidae. With only a single species, this taxon was previously, unsurprisingly, placed in Viverridae (Veron, this volume, Chapter 3). Recent molecular analyses confirm its basal position among extant feliform clades, although its primitive bullar and basicranial morphology had already hinted to many workers that it did not belong with viverrids (Hunt, 1987). Neither Nandiniidae nor Eupleridae have a pre-Recent fossil record.

Nimravidae

Nimravidae is a wholly extinct clade of large, cat-like predators that have often been identified as basal feliforms, but alternatively as stem carnivorans or stem caniforms (Flynn *et al.*, this volume, Chapter 2). Commonly called 'false sabre-toothed cats', nimravids are distinguished by their long, laterally compressed upper canines, mandibular flange, and reduced or absent m2, similar to sabre-toothed felids. With approximately nine genera, nimravids are well represented in the fossil record from the late Eocene in Asia and North America, invading Europe by the Oligocene (Bryant, 1991; McKenna and Bell, 1997; Peigné, 2003;

Flynn and Wesley-Hunt, 2005). Nimravids persist in these three regions until the late Oligocene.

Barbourofelines, another clade of sabre-toothed forms with approximately five named genera, have recently been removed from Nimravidae, with suggestions that they are more closely related to Felidae (Morlo *et al.*, 2004). These specialised carnivorans are geographically widespread but temporally restricted to the Miocene. They first appear in Africa and Europe in the early Miocene, but spread to Asia and North America before going extinct at the end of the Miocene (Bryant, 1991).

Caniformia (Figure 1.2)

Turning to the other major branch of Carnivora, we encounter a few clades that are far more speciose than their feliform relatives.

Mustelidae

Mustelidae is the most taxonomically diverse carnivoran family-level clade, with 107 recognised genera, even after the exclusion of Mephitidae (skunks and stink badgers). Mustelidae presently includes many familiar and fascinating animals, including otters (Figure 1.2a), sea otters, martens, weasels, ferrets, polecats, honey badgers, wolverines, and New and Old World badgers (Myers *et al.*, 2008). Mustelids are well-represented in the fossil record from the early Oligocene, with at least 84 extinct genera. They first appear in Eurasia, spreading to North America and Africa by the late Oligocene or early Miocene (Wolsan, 1993; McKenna and Bell, 1997). Unlike raccoons, mustelids do not enter South America prior to the formation of the Panamanian land bridge in the late Pliocene (Marshall *et al.*, 1982). Studies of mustelid evolution suggest that most of their diversification has occurred in Eurasia, with multiple invasions of the other continents from that region (Koepfli *et al.*, 2008).

While most mustelids are small- to medium-sized animals, there are several large species that reach 30–40 kg, and the clade displays an order of magnitude range in body size (Finarelli and Flynn, 2006). Mustelids are generally short-faced and elongate, with short limbs. They have successfully invaded arboreal, riverine, and marine habitats, but few mustelids deviate from a carnivorous diet. They do, however, demonstrate remarkable specialisations in the acquisition and consumption of prey, with one of the most interesting being sea otters, which regularly use rocks to break open shells of their prey (Myers *et al.*, 2008).

Relationships among mustelids and other arctoid caniforms have been revised extensively in recent years, as discussed by Flynn *et al.* (Chapter 2). Mephitidae, Procyonidae, and Phocidae have all been suggested as either

subclades within Mustelidae or close relatives, although several of these controversies have been settled with new molecular data (Flynn and Nedbal, 1998;
Flynn *et al.*, 2000, 2005; Koepfli *et al.*, 2008) or with total evidence analyses that
incorporate the many problematic fossil taxa previously described as basal
arctoids or basal musteloids (Finarelli, 2008).

Procyonidae

The closest relatives to Mustelidae appear to be raccoons (Procyonidae). This
clade consists of approximately 18 genera, although only 6 are extant and the
affinities of the fossil forms are highly disputed. The earliest uncontested record
of procyonids comes from the early Miocene of Europe, with their appearance
in North America soon after. In the late Miocene, procyonids invade South
America, where they are one of the few mammalian clades to invade that island
continent prior to the closure of the Panamanian isthmus (Marshall *et al.*, 1982).
While the extinct Simocyoninae, from the Miocene of North America,
Europe, and Asia, have been placed in Procyonidae, some argue for a closer
relationship to Ailuridae (Morlo and Peigné, this volume, Chapter 4), which
suggests that true procyonids never colonised Asia. Procyonids disappear from
Europe by the end of the Miocene, after which they are strictly a New World
clade (McKenna and Bell, 1997).

Although they are not particularly taxonomically diverse and are all medium-
sized, primarily nocturnal, and at least partially arboreal, living procyonids do
display interesting variation in both morphology and ecology. Perhaps the
most familiar forms are the omnivorous raccoons of North America, but the
South American species in particular have more specialised diets. Olingos
(*Bassaricyon*) and kinkajous (*Potos*) are primarily frugivorous and highly arboreal, while coatis (*Nasua*; Figure 1.2b) are more terrestrial and insectivorous
(Myers *et al.*, 2008). They also display great variation in skull shape, from very
short-snouted forms like kinkajous to the long-snouted coatis. As noted above,
kinkajous are also one of the two living carnivorans to bear a prehensile tail,
demonstrating their highly arboreal nature.

Ailuridae

Red pandas (Figure 1.2c) and allies (Ailuridae) have often been placed in
Procyonidae, and the superficial resemblance in size, general shape, and pelage
is striking. However, as discussed by Morlo and Peigné (this volume, Chapter 4),
molecular, morphological, and fossil evidence strongly supports ailurids as a
distinct clade, and many molecular studies place Ailuridae as the sister clade to
other musteloids (Mephitidae, Procyonidae, and Mustelidae; Flynn *et al.*, this
volume, Chapter 2). As mentioned above, simocyonines may represent the

extinct sister clade to ailurines, which would extend the temporal range of this family to the middle Miocene. Both clades are found in Europe, North America, and Asia, although simocyonines do not extend beyond the late Miocene or early Pliocene. Ailurines are first observed in the middle Miocene of Europe and early Pliocene of North America, but are extinct in both regions by the late Pliocene and are currently found only in Asia. The fossil record of ailurids has grown extensively in recent years, with multiple new genera identified along with great extension of their geographic and temporal range (Morlo and Peigné, this volume, Chapter 4). As only one species of Ailuridae survives today, *Ailurus fulgens*, and is quite specialised for bamboo-feeding, these new fossil forms will have important implications for understanding the evolution of their unusual living relative.

Mephitidae

The last of the musteloid clades, Mephitidae (skunks and stink badgers), was only recently recognised as a separate clade from Mustelidae (Dragoo and Honeycutt, 1997; Flynn and Nedbal, 1998). Mephitids are known from the early Miocene of Europe and the late Miocene of North America, with a single genus, *Promephitis*, known from the late Miocene to early Pliocene of Asia. Skunks persist only in the New World today, but stink badgers (*Mydaus*) are found in Indonesia and the Philippines (Myers *et al.*, 2008). Like most carnivorans, mephitids invade South America as part of the Great American Biotic Interchange after the formation of the Panamanian land bridge in the late Pliocene (Marshall *et al.*, 1982; McKenna and Bell, 1997). Represented today by only 4 genera, there are 11 recognised fossil mephitid genera, primarily from Europe and North America.

Like most musteloids, mephitids are small- to medium-sized, but they generally have fairly stocky bodies, pointed snouts, and large digging claws (Figure 1.2d). Of course, mephitids are best known for the noxious odours that they produce from their anal glands when threatened, and they all bear conspicuous markings, usually white or yellow stripes or spots on a brown or black coat, to warn potential predators (Myers *et al.*, 2008). Most mephitids are omnivorous, but several species, particularly stink badgers, are primarily insectivorous, using their strong claws to dig for prey.

Pinnipedia

These four musteloid families are united with Pinnipedia (seals, sea lions, and walruses) and Ursidae (bears) in Arctoidea, although historically the exact interrelationships among arctoids have been highly contentious. A long debate has raged on the monophyly of pinnipeds, with some arguing that Phocidae (seals) were more closely related to mustelids, while Otariidae (sea lions and fur seals) and Odobenidae (walruses) were closer to bears. More recent studies, including several molecular analyses, demonstrate that pinnipeds are monophyletic and are likely the sister clade to Musteloidea (see Flynn *et al.*, this volume, Chapter 2). Pinnipeds are a fascinating group, representing a major transition to a primarily aquatic life that is accompanied by a large radiation (Jones and Goswami, this volume, Chapter 12). Extant pinnipeds comprise 21 genera, but there are at least 48 extinct genera recognised from the late Oligocene of North America (McKenna and Bell, 1997; Deméré *et al.*, 2003). Perhaps the best known is *Enaliarctos*, which already has well-developed flippers, from the late Oligocene to early Miocene of western North America and Asia (Berta *et al.*, 1989), but a recent discovery of an early Miocene pinniped from the Canadian Arctic provides an exceptionally preserved transitional fossil. *Puijila darwini* shows several skull characters linking it to pinnipeds but has large, possibly webbed feet, and an unspecialised tail (Rybczynski *et al.*, 2009). The precise relationships among fossil forms and even extant clades are highly debated, with disagreement on the affinities of odobenids, in particular, but *Puijila* and *Enaliarctos* both provide morphological support that quadrapedal swimming was likely the primitive condition for all pinnipeds. Today, odobenids continue to use quadrapedal locomotion in the water, while phocids use hindlimb-powered swimming and otariids rely instead on their forelimbs for propulsion and manoeuvering in the water.

Odobenidae

Odobenids are today represented by only a single species, the walrus (Figure 1.2e), but there are as many as 14 extinct genera of odobenids, ranging back to the middle Miocene of Asia and North America and the early Pliocene of Europe (Deméré *et al.*, 2003). Most early odobenids do not show greatly enlarged canines and appear to have retained a more typical piscivorous diet, rather than sharing the specialisations for suction-feeding of molluscs observed in extant walruses (Berta *et al.*, 2006). Walruses are divided into two monophyletic clades: odobenines, including the extant walrus, and dusignathines. The extinct dusignathines are known only from the late Miocene to the early Pliocene of North America. Unlike modern odobenids, dusignathines show enlargement of both upper and lower canines and likely evolved suction feeding independently from odobenines (Adam and Berta, 2002). Modern walruses are highly gregarious animals confined to the Arctic region. They live primarily on ice floes and are large, conspicuous animals, where both genders bear large canines for fighting, cutting ice, and even tearing apart occasional vertebrate prey (Myers *et al.*, 2008).

Otariidae

Molecular analyses often ally Odobenidae with Otariidae (sea lions and fur seals), in the clade Otarioidea (Flynn *et al.*, 2005), although morphological analyses often prefer a topology uniting Odobenidae with Phocidae (seals) in Phocamorpha (Berta *et al.*, 2006). Otariids (Figure 1.2f) retain the most terrestrial morphology among extant pinnipeds, in that they are able to rotate their hind flippers under their bodies while on land and can only mate and breed on land, unlike phocids and odobenids. Otariids are large and gregarious, with most species displaying great sexual size dimorphism, but they are relatively uniform in ecology. Most species are generalists, eating fish, small vertebrates, and cephalopods, and, while they inhabit a broad geographic range, their requirements of land for breeding restricts them from parts of the Arctic and Antarctic where phocids flourish. With only seven extant and three extinct genera known, otariids are the least speciose of the three pinniped families. Otariids also have the latest appearance of the extant pinnipeds, with the first unambiguous otariid from the late Miocene of California. The record for crown Otariidae is even worse, with no unambiguous representatives prior to the late Pliocene or early Pleistocene. Otariids are generally split into Arctocephalinae (fur seals) and Otariinae (sea lions), although the monophyly of these groups is debated (Deméré *et al.*, 2003).

Phocidae

The last of the extant pinniped clades is Phocidae, which is the most diverse and well-represented in the fossil record. If desmatophocines are accepted as phocids (Berta *et al.*, 2006), there are approximately 24 extinct and 10 extant genera in Phocidae, with a first appearance in the early Miocene. Phocids are generally split into two clades – phocines and monachines (Davis *et al.*, 2004) – although other groupings have also been suggested (Wyss, 1988). Phocids are highly derived for aquatic life, with several species able to mate at sea and breed on ice, freeing them from the terrestrial realm. Several phocids display exceptional diving abilities, and many have evolved specialised diets, such as krill-feeding in *Lobodon carcinophaga*, large vertebrate carnivory in *Hydrurga leptonyx*, and suction feeding in *Erignathus barbatus* (Adam and Berta, 2002; Jones and Goswami, this volume, Chapter 12). Unlike otariids, both of the phocid subclades are represented in the fossil record as early as the middle Miocene of Europe and North America, although none of the extant genera appear prior to the late Pliocene. Phocids are currently distributed in all of the world's oceans, including several species that are exclusively polar, and one freshwater species in Lake Baikal.

Ursidae

The last of the arctoid caniform families is Ursidae. Bears are not a particularly speciose clade, with only five genera and eight species (Myers *et al.*, 2008). However, bears have a long and interesting fossil record, with approximately 20 extinct genera ranging back to the late Eocene of Europe and North America. While relationships are, as usual, contentious, bears and their fossil relatives are typically divided into three chronologically distinct groups: Amphicynodontinae, a likely paraphyletic group from the late Eocene to the early Oligocene of Europe, Asia, and North America; Hemicyoninae, from the early Oligocene to the late Pliocene of Asia, Europe, and North America; and Ursinae, from the early Miocene of Asia, Europe, and North America, spreading to Africa in the late Miocene and to South America in the early Pleistocene (Marshall *et al.*, 1982; Hunt, 1998).

Amphicynodontines (not to be confused with amphicyonids) are relatively small- to medium-sized dog-like animals, displaying some arboreality, but remaining relatively generalised. Hemicyonines, in contrast, evolve a larger body size and more predatory morphology and ecology, with a digitigrade stance that suggests that they were capable runners and hunted down large vertebrate prey. While the hemicyonines successfully invaded North America from Eurasia in the Miocene, potentially displacing other carnivores, such as creodonts and nimravids, these carnivorous bears disappear by the end of the Miocene, leaving their more generalised relatives to continue the bear lineage (Hunt, 1998). Ursinae is an unusual clade of large-bodied, primarily omnivorous forms (Figure 1.2g), with many species with extreme specialisations. Pandas, of course, are well known for bamboo feeding, while sloth bears feed primarily on ants and termites. Spectacled bears are more frugivorous, and polar bears are entirely carnivorous. Thus, for a clade of only eight living species, bears show exceptional ecological diversity.

Among crown ursids, pandas (*Ailuropoda* and allies) are the first to diverge, with molecular clock estimates dating this split at 12 Mya (Wayne *et al.*, 1991), while the first fossil evidence of the distinct ailuropodine lineage is in the late Miocene. The first tremarctine bears, including *Arctodus*, the giant short-faced bear, and *Tremarctos*, the modern spectacled bear, also appear in the late Miocene (McKenna and Bell, 1997), with molecular clock estimates dating the split between tremarctines and ursines also in the Miocene, approximately 6 Mya. Ursini, including the rest of the extant bears (sloth bears, sun bears, polar bears, and black and brown bears), experiences its major radiation into the modern forms around the Miocene–Pliocene boundary (Wayne *et al.*, 1991; Krause *et al.*, 2008).

Canidae

The last of the extant caniform clades, Canidae (Figure 1.2g), is one of the most diverse, with approximately 47 named genera and one of the best fossil records, dating from the middle Eocene (Munthe, 1998). Canids are perhaps the most familiar of all of carnivorans, as they have invaded human homes as successfully as they have invaded every continent except Antarctica (albeit Australia with human help). Canids are a well-studied group, with three major clades providing an ideal system for studying macroevolutionary patterns (Finarelli, 2007). The earliest canids are the hesperocyonines, with at least 10 genera known from the middle Eocene to the middle Miocene of North America (Wang, 1994). While the earliest forms are small- to medium-sized, large, hypercarnivorous forms evolve during the Oligocene, with hesperocyonines achieving their maximum diversity in the late Oligocene.

The second major radiation is that of the borophagine dogs. Borophagines are also exclusively North American, with the earliest members appearing in the early Oligocene (Wang et al., 1999). However, borophagines exhibit their maximum diversity in the Miocene, during which 13 of the approximately 15 recognised genera exist. Although borophagines are typically thought of as bone-crackers, similar to modern hyaenas, this morphology really characterises the later forms that dominated in the late Miocene and Pliocene. Among the early to middle Miocene forms, several taxa show signs of hypocarnivory or omnivory, with some even suggested as primarily frugivorous. Indeed, the small-bodied borophagine *Cynarctus* was originally placed in Procyonidae based on its hypocarnivorous dentition (Wang et al., 1999). However, by the late Miocene and into the Pliocene, borophagines decline quickly, likely due to competition with canines, and the large-bodied carnivorous or bone-cracking forms are the last of the borophagine radiation to go extinct at the end of the Pliocene (Munthe, 1998).

Canines are the last and only extant canid radiation, with approximately 13 extant and 7 extinct genera, and the only ones to expand beyond North America. Canines first appear in the early Miocene of North America, spreading to Europe in the late Miocene, then to Africa and Asia in the Pliocene. They do not colonise South America until the late Pliocene or early Pleistocene, after the emergence of the Panamanian land bridge (Marshall et al., 1982). Canine generic diversity remains low for most of the Miocene, with a pulse of diversification in the late Miocene, correlated with a decline in borophagine diversity in North America and hyaenid diversity in Eurasia, and a larger pulse, particularly in species diversification, in the early Pliocene (Munthe, 1998; Van Valkenburgh, 1999; Finarelli, 2007). Modern canids are generally medium-sized and relatively omnivorous, with long rostra and a digitigrade stance (Figure 1.2h). They are specialised for long-distance pursuit and are generally gregarious, forming packs with complex social systems.

Amphicyonidae

The last caniform clade is the problematic Amphicyonidae, or 'bear-dogs'. This extinct clade is taxonomically diverse, with 34 genera, and a long fossil record spanning the Eocene to the Miocene (McKenna and Bell, 1997). Amphicyonids first appear in North America and Eurasia in the Eocene, only invading Africa in the Miocene. These medium- to large-bodied predators show a range of dental and locomotor morphologies similar to both canids and ursids, driving the confusion on their phylogenetic placement (Hunt, 1996). While early forms appear to be more cursorial, like canids, later amphicyonids display a more bear-like, semi-plantigrade stance, perhaps related to a trend of increasing body size that is well documented in this clade (Finarelli and Flynn, 2006).

Non-carnivoran carnivores

It is worth noting here that many other clades of mammals have also evolved carnivorous forms, allowing for many interesting studies of ecomorphology and convergence. An extinct group with particular relevance to carnivoran evolution is the order Creodonta, composed of two families, Oxyaenidae and Hyaenodontidae (McKenna and Bell, 1997). Creodonts were carnivorous mammals that were the dominant predators for much of the early Cenozoic, before going extinct in the late Miocene (~8 Mya). The largest terrestrial, mammalian carnivore was a hyaenodontid creodont, *Megistotherium osteothlastes*, with a skull length of over a metre and an estimated body size of over 800 kg (Rasmussen *et al.*, 1989). Creodonts share carnassials with carnivorans, suggesting common ancestry, although this interpretation is heavily debated. However, the molars of creodonts became carnassials, with no premolar carnassials, as seen in Carnivora, leaving creodonts without grinding ability on the molars. Because of the great temporal and geographic overlap between creodonts and carnivorans, one might suspect competition. However, given the rapid diversification of carnivorans into their modern range of niches, noted above, there is little evidence that creodonts suppressed early carnivoran evolution through competition (Wesley-Hunt, 2005). Instead, carnivorans may well have outcompeted creodonts (Friscia and Van Valkenburgh, this volume, Chapter 11).

Carnivory has also evolved at least three times in marsupial mammals, with perhaps even more extreme specialisations than are observed in any placental carnivoran. The South American borhyaenid marsupials evolved forms that converge on the morphology of mustelids, bears, dogs, hyaenas, and perhaps most strikingly, sabre-toothed cats. *Thylacosmilus atrox*, the sabre-toothed marsupial, goes even further than sabre-toothed felids, in evolving massive carnassials and open-rooted, evergrowing canines (Riggs, 1934). *Thylacoleo carnifex*, the marsupial lion of Australia, also shows unique specialisations, with the largest carnassials of any carnivorous mammal, and enlarged, procumbent incisors acting as canines (Argot, 2004). Unfortunately, Australia's marsupial carnivores have not fared well since the arrival of humans, with the marsupial wolf, *Thylacinus cynocephalus*, going extinct in the twentieth century. In the last chapter of this volume, Wroe (this volume, Chapter 15) uses finite element analysis to compare these marsupial predators to the more familiar placental carnivorans.

Ecomorphology and macroevolutionary patterns

Because of their living diversity and excellent fossil record, Carnivora has been the focus of many studies in recent years. As described in several chapters in this volume, some of the greatest advances in the understanding of carnivoran evolution involve resolving the relationships of the living and extinct species, providing a framework for more detailed study of their evolutionary history. These phylogenetic studies provide a solid foundation for studies of carnivoran evolution. A strong phylogenetic framework is essential to rigorous assessment of evolutionary trends, to isolate the effects of external influences from patterns that simply reflect ancestral conditions. Several studies in this volume employ recent phylogenies to assess, for example: patterns of body and brain size evolution in Carnivora (Flynn *et al.*, Chapter 2); the effects of character correlations on phylogenetic analyses (Goswami and Polly, Chapter 5); the influence of specialisation on subsequent morphological evolution (Holliday, Chapter 7); the relationship between ecology and cranial shape in aquatic carnivorans (Jones and Goswami, Chapter 12); and the relationship between habitat and limb morphology in terrestrial carnivorans (Polly, Chapter 13).

Ecomorphology and competition in particular have been studied extensively in the fossil record of carnivorans (Van Valkenburgh, 1985, 1989, 1999; Werdelin, 1996a; Wesley-Hunt, 2005). Teeth reflect diet and ecology (Lucas, 1979), and studies of fossil teeth reveal much about paleoecology and its relationship to evolutionary diversity, as many chapters in this volume discuss in detail (Werdelin and Wesley-Hunt, Chapter 8; Wesley-Hunt *et al.*, Chapter 9; Morlo *et al.*, Chapter 10; Friscia and Van Valkenburgh, Chapter 11). The early fossil record of carnivoran dentition shows that diversity increased rapidly in the early Cenozoic (Wesley-Hunt, 2005). Interestingly, by the late Eocene–early Oligocene, the early carnivorans had filled most of the same ecological niches occupied by living species. Although different clades are dominant at different times, entirely new forms and consequently entirely new ecological niches are rare. Even what we think of as a highly specialised morphology, sabre-toothery, evolved independently in both Felidae and Nimravidae, as well as in marsupials, discussed further below. This lack of novelty in the carnivoran record perhaps reflects the stability of prey as a food source, in contrast to the environment-driven shifts affecting herbivore diets (Van Valkenburgh, 1999).

Large hypercarnivorous forms in particular have evolved several times. Large cat-like forms have evolved in at least six different families, from short-faced bear-dogs to leopard-sized mustelids. Bone-cracking forms have evolved at least twice, in hyaenas and dogs. Wolf-like forms have evolved at least five times, in dogs, bears, red pandas, bear-dogs, and hyaenas (Van Valkenburgh, 1999, 2007). However, despite the repeated evolution of hypercarnivorous forms, it has been demonstrated that hypercarnivory is often an evolutionary dead end. Large hypercarnivores diversify quickly, but also decline and go extinct relatively quickly, often being replaced by another hypercarnivorous group. It has been suggested that this pattern is due to the increasing special-isation limiting the group's ability to generalise or expand into other niches, thus increasing their extinction risk (Van Valkenburgh, 1999; Van Valkenburgh *et al.*, 2004; Holliday, this volume, Chapter 7). Correspondingly, recent studies have shown that hypercarnivores are always less morphologically diverse than their closest non-hypercarnivorous relatives (Holliday and Steppan, 2004). Thus, while the sabre-toothed cat may be the classic image of the carnivoran radiation, the raccoon may well be the better model for success in carnivoran evolution.

Locomotor styles also reflect diversity in carnivoran paleoecology, espe-cially when there is significant dietary overlap among coexisting predators (Morlo *et al.*, this volume, Chapter 10; Polly, this volume, Chapter 13). Coexisting carnivorans in modern ecosystems can partition resources by inhabiting different locomotor niches defined by habitat (arboreal or terres-trial) or hunting style (pursuit or ambush). Studies of fossil carnivoran ecomorphology have shown that the locomotor diversity of coexisting carni-vorans is similar in fossil and Recent ecosystems (Van Valkenburgh, 1985; Andersson and Werdelin, 2003). Although the species are different, the ecological structure is similar, demonstrating that extinct taxa partitioned resources similarly to living species.

As described above, these ecological niches are not exclusive to Carnivora; several other mammalian clades have evolved carnivorous forms. Yet, while these other ecological carnivores dominate for long periods on some continents, none approach the taxonomic and ecological diversity and temporal persistence of Carnivora. Why some clades diversify and flourish while others wither is a question of interest not only for evolutionary biology, but also for conservation, and Holliday (this volume, Chapter 7) and Friscia and Van Valkenburgh (this volume, Chapter 11) touch on the answer. Specialisation for hypercarnivory in members of the order Carnivora often involves narrowing and lengthening of the carnassials into shearing blades and reduction or complete loss of post-carnassial molars. In creodonts and marsupial carnivores, all of the molars are specialised for carnivory, either through reduction of all post-carnassial denti-tion or, more often, modification of all of the molar teeth into carnassials. In contrast, most carnivorans retain at least some post-carnassial grinding denti-tion, and many of the herbivorous carnivorans, most notably the giant panda, greatly expand the grinding surface of their molars and reduce their carnassials. While all of their competitors specialised further and further towards hyper-carnivory, carnivorans never develop shearing dentition beyond the original P4/m1 carnassial pair, and this combination of shearing and grinding dentition has served Carnivora well (Van Valkenburgh, 1999). The dental flexibility conferred by the carnivoran dental arrangement may well be the secret to its success. While many carnivoran lineages have gone down the path of greater specialisation, through reduction of the post-carnassial dentition, the greater diversity of carnivorans rests with those that, morphologically and ecologically, keep their options open (Holliday, this volume, Chapter 7).

Conclusions

In closing, there are many reasons why carnivorans are one of the most interesting clades for studies of evolutionary biology. With their great taxonomic, morphological, and ecological diversity, excellent fossil record and well-studied phylogeny, they provide an ideal system for studying conver-gence and ecomorphology, macroevolutionary patterns, and even life history evolution. This volume brings together some of the most exciting and broad studies, using an array of methods, to examine the evolutionary history of Carnivora and, in doing so, displays the cutting edge of vertebrate palaeon-tology. While their obvious charisma may lead people to dismiss the focus on carnivoran evolution as better suited to the popular media, the studies in this volume provide ample evidence that Carnivora truly is a model clade for macroevolutionary studies.

Acknowledgements

J. A. Finarelli, T. Friscia, V. Weisbecker, and A. Crosby all read drafts of this introduction, and their suggestions helped improve this draft greatly. P. Goswami, P. D. Polly, and S. Peigné provided some of the images of carnivorans used in this volume.

REFERENCES

Adam, P. J. and Berta, A. (2002). Evolution of prey capture strategies and diet in the pinnipedimorpha (mammalia, carnivora). *Oryctos*, **4**, 83–107.

Andersson, K. and Werdelin, L. (2003). The evolution of cursorial carnivores in the Tertiary: implications of elbow-joint morphology. *Proceedings of the Royal Society of London, B*, **270**, s163–65.

Argot, C. (2004). Evolution of South American mammalian predators (Borhyaenoidea): anatomical and palaeobiological implications. *Zoological Journal of the Linnean Society*, **140**, 487–521.

Berta, A. (1981). The Plio-Pleistocene hyaena *Chasmaporthetes ossifragus* from Florida. *Journal of Vertebrate Paleontology*, **1**, 341–56.

Berta, A., Ray, C. E. and Wyss, A. R. (1989). Skeleton of the oldest known pinniped, *Enaliarctos mealsi. Science*, **244**, 60–62.

Berta, A., Sumich, J. L. and Kovacs, K. (2006). *Marine Mammals: Evolutionary Biology*, 2nd ed. Amsterdam: Elsevier.

Bryant, H. N. (1991). Phylogenetic relationships and systematics of the Nimravidae (Carnivora). *Journal of Mammalogy*, **72**, 56–78.

Davis, C. S., Delisle, I., Stirling, I., Siniff, D. B. and Strobeck, C. (2004). A phylogeny of the extant Phocidae inferred from complete mitochondrial DNA coding regions. *Molecular Phylogenetics and Evolution*, **33**, 363–77.

Deméré, T. A., Berta, A. and Adam, P. J. (2003). Pinnipedimorph evolutionary biogeography. *Bulletin of the American Museum of Natural History*, **279**, 32–76.

Dragoo, J. W. and Honeycutt, R. L. (1997). Systematics of mustelid-like carnivores. *Journal of Mammalogy*, **78**, 426–43.

Finarelli, J. A. (2007). Mechanisms behind active trends in body size evolution in the Canidae (Carnivora: Mammalia). *American Naturalist*, **170**, 876–85.

Finarelli, J. A. (2008). A total evidence phylogeny of the Arctoidea (Carnivora: Mammalia): relationships among basal taxa. *Journal of Mammalian Evolution*, **15**, 231–54.

Finarelli, J. A. and Flynn, J. J. (2006). Ancestral state reconstruction of body size in the Caniformia (Carnivora, Mammalia): the effects of incorporating data from the fossil record. *Systematic Biology*, **55**, 301–13.

Flynn, J. J. and Nedbal, M. A. (1998). Phylogeny of the Carnivora (Mammalia): congruence versus incompatability among multiple data sets. *Molecular Phylogenetics and Evolution*, **9**, 414–26.

Flynn, J. J. and Wesley-Hunt, G. D. (2005). Carnivora. In *The Rise of Placental Mammals: Origins and Relationships of the Major Extant Clades*, ed. D. Archibald and K. Rose. Baltimore, MD: Johns Hopkins University Press, pp. 175–98.

Flynn, J. J., Nedbal, M. A., Dragoo, J. W. and Honeycutt, R. L. (2000). Whence the red panda? *Molecular Phylogenetics and Evolution*, **17**, 190–99.

Flynn, J. J., Finarelli, J. A., Zehr, S., Hsu, J. and Nedbal, M. A. (2005). Molecular phylogeny of the Carnivora (Mammalia): assessing the impact of increased sampling on resolving enigmatic relationships. *Systematic Biology*, **54**, 317–37.

Fox, R. C. and Youzwyshyn, G. P. (1994). New primitive Carnivorans (Mammalia) from the Paleocene of Western Canada, and their bearing on relationships of the order. *Journal of Vertebrate Paleontology*, **14**, 382–404.

Holliday, J. A. and Steppan, S. J. (2004). Evolution of hypercarnivory: the effect of specialization on morphological and taxonomic diversity. *Palaeobiology*, **30**, 108–28.

Hunt, R. M. (1987). Evolution of Aelurioidea Carnivora: significance of the ventral promontorial process of the petrosal and the origin of basicranial patterns in the living families. *American Museum Novitates*, **2930**, 1–32.

Hunt, R. M. (1996). Amphicyonidae. In *The Terrestrial Eocene–Oligocene Transition in North America*, ed. D. R. Prothero and R. J. Emry. Cambridge: Cambridge University Press, pp. 476–85.

Hunt, R. M. (1998). Ursidae. In *Evolution of Tertiary Mammals of North America*, ed. C. M. Janis, K. M. Scott and L. Jacobs. Cambridge: Cambridge University Press, pp. 174–95.

Koepfli, K.-P., Jenks, S. M., Eizirik, E., Zahirpour, T., Van Valkenburgh, B. and Wayne, R. K. (2006). Molecular systematics of the Hyaenidae: relationships of a relictual lineage resolved by a molecular supermatrix. *Molecular Phylogenetics and Evolution*, **38**, 603–20.

Koepfli, K.-P., Deere, K. A., Slater, G. J., *et al.* (2008). Multigene phylogeny of the Mustelidae: resolving relationships, tempo and biogeographic history of a mammalian adaptive radiation. *BMC Biology*, **6**, 10.

Krause, J., Unger, T., Nocon, A., *et al.* (2008). Mitochondrial genomes reveal an explosive radiation of extinct and extant bears near the Miocene–Pliocene boundary. *BMC Evolutionary Biology*, **8**, 220.

Lucas, P. W. (1979). The dental–dietary adaptations of mammals. *Neues Jahrbuch fur Geologie und Palaeontologie, Monatshefte*, **8**, 486–512.

Marshall, L. G., Webb, S. G., Sepkoski, J. J. and Raup, D. M. (1982). Mammalian evolution and the great American biotic interchange. *Science*, **215**, 1351–57.

McKenna, M. C. and Bell, S. K. (1997). *Classification of Mammals above the Species Level*. New York: Columbia University Press.

Morlo, M., Peigné, S. and Nagel, D. (2004). A new species of *Prosansansosmilus*: implications for the systematic relationships of the family Barbourofelidae new rank (Carnivora, Mammalia). *Zoological Journal of the Linnean Society*, **140**, 43–61.

Munthe, K. (1998). Canidae. In *Evolution of Tertiary Mammals of North America: Terrestrial Carnivores, Ungulates, and Ungulatelike Mammals*, ed. C. M. Janis, L. Jacobs and K. M. Scott. Cambridge: Cambridge University Press, pp. 124–43.

Murphy, W. J., Eizirik, E., Johnson, W. E., Zhang, Y. P., Ryder, O. A. and O'Brien, S. J. (2001). Molecular phylogenetics and the origins of placental mammals. *Nature*, **409**, 614–18.

Murphy, W. J., Pringle, T. H., Crider, T. A., Springer, M. S. and Miller, W. (2007). Using genomic data to unravel the root of the placental mammal phylogeny. *Genome Research*, **17**, 413–21.

Myers, P., Espinosa, R., Parr, C. S., Jones, T., Hammond, G. S. and Dewey, T. A. (2008). The animal diversity web (http://animaldiversity.ummz.umich.edu/).

Peigné, S. (2003). Systematic review of European Nimravinae (Mammalia, Carnivora, Nimravidae) and the phylogenetic relationships of Palaeogene Nimravidae. *Zoologica Scripta*, **32**, 199–229.

Polly, P. D., Wesley-Hunt, G. D., Heinrich, R. E., Davis, G. and Houde, P. (2006). Earliest know carnivoran auditory bulla and support for a recent origin of crown-group carnivora (Eutheria, Mammalia). *Palaeontology*, **49**, 1019–27.

Rasmussen, D. T., Tilden, C. D. and Simons, E. L. (1989). New specimens of the giant creodont *Megistotherium* (Hyaenodontidae) from Moghara, Egypt. *Journal of Mammalogy*, **70**, 442–47.

Riggs, E. S. (1934). A new marsupial saber-tooth from the pliocene of Argentina and its relationships to other South American predaceous marsupials. *Transactions of the American Philosophical Society*, **24**, 1–32.

Rybczynski, N., Dawson, M. R. and Tedford, R. H. (2009). A semi-aquatic mammalian carnivore from the Miocene epoch and origin of Pinnipedia. *Nature*, **458**, 1021–24.

Sears, K. E., Goswami, A., Flynn, J. J. and Niswander, L. (2007). The correlated evolution of *runx2* tandem repeats and facial length in carnivora. *Evolution and Development*, **9**, 555–65.

Van Valkenburgh, B. (1985). Locomotor diversity within past and present guilds of large predatory mammals. *Palaeobiology*, **11**, 406–28.

Van Valkenburgh, B. (1989). Carnivore dental adaptations and diet: a study of trophic diversity within guilds. In *Carnivore Behavior, Ecology, and Evolution*, ed. J. L. Gittleman. Ithaca, NY: Comstock Publishing Associates, pp. 410–36.

Van Valkenburgh, B. (1999). Major patterns in the history of carnivorous mammals. *Annual Review of Earth and Planetary Science*, **27**, 463–93.

Van Valkenburgh, B. (2007). Deja vu: the evolution of feeding morphologies in the Carnivora. *Integrative and Comparative Biology*, **47**, 147–63.

Van Valkenburgh, B., Wang, X. M. and Damuth, J. (2004). Cope's rule, hypercarnivory, and extinction in North American canids. *Science*, **306**, 101–04.

Wang, X. M. (1994). Phylogenetic systematics of the Hesperocyoninae (Carnivora: Canidae). *Bulletin of the American Museum of Natural History*, **221**, 1–207.

Wang, X. M., Tedford, R. H. and Taylor, B. E. (1999). Phylogenetic systematics of the Borophaginae (Carnivora: Canidae). *Bulletin of the American Museum of Natural History*, **243**, 1–391.

Wayne, R. K. (1986). Cranial morphology of domestic and wild canids: the influence of development on morphological change. *Evolution*, **40**, 243–61.

Wayne, R. K., Van Valkenburgh, B. and O'Brien, S. J. (1991). Molecular distance and divergence time in carnivores and primates. *Molecular Biology and Evolution*, **8**, 297–319.

Werdelin, L. (1996a). Carnivoran ecomorphology: a phylogenetic perspective. In *Carnivore Behavior, Ecology, and Evolution*, ed. J. L. Gittleman. Ithaca, NY: Cornell University Press, pp. 582–624.

Werdelin, L. (1996b). Community-wide character displacement in Miocene hyaenas. *Lethaia*, **29**, 97–106.

Werdelin, L. and Solounias, N. (1991). The Hyaenidae: taxonomy, systematics, and evolution. *Lethaia*, **30**, 1–105.

Wesley-Hunt, G. D. (2005). The morphological diversification of carnivores in North America. *Palaeobiology*, **31**, 35–55.

Wesley-Hunt, G. D. and Flynn, J. J. (2005). Phylogeny of the Carnivora: basal relationships among the carnivoramorphans, and assessment of the position of 'Miacoidea' relative to crown-clade carnivora. *Journal of Systematic Palaeontology*, **3**, 1–28.

Wolsan, M. (1993). Phylogeny and classification of early European Mustelida (Mammalia: Carnivora). *Acta Theriologica*, **38**, 345–84.

Wozencraft, W. C. (2005). Order Carnivora. In *Mammal Species of the World – A Taxonomic and Geographic Reference*, ed. D. E. Wilson and D. M. Reeder. Baltimore, MD: Johns Hopkins University Press, pp. 532–628.

Wyss, A. R. (1988). On 'retrogression' in the evolution of the Phocinae and phylogenetic affinities of the monk seals. *American Museum Novitates*, **2924**, 1–38.

Wyss, A. R. and Flynn, J. J. (1993). A phylogenetic analysis and definition of the Carnivora. In *Mammal Phylogeny*, ed. F. S. Szalay, M. J. Novacek and M. C. McKenna. New York: Springer Verlag, pp. 32–52.

Yoder, A. D., Burns, M. M., Zehr, S., *et al.* (2003). Single origin of Malagasy Carnivora from an African ancestor. *Nature*, **421**, 734–37.

2

Phylogeny of the Carnivora and Carnivoramorpha, and the use of the fossil record to enhance understanding of evolutionary transformations

JOHN J. FLYNN, JOHN A. FINARELLI, AND MICHELLE SPAULDING

Introduction

Phylogeny of the Carnivora – molecules, fossils, and total evidence

Fossil taxa are inherently at a disadvantage in resolving phylogenetic relationships, relative to living forms, as soft anatomy, DNA, physiology, and most life-history attributes are not readily available for the vast majority of these taxa, other than some fascinating new sequences available for Pleistocene fossil taxa (e.g. *Smilodon*, *Homotherium*, *Miracinonyx*, *Ursus spelaeus*, etc.; Loreille *et al.*, 2001; Barnett *et al.*, 2005). Nevertheless, fossil data possess several key advantages in phylogenetic analyses, including the ability to break-up 'long branches' in phylogenies, where the divergence between modern-day clades occurred deep in geological time. Fossils preserve morphologies that can become obscured along these long branches, and also provide temporal context for the evolution of living clades that may be crucial for accurately reconstructing ancestral conditions and partitioning synapomorphic versus homoplasious resemblances among modern-day taxa. Some workers feel that molecular data are inherently superior for reconstructing phylogeny than morphological characters (see for example: Scotland *et al.*, 2003; but see Jenner, 2004), and as a consequence, phylogenies for many clades, particularly those that are not well represented in the fossil record, often are based solely on molecular sequence data. Within Carnivora, for example, the most recent studies reconstructing phylogenetic relationships among living taxa have relied principally on molecular

Carnivoran Evolution: New Views on Phylogeny, Form, and Function, ed. A. Goswami and A. Friscia. Published by Cambridge University Press. © Cambridge University Press 2010.

sequences (e.g. Flynn *et al.*, 2000, 2005). In contrast, those studies seeking to integrate fossil taxa into an evolutionary context together with living species de facto have tended to emphasise morphological data (e.g. Wesley-Hunt and Flynn, 2005). 'Total evidence' phylogenetic analyses, incorporating all available data into a single data set, are becoming more common in analyses of extant taxa (Vrana *et al.*, 1994; Flynn and Nedbal, 1998; Wheeler and Hayashi, 1998; Giribet *et al.*, 2001; Nylander *et al.*, 2004; Bond and Hedin, 2006; Grant *et al.*, 2006; Sanders *et al.*, 2006), as well as analyses incorporating fossil and extant taxa (O'Leary, 1999, 2001; Gatesy and O'Leary, 2001; Gatesy *et al.*, 2003; Rothwell and Nixon, 2006; Arango and Wheeler, 2007; Magallon, 2007; Manos *et al.*, 2007; Finarelli, 2008b). This approach arguably is preferable, on both methodological and philosophical grounds, as it incorporates all potentially phylogenetically significant data within a single analysis. Indeed, empirical studies have demonstrated that data combination can be superior to congruence-based approaches (Allard and Carpenter, 1996; Baker and DeSalle, 1997; Flynn and Nedbal, 1998; Asher *et al.*, 2005; Gatesy and Baker, 2005).

In this chapter we provide a summary of recent phylogenetic analyses based on primary character data for the Carnivora and their closest extinct relatives (basal members of the more inclusive clade Carnivoramorpha), with brief consideration of their nearest sister clades among the Eutheria. While preference is given to synthetic, 'total evidence' analyses when available, we review the spectrum of most inclusive analyses currently available for a broad array of higher-level clades within the Carnivoramorpha. In addition, we name and propose a phylogenetic definition and diagnosis for a now consistently recovered major clade within Carnivoramorpha, the Carnivoraformes (see other phylogenetic definitions in Wesley-Hunt and Flynn, 2005, as well as Wyss and Flynn, 1993, and discussions in both papers; see also the Phylocode, Cantino and deQuiroz, 2007).

For clades in which the interrelationships of major groups are well supported, a variety of interesting evolutionary questions can be addressed within this framework. In addition, those studies that can incorporate fossil taxa permit direct assessment and evaluation of the additional information that can be provided by such fossils, including constraints on divergence ages, morphological attributes of basal members of clades (and enhanced understanding of ancestral reconstructions for these groups), biogeography through time, and so on. The intent of this paper is to provide exemplars of the types of interesting evolutionary studies that can be developed within a robust phylogenetic context, and we present discussions of some recently published analyses and works in progress by the authors.

Phylogeny of the Carnivoramorpha and Carnivora

Various aspects of the phylogeny of crown-clade Carnivora and the more inclusive Carnivoramorpha have been well resolved by morphological data for some time. Some higher-level interrelationships have remained controversial or poorly resolved, however, including determination of the closest relatives of Carnivora among living eutherians. The increased prominence of molecular phylogenies, as well as 'total evidence' analyses, has led to even more robust support for these generally accepted relationships, resolution of many areas of ambiguity or conflict, and uncovering of some novel hypotheses of interrelationships. In addition, there has been long-standing uncertainty about the placement of key fossil taxa, such as the closest early Cenozoic relatives (traditionally the 'Miacoidea' or basal Carnivoramorpha) and the extinct 'Creodonta', some of which is beginning to be resolved through more comprehensive sampling of taxa and characters and broader integrative analyses. Flynn and Wesley-Hunt (2005) provided the most recent extensive discussion of these phylogenetic issues and results, which we summarise and expand upon here.

Morphological analyses (Wyss and Flynn, 1993; Wesley-Hunt and Flynn, 2005) provide support for the long-standing notion (e.g. Matthew, 1909) that 'Creodonta' are the nearest relatives of Carnivoramorpha, although monophyly of the two major lineages still assigned to the 'Creodonta' (Hyaenodontidae and Oxyaenidae) remains uncertain (Polly, 1996; Gunnell, 1998). Molecular results generally indicate that Pholidota are the nearest living relatives of Carnivora (e.g. Murphy *et al.*, 2001, 2007; Delsuc *et al.*, 2002), although there is disagreement among molecular studies, variability in taxon sampling and topologies among various molecular studies, analytical algorithm applied, and only weak morphological support for this hypothesis (other than the osseous tentorium shared by Pholidota, Carnivoramorpha and creodonts; Wyss and Flynn, 1993) (e.g. Asher, 2007; Kjer and Honeycutt, 2007).

Virtually all recent cladistic higher level phylogenetic studies (e.g. Flynn *et al.*, 1988, 2000, 2005; Wyss and Flynn, 1993; Vrana *et al.*, 1994; Flynn and Nedbal, 1998; Sato *et al.*, 2004, 2006; Yu *et al.*, 2004a; Wesley-Hunt and Flynn, 2005; Wesley-Hunt and Werdelin, 2005; Fulton and Strobeck, 2006; Yu and Zhang, 2006; Arnason *et al.*, 2007; Finarelli, 2008b) provide strong measures of nodal support for monophyly of the Carnivora/Carnivoramorpha and numerous monophyletic subclades (Figures 2.1 and 2.2). Examples include monophyly of the crown-clade Carnivora and the division between its two major subclades Caniformia and Feliformia; the caniform subclades Arctoidea, Pinnipedimorpha and Pinnipedia, and Musteloidea; the feliform subclades Feloidea and Herpestoidea; and monophyly of all traditional

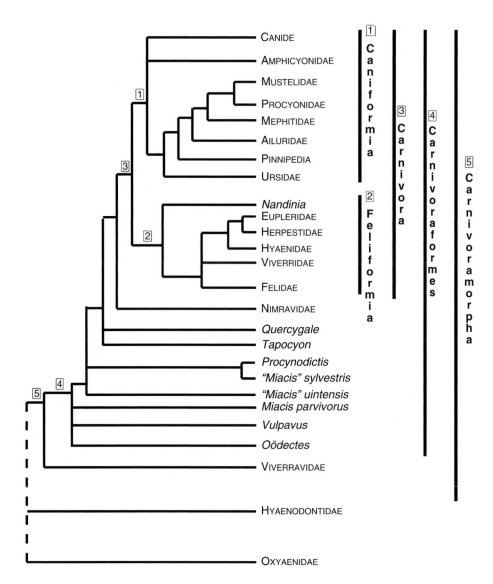

Figure 2.1 Phylogeny of the Carnivoramorpha. Summary tree schematically portraying the congruent topologies from the phylogenetic analyses of Flynn *et al.* (2005), Wesley-Hunt and Flynn (2005), Spaulding (2007), and Finarelli (2008).

modern families except Viverridae and Mustelidae. Many lower-level sub-clades of those groups also are well resolved, but others remain the focus of intensive study.

Among an array of either previously controversial or unexpected hypotheses for living clades, several phylogenetic relationships now appear to be well

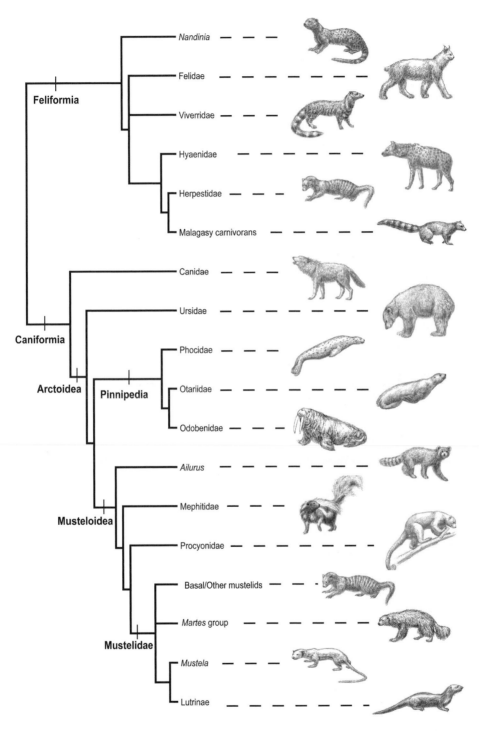

Figure 2.2 See Plate 1 for colour. Diagrammatic summary of the molecular phylogeny of major clades of living Carnivora (from Flynn *et al.*, 2005). Illustrations of

supported by molecular, morphological, or 'total evidence' analyses, using various nodal support metrics (Figures 2.1 and 2.2); these include:

1. Monophyly of the Pinnipedia (Wyss, 1987; Arnason *et al.*, 1995, 2006; Lento *et al.*, 1995; Flynn and Nedbal, 1998; Flynn *et al.*, 2000, 2005; Fulton and Strobeck, 2006; Sato *et al.*, 2006; Higdon *et al.*, 2007), with the Pinnipedia initially of uncertain relationships within the Arctoidea but now definitively shown in several studies to be more closely related to the Musteloidea than to Ursidae (Flynn *et al.*, 2005; Fulton and Strobeck, 2006; Sato *et al.*, 2006; Finarelli, 2008b). In addition, the Odobenidae are consistently resolved as more closely related to Otariidae than to Phocidae (Arnason *et al.*, 1995, 2007; Lento *et al.*, 1995; Flynn and Nedbal, 1998; Flynn *et al.*, 2000, 2005; Delisle and Strobeck, 2005; Fulton and Strobeck, 2006; Higdon *et al.*, 2007; but see Berta *et al.*, 2006 for Odobenidae/Phocidae link).
2. Skunks and stink badgers (Mephitidae) are not included within the Mustelidae (*sensu stricto*), contrary to their traditional taxonomic placement (e.g. Mivart, 1885; Simpson, 1945), but rather form a distinct clade unto themselves (Dragoo and Honeycutt, 1997; Flynn and Nedbal, 1998; Flynn *et al.*, 2000, 2005; Arnason *et al.*, 2007).
3. The red panda (*Ailurus*) is placed within a family (Ailuridae) that is monotypic for living taxa (but which has fossil representatives), and this clade remains in an unresolved polytomy between Ailuridae, Mephitidae, and Musteloidea (Flynn *et al.*, 2000, 2005; Yu *et al.*, 2004a; Fulton and Strobeck, 2006; Yu and Zhang, 2006; Arnason *et al.*, 2007; see also the detailed discussions in Yonezawa *et al.*, 2007 and Morlo and Peigné, this volume [the latter includes detailed discussion of 26 species (in 9 genera) of simocyonine and ailurine ailurids, morphological and molecular phylogeny evidence, and the oldest ailurid, the 25 million year old *Amphictis*]).
4. *Nandinia* is the sister group to all other living feliforms (not a 'viverrid') (Hunt, 1987; Flynn *et al.*, 1988, 2005; Veron, 1995; Flynn, 1996; Flynn and

Caption for figure 2.2 (*cont.*)
representative taxa for major lineages include (from top): *Nandinia binotata*; Felidae (*Lynx rufus*); Viverridae (*Viverra zibetha*); Hyaenidae (*Crocuta crocuta*); Herpestidae (*Mungos mungo*); Malagasy carnivorans (*Eupleres goudotii*); Canidae (*Canis lupus*); Ursidae (*Ursus americanus*); Phocidae (*Phoca vitulina*); Otariidae (*Zalophus californianus*); Odobenidae (*Odobenus rosmarus*); *Ailurus fulgens*; Mephitidae (*Mephitis mephitis*); Procyonidae (*Potos flavus*); Mustelidae, basal/other mustelids (generalised schematic representing diverse taxa [African polecat and striped marten, badger, etc.]); Mustelidae, *Martes*-group (*Gulo gulo*); Mustelidae, *Mustela* (*Mustela frenata*); Mustelidae, Lutrinae (*Lontra canadensis*).

Nedbal, 1998; Gaubert and Veron, 2003; Yoder and Flynn, 2003; Yoder *et al.*, 2003; Yu *et al.*, 2004a; Gaubert and Cordeiro-Estrela, 2006; Veron, this volume).

5. Linsangs (Asian *Prionodon*, and possibly African *Poiana*) also are not viverrids, but instead appear to form a clade (Prionodontidae) that is the sister taxon to the Felidae (Gaubert and Veron, 2003; Gaubert and Cordeiro-Estrela, 2006; Veron, this volume).

6. Madagascar's feliform carnivorans, species of which have been variably included within Felidae, Herpestidae, and Viverridae, instead comprise a monophyletic clade (Eupleridae) that is the sister clade to the Herpestidae, bearing significantly on interpretations of their biogeography (Yoder and Flynn, 2003; Yoder *et al.*, 2003; Flynn *et al.*, 2005; Gaubert and Cordeiro-Estrela, 2006).

7. Within Herpestoidea, Hyaenidae is the nearest relative to the clade Herpestidae + Eupleridae, although the interrelationships among Herpestoidea, Felidae (plus Prionodontidae) and Viverridae (*sensu stricto*) remain controversial, and are best represented as an unresolved polytomy at this time (Yoder and Flynn, 2003; Yoder *et al.*, 2003; Flynn *et al.*, 2005; Gaubert and Cordeiro-Estrela, 2006).

8. The phylogenetic position of the giant panda *Ailuropoda*, which has been considered problematic taxonomically (although morphological evidence has long supported its position as an ursid; Davis, 1964; see discussion in Flynn and Wyss, 1988), is strongly supported as basal to all other living ursids in recent morphological, molecular and 'total evidence' analyses (Flynn and Nedbal, 1998; Yu *et al.*, 2004b; Flynn *et al.*, 2005; Fulton and Strobeck, 2006; Yu and Zhang, 2006; Arnason *et al.*, 2007).

Flynn *et al.* (2005) presented the most comprehensive molecular phylogenetic analysis to date across the Carnivora, including 42 extant caniform and 32 extant feliform taxa, and incorporating more than 6 kbp of concatenated sequence data from 6 loci (3 nuclear [*TR-i-I, TBG, IRBP*]) and 3 mitochondrial [*ND2, CYTB, 12S*]). Finarelli (2008b) subsequently performed a 'total evidence' phylogenetic analysis for the subclade Caniformia, integrating about 5.6 kbp of the molecular sequence data from the Flynn *et al.* (2005) analysis, combined with a morphological matrix consisting of 80 craniodental characters for 32 caniform (14 extant and 18 extinct) taxa. That analysis yielded a single most parsimonious phylogeny, congruent with the molecular phylogeny of Flynn *et al.* (2005) for living taxa, but substantially revising traditional notions of basal arctoid phylogeny, as many fossil taxa that have typically been placed either with the bears (Ursidae), raccoons (Procyonidae) or mustelids

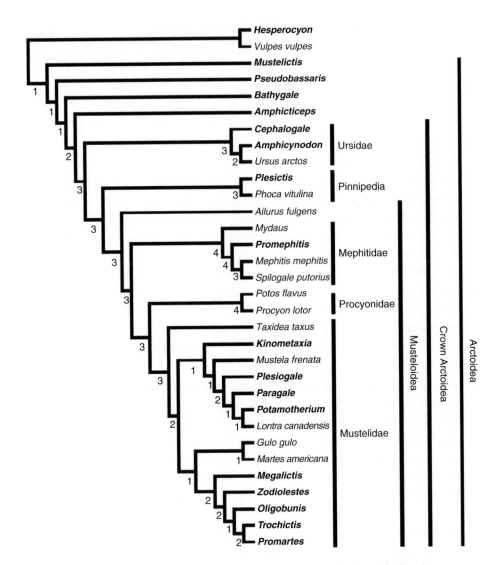

Figure 2.3 Arctoid 'total evidence' phylogeny, from Finarelli (2008b). Single-most parsimonious cladogram; 5.6 kbp of molecular sequence data and 80-character craniodental morphological data; fossil taxa are shown in bold; nodal Bremer Decay Indices indicated under branches.

(Mustelidae, *sensu lato*, including skunks and 'basal mustelids') instead appear to be basal arctoids or stem taxa lying entirely outside the crown clade Arctoidea (Figure 2.3). For example, *Bathygale* and *Plesictis* appear to be stem arctoids rather than members of the Mustelidae (Wolsan, 1993). Of particular interest has been resolution of the phylogenetic relationships of the 'Paleomustelidae' (see discussion in Finarelli and Flynn, 2006). While this is almost certainly a polyphyletic assemblage of taxa, a subset of 'paleomustelids', previously assigned to the Oligobuninae, clearly forms a monophyletic clade nested within the radiation of Mustelidae *sensu stricto* (that is, mustelids excluding skunks), reconstructed as the sister clade to *Gulo* + *Martes*. Finarelli's (2008b) results highlight the importance of including both molecular and morphological data in reconstructing phylogenetic topologies.

Recent phylogenetic work among extant taxa has emphasised molecular data, although the number of loci (and total number of base pairs) and the proportion of living species sampled vary substantially across studies. For Caniformia, these studies include the Canidae (Wayne *et al.*, 1997; Bardeleben *et al.*, 2005), Ursidae (Yu *et al.*, 2004b; Yu and Zhang, 2006; Arnason *et al.*, 2007), Phocidae (Delisle and Strobeck, 2005; Fyler *et al.*, 2005; Higdon *et al.*, 2007), Otariidae (Wynen *et al.*, 2001; Higdon *et al.*, 2007), Mephitidae (Dragoo *et al.*, 1993), Mustelidae (Marmi *et al.*, 2004; Yonezawa *et al.*, 2007), and Procyonidae (Fulton and Strobeck, 2007; Koepfli *et al.*, 2007). Although the Feliformia as a whole have been relatively less well-studied, an increasing number of analyses have extensively sampled feliform family-level clades, including Viverridae *sensu stricto* (that is, not including taxa such as *Nandinia*, *Prionodon*, or Malagasy eupllerids) (Gaubert and Veron, 2003; Gaubert and Cordeiro-Estrela, 2006; Gaubert and Begg, 2007; Veron, this volume [now limited to 34 species grouped into 4 subfamilies, the Asian Hemigalinae and Paradoxurinae, African Genettinae, and Afro-Asian Viverrinae, with the perineal gland as a morphological synapomorphy for the clade, and the oldest fossil being *Herpestides* and *Semigenetta* at about 23 Mya]), Felidae (and Prionodontidae; Gaubert and Veron, 2003; Johnson *et al.*, 2006; Veron, this volume), Hyaenidae (Koepfli *et al.*, 2006), Herpestidae (Yoder *et al.*, 2003; Veron *et al.*, 2004; Flynn *et al.*, 2005), and the newly recognised clade of Eupleridae containing all of the carnivorans endemic to Madagascar (Yoder and Flynn, 2003; Yoder *et al.*, 2003; Flynn *et al.*, 2005).

Phylogeny of basal Carnivoramorpha, and the problematic fossil groups Nimravidae and Amphicyonidae

Studies of relationships among living Carnivora alone hinders development of the most meaningful and comprehensive view of their evolutionary history, as it ignores key character states and transformations, temporal and biogeographic data associated with the rich diversity of early carnivoramorphan fossil taxa (see Finarelli and Flynn, 2006, 2007). As patterns of morphological transformation during the initial diversification of Carnivoramorpha can only be accurately determined through enhanced understanding of the relationships of basal taxa, specifically early Cenozoic members of this group, we provide a more detailed discussion of them here. Matthew (1909) provided the first 'modern' consideration of the phylogenetic relationships of Carnivora and their closest extinct relatives, although his classifications were explicitly gradal in nature and excluded early Cenozoic fossil forms (his 'Miacidae', later 'Miacoidea' = 'Miacidae' in the modern sense, and Viverravidae) from the 'Carnivora Fissipedia' or terrestrial carnivorans. Simpson's (1945) influential classification clearly separated the 'Miacoidea' as group distinct from the clades of extant Carnivora.

The Nimravidae ('false sabretooths') are an archaic group of sabre-toothed carnivoramorphans, initially considered to be close relatives of the Felidae but now of uncertain placement as basal feliforms, basal caniforms, or even outside the carnivoran crown-clade. In addition, it is unclear whether the hypercarnivorous barbourofelines are members of the Nimravidae or instead represent another independent evolution of a sabre-toothed carnivoramorphan clade (Morlo et al., 2004). The Amphicyonidae are an extinct lineage often referred to as the 'bear-dogs' because of anatomical resemblances to members of both groups. They are of uncertain phylogenetic position; while usually allied with either the canids or the ursids, they might represent basal caniforms or basal arctoids (Hunt, 1977, 1996; Wolsan, 1993; Wyss and Flynn, 1993; Viranta, 1996; see discussion in Finarelli and Flynn, 2006).

Most early, non-computer-based cladistic analyses of the group recognised monophyly of the crown-clade Carnivora, and placed some or all members of the 'Miacidae' and Viverravidae within the two main carnivoran subclades Caniformia and Feliformia, respectively (e.g. Flynn and Galiano, 1982; Flynn et al., 1988; Wang and Tedford, 1994). There is now substantial evidence from computer-based parsimony analyses that Early Cenozoic carnivoramorphans form a series of stem groups to crown Carnivora (Wyss and Flynn, 1993; Wesley-Hunt and Flynn, 2005; Wesley-Hunt and Werdelin, 2005; Polly et al., 2006; Spaulding and Flynn, 2009; see also discussion in Bryant, 1991). Initially, there was uncertainty about whether viverravids or 'miacids' were closer to Carnivora. While most prior studies treated the 'Miacidae' and Viverravidae as two different composite, familial Operational Taxonomic Units (OTUs), Wesley-Hunt and Flynn (2005) dealt with taxa at the species-level, providing the most comprehensive phylogenetic analysis of taxa and characters for the basal Carnivoramorpha (40 taxa, 99 characters), which yielded 3 most-parsimonious trees (Figure 2.4). That analysis clearly documented monophyly of the Viverravidae, placed that clade as the sister clade to all other Carnivoramorpha, and recovered 'Miacidae' as a paraphyletic stem group to crown-clade Carnivora. In addition, Nimravidae were

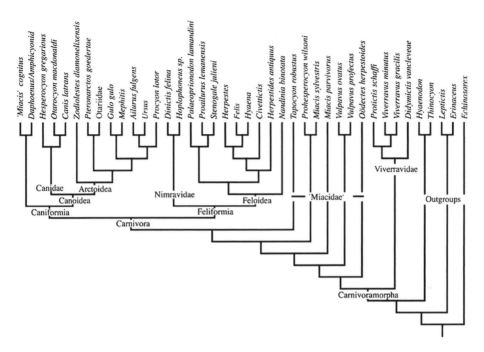

Figure 2.4 Consensus morphological phylogeny for Carnivoramorpha, emphasising
the interrelationships of basal carnivoramorphans, from Wesley-Hunt and Flynn
(2005). Fifty per cent majority rule consensus of 3 trees; tree length = 439 steps;
Consistency Index = 0.318 (excluding uninformative characters); Retention Index
(RI) = 0.659.

recovered as basal feliforms and Amphicyonidae as basal caniforms, weakly
supported as the sister clade to crown-clade caniforms (Canoidea, see: Flynn
and Wesley-Hunt, 2005).

When the hypothesised relationships of Wesley-Hunt and Flynn (2005) are
compared to those generated by subsequent studies, that either expand taxon
sampling (Wesley-Hunt and Werdelin, 2005; Polly *et al.*, 2006; Spaulding and
Flynn, 2009; Spaulding *et al.*, in press) or character sampling (Spaulding, 2007),
several aspects remain consistent. For example, while the study of Wesley-Hunt
and Werdelin (2005), which added one additional taxon beyond that analysed
by Wesley-Hunt and Flynn (2005: *Quercygale*, a European 'miacoid'), yielded
many of the same phylogenetic relationships for basal carnivormorphans, it
differed in resolving several nodes that previously received only weak support,
emphasising the potential importance of increased taxon sampling for resolving
ambiguous or weakly supported hypotheses of interrelationships. Relationships
among the basal taxa indicated that *Quercygale* is the sister taxon to a clade of

the Nimravidae plus crown-clade Carnivora, and allied Amphicyonidae with Ursidae. Spaulding and Flynn (2009) provide the first detailed description of the only known postcranial skeletal elements of '*Miacis*' *uintensis* and include substantial new postcranial data bearing on the resolution of early carnivor-amorphan interrelationships. Analysis of that material points to important new postcranial character variation that can be incorporated into phylogenetic studies, helping to resolve early Carnivoramorpha phylogeny, and highlights a previously unrecognised diversity in postcranial morphology and locomotor styles in these early taxa (see below).

In recent studies, the Viverravidae is always found to be monophyletic, and represents the nearest sister group to the remainder of Carnivoramorpha (Wesley-Hunt and Flynn, 2005; Spaulding and Flynn, 2009). These are the earliest carnivoramorphans in the fossil record, first occurring in the early Paleocene of North America and ranging to the Late Eocene across Europe and North America, with the last known specimens also found in North America (Flynn and Wesley-Hunt, 2005). However, relatively few viverravid taxa occur in rocks younger than the Paleocene (Flynn, 1998). Body sizes of those taxa ranged from the roughly small weasel sized *Viverravus minutus* (slightly >1 kg) to the coyote-sized *Didymictis vancleaveae* (possibly >20 kg). All species appear to have been highly carnivorous, with the smaller taxa perhaps more specialised for insectivory. Dental and cranial features that support this clade include the loss of the M3/m3, a well-defined parastylar cusp on the P4, M1 with a protocone larger than the paracone, and an almond-shaped promontorium (Wesley-Hunt and Flynn, 2005).

A paraphyletic series of taxa is typically found between the Viverravidae and the crown Carnivora (plus Nimravidae, whenever this family is not found within crown-clade Carnivora). These taxa were once thought to make up a group known as the 'Miacidae', originally proposed by Cope (1880), but this assemblage has received no support as a natural group in recent studies (e.g. Wyss and Flynn, 1993; Wesley-Hunt and Flynn, 2005; Spaulding and Flynn, 2009). Relationships between various species usually assigned to the 'Miacidae' are unstable among studies; however, all relevant taxa are always found to be more closely related to crown Carnivora than to the Viverravidae. Since the Miacidae clearly are not a monophyletic group, and referring to this suite of taxa via terms such as 'non-carnivoran, non-viverravid carnivoramorphans' is unwieldy, we propose the formal phylogenetic taxonomic name (see the Phylocode, Cantino and deQuiroz, 2007) Carnivoraformes for this consistently recovered grouping of most basal carnivoramorphans ('Miacidae', or all taxa other than the monophyletic Viverravidae) plus the crown-clade Carnivora (node 4, Figure 2.1).

Goswami and Polly (this volume) tested the potential influence of cranial character correlations on morphologically based phylogenetic analyses, concluding that although these characters primarily derive from the basicranium and molars and that there are high character correlations among modules and within clades, the emphasis on characters from only two cranial 'modules' did not lead to significant errors (as compared to phylogenies generated from molecular data). Interestingly, they determined that basicranial anatomy, long used for distinguishing major carnivoran groups, retains strong phylogenetic signal across the clade.

Phylogenetic taxonomic definition of a newly named clade: Carnivoraformes

CARNIVORAMORPHA Wyss and Flynn, 1993
CARNIVORAFORMES new clade

Phylogenetic Taxonomic Definition (stem-based): Carnivora and all taxa that are more closely related to Carnivora (represented by *Canis lupus*) than to *Viverravus gracilis* (the holotype species of *Viverravus*, and representative of the Viverravidae).

Diagnosis: Carnivoraformes are distinguished from the Viverravidae by the presence of the following features: round infraorbital foramen; the mastoid process is blunt, rounded, and does not protrude significantly; the presence of a rostral entotympanic or evidence of a rostral entotympanic in life; the fossa for the tensor tympanic muscle is well-defined and deep; and the m2 talonid is not elongate, and does not possess an enlarged hypoconulid.

Discussion: The literature is filled with names that will see modest or no use. Often these are based upon weakly supported nodes in an individual analysis or lack of strong support for monophyly of the included taxa. While there is little benefit in naming every node in an analysis, naming the clade represented by this node is well justified and will be useful, as it is recovered with high nodal support in all recent phylogenetic studies of the Carnivoramorpha. Further, it would permit abandonment of the wastebasket taxonomic terms 'Miacidae' and 'Miacoidea' (or 'miacids' and 'miacoids'). At this point, the inclusive informal terms 'basal carnivoramorphan' or 'early diverging carnivoramorphan' refer to a paraphyletic assemblage of both the Viverravidae and other early diverging taxa (traditionally referred to as 'Miacidae'), representing the spectrum of early fossil taxa that lie outside a clearly monophyletic clade including all of the living carnivorans (crown-clade Carnivora). The new clade name permits recognition that many of these taxa are more closely related to Carnivora than to Viverravidae or carnivoramorphan outgroups.

Pending recognition of additional well-supported subclades along the carnivoraform backbone phylogeny, the more restrictive informal terms 'basal carnivoraform' or 'early diverging carnivoraform' can be used to refer to all of the non-carnivoran Carnivoramorpha taxa other than the early diverging clade of Viverravidae, instead of more unwieldy terms such as 'basal non-viverravid carnivoramorphan'. We stress that erecting a new name that excludes Viverravidae does not imply that they are irrelevant to the history of the Carnivoraformes (see discussion below), and we examine locomotor reconstructions of all of these basal taxa (and consider the importance of sampling the spectrum of taxa in Viverravidae for accurate reconstructions of the ancestral states for Carnivoramorpha).

As mentioned earlier, relationships among some individual species of basal carnivoraforms are unstable across analyses, most notably when additional characters or taxa are added. Some groupings are generally consistent across studies, however. For example, either *Tapocyon* or *Quercygale* (or both, in an unresolved polytomy) are always the early diverging carnivoraforms most closely related to (or in a polytomy with) the Carnivora. The species of *Vulpavus* always form a monophyletic group, and *Vulpavus* and *Oödectes* both typically lie near the base of the Carnivoraformes diversification. '*Miacis*' *uintensis*, *Procynodictis vulpiceps*, and '*Miacis*' *sylvestris* consistently group together. One taxon that is never found to be monophyletic is the genus *Miacis*; there is no compelling evidence of a close relationship of the type species *Miacis parvivorus* to any other putative species assigned to *Miacis*. A thorough discussion of *Miacis* is beyond the scope of this review, although it is obvious that the taxon is badly in need of taxonomic revision (Spaulding *et al.*, in press). In addition, there are several other basal taxa such as *Uintacyon* or *Procynodictis* that have not yet had their monophyly tested in a rigorous phylogenetic study.

Basal Carnivoraformes appear later in the stratigraphic record than any species of Viverravidae, and are known from the Late Paleocene to the Late Eocene. Like the Viverravidae, they are first known from North America and then quickly spread to the rest of Laurasia (Flynn and Wesley-Hunt, 2005). There is a brief burst of diversification in the late-middle Eocene (Flynn, 1998). These animals generally had a smaller body mass range than the Viverravidae, with most approximating the size a small house cat (∼2 kg), although some grew as large as a coyote (∼20 kg). Their diet was predominantly carnivorous, although the dentition of some taxa (e.g. *Vulpavus*) suggests a tendency towards hypocarnivory and a more generalised diet.

Reconstructing the relationships and patterns of evolutionary diversification among the basal Carnivoraformes is a stimulating and active area of research, but many more taxa and characters await inclusion in future analyses.

Character transformations in a phylogenetic framework

Phylogenetic reconstruction within a clade forms the fundamental evolutionary frame of reference for further analysis into the evolution of character transformations and correlations within that group (Felsenstein, 1985; Swofford and Maddison, 1987; Maddison, 1991; Garland et al., 1999; Garland and Ives, 2000; Webster and Purvis, 2002). As such, phylogenetic reconstruction is a critical first step in a comprehensive study of the evolution of Carnivora, or any other group. Analyses of character evolution and comparative analyses often make use of well-supported phylogenies of extant taxa, but do so without reference to potentially relevant character information that is documented in the fossil record (Gittleman, 1993; Gittleman and Purvis, 1998; Webster and Purvis, 2002; Webster et al., 2004). Reconstructions of ancestral character states have greater associated errors as one proceeds from the tips to the root of a presumed ultrametric phylogeny, even in exceptional cases where the phylogeny is completely and perfectly known (Oakley and Cunningham, 2000). It is in this context that fossil taxa provide two important sources of information for ancestral character reconstruction. First, fossil data provide a more complete sampling of the entire distribution of character states through the evolutionary history of a clade, documenting morphologies that through extinctions are no longer represented in the extant sample. Second, fossils provide temporal information associated with the specimens, which can be used to weight character state observations in the reconstructions. Adding fossil taxa has been shown to positively affect both accuracy and precision of ancestral character state reconstructions (Oakley and Cunningham, 2000; Polly, 2001; Finarelli and Flynn, 2006, 2007).

In the following sections we present three detailed examples where an enhanced understanding of both the phylogeny and fossil record of the Carnivora have led to improved reconstructions of the patterns and underlying processes in character evolution. In the first, we discuss body size evolution. In the second, we examine the evolution of relative brain volume (brain volume scaled to body mass). In the third, we evaluate new information from the postcranial skeleton and previously unstudied basal fossil taxa, to more accurately infer the locomotor habitus of early carnivoran and carnivoramorphan taxa.

Evolution of body mass

Adult body mass has been described as a fundamental organismal variable in mammalian biology (Schmidt-Nielsen, 1984), as it is directly related to the energetics and physiology of the organism (McNab, 1988; Eisenberg,

1990; Harvey *et al.*, 1991; Carbone *et al.*, 1999, 2007). The tight interrelationship of basic energy requirements and body mass manifests itself in high correlations between body mass and many aspects of carnivoran life history (Gittleman and Harvey, 1982; Gittleman, 1986b, 1991) and ecology (Gittleman and Van Valkenburgh, 1997; Gittleman and Purvis, 1998; Meiri *et al.*, 2004a,b; Webster *et al.*, 2004; Friscia *et al.*, 2007). Not surprisingly, body mass has received considerable attention from paleobiologists. Estimation of body mass in extinct taxa remains an important, if contentious, area of research, with numerous morphometric proxies examined at different phylogenetic scales (e.g. Gingerich, 1977, 1990; Gingerich *et al.*, 1982; Legendre, 1986; Conroy, 1987; Legendre and Roth, 1988; Damuth and MacFadden, 1990; Jungers, 1990; Van Valkenburgh, 1990; Dagosto and Terranova, 1992; Anyonge, 1993; Delson *et al.*, 2000; Ruff, 1990; Sears *et al.*, 2008). With body mass estimates for fossil taxa becoming increasingly available, the evolution and timing, as well as potential mechanisms responsible for observed patterns, of body size evolution have been investigated for many mammalian clades (e.g. Stanley, 1973; Bookstein *et al.*, 1978; Van Valkenburgh, 1989, 1991; Alroy, 1998; Van Valkenburgh *et al.*, 2004; Finarelli and Flynn, 2006; Finarelli, 2007).

Body masses among extant carnivorans span more than four orders of magnitude, with this entire range realised in the Caniformia (least weasel, *Mustela nivalis*, ~100 g to southern elephant seal, *Mirounga leonina*, ~1600 kg) (Smith *et al.*, 2003). Among the major caniform clades, extant ursids and pinnipeds are large-bodied, with median body sizes of ~104 kg and ~145 kg, respectively, whereas musteloids are generally small-bodied, with a median body size of only ~1.5 kg (Finarelli and Flynn, 2006). Given the branching order observed among the family-level clades (with pinnipeds allied with musteloids), the body mass of the last common ancestor (LCA) of all Caniformia and the LCA of Arctoidea are reconstructed as large-bodied organisms (~10–50 kg) when only extant taxa are considered.

Finarelli and Flynn (2006) gathered body mass data for 149 extant caniforms and body mass estimates for 367 fossil caniforms. Using the molecular phylogeny of Flynn *et al.* (2005) as a backbone, they incorporated fossil taxa using overlapping taxa from morphological phylogenies, and reconstructed ancestral body masses for four deep nodes in the evolutionary history of the Caniformia: the LCAs of Caniformia, Arctoidea, Pinnipedia + Musteloidea node, and Musteloidea, using weighted, squared-change parsimony. The LCAs of Caniformia and Arctoidea were reconstructed as small-bodied (~1–5 kg) when fossil taxa and temporal information were included in the analysis. These results were robust to ambiguity in the phylogenetic position of problematic fossil groups and to a current lack of accurate body mass estimates for fossil pinnipeds. This order of magnitude difference in reconstructed ancestral mass has a profound impact on our interpretations of the biology of the caniform LCA, and will influence our reconstructions on a wide range of life-history and ecological attributes, such as diet, energy expenditure, home range size, reproductive biology, and sociality (Gittleman and Harvey, 1982; Gittleman, 1986a, b; Carbone *et al.*, 1999, 2007; Van Valkenburgh *et al.*, 2003, 2004; Muñoz-Garcia and Williams, 2005; Dalerum, 2007; Friscia *et al.*, 2007). Additionally, this indicates that caniform clades that are today represented by large forms (Ursidae and Pinnipedia) and several clades that achieved large body size in the fossil record (Borophaginae, Hesperocyoninae, Amphicyonidae) did so independently.

Among caniforms, the fossil record of the Canidae is exceptionally well sampled, spanning approximately the last 40 million years (Wang, 1994; Tedford *et al.*, 1995; Wang *et al.*, 1999). The Canidae form a model clade for examining character evolution across phylogeny and through time. Pronounced trends towards increased body size have been documented in each of the three canid subfamilies (as they were in several other caniform clades [Finarelli and Flynn, 2006]): Hesperocyoninae, Borophaginae, and Caninae (Figure 2.5). Furthermore, Canidae often has been used to document 'Cope's Rule' of progressive body size increase through time (Wang, 1994; Wang *et al.*, 1999; Van Valkenburgh *et al.*, 2004; Finarelli and Flynn, 2006). Body size evolution in the Hesperocyoninae and Borophaginae appears to be an active replacement of smaller taxa with larger forms through time. The pattern in Caninae is qualitatively different, with an increase in maximum and mean body sizes but without the disappearance of small body sizes.

Finarelli (2007) investigated mechanisms responsible for the patterns observed in canid body size evolution for a data set of 151 species. First and last appearance events were documented for each, and species were binned into 2-million-year time slices. Proportions of taxon origination and extinction events within each time slice were classified as 'large' or 'small', that is, greater or less than median bin value. Several evolutionary models were then tested against the origination and extinction data. Extinction events never significantly deviated from a model that was unbiased with respect to larger or smaller species going extinct (Finarelli, 2007), implying that canid body size evolution is driven by bias in originations, not selective culling of taxa. Body size evolution in the Canidae is characterised by a background rate of size increase across the entire Canidae, interrupted by more extreme biases towards large originations during periods of increased diversification in each canid subfamily (Finarelli, 2007), validating the concept of 'Cope's Rule' among canids (Van Valkenburgh *et al.*, 2004), as it documents continuous, driven trend in body size (McShea, 1994). However, a final feature of canid body size evolution

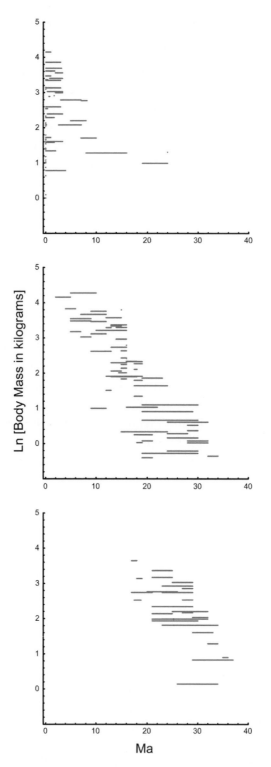

Figure 2.5 Canidae body size evolution. Range plots of body mass through time for the Canidae. Lines connect first and last appearances for canid

is a reversal promoting biases towards smaller forms in the Quaternary, or a counter to 'Cope's Rule' (Finarelli, 2007). This is associated with taxonomic diversification among small-bodied canids typically referred to as 'foxes', although these species are distributed among each of the three main clades of extant dogs (Bardeleben et al., 2005). This bias toward smaller forms is not a 'pull of the Recent' bias, as the pattern remains even if all taxa not possessing a fossil record are removed from the analysis. This documents a reoccupation of small-body space by canine dogs, and explains the qualitative differences observed in canid body size evolution.

Evolution of relative brain volume

The study of the evolution of relative brain size is a classic theme in vertebrate paleobiology, and the literature is replete with analyses demonstrating progressive increase in relative brain size across phylogenetic scales: among vertebrates (Jerison, 1973; Martin, 1981, 1996), across mammalian orders (Jerison, 1970; Radinsky, 1978; Pagel and Harvey, 1988a,b, 1989; Marino, 1998; Martin et al., 2005), and within mammalian clades (Radinsky, 1977a; Marino et al., 2004). Many studies of relative brain size evolution are primatological (e.g. Jerison, 1973; Radinsky, 1973, 1977b; Martin, 1984, 1990; Simons, 1993; Elton et al., 2001; Sears et al., 2008), a taxonomic bias likely imparted by their large-brained human authors. As such, there often is an implicit assumption that increased encephalisation through the evolutionary history of a clade is somehow causally connected to increased 'gross intelligence' (Jerison, 1970, 1991), although this is difficult, and maybe even impossible, to define rigorously.

Brain volume scales to body mass with negative allometry, such that progressively larger animals have a progressively smaller brain volume to body mass ratios (Jerison, 1961). As such, encephalisation (volume scaled to mass) is usually studied as the metric for the evolution of brain size, with encephalisation

Figure 2.5 (*cont.*)
species against the natural logarithm of body mass in kilograms. The bottom panel is the extinct subfamily Hesperocyoninae. Middle panel is the extinct subfamily Borophaginae. The top panel is the subfamily Caninae, to which all extant dog species belong. The pattern of body size evolution in the two extinct subfamilies appears to be an active replacement of smaller body sizes with larger body sizes. The pattern for the Caninae is different in that, while there is increase in both maximum and mean body mass through time, the region of small body sizes is not evacuated.

typically measured with the Encephalisation Quotient (EQ) (e.g. Jerison, 1973; Radinsky, 1977a), which is the ratio of observed to expected brain volume, relative to the brain–body size allometry. However, we generally are not concerned with the ratio of observed to expected volume, but rather proportional change in relative volume, or logEQ (= the natural logarithm of EQ, or the deviation from the allometry regression in log-space) (Finarelli and Flynn, 2007; Finarelli, 2008a), that is the measure of interest (see also: Marino *et al.*, 2004).

Brain–body size relationships have been investigated among extant Carnivora with respect to the correlation between encephalisation and a range of life-history and ecological variables (Gittleman and Harvey, 1982; Gittleman, 1986a,b, 1991, 1994; Dunbar and Bever, 1998; Iwaniuk *et al.*, 1999). However, until recently, our only knowledge of brain volumes for fossil carnivorans was restricted to preserved endocasts (sediment-filled endocranial cavities, where subsequent weathering has removed the fossilised bone exposing a natural cast) (Jerison, 1970, 1973; Radinsky, 1977a, 1978). The exceptional preservation required to preserve a complete endocast from which a volume can be derived has severely limited the sample size of endocranial volumes for fossil carnivorans. For example, within Caniformia, endocranial volumes from endocasts had been reported for only eleven fossil taxa (Jerison, 1970, 1973; Radinsky, 1977a, 1978), which contrasts markedly with the 367 fossil taxa incorporated in an analysis of body size evolution (Finarelli and Flynn, 2006). Finarelli (2006) recently employed multiple linear regression and a model-averaging technique using the Akaike Information Criterion (AIC) (Burnham and Anderson, 2002) to accurately estimate endocranial volume for extant carnivorans from three external measurements of the cranium approximating braincase length, width and height. Finarelli and Flynn (2007) verified that this model also accurately predicted brain volumes for fossil taxa, using a set of fossil taxa with both reported endocast volumes and sufficient cranial material to apply the model. The crania of 123 fossil specimens were then measured, expanding the data set of fossil caniforms with endocranial volume estimates from 11 to 60 species (Finarelli and Flynn, 2007), and the list of fossil taxa with brain volume estimates continues to expand using this method.

Finarelli and Flynn (2007) calculated logEQs for all caniforms relative to an allometry defined by the extant taxa, relating encephalisation of fossil taxa to an extant benchmark. From this, they observed that for much of caniform evolutionary history, logEQs are at or below the modern median, indicating that fossil caniforms had consistently smaller brains relative to their body mass. When binned into time slices, logEQs show only a single significant change in the median, an upward shift to a distribution with the modern median value, corresponding approximately with the Miocene/Pliocene transition (Figure 2.6)

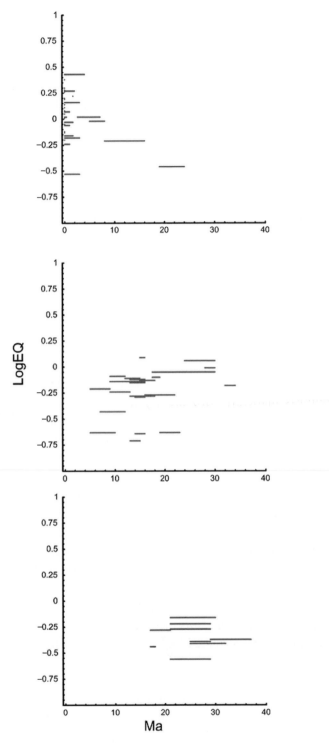

Figure 2.6 Canidae brain size evolution. Range plots of relative brain size through time for the Canidae. Lines connect first and last appearances for

(Finarelli and Flynn, 2007). However, almost half of the fossil taxa in that analysis were canids, and Canidae underwent substantial taxonomic diversification at this time (Munthe, 1998; Finarelli, 2007). As such, it is possible that over-sampling in this clade drove the observed shift. Applying the ancestral reconstruction methods for body size (Finarelli and Flynn, 2006) to relative brain size, ancestral logEQs were calculated including all fossil taxa and excluding fossil canids. If the Canidae were responsible for the pattern, then removal of fossil canids should impact reconstructions of relative brain size for the LCAs of the Musteloidea and the families within the Musteloidea. This was not the case, however; reconstructions without canids were nearly identical to the all-taxa reconstructions, and increases in encephalisation to the modern distribution must have occurred independently in the extant families.

Returning to the Canidae as a model system, Finarelli (2008a) expanded the sample of canid endocranial volume estimates, such that 28 species of extant canids and 44 extinct taxa were sampled, allowing detailed examination of the pattern of brain size evolution with respect to the canid phylogeny. Finarelli (2008a) compared several encephalisation models, including a single allometry for the Canidae and models proposing different allometries among the three subfamilies. However, it was found that the most supported model actually grouped the subfamilies Hesperocyoninae and Borophaginae with the extinct stem canine genus *Leptocyon*, in contrast to the crown clade of the subfamily Caninae. Rank-order comparison of logEQs for canids in these two groups indicates a shift to increased relative brain size that can be localised to the branch of the canid phylogeny representing the crown clade. To test whether this shift represented a single shift to higher encephalisation for all members of the crown clade, or if there were differences among the three distinct subgroup radiations of modern canids (*Canis*-like, *Vulpes*-like, and South American canids) (Wayne *et al.*, 1997; Bardeleben *et al.*, 2005), Finarelli (2008a) also compared the rank logEQs among crown-clade taxa, and there were no

Caption for figure 2.6 (*cont.*)

canid species against logEQ (the natural logarithm of the encephalisation quotient, measured against extant caniforms). The bottom panel is the extinct subfamily Hesperocyoninae. The middle panel is the extinct subfamily Borophaginae. The top panel is the subfamily Caninae, to which all extant dog species belong. Both Hesperocyoninae and Borophaginae show no tendency for increase in relative brain size through time. The median logEQ for both clades is below the modern value (logEQ = 0). Caninae shows a pronounced increase in both maximum and median logEQ. This shift can be localised to the branch of the phylogeny that defines the crown radiation of the Caninae.

significant differences among any of the extant clades, indicating a single, apomorphic shift for the entire crown clade.

Reconstructing locomotor styles and habitus in early Carnivoramorpha

The postcranial anatomy of basal carnivoramorphans has received little attention. As for many other mammalian groups, prior studies have predominately focused on dental and cranial material, even when well-preserved post-cranial skeletons were recovered with those craniodental specimens. Historical exceptions to this are found in Matthew (1909) and Clark (1939), although these descriptions are very brief when compared with later works. Jenkins and Camazine (1977) examined several non-Viverravidae taxa, primarily utilising ratios, and it was not until the work of Heinrich and Rose (1995, 1997) that detailed examinations of discrete morphological features in basal carnivoramorphan postcranial skeletons were first undertaken.

These initial studies of basal carnivoramorphans noted extreme differences between the two supposed monophyletic families of basal carnivoramorphans, the Viverravidae and the basal Carnivoraformes ('miacids'). The viverravids, exemplified by *Didymictis*, were considered terrestrially adapted, perhaps incipiently cursorial or with some fossorial tendencies, whereas the basal carnivoraforms were portrayed as arboreal, with *Vulpavus* the best represented taxon. It should be stressed that this apparent locomotor dichotomy between the two groups was not the view of the original authors, but rather how others have generally perceived the taxa due to the sparse sampling of basal carnivoramorphans.

Re-examination of 'Miacidae' (Wesley and Flynn, 2003; Spaulding, 2007; Spaulding and Flynn, 2009) postcranial elements has documented a wide variety of morphological diversity (Figure 2.7). This diversity emphasises the inadvisability of assuming that conditions observed in *Vulpavus* would necessarily extend to other early carnivoramorphans, as was the case when 'Miacidae' was viewed as monophyletic. Perhaps the most extreme differences from previously described early carnivoraforms can be seen in '*Miacis*' *uintensis* (Spaulding and Flynn, 2009), which has many features that are more in line with a scansorial way of life than the arboreal one inferred previously (e.g. for *Vulpavus*). It is noteworthy that this specimen was referred to by Matthew (1909) as an 'aberrant form' and disregarded, remaining undescribed until recently. However, it now is clear that this species is not an anomaly in a sea of more arboreal taxa, as *Procynodictis vulpiceps* (Clark, 1939) and *Tapocyon robustus* also possess postcranial skeletons that appear to be ill-adapted for a

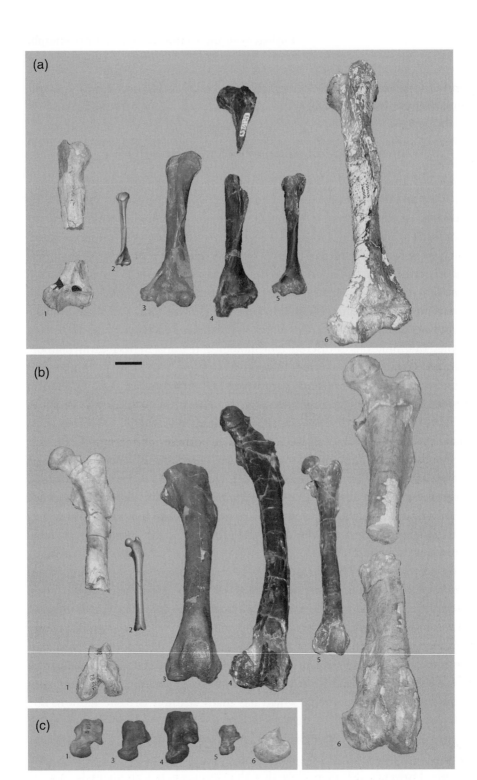

Figure 2.7 Postcranial elements of representative basal carnivoramorphans.
a, Anterior view of humeri; b, anterior view of femora; c, dorsal view of astragali.

primarily arboreal way of life. It is likely, in fact, that the well-preserved and most intensively studied taxon, *Vulpavus*, instead could be the most arboreally specialised and distinctive basal carnivoraform known.

The Viverravidae also vary substantially in their postcranial morphology, as shown by the recent analysis of *Viverravus acutus* (Heinrich and Houde, 2006). This taxon has features that suggest a considerably less terrestrial and more scansorial locomotor style than in *Didymitcis*. Furthermore, there appears to be locomotor variation even among the species of *Didymticis* (Spaulding, pers. observ.). Unfortunately, well-preserved viverravid remains are less common than for basal carnivoraforms, but there are additional specimens yet to be studied which will better elucidate the extent of the variation in locomotor specialisations among viverravids.

Studying these ancient fossil carnivoramorphans allows us to determine more than just the variety of locomotor modes of the extinct taxa themselves. Rather, inclusion of more fossil taxa will lead to enhanced understanding of the locomotor habits of the Viverravidae, the ancestral condition for Carnivora-formes, potential shared locomotor habitus for subclades of basal carnivora-forms, transformations leading to the Carnivora, and assessment of the stability or lability of locomotor habitus during evolution of the group. This also will allow us to develop more rigorously tested hypotheses of ancestral locomotor conditions for the Carnivora than could be attained solely from examining crown taxa. To accomplish this goal, of course we must first examine all the relevant taxa in a phylogenetic context. The trees that have been constructed for the initial analysis presented here are based on preliminary data (Spaulding *et al.*, in press; Spaulding PhD, in progress), but still allow us to build a better view of the evolution of locomotor habits than has been possible in past studies.

Figure 2.8 shows the hypotheses of relationships generated by morphological character-based phylogenetic analyses of early carnivoramorphans and repre-sentative Carnivora (Spaulding *et al.*, in press). Locomotor conditions have been mapped upon the tree and ancestral conditions have been reconstructed utilising MacClade (Maddison and Maddison, 2007), implementing ACCTRAN and DELTRAN optimisations to generate unambiguous and robust results. We reconstruct ambiguous ancestral conditions for Viverravidae

Figure 2.7 (*cont.*)

1: *Didymictis protenus* USGS 27585, 2: *Viverravus acutus* USNM 489122, 3: *Vulpavus* AMNH 12626 and 11497, 4: '*Miacis*' *uintensis* AMNH 1964, 5: *Oödectes herpestoides* 140008, 6: *Tapocyon robustus* SDSNH 36000. Drawings of *Viverravus acutus* from Heinrich and Houde (2006). Scale bar = 1 cm.

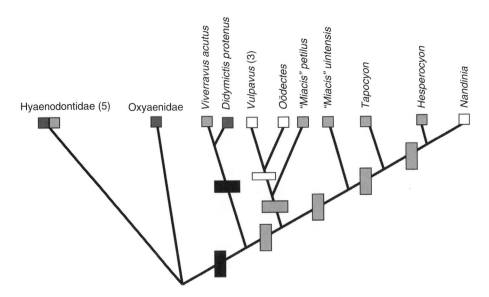

Figure 2.8 Inferred ancestral locomotor reconstructions for Carnivoramorpha. Locomotor conditions mapped upon a hypothesis of relationships for basal carnivoramorphans and their putative nearest outgroups, the Hyaenodontidae and Oxyaenidae (Creodonta). Light grey = scansorial, dark grey = terrestrial, white = arboreal; black = ambiguous reconstruction. A scansorial basal Carnivoraformes is reconstructed in this analysis, with an ambiguous ancestral locomotor condition for the Carnivoramorpha. The limited sample of Carnivora yields a basal condition of scansorial for this clade.

and Carnivoramorpha, with an unambiguous reconstruction of an ancestral scansorial locomotor style for Carnivoraformes and Carnivora. It is important to note that this preliminary analysis emphasised reconstructions of ancestral locomotor states for clades in the early diversification of the Carnivoramorpha, and therefore included only representative members of the crown-clade Carnivora, and thus the ancestral reconstruction for Carnivora must be viewed as tentative. It is noteworthy that even though relationships among basal carnivoraform species has changed substantially as taxon and character sampling has increased, the backbone topology of interrelationships has remained stable, with a monophyletic basal Viverravidae, paraphyletic array of species previously assigned to 'Miacidae', then Nimravidae (false sabre-cats) as nearest sister-group to crown-clade Carnivora.

   This preliminary analysis emphasises how sensitive such reconstructions are to taxon sampling (pertinent to both extinct and extant taxa), both in terms of how they affect the topology of the framework phylogeny that is required for rigorous assessments of ancestral conditions, and how sampling variation in locomotor styles among close relatives also can substantially alter ancestral reconstructions. In particular, it is clear that a better understanding of the evolution of locomotor styles in the Viverravidae is needed. Only two relatively complete skeletons of this clade have been studied in depth previously, and they differ substantially from one another in their postcranial morphology. However, many more skeletons remain unstudied in museum collections, and more work is required to determine the primitive anatomical features and inferred locomotor condition for the Viverravidae clade. Whatever these conditions might be, they will have a major impact on reconstructing the phylogeny and ancestral conditions of the Carnivoramorpha, Carnivoraformes, and Carnivora.

Based on the most conservative assessment of currently available information, the ancestral locomotor condition for both the Carnivoraformes and Carnivora is scansorial (not terrestrial), although including more crown taxa is essential for generating a more reliable reconstruction for the Carnivora node.

Conclusions

Recent morphological, molecular, and combined primary character analyses yield a well-resolved and stable higher-level phylogeny of Carnivora and parts of Carnivoramorpha; current studies will soon permit similar refinements at lower levels. Morphological analyses indicate that 'Creodonta' are the nearest relatives of Carnivoramorpha, while molecular results suggest that Pholidota are the nearest living relatives of Carnivora. There is strong support for monophyletic Carnivoramorpha, Carnivora, Caniformia, Feliformia, Arctoidea, Pinnipedia, Musteloidea, Feloidea, Herpestoidea, and all traditional modern families (except Viverridae and Mustelidae); many subclades also are well resolved. Early Cenozoic carnivoramorphans (a monophyletic Viverravidae plus a paraphyletic series of stem 'miacids') form basal outgroups to crown Carnivora. Based on those results, we name and provide a phylogenetic definition and diagnosis for the clade including all carnivoramorphans except Viverravidae (the Carnivoraformes).

Fossils, placed within a robust phylogeny, are crucial to interpretations of many evolutionary transformations. Studies of Recent taxa alone indicate a large ancestral body size for Arctoidea, whereas including fossils documents small-bodied ancestors for Caniformia and Arctoidea, and a moderate-sized ancestral musteloid with size reduction in some mustelids. Brain volumes can be accurately estimated for fossil carnivoran taxa using three simple external cranial measures, allowing the number of fossil caniform endocranial volume estimates to be increased by ~700%. Body size and brain volume estimates for fossil caniforms suggest their encephalisation was at or below the modern median from the Oligocene–late Miocene and that modern encephalisation levels were achieved independently in caniform 'familial' clades. Canidae encephalisation appears to follow a simple phase shift of increased brain volume restricted to the crown radiation, coinciding with both taxonomic diversification and anatomical reorganisation of the neocortex.

Prior studies of 'miacoid' postcranial anatomy generally used exemplars. Detailed analyses of more (and more varied) taxa document mosaic distributions of features, with greater disparity in anatomy and inferred habitus (terrestrial, scansorial, arboreal). These results indicate that reliable ancestral locomotor reconstructions for major carnivoramorphan clades and understanding of transformations during the evolution of the group will require enhanced taxon sampling, which is feasible given the large number of early carnivoramorphan taxa now recognised as having preserved postcrania.

Acknowledgements

L. Meeker and C. Tarka provided the excellent photographs in Figure 2.7; E. Peterson and J. Kelly assisted with specimen preparation. We thank W. Simpson, W. Stanley, J. Meng, C. Norris, J. Galkin, R. Purdy, W. Joyce, D. Brinkman, M. Benoit, P. Tassy, C. Sagne, A. Currant, X. Wang, S. McLeod, and G. Takeuchi for access to collections. This project was supported by an AMNH Collections Study Grant, the National Science Foundation (DEB-0608208 to JAF; DEB-0614098 to JJF; and AToL Mammalia Morphology grant BIO EF– 0629811 to JJF and colleagues); a Brown Family Foundation Graduate Fellowship and the University of Michigan, Society of Fellows (to JAF); and an NSF Graduate Research Fellowship, the Frick Fund (AMNH), and Columbia University Graduate Fellowship (to MS).

REFERENCES

Allard, M. W. and Carpenter, J. M. (1996). On weighting and congruence. *Cladistics*, **12**, 183–98.
Alroy, J. (1998). Cope's rule and the dynamics of body mass evolution in North American fossil mammals. *Science*, **280**, 731–34.
Anyonge, W. (1993). Body-mass in large extant and extinct carnivores. *Journal of Zoology*, **231**, 339–50.
Arango, C. P. and Wheeler, W. C. (2007). Phylogeny of the sea spiders (Arthropoda, Pycnogonida) based on direct optimization of six loci and morphology. *Cladistics*, **23**, 255–93.

Arnason, U., Bodin, K., Gullberg, A., Ledje, C. and Mouchaty, S. (1995). A molecular view of pinniped relationships with particular emphasis on the true seals. *Journal of Molecular Evolution*, **40**, 78–85.

Arnason, U., Gullberg, A., Janke, A., *et al.* (2006). Pinniped phylogeny and a new hypothesis for their origin and dispersal. *Molecular Phylogenetics and Evolution*, **41**, 345–54.

Arnason, U., Gullberg, A., Janke, A. and Kullberg, M. (2007). Mitogenomic analyses of caniform relationships. *Molecular Phylogenetics and Evolution*, **45**, 863–74.

Asher, R. J. (2007). A web-database of mammalian morphology and a reanalysis of placental phylogeny. *BMC Evolutionary Biology 2007*, **7**, 108 (10 pages; doi 10.1186/1471–2148–7–108).

Asher, R. J., Emry, R. J. and McKenna, M. C. (2005). New material of *Centetodon* (Mammalia, Lipotyphla) and the importance of (missing) DNA sequences in systematic paleontology. *Journal of Vertebrate Paleontology*, **25**, 911–23.

Baker, R. H. and DeSalle, R. (1997). Multiple sources of character information and the phylogeny of Hawaiian drosophilids. *Systematic Biology*, **46**, 654–73.

Bardeleben, C., Moore, R. L. and Wayne, R. K. (2005). A molecular phylogeny of the Canidae based on six nuclear loci. *Molecular Phylogenetics and Evolution*, **37**, 815–31.

Barnett, R., Barnes, I., Phillips, M. J., *et al.* (2005). Evolution of the extinct sabretooths and the American cheetah-like cat. *Current Biology*, **15**(15), R589–90.

Berta, A., Sumich, J. L. and Kovacs, K. M. (2006). *Marine Mammals: Evolutionary Biology*, 2nd edn. San Diego, CA: Academic Press, 560 pp.

Bond, J. E. and Hedin, M. (2006). A total evidence assessment of the phylogeny of North American euctenizine trapdoor spiders (Araneae, Mygalomorphae, Cyrtaucheniidae) using Bayesian inference. *Molecular Phylogenetics and Evolution*, **41**, 70–85.

Bookstein, F. L., Gingerich, P. D. and Kluge, A. G. (1978). Hierarchical linear modeling of the tempo and mode of evolution. *Paleobiology*, **4**, 120–34.

Bryant, H. N. (1991). Phylogenetic relationships and systematics of the Nimravidae (Carnivora). *Journal of Mammalogy*, **72**, 56–78.

Burnham, K. P. and Anderson, D. R. (2002). *Model Selection and Multimodel Inference: A Practical Information-Theoretic Approach*. New York, NY: Springer, 488 pp.

Cantino, P. D. and deQuiroz, K. (eds.) (2007). *Phylocode. International Code of Phylogenetic Nomenclature*, version 4b (12 September 2007; http://www.ohio.edu/phylocode/PhyloCode4b.pdf).

Carbone, C., Mace, G. M., Roberts, S. C. and Macdonald, D. W. (1999). Energetic constraints on the diet of terrestrial carnivores. *Nature*, **402**, 286–88.

Carbone, C., Teacher, A. and Rowcliffe, J. M. (2007). The cost of carnivory. *PLoS Biology*, **5**, 363–68.

Clark, J. (1939) *Miacis gracilis*, a new carnivore from the Uinta Eocene. *Annals of Carnegie Museum*, **27**, 349–70.

Conroy, G. C. (1987). Problems of body weight estimation in fossil Primates. *International Journal of Primatology*, **8**, 115–38.

Cope, E. D. (1880). On the genera of the Creodonta. *Proceedings of the American Philosophical Society*, **19**, 76–82.

Dagosto, M. and Terranova, C. J. (1992). Estimating the body size of Eocene Primates: A comparison of results from dental and postcranial variables. *International Journal of Primatology*, **13**, 307–44.

Dalerum, F. (2007). Phylogenetic reconstruction of carnivore social organizations. *Journal of Zoology*, **273**, 90–97.

Damuth, J. and MacFadden, B. J. (1990). Introduction: body size and its estimation. In *Body Size in Mammalian Paleobiology*, ed. J. Damuth and B. J. MacFadden. New York, NY: Cambridge University Press, pp. 1–9.

Davis, C. S., Delisle, I., Stirling, I., Siniff, D. B. and Strobeck, C. (2004). A phylogeny of the extant Phocidae inferred from complete mitochondrial DNA coding regions. *Molecular Phylogenetics and Evolution*, **33**, 363–77.

Davis, D. D. (1964). The giant panda: a morphological study of evolutionary mechanisms. *Fieldiana Zoology Memoirs*, **3**: 1–339.

Delisle, I. and Strobeck, C. (2005). A phylogeny of the Caniformia (order Carnivora) based on 12 complete protein-coding mitochondrial genes. *Molecular Phylogenetics and Evolution*, **37**, 192–201.

Delson, E., Terranova, C. J., Jungers, W. L., Sargis, E. J., Jablonski, N. G. and Dechow, P. C. (2000). Body mass in Cercopithecidae (Primates, Mammalia): estimation and scaling in extinct and extant taxa. *Anthropological Papers of the American Museum of Natural History*, **83**, 1–159.

Delsuc, F., Scally, M., Madsen, O., *et al.* (2002). Molecular phylogeny of living xenarthrans and the impact of character and taxon sampling on the placental tree rooting. *Molecular Biology and Evolution*, **19**, 1656–71.

Dragoo, J. W., Bradley, R. D., Honeycutt, R. L. and Templeton, J. W. (1993). Phylogenetic relationships among the skunks: a molecular perspective. *Journal of Mammalian Evolution*, **1**(4), 255–67.

Dragoo, J. W. and Honeycutt, R. L. (1997). Systematics of mustelid-like carnivores. *Journal of Mammalogy*, **78**, 426–43.

Dunbar, R. I. M. and Bever, J. (1998). Neocortex size predicts group size in carnivores and some insectivores. *Ethology*, **104**, 695–708.

Eisenberg, J. F. (1990). The behavioral/ecological significance of body size in the Mammalia. In *Body Size in Mammalian Paleobiology*, ed. J. Damuth and B. J. MacFadden. New York, NY: Cambridge University Press, pp. 25–37.

Elton, S., Bishop, L. C. and Wood, B. (2001). Comparative context of Plio-Pleistocene hominin brain evolution. *Journal of Human Evolution*, **41**, 1–27.

Felsenstein, J. (1985). Phylogenies and the comparative method. *American Naturalist*, **125**, 1–15.

Finarelli, J. A. (2006). Estimation of endocranial volume through the use of external skull measures in the Carnivora (Mammalia). *Journal of Mammalogy*, **87**, 1027–36.

Finarelli, J. A. (2007). Mechanisms behind active trends in body size evolution in the Canidae (Carnivora: Mammalia). *American Naturalist*, **170**, 876–85.

Finarelli, J. A. (2008a). Testing hypotheses of the evolution of brain–body size scaling in the Canidae (Carnivora, Mammalia). *Paleobiology*, **34**, 48–58.

Finarelli, J. A. (2008b). A total evidence phylogeny of the Arctoidea (Carnivora: Mammalia): relationships among basal taxa. *Journal of Mammalian Evolution*, **15**, 231–59.

Finarelli, J. A. and Flynn, J. J. (2006). Ancestral state reconstruction of body size in the Caniformia (Carnivora, Mammalia): the effects of incorporating data from the fossil record. *Systematic Biology*, **55**, 301–13.

Finarelli, J. A. and Flynn, J. J. (2007). The evolution of encephalization in caniform carnivorans. *Evolution*, **61**, 1758–72.

Flynn, J. J. (1996). Phylogeny and rates of evolution: morphological, taxic and molecular. In *Carnivore Behavior, Ecology, and Evolution, Vol. 2*, ed. J. Gittleman. Ithaca, NY: Cornell University Press, pp. 542–81.

Flynn, J. J. (1998). Early Cenozoic Carnivora ('Miacoidea'). In *Evolution of Tertiary Mammals of North America (Vol. 1: Terrestrial Carnivores, Ungulates, and Ungulatelike Mammals)*, ed. C. M. Janis, K. M. Scott and L. L. Jacobs. Cambridge: Cambridge University Press, pp. 110–23.

Flynn, J. J. and Galiano, H. (1982). Phylogeny of Early Tertiary Carnivora, with a description of a new species of *Protictis* from the Middle Eocene of Northwestern Wyoming. *American Museum Novitates*, **2725**, 1–64.

Flynn, J. J. and Wyss, A. R. (1988). Letter to the Editor [re: S. J. O'Brien, 'The ancestry of the giant panda', *Scientific American*, November, 1987]. *Scientific American*, **June 1988**, 8.

Flynn, J. J. and Nedbal, M. A. (1998). Phylogeny of the Carnivora (Mammalia): congruence vs. incompatibility among multiple data sets. *Molecular Phylogenetics and Evolution*, **9**, 414–26.

Flynn, J. J. and Wesley-Hunt, G. D. (2005). Carnivora. In *Origin, Timing, and Relationships of the Major Clades of Extant Placental Mammals*, ed. D. Archibald, and K. D. Rose. Baltimore, MD: Johns Hopkins University Press, pp. 175–98.

Flynn, J. J., Neff, N. A. and Tedford, R. H. (1988). Phylogeny of the Carnivora. In *Phylogeny and Classification of the Tetrapods*, ed. M. J. Benton. Oxford: Clarendon Press, pp. 73–116.

Flynn, J. J, Nedbal, M. A., Dragoo, J. W. and Honeycutt, R. L. (2000). Whence the red panda? *Molecular Phylogenetics and Evolution*, **17**, 190–99.

Flynn, J. J., Finarelli, J. A., Zehr, S., Hsu, J. and Nedbal, M. A. (2005). Molecular phylogeny of the Carnivora (Mammalia): assessing the impact of increased sampling on resolving enigmatic relationships. *Systematic Biology*, **54**, 317–37.

Friscia, A. R., Van Valkenburgh, B. and Biknevicius, A. R. (2007). An ecomorphological analysis of extant small carnivorans. *Journal of Zoology*, **272**, 82–100.

Fulton, T. L. and Strobeck, C. (2006). Molecular phylogeny of the Arctoidea (Carnivora): effect of missing data on supertree and supermatrix analyses of multiple gene data sets. *Molecular Phylogenetics and Evolution*, **41**, 165–81.

Fulton, T. L. and Strobeck, C. (2007). Novel phylogeny of the raccoon family (Procyonidae: Carnivora) based on nuclear and mitochondrial DNA evidence. *Molecular Phylogenetics and Evolution*, **43**, 1171–77.

Fyler, C. A., Reeder, T. W., Berta, A., Antonelis, G., Aguilar, A. and Androukaki, E. (2005). Historical biogeography and phylogeny of monachine seals (Pinnipedia: Phocidae) based on mitochondrial and nuclear DNA data. *Journal of Biogeography*, **32**, 1267–79.

Garland, T. and Ives, A. R. (2000). Using the past to predict the present: confidence intervals for regression equations in phylogenetic comparative methods. *American Naturalist*, **155**, 346–64.

Garland, T., Midford, P. E. and Ives, A. R. (1999). An introduction to phylogenetically based statistical methods, with a new method for confidence intervals on ancestral values. *American Zoologist*, **39**, 374–88.

Gatesy, J. and Baker, R. H. (2005). Hidden likelihood support in genomic data: can forty-five wrongs make a right? *Systematic Biology*, **54**, 483–92.

Gatesy, J. and O'Leary, M. A. (2001). Deciphering whale origins with molecules and fossils. *Trends in Ecology & Evolution*, **16**, 562–70.

Gatesy, J., Amato, G., Norell, M., DeSalle, R. and Hayashi, C. (2003). Combined support for wholesale taxic atavism in gavialine crocodylians. *Systematic Biology*, **52**, 403–22.

Gaubert, P. and Begg, C. M. (2007). Re-assessed molecular phylogeny and evolutionary scenario within genets (Carnivora, Viverridae, Genettinae). *Molecular Phylogenetics and Evolution*, **44**, 920–27.

Gaubert, P. and Cordeiro-Estrela, P. (2006). Phylogenetic systematics and tempo of evolution of the Viverrinae (Mammalia, Carnivora, Viverridae) within feliformians: implications for faunal exchanges between Asia and Africa. *Molecular Phylogenetics and Evolution*, **41**, 266–78.

Gaubert, P. and Veron, G. (2003). Exhaustive sample set among Viverridae reveals the sister-group of felids: the linsangs as a case of extreme morphological convergence within Feliformia. *Proceedings of the Royal Society London B Biological Sciences*, **270**, 2523–30.

Gingerich, P. D. (1977). Correlation of tooth size and body size in living hominoid primates, with a note on relative brain size in *Aegyptopithecus* and *Proconsul*. *American Journal of Physical Anthropology*, **47**, 395–98.

Gingerich, P. D. (1990). Prediction of body mass in mammalian species from long bone lengths and diameters. *Contributions from the Museum of Paleontology, The University of Michigan*, **28**, 79–92.

Gingerich, P. D., Smith, H. B. and Rosenberg, K. (1982). Allometric scaling in the dentition of Primates and prediction of body weight from tooth size in fossils. *American Journal of Physical Anthropology*, **58**, 81–100.

Giribet, G., Edgecombe, G. D. and Wheeler, W. C. (2001). Arthropod phylogeny based on eight molecular loci and morphology. *Nature*, **413**, 157–61.

Gittleman, J. L. (1986a). Carnivore brain size, behavioral ecology, and phylogeny. *Journal of Mammalogy*, **67**, 23–36.

Gittleman, J. L. (1986b). Carnivore life history patterns: allometric, phylogenetic, and ecological associations. *American Naturalist*, **127**, 744–71.

Gittleman, J. L. (1991). Carnivore olfactory bulb size: allometry, phylogeny and ecology. *Journal of Zoology*, **225**, 253–72.

Gittleman, J. L. (1993). Carnivore life histories: a reanalysis in light of new models. *Symposia of the Zoological Society of London*, **65**, 65–86.

Gittleman, J. L. (1994). Female brain size and parental care in carnivores. *Proceedings of the National Academy of Sciences USA*, **91**, 5495–97.

Gittleman, J. L. and Harvey, P. H. (1982). Carnivore home-range size, metabolic needs and ecology. *Behavioral Ecology and Sociobiology*, **10**, 57–63.

Gittleman, J. L. and Purvis, A. (1998). Body size and species-richness in carnivores and primates. *Proceedings of the Royal Society of London Series B Biological Sciences*, **265**, 113–19.

Gittleman, J. L. and Van Valkenburgh, B. (1997). Sexual dimorphism in the canines and skulls of carnivores: effects of size, phylogeny, and behavioral ecology. *Journal of Zoology*, **242**, 97–117.

Grant, T., Frost, D. R., Caldwell, J. P., *et al.* (2006). Phylogenetic systematics of dart-poison frogs and their relatives (Amphibia: Athesphatanura: Dendrobatidae). *Bulletin of the American Museum of Natural History*, **299**, 1–262.

Gunnell, G. F. (1998). Creodonta. In *Evolution of Tertiary Mammals of North America (Vol. 1: Terrestrial Carnivores, Ungulates, and Ungulatelike Mammals)*, ed. C. M. Janis, K. M. Scott and L. L. Jacobs. Cambridge: Cambridge University Press, pp. 91–109.

Harvey, P. H., Pagel, M. D. and Rees, J. A. (1991). Mammalian metabolism and life histories. *American Naturalist*, **137**, 556–66.

Heinrich, R. E. and Houde, P. (2006). Postcranial anatomy of *Viverravus* (Mammalia, Carnivora) and implications for substrate use in basal Carnivora. *Journal of Vertebrate Paleontology*, **26**, 422–35.

Heinrich, R. E. and Rose, K. D. (1995). Partial skeleton from the primitive carnivoran *Miacis petilus* from the early Eocene of Wyoming. *Journal of Mammalogy*, **76**, 148–62.

Heinrich, R. E. and Rose, K. D. (1997). Postcranial morphology and locomotor behaviour of two early Eocene miacoid carnivorans, *Vulpavus* and *Didymictis*. *Palaeontology*, **40**, 279–305.

Higdon, J. W., Bininda-Emonds, O. R. P., Beck, R. M. D. and Ferguson, S. H. (2007). Phylogeny and divergence of the pinnipeds (Carnivora: Mammalia) assessed using a multigene dataset. *BMC (BioMed Central) Evolutionary Biology*, 7, **216** (doi:10.1186/1471-2148-7-216, 19 pages).

Hunt, R. M., Jr. (1977). Basicranial anatomy of *Cynelos* Jourdan (Mammalia: Carnivora), an Aquitanian amphicyonid from the Allier Basin, France. *Journal of Paleontology*, **51**, 826–43.

Hunt, R. M., Jr. (1987). Evolution of the aeluroid Carnivora: significance of auditory structure in the nimravid cat *Dinictis*. *American Museum Novitates*, **2886**, 1–74.

Hunt, R. M., Jr. (1996). Amphicyonidae. In *The Terrestrial Eocene–Oligocene Transition in North America*, ed. D. Prothero and R. J. Emry. Cambridge: Cambridge University Press, pp. 476–85.

Iwaniuk, A. N., Pellis, S. M. and Whishaw, I. Q. (1999). Brain size is not correlated with forelimb dexterity in fissiped carnivores (Carnivora): a comparative test of the principle of proper mass. *Brain Behavior and Evolution*, **54**, 167–80.

Jenkins, F. A. and Camazine, S. M. (1977). Hip structure and locomotion in ambulatory and cursorial carnivores. *Journal of Zoology*, **181**, 351–70.

Jenner, R. A. (2004). Accepting partnership by submission? Morphological phylogenetics in a molecular millennium. *Systematic Biology*, **52**, 333–42.

Jerison, H. (1961). Quantitative analysis of evolution of the brain in mammals. *Science*, **133**, 1012–14.

Jerison, H. (1970). Brain evolution: new light on old principles. *Science*, **170**, 1224–25.

Jerison, H. (1973). *Evolution of the Brain and Intelligence*. New York, NY: Academic Press, 482 pp.

Jerison, H. (1991). *Brain Size and the Evolution of Mind*. New York, NY: American Museum of Natural History, 99 pp.

Johnson, W. E., Eizirik, E., Pecon-Slattery, J., *et al.* (2006). The Late Miocene radiation of modern Felidae: a genetic assessment. *Science*, **311**, 73–77.

Jungers, W. L. (1990). Problems and methods in reconstructing body size in fossil Primates. In *Body Size in Mammalian Paleobiology*, ed. J. Damuth and B. J. MacFadden. New York, NY: Cambridge University Press, pp. 103–18.

Kjer, K. M. and Honeycutt, R. L. (2007). Site specific rates of mitochondrial genomes and the phylogeny of eutheria. *BMC Evolutionary Biology*, **7**, 8 (doi:10.1186/1471-2148-7-8).

Koepfli, K.-P., Jenks, S. M., Eizirik, E., Zahirpour, T., Van Valkenburgh, B. and Wayne, R. K. (2006). Molecular systematics of the Hyaenidae: relationships of a relictual lineage resolved by a molecular supermatrix. *Molecular Phylogenetics and Evolution*, **38**, 603–20.

Koepfli, K.-P., Gompper, M. E., Eizirik, E., *et al.* (2007). Phylogeny of the Procyonidae (Mammalia: Carnivora): molecules, morphology and the Great American Interchange. *Molecular Phylogenetics and Evolution*, **43**, 1076–95.

Legendre, S. (1986). Analysis of mammalian communities from the late Eocene and Oligocene of southern France. *Palaeovertebrata*, **16**, 191–212.

Legendre, S. and Roth, C. (1988). Correlation of carnassial tooth size and body weight in Recent carnivores (Mammalia). *Historical Biology*, **1**, 85–98.

Lento, G. M., Hickson, R. E., Chambers, G. K. and Penny, D. (1995). Use of spectral analysis to test hypotheses on the origin of pinnipeds. *Molecular Biology and Evolution*, **12**, 28–52.

Loreille, O., Orlando, L., Patou-Mathis, M., Philippe, M., Taberlet, P. and Hänni, C. (2001). Ancient DNA analysis reveals divergence of the cave bear, *Ursus spelaeus*, and brown bear, *Ursus arctos*, lineages. *Current Biology*, **11**, 200–03.

Maddison, W. P. (1991). Squared-change parsimony reconstructions of ancestral states for continuous-valued characters on a phylogenetic tree. *Systematic Zoology*, **40**, 304–14.

Maddison, W. P. and Maddison, D. R. (2007). Mesquite: a modular system for evolutionary analysis. Version 2.01. http://mesquiteproject.org

Magallon, S. (2007). From fossils to molecules: phylogeny and the core eudicot floral groundplan in Hamamelidoideae (Hamamelidaceae, Saxifragales). *Systematic Botany*, **32**, 317–47.

Manos, P. S., Soltis, P. S., Soltis, D. E., *et al.* (2007). Phylogeny of extant and fossil Juglandaceae inferred from the integration of molecular and morphological data sets. *Systematic Biology*, **56**, 412–30.

Marino, L. (1998). A comparison of encephalization between odontocete cetaceans and anthropoid primates. *Brain Behavior and Evolution*, **51**, 230–38.

Marino, L., McShea, D. W. and Uhen, M. D. (2004). Origin and evolution of large brains in toothed whales. *Anatomical Record Part A – Discoveries in Molecular Cellular and Evolutionary Biology*, **281A**, 1247–55.

Marmi, J., López-Giráldez, J. F. and Domingo-Roura, X. (2004). Phylogeny, evolutionary history and taxonomy of the Mustelidae based on sequences of the cytochrome *b* gene and a complex repetitive flanking region. *Zoologica Scripta*, **33**(6), 481–99.

Martin, R. D. (1981). Relative brain size and basal metabolic-rate in terrestrial vertebrates. *Nature*, **293**, 57–60.

Martin, R. D. (1984). Body size, brain size and feeding strategies. In *Food Acquisition and Processing in Primates*, ed. D. J. Chivers, B. A. Wood and A. Bilsborough. New York, NY: Plenum Press, pp. 73–103.

Martin, R. D. (1990). *Primate Origins and Evolution: A Phylogenetic Reconstruction*. London: Chapman and Hall, 828 pp.

Martin, R. D. (1996). Scaling of the mammalian brain: the maternal energy hypothesis. *News in Physiological Sciences*, **11**, 149–56.

Martin, R. D., Genoud, M. and Hemelrijk, C. K. (2005). Problems of allometric scaling analysis: examples from mammalian reproductive biology. *Journal of Experimental Biology*, **208**, 1731–47.

Matthew, W. D. (1909). The Carnivora and Insectivora of the Bridger Basin, middle Eocene. *Memoirs of the American Museum of Natural History*, **9**, 289–567.

McNab, B. K. (1988). Complications inherent in scaling the basal rate of metabolism in mammals. *Quarterly Review of Biology*, **63**, 25–54.

McShea, D. W. (1994). Mechanisms of large-scale evolutionary trends. *Evolution*, **48**, 1747–63.

Meiri, S., Dayan, T. and Simberloff, D. (2004a). Body size of insular carnivores: little support for the island rule. *American Naturalist*, **163**, 469–79.

Meiri, S., Dayan, T. and Simberloff, D. (2004b). Carnivores, biases and Bergmann's rule. *Biological Journal of the Linnean Society*, **81**, 579–88.

Mivart, S. G. J. (1885). On the anatomy, classification and distribution of the Arctoidea. *Proceedings of the Zoological Society of London*, **23**, 340–404.

Morlo, M., Peigné, S. and Nagel, D. (2004). A new species of *Prosansanosmilus*: implications for the systematic relationships of the family Barbourofelidae new rank (Carnivora, Mammalia). *Zoological Journal of the Linnean Society*, **140**, 43–61.

Muñoz-Garcia, A. and Williams, J. B. (2005). Basal metabolic rate in carnivores is associated with diet after controlling for phylogeny. *Physiological and Biochemical Zoology*, **78**, 1039–56.

Munthe, K. (1998). Canidae. In *Evolution of Tertiary Mammals of North America (Vol. 1: Terrestrial Carnivores, Ungulates, and Ungulatelike Mammals)*, ed. C. M. Janis, K. M.Scott and L. L. Jacobs. New York, NY: Cambridge University Press, pp. 124–43.

Murphy, W. J., Eizirik, E., Johnson, W. E., Zhang, Y. P., Ryder, O. A. and O'Brien, S. J. (2001). Molecular phylogenetics and the origins of placental mammals. *Nature*, **409**, 614–18.

Murphy, W. J., Pringle, T. H., Crider, T. A., Springer, M. S. and Miller, W. (2007). Using genomic data to unravel the root of the placental mammal phylogeny. *Genome Research*, **17**, 413–21.

Nylander, J. A. A., Ronquist, F., Huelsenbeck, J. P. and Nieves-Aldrey, J. L. (2004). Bayesian phylogenetic analysis of combined data. *Systematic Biology*, **53**, 47–67.

O'Leary, M. A. (1999). Parsimony analysis of total evidence from extinct and extant taxa and the cetacean–artiodactyl question (Mammalia, Ungulata). *Cladistics*, **15**, 315–30.

O'Leary, M. A. (2001). The phylogenetic position of cetaceans: further combined data analyses, comparisons with the stratigraphic record and a discussion of character optimization. *American Zoologist*, **41**, 487–506.

Oakley, T. H. and Cunningham, C. W. (2000). Independent contrasts succeed where ancestor reconstruction fails in a known bacteriophage phylogeny. *Evolution*, **54**, 397–405.

Pagel, M. D. and Harvey, P. H. (1988a). The taxon-level problem in the evolution of mammalian brain size: facts and artifacts. *American Naturalist*, **132**, 344–59.

Pagel, M. D. and Harvey, P. H. (1988b). How mammals produce large-brained offspring. *Evolution*, **42**, 948–57.

Pagel, M. D. and Harvey, P. H. (1989). Taxonomic differences in the scaling of brain on body-weight among mammals. *Science*, **244**, 1589–93.

Polly, P. D. (1996). The skeleton of *Gazinocyon vulpeculus* Gen. et Comb. Nov. and the cladistic relationships of Hyaenodontidae (Eutheria, Mammalia). *Journal of Vertebrate Paleontology*, **16**, 303–19.

Polly, P. D. (2001). Paleontology and the comparative method: ancestral node reconstructions versus observed node values. *American Naturalist*, **157**, 596–609.

Polly, P. D., Wesley-Hunt, G. D., Heinrich, R. E., Davis, G. and Houde P. (2006). Earliest known carnivoran auditory bulla and support for a recent origin of crown-group Carnivora (Eutheria, Mammalia). *Palaeontology* **49**, 1019–27.

Radinsky, L. B. (1973). *Aegyptopithecus* endocasts: oldest records of a pongid brain. *American Journal of Physical Anthropology*, **39**, 239–47.

Radinsky, L. B. (1977a). Brains of early carnivores. *Paleobiology*, **3**, 333–49.

Radinsky, L. B. (1977b). Early primate brains: facts and fiction. *Journal of Human Evolution*, **6**, 79–86.

Radinsky, L. B. (1978). Evolution of brain size in carnivores and ungulates. *American Naturalist*, **112**, 815–31.

Rothwell, G. W. and Nixon, K. C. (2006). How does the inclusion of fossil data change our conclusions about the phylogenetic history of euphyllophytes? *International Journal of Plant Sciences*, **167**, 737–49.

Ruff, C. (1990). Body mass and hindlimb bone cross-sectional and articular dimensions in anthropoid Primates. In *Body Size in Mammalian Paleobiology*, ed. J. Damuth and B. J. MacFadden. New York, NY: Cambridge University Press, pp. 119–50.

Sanders, K. L., Malhotra, A. and Thorpe, R. S. (2006). Combining molecular, morphological and ecological data to infer species boundaries in a cryptic tropical pitviper. *Biological Journal of the Linnean Society*, **87**, 343–64.

Sato, J. J., Hosoda, T., Wolsan, M. and Suzuki, H. (2004). Molecular phylogeny of arctoids (Mammalia: Carnivora) with emphasis on phylogenetic and taxonomic positions of the ferret-badgers and skunks. *Zoological Science*, **21**, 111–18.

Sato, J. J., Wolsan, M., Suzuki, H., *et al.* (2006). Evidence from nuclear DNA sequences sheds light on the phylogenetic relationships of Pinnipedia: single origin with affinity to Musteloidea. *Zoological Science*, **23**, 125–46.

Schmidt-Nielsen, K. (1984). *Scaling: Why is Animal Size so Important?* Cambridge: Cambridge University Press, 239 pp.

Scotland, R. W., Olmstead, R. G. and Bennett, J. R. (2003). Phylogeny reconstruction: the role of morphology. *Systematic Biology*, **52**, 539–48.

Sears, K. E., Finarelli, J. A., Flynn, J. J. and Wyss, A. R. (2008). Estimating body mass in New World 'monkeys' (Platyrrhini, Primates) from craniodental measurements, with a consideration of the Miocene platyrrhine, *Chilecebus carrascoensis*. *American Museum Novitates*, **3617**, 1–29.

Simons, E. L. (1993). New endocasts of *Aegyptopithecus*: oldest well-preserved record of the brain in Anthropoidea. *American Journal of Science*, **293–A**, 383–90.

Simpson, G. G. (1945). The principles of classification and a classification of mammals. *Bulletin of the American Museum of Natural History*, **85**, 1–350.

Smith, F. A., Lyons, S. K., Ernest, S. K. M., *et al.* (2003). Body mass of late Quaternary mammals. *Ecology*, **84**, 3403.

Spaulding, M. (2007). The impact of postcranial characters on reconstructing the phylogeny of Carnivoramorpha. *Journal of Vertebrate Paleontology*, **27**(3 suppl.), 151A.

Spaulding, M. and Flynn, J. J. (2009). Anatomy of the postcranial skeleton of '*Miacis*' *uintensis* (Mammalia: Carnivoramorpha). *Journal of Vertebrate Paleontology*, **29**(4), 1212–23.

Spaulding, M., Flynn, J. J. and Stucky, R. (in press). A new basal carnivoramorphan (Mammalia) from the 'Bridger B' (Bridger Formation, Bridgerian NALMA, Middle Eocene) of Wyoming. *Palaeontology*.

Stanley, S. M. (1973). An explanation for Cope's rule. *Evolution*, **27**, 1–26.

Swofford, D. L. and Maddison, W. P. (1987). Reconstructing ancestral character states under Wagner parsimony. *Mathematical Biosciences*, **87**, 199–229.

Tedford, R. H., Taylor, B. E. and Wang, X. M. (1995). Phylogeny of the Caninae (Carnivora: Canidae): the living taxa. *American Museum Novitates*, **3146**, 1–37.

Van Valkenburgh, B. (1989). Carnivore dental adaptations and diet: a study of trophic diversity within guilds. In *Carnivore Behavior, Ecology, and Evolution*, ed. J. L. Gittleman. Ithaca, NY: Cornell University Press, pp. 410–36.

Van Valkenburgh, B. (1990). Skeletal and dental predictors of body mass in carnivores. In *Body Size in Mammalian Paleobiology*, ed. J. Damuth and B. J. MacFadden. New York, NY: Cambridge University Press, pp. 181–206.

Van Valkenburgh, B. (1991). Iterative evolution of hypercarnivory in canids (Mammalia, Carnivora) – evolutionary interactions among sympatric predators. *Paleobiology*, **17**, 340–62.

Van Valkenburgh, B., Sacco, T. and Wang, X. M. (2003). Pack hunting in Miocene borophagine dogs: evidence from craniodental morphology and body size. *Bulletin of the American Museum of Natural History*, **279**, 147–62.

Van Valkenburgh, B., Wang, X. M. and Damuth, J. (2004). Cope's Rule, hypercarnivory, and extinction in North American canids. *Science*, **306**, 101–04.

Veron, G. (1995). La position systématique de *Cryptoprocta ferox* (Carnivora). Analyse cladistique des caractères morphologiques de carnivores Aeluroidea actuels et fossiles. *Mammalia*, **59**, 551–82.

Veron, G., Colyn, M., Dunham, A. E., Taylor, P. and Gaubert, P. (2004). Molecular systematics and origin of sociality in mongooses (Herpestidae, Carnivora). *Molecular Phylogenetics and Evolution*, **30**, 582–98.

Viranta, S. (1996). European Miocene Amphicyonidae – taxonomy, systematics and ecology. *Acta Zoologica Fennica*, **204**, 1–61.

Vrana, P. B., Milinkovitch, M. C., Powell, J. R. and Wheeler, W. C. (1994). Higher level relationships of the arctoid Carnivora based on sequence data and total evidence. *Molecular Phylogenetics and Evolution*, **3**, 47–58.

Wang, X. M. (1994). Phylogenetic systematics of the Hesperocyoninae (Carnivora: Canidae). *Bulletin of the American Museum of Natural History*, **221**, 1–207.

Wang, X. M. and Tedford, R. H. (1994). Basicranial anatomy and phylogeny of primitive canids and closely related miacids (Carnivora: Mammalia). *American Museum Novitates*, **3092**, 1–34.

Wang, X. M., Tedford, R. H. and Taylor, B. E. (1999). Phylogenetic systematics of the Borophaginae (Carnivora: Canidae). *Bulletin of the American Museum of Natural History*, **243**, 1–391.

Wayne, R. K., Geffen, E., Girman, D. J., Koepfli, K.-P., Lau, L. M. and Marshall, C. R. (1997). Molecular systematics of the Canidae. *Systematic Biology*, **46**, 622–53.

Webster, A. J. and Purvis, A. (2002). Ancestral states and evolutionary rates of continuous characters. In *Morphology, Shape and Phylogeny*, ed. N. McLeod and P. L. Forey. New York, NY: Taylor and Francis, pp. 247–68.

Webster, A. J., Gittleman, J. L. and Purvis, A. (2004). The life history legacy of evolutionary body size change in carnivores. *Journal of Evolutionary Biology*, **17**, 396–407.

Wesley-Hunt, G. D. and Flynn, J. J. (2005). Phylogeny of the Carnivora: basal relationships among the carnivoramorphans, and assessment of the position of 'Miacoidea' relative to crown-clade Carnivora. *Journal of Systematic Palaeontology*, **3**, 1–28.

Wesley-Hunt, G. D. and Werdelin, L. (2005). Basicranial morphology and phylogenetic position of the upper Eocene carnivoramorphan *Quercygale*. *Acta Palaeontologica Polonica*, **50**, 837–46.

Wheeler, W. C. and Hayashi, C. Y. (1998). The phylogeny of the extant chelicerate orders. *Cladistics*, **14**, 173–92.

Wolsan, M. (1993). Phylogeny and classification of early European Mustelida (Mammalia: Carnivora). *Acta Theriologica*, **38**, 345–84.

Wynen, L. P., Goldsworthy, S. D., Insley, S. J., *et al.* (2001). Phylogenetic relationships within the eared seals (Otariidae: Carnivora): implications for the historical biogeography of the family. *Molecular Phylogenetics and Evolution*, **21**(2), 270–84.

Wyss, A. R. (1987). The walrus auditory region and the monophyly of pinnipeds. *American Museum Novitates*, **2871**, 1–31.

Wyss, A. R. and Flynn, J. J. (1993). A phylogenetic analysis and definition of the Carnivora. In *Mammal Phylogeny: Placentals*, ed. F. Szalay, M. Novacek, and M. McKenna. New York, NY: Springer-Verlag, pp. 32–52.

Yoder, A. D. and Flynn, J. J. (2003). Origin of Malagasy Carnivora. In *The Natural History of Madagascar*, ed. S. M. Goodman and J. Benstead. Chicago, IL: University of Chicago Press, pp. 1253–56.

Yoder, A. D., Burns, M. M., Zehr, S., *et al.* (2003). Single origin of Malagasy Carnivora from an African ancestor. *Nature*, **421**, 734–37.

Yonezawa, T., Nikaido, M., Kohno, N., Fukumoto, Y., Okada, N. and Hasegawa, M. (2007). Molecular phylogenetic study on the origin and evolution of Mustelidae. *Gene*, **396**, 1–12.

Yu, L. and Zhang, Y. P. (2006). Phylogeny of the caniform carnivora: evidence from multiple genes. *Genetica*, **127**, 65–79.

Yu, L., Li, Q., Ryder, O. A. and Zhang, Y. (2004a). Phylogenetic relationships within mammalian Order Carnivora indicated by sequences of two nuclear DNA genes. *Molecular Phylogenetics and Evolution*, **33**, 694–705.

Yu, L., Li, Q., Ryder, O. A. and Zhang, Y. (2004b). Phylogeny of the bears (Ursidae) based on nuclear and mitochondrial genes. *Molecular Phylogenetics and Evolution*, **32**, 480–94.

3

Phylogeny of the Viverridae and 'Viverrid-like' feliforms

GERALDINE VERON

Introduction

The phylogenetic relationships of the extant feliform carnivores, Felidae (cats), Herpestidae (mongooses), Hyaenidae (hyenas and aardwolf), and Viverridae (civets, genets, and oyans), have been debated for a long time, with several proposed hypotheses for the relationships of these families (Flower, 1869; Gregory and Hellman, 1939, see Figure 3.1; Simpson, 1945; Hunt, 1987; Flynn *et al.*, 1988; Wayne *et al.*, 1989; Wozencraft, 1989a; Hunt and Tedford, 1993; Wyss and Flynn, 1993; Veron, 1994). The position of the Viverridae family is still unresolved (see e.g. Gaubert and Veron, 2003; Flynn *et al.*, 2005; Koepfli *et al.*, 2006; Holliday, 2007).

The mongooses were initially included within the Viverridae (Flower, 1869; Mivart, 1882) until Pocock (1916a, 1919) advocated for a family rank, to which he gave the name Mungotidae. Gregory and Hellman (1939) also placed them in a separate family, the Herpestidae Bonaparte, 1845. This separation was not followed by Simpson (1945) and several other authors (e.g. Albignac, 1973; Ewer, 1973; Petter, 1974; Rosevear, 1974; Coetzee, 1977; Kingdon, 1977; Payne *et al.*, 1985; Stains, 1987; Taylor, 1988; Schreiber *et al.*, 1989; Dargel, 1990; Skinner and Smithers, 1990). However, this split has been supported by further studies, based on morphology, chromosomes and molecular data (e.g. Wurster, 1969; Fredga, 1972; Radinsky, 1975; Bugge, 1978; Neff, 1983; Hunt, 1987; Wozencraft, 1984; Hunt and Tedford, 1993; Veron and Catzeflis, 1993; Wyss and Flynn, 1993; Veron, 1994, 1995; Flynn and Nedbal, 1998; Veron and Heard, 2000; Gaubert and Veron, 2003; Veron *et al.*, 2004a; Flynn *et al.*, 2005), and it is now generally accepted that the mongooses should be placed in a separate family, the Herpestidae (see Honacki *et al.*, 1982; Wozencraft, 1989b, 1993, 2005; Gilchrist *et al.*, 2009).

Carnivoran Evolution: New Views on Phylogeny, Form, and Function, ed. A. Goswami and A. Friscia. Published by Cambridge University Press. © Cambridge University Press 2010.

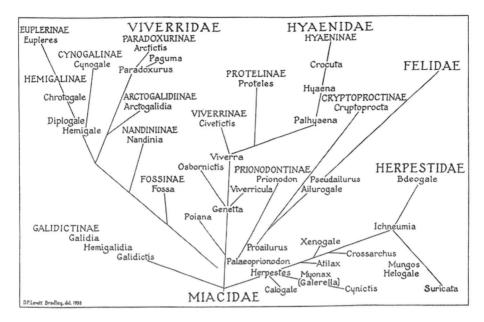

Figure 3.1 Phylogeny of the feliform carnivores reproduced from Gregory and Hellman (1939).

The Viverridae (Viverridae Gray, 1821; type genus *Viverra* Linnaeus, 1758) occur in Africa and Asia; one species, the common genet (*Genetta genetta*), also occurs in Europe, but it may have been introduced to this region during historical times (Amigues, 1999). Viverrids are small carnivores that range from 0.6 kg (African oyans, *Poiana*) to 20 kg (the African civet, *Civettictis civetta*). They have a long and slender body, with a pointed face, small ears, and a long tail; most species have a spotted coat and a banded tail. This family includes digitigrade terrestrial species (civets), semi-digitigrade semi-arboreal species (genets), and plantigrade arboreal species (palm civets) (Ewer, 1973; Wozencraft, 1984; Taylor, 1988; Veron, 1994, 1999; Nowak, 2005; Jennings and Veron, 2009). Their systematics was based mainly on the morphology of the basicranium (size and shape of the anterior and posterior chambers of the auditory bullae, position of the foramens at the base of the skull), the dentition, the feet, and the perineal gland (a scent pouch that lies between the genitals and the anus) (Gray, 1864; Flower, 1869; Mivart, 1882; Pocock, 1915a,b,c,d, 1916b,c, 1929, 1933a,b,c, 1934a,b,c; 1939; Gregory and Hellman, 1939; Hunt, 1974, 1987, 1989, 1991, 2001; Wozencraft, 1984, 1989a; Veron, 1994, 1995). The perineal gland is specific to the Viverridae, and its product, called civet, is used in scent marking (see Jennings and Veron, 2009).

The relationships within the Viverridae have been long debated, largely due to their large diversity of forms and also because this family was a dumping ground for all 'viverrid-like' species of feliforms (Pocock, 1916c; and see reviews in Wozencraft, 1984; Veron, 1994, 1995; Bininda-Emonds *et al.*, 1999; Veron and Heard, 2000; Hunt, 2001; Gaubert *et al.*, 2002; Gaubert and Veron, 2003; Jennings and Veron, 2009). Until recently, the Viverridae was divided into six subfamilies (Wozencraft, 1993): two endemic to Madagascar, the Cryptoproctinae Gray, 1864 and the Euplerinae Chenu, 1852; two Asian subfamilies, the Hemigalinae Gray, 1864 and the Paradoxurinae Gray, 1864, one monospecific African subfamily, the Nandiniinae Pocock, 1929, and one subfamily, the Viverrinae Gray, 1821, with both Asian and African representatives. However, the systematic position of several viverrid species were questioned, in particular the African palm civet (*Nandinia binotata*), the Malagasy viverrids (the fossa *Cryptoprocta ferox*, falanouc *Eupleres goudotii*, and Malagasy civet *Fossa fossana*), and the Asian linsangs (the spotted linsang *Prionodon pardicolor* and banded linsang *Prionodon linsang*) (see Gray, 1864; Milne-Edwards and Grandidier, 1867; Flower, 1869; Filhol, 1879, 1894; Mivart, 1882; Carlsson, 1911; Gregory and Hellman, 1939; Pocock, 1940; Beaumont, 1964; Petter, 1974; Hunt, 1987, 2001; Flynn *et al.*, 1988; Wozencraft, 1984, 1989a; Hunt and Tedford, 1993; Wyss and Flynn, 1993; Veron, 1994, 1995; Gaubert, 2003a). The first molecular phylogenies revealed a conflict with the traditional taxonomy (e.g. Veron and Catzeflis, 1993; Flynn and Nedbal, 1998), which prompted re-evaluations of the relationships within the family. Also, these studies raised questions about the morphological homoplasy within this group.

This chapter reviews the results of recent molecular and morphological studies and provides an up-to-date classification of the Viverridae. It also discusses some morphological features in the light of the phylogenetic relationships recently established.

'Viverrid-like' feliforms

African palm civet (*Nandinia binotata*)

The African palm civet, *N. binotata*, was placed by Gray (1864) in the Viverridae, within the subfamily Paradoxurinae (Asian palm civets). A similar classification was suggested by Flower (1869), although this author noticed that *N. binotata* has a peculiar characteristic, a cartilaginous posterior chamber of the tympanic bullae. On the basis of this morphological trait, Pocock (1929) advocated that *N. binotata* should be placed in its own separate family, the Nandiniidae. Gregory and Hellman (1939) disagreed with this opinion and placed *Nandinia* in the Viverridae, within the subfamily Nandiniinae. They considered the African palm civet close to the Paradoxurinae, and viewed its peculiar morphological trait as a specialisation rather than a primitive feature. The classification by Simpson (1945) placed *Nandinia* in the Paradoxurinae. Novacek (1977) argued that the cartilaginous tympanic bone in this species is a derivative condition and that in all other aspects its anatomy is the same as other viverrids. Hunt (1987), however, considered *Nandinia*'s bullae as primitive relative to other feliforms and suggested that this species may be a sister group to the rest of the feliforms. Flynn *et al.* (1988) placed it *incertae sedis* due to these conflicting opinions.

The molecular results of Flynn and Nedbal (1998) revealed that the African palm civet is the sister taxon of the other Feliformia. This is congruent with the morphological studies of Hunt (1987) and Hunt and Tedford (1993), and was also confirmed by other molecular studies (e.g. Gaubert and Veron, 2003; Gaubert *et al.*, 2004a; Yoder *et al.*, 2003; Flynn *et al.*, 2005; Koepfli *et al.*, 2006). The viverrid-like morphological features of *Nandinia* were thus primitive or resulting from convergence (Gaubert *et al.*, 2005c). The African palm civet is now believed to be an early offshoot within the feliforms and is placed in a separate family, the Nandiniidae (McKenna and Bell, 1997; Hunt, 2001; Wozencraft, 2005).

The Malagasy carnivores (*Cryptoprocta ferox, Eupleres goudotii, Fossa fossana, Galidia elegans, Galidictis fasciata, Galidictis grandidieri, Mungotictis decemlineata,* and *Salanoia concolor*)

Three 'civet' species are endemic to Madagascar: the fossa (*Cryptoprocta ferox*), falanouc (*Eupleres goudotii*), and Malagasy civet (*Fossa fossana*). The presence of another viverrid species on Madagascar, the small Indian civet (*Viverricula indica*), resulted from human introduction (see Albignac, 1973).

The fossa (*C. ferox*) has cat-like dentition, puma-like external morphology, and plantigrade feet with wide plantar pads that are very similar to those of the Asian palm civets (Pocock, 1916b, Albignac, 1973; Petter, 1974; Wozencraft, 1984; Veron, 1994). Bennett (1833) described this species as a viverrid, on the basis of its bare, webbed feet and retractile claws. Milne-Edwards and Grandidier (1867) believed that its general morphology and cranial features showed a close relationship to the Felidae. On the basis of the dentition, the shape of the tympanic bullae, and the disposition of the basicranial foramina, Flower (1869) placed *C. ferox* in a separate family, the Cryptoproctidae, which he believed to be intermediate between the Viverridae and Felidae. Mivart (1882) and Gregory and Hellman (1939) pointed out that the presence of an anal pouch, the absence of a perineal gland, and the size of the baculum are features shared by *C. ferox* and the Herpestidae, but not by the Viverridae. However, Gregory and Hellman (1939) said that this species is a 'little modified survivor of *Proailurus*' (an Oligocene–Miocene felid; see Hunt, 1989), and 'may be assumed to stand in a border zone between the Viverridae and the Felidae'. They suggested that *C. ferox* should be placed in a distinct subfamily, Cryptoproctinae, within the Felidae. Simpson (1945) considered *C. ferox* closer to the Viverridae rather than to the Felidae, and he included it in the Viverridae, in the subfamily Cryptoproctinae. The fossa has thus been placed in the Viverridae (Lesson, 1842; Gray, 1864; Pocock, 1916b, 1940; Simpson, 1945; Albignac, 1973; Honacki *et al.*, 1982; Stains, 1987; Wozencraft, 1989b, 1993), sometimes in the Felidae (Gregory and Hellman, 1939), or in its own separate family, the Cryptoprocti-dae (Flower, 1869).

The first molecular study to tackle the relationships of the Malagasy carnivores, using DNA hybridisation experiments, suggested that *C. ferox* is closer to the Herpestidae than to the Viverridae (Veron and Catzeflis, 1993, see Figure 3.2). A cladistic morphological analysis by Veron (1994, 1995) clustered *C. ferox* with the Felidae, on the basis of shared dental features. However, when fossils were included in the analysis, it was clear that the dental features related to hypercarnivory had occurred by convergence in *C. ferox*, the Felidae, and the Hyaenidae (as already suggested by Petter, 1974). In fact, *C. ferox* shares derived features with the Herpestidae (the reduction of *presylvian sulcus*), and with the Herpestidae and Hyaenidae (the presence of an anal pouch) (Veron, 1994, 1995).

The falanouc (*Eupleres goudotii*) was first described as an insectivore (Doyère, 1835) because of its long and thin snout, and its very small and sharp teeth, but it was very soon after included in the feliform carnivores, within the Viverridae (Gray, 1864). It was believed to be close to the Hemigalinae (Asian palm civets), on the basis of skull and teeth similarities with Owston's palm civet (*Chrotogale owstoni*) and on its foot morphology (Thomas, 1912; Gregory and Hellman, 1939). The falanouc was, however, placed in a separate subfamily, the Euplerinae Chenu, 1852, by Gregory and Hellman (1939). Simpson (1945) placed *Eupleres* in the subfamily Hemigalinae, within the Viverridae.

The Malagasy civet (*Fossa fossana*) has a general morphology quite similar to the terrestrial civets and a viverrid-like basicranium (see Mivart, 1882), so its inclusion in the Viverridae was not much debated. Gregory and Hellman (1939) suggested some affinities with the Hemigalinae, while Pocock (1915b) pointed out some shared morphological characteristics with the linsangs (*Poiana* and *Prionodon*). *F. fossana* was either placed in its own subfamily, the Fossinae Pocock, 1915 (Pocock, 1915b; Gregory and Hellman, 1939), in the Hemigalinae

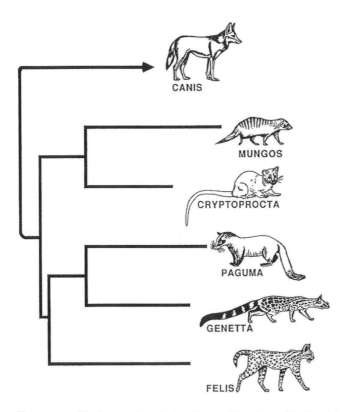

CANIS

MUNGOS

CRYPTOPROCTA

PAGUMA

GENETTA

FELIS

Figure 3.2 Phylogenetic relationships of the fossa (*C. ferox*) based on DNA–DNA hybridisation (dominant tree (94.6%) derived from the bootstrap procedure (5000 replications) on Mode values; modified from Veron and Catzeflis, 1993). Dating estimate for the divergence *Cryptoprocta*–Herpestidae (*Mungos*) was 24 million years.

(Simpson, 1945), or in the Euplerinae, together with *Eupleres* (Wozencraft, 1993). Petter (1974) underlined that several morphological features pointed to a close affinity between *Fossa* and *Eupleres*, and that the resemblances of these taxa to the Hemigalinae are likely to be due to convergence. Wozencraft (1989b) placed *C. ferox*, *E. goudotii* and *F. fossana* in the subfamily Cryptoproctinae within the Viverridae.

The Malagasy 'mongooses' (*Galidia elegans*, *Galidictis fasciata*, *Galidictis grandidieri*, *Mungotictis decemlineata*, and *Salanoia concolor*) were grouped in the subfamily Galidiinae Gray, 1864 and included in the Viverridae (along with the other mongooses, placed in the subfamily Hespestinae; Gray, 1864; Simpson, 1945; Albignac, 1973). Once the Herpestinae was considered separate from the Viverridae and placed in its own family, the Herpestidae (Gregory and Hellman, 1939; Honacki *et al.*, 1982; Wozencraft, 1989b, 1993), the Galidiinae was also placed in the Herpestidae by most authors, based on shared morphological features (see Pocock, 1915e; Albignac, 1973; Petter, 1974). However, on the basis of some soft anatomy features, Gregory and Hellman (1939) believed that the Galidiinae were an 'offshoot from the base of the viverrid stem where it joins the herpestid branch' and placed them in the subfamily Galidictinae, within the Viverridae.

A recent molecular phylogenetic analysis, based on two mitochondrial genes (cytochrome *b* and ND2) and two nuclear genes (exon 1 of the interphotoreceptor retinoid-binding protein and intron 1 of the transthyretin gene), revealed that the Malagasy carnivores form a monophyletic group, which is the sister group to the Herpestidae (Yoder *et al.*, 2003; Figure 3.3). Within the Malagasy carnivore clade, all the Galidiinae species group together, and the fossa (*C. ferox*) and the Malagasy civet (*F. fossana*) are sister taxa (which agrees with the suggestion of Petter (1974) of a close affinity between these two species). However, the relationship of the falanouc (*E. goudotii*) to the other Malagasy carnivores remains uncertain (no fresh sample was obtained for this species, and the sequence of cytochrome *b* obtained from a museum skin gave an unresolved position within the Malagasy clade; Yoder *et al.*, 2003).

This close relationship of the Malagasy carnivores to the Herpestidae has now been confirmed by other studies (e.g. Veron *et al.*, 2004a; Flynn *et al.*, 2005). The divergence of the Malagasy carnivores from their African relatives has been estimated at 18–24 million years ago (Mya), using Bayesian methods (Yoder *et al.*, 2003), which is congruent with the previous divergence date for Herpestidae–*Cryptoprocta* at 24 Mya, inferred from DNA–DNA hybridisation results (Veron and Catzeflis, 1993). Thus, an African ancestor (closely related to the mongooses) colonised Madagascar around 18–24 Mya, and, in the absence of any other representatives of the Carnivora, diversified into mongoose-like, civet-like, and cat-like carnivores on the island.

The Malagasy carnivores are now placed in a separate family, the Eupleridae Chenu, 1852, with two subfamilies, the Euplerinae (*Cryptoprocta*, *Eupleres*, and *Fossa*) and the Galidiinae (*Galidia*, *Galidictis*, *Mungotictis*, and *Salanoia*) (Wozencraft, 2005). The mongoose family (Herpestidae) now no longer includes the Malagasy 'mongooses' (Galidiinae). Recent molecular studies have shown that the Herpestidae should be split into two subfamilies: the Mungotinae Gray 1864 (11 small, social species) and the Herpestinae Bonaparte 1845 (23 large, solitary species) (Veron *et al.*, 2004a; Perez *et al.*, 2006; Gilchrist *et al.*, 2009).

The new systematic position of the Malagasy 'civets' suggests that some of the morphological features that were used to place them previously within the Viverridae are either plesiomorphic characters or convergences. In fact, it was

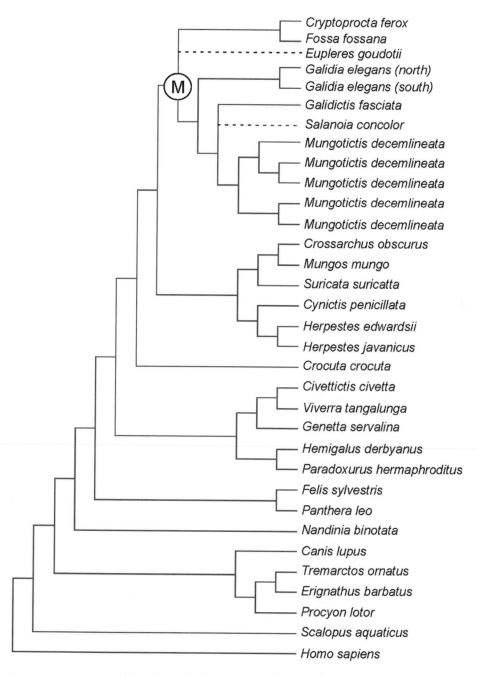

Figure 3.3 Phylogenetic relationships of Malagasy carnivores (maximum likelihood tree based on two mitochondrial and two nuclear genes; dashed lines indicate taxa sampled from museum specimens for which only the cytochrome *b* gene has been sequenced; M: Malagasy carnivore node – divergence was estimated *ca.* 18–24 Mya; from Yoder *et al.*, 2003).

shown that they share no synapomorphy with the Viverridae (Flynn *et al.*, 1988; Veron and Catzeflis, 1993; Veron, 1994, 1995; Gaubert *et al.*, 2005c). For instance, the felid-like teeth of *C. ferox* (slicing large carnassials, reduction of postcarnassial molars) can be considered as adaptive features to predation, which has occurred by convergence in *C. ferox*, the Felidae, and the Hyaenidae (Petter, 1974; Veron, 1994, 1995). Beside these derived features, *C. ferox* exhibits several primitive features (such as a viverrid-like basicranium; see Hunt, 1974, 1987; Flynn *et al.*, 1988; Veron, 1994, 1995), which were wrongly used to include it the Viverridae. Even though some authors had noted several shared morphological features between the Malagasy 'civets' and the Herpestidae and Galidiinae (see for instance Mivart, 1882 for *Cryptoprocta*; Petter, 1974; Veron, 1994, 1995), the close relationship between these groups was not clearly suggested.

Further studies are needed to understand the evolution of the Malagasy carnivores. Unfortunately, no African fossils have yet been found which may represent a 'proto Malagasy carnivore', and there is no fossil record of early Malagasy carnivores.

The Asian linsangs (*Prionodon*)

The two Asian linsang species (the banded linsang *Prionodon linsang* and the spotted linsang *Prionodon pardicolor*) have been traditionally included in the Viverridae (Gray, 1864; Mivart, 1882; Simpson, 1945; Corbet and Hill, 1992), within the subfamily Viverrinae (Ewer, 1973; Honacki *et al.*, 1982; Wozencraft, 1989b, 1993) or in the subfamily Prionodontinae (Pocock, 1933a; Gregory and Hellman, 1939; Hunt, 2001; Wozencraft, 2005). The Asian linsangs share many features with the African linsangs or oyans (Central African oyan *Poiana richardsoni* and Leighton's oyan *Poiana leightoni*): they are small, arboreal predators that live in tropical forests and have a small, rounded skull, hypercarnivorous dentition, absence of M^2, similar body proportions, a spotted pelage, and a long banded tail (Nowak, 2005; Jennings and Veron, 2009). However, the Asian linsangs lack a perineal gland (Pocock, 1915b), which is a characteristic of the Viverridae (see Jennings and Veron, 2009). The presence of the perineal gland in the African oyans is debated, as some authors state that it may be present in a rudimentary state (Pocock, 1933a, Rosevear, 1974; Wozencraft, 1984), whereas others suggest this gland is absent (Gaubert *et al.*, 2005c).

The morphological similarities between the linsangs and oyans were recognised by many authors (e.g. Gray, 1864; Mivart, 1882; Pocock, 1915b; Gregory and Hellman, 1939; Wozencraft, 1984; Hunt, 2001), and on this basis they were placed together within the Viverridae, either in the 'Prionodontina'

(Gray, 1864) or in the subfamily Prionodontinae (including also *Genetta*; Hunt, 2001). The linsangs and oyans have also been considered congeneric by Wozencraft and Grubb (in Honacki *et al.*, 1982). However, several authors considered the oyans (*Poiana*) close to the genets (*Genetta*) rather than to *Prionodon* (e.g. Rosevear, 1974; Crawford-Cabral, 1993), and suggested that their similarities to the Asian linsangs were due to convergences (e.g. loss of M^2) or plesiomorphic characters (e.g. no median dorsal stripe and a large number of rings on the coat pattern) (Crawford-Cabral, 1993).

The Asian linsangs share morphological features with the Felidae (e.g. absence of the perineal gland, close proximity of the genitalia to the anus, digitigrade feet, hairy metapodes, retractile claws, hypercarnivorous dentition; Horsfield, 1821; Mivart, 1882; Pocock, 1915b; Gregory and Hellman, 1939; Veron, 1994, 1995) and their basicranium is very similar to that of some fossils from the Oligocene and Miocene, *Paleoprionodon* (feliform) and *Proailurus* (early felid) (Teilhard de Chardin, 1915; Gregory and Hellman, 1939; Veron, 1994, 1995; Hunt, 1987, 2001). Horsfield (1821) believed that the Asian linsangs are related to the cats and placed them in the 'Prionodontidae', a section within the genus *Felis*. From a study of the basicranial region of the linsangs, Hunt (2001) suggested that the auditory region of *Prionodon* was a 'transitional state' between the Oligocene feliform *Paleoprionodon* and *Poiana* and, coupled with other morphological traits, he grouped *Prionodon*, *Poiana* and *Genetta* together with *Paleoprionodon* within the Viverridae, in the subfamily Prionodontinae.

The first molecular systematic study of the Viverridae (Veron and Heard, 2000), using a mitochondrial gene (cytochrome *b*), revealed for the first time that the Asian linsangs are not closely related to the Viverridae. The molecular results of Gaubert and Veron (2003), based on both mitochondrial and nuclear genes (cytochrome *b* and intron 1 of the transthyretin gene), suggested that the Asian linsangs are in fact the sister group of the Felidae (Figure 3.4). This was later confirmed by another molecular study, using 22,789 base pairs of nuclear and mitochondrial genes (Johnson *et al.*, 2006). Based on these findings, the Asian linsangs are now placed in a separate family, the Prionodontidae (first proposed by Horsfield, 1821). Gaubert and Veron (2003) also estimated that the divergence time of the Asian linsang lineage from the cat lineage was *ca.* 33 Mya. The African oyans have now been shown to be the sister group of the genets (*Genetta*), as was proposed by Rosevear (1974) and Crawford-Cabral (1993) (see Figure 3.4).

These molecular studies now suggest that the morphological features shared by the Asian linsangs and African oyans are primitive features and/or convergences. This includes hair ultrastructure features (Gaubert *et al.*, 2002), and also the basicranium. Although the Asian linsangs have a typical viverrid

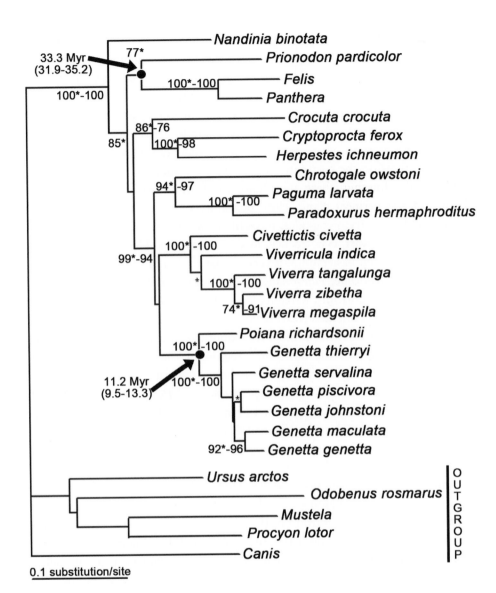

0.1 substitution/site

Figure 3.4 Systematic relationships of the linsangs, with dating estimates for *Prionodon*–Felidae and *Poiana*–*Genetta* (maximum likelihood tree based on cytochrome *b* and transthyretin intron1 genes; boostrap values aboves 70% are shown for ML analyses and MP analyses; calibration point used for the estimation of divergence time was the divergence *Felis*–*Panthera* 3 Mya; from Gaubert and Veron, 2003).

basicranium (Hunt, 2001), it is also very similar to that of the early feliform *Paleoprionodon* (Veron, 1994; Hunt, 2001) and thus may be a primitive feature. The resemblances between the Asian linsangs and African oyans, which are now known to be homoplasic, have raised the question as to whether they have retained ancestral feliform characteristics or they have evolved to the same ecomorphotype by convergence.

From the newly established relationships of the Asian linsangs, it can be seen that several morphological features that were considered as diagnostic features within the feliforms, such as the structure of the basicranium, can be homoplasic. It now appears that some characters, notably of the basicranium, are plesiomorphic, and others, like dentition, may be convergent (Gaubert *et al.*, 2005c). Thus, establishing the relationships of the extant or fossil feliforms on the basis of basicranium features or dental characters can be problematic, as it has been for the Asian linsangs.

Systematics and evolution of the Viverridae

The Viverridae, now redefined, forms a monophyletic group that comprises 34 species, grouped into 4 subfamilies (Gaubert and Veron, 2003; Gaubert *et al.*, 2005c, Gaubert and Cordeiro-Estrela, 2006; Patou *et al.*, 2008; Jennings and Veron, 2009): 2 Asian subfamilies (Hemigalinae and Paradoxurinae), 1 African subfamily (Genettinae) and 1 Afro-Asian subfamily (Viverrinae) (see Table 3.1; Figure 3.5). Now that several 'viverrid-like' taxa have been excluded from this family, the major synapomorphy of the Viverridae appears to be the presence of a perineal gland (Gaubert *et al.*, 2005c; Patou *et al.*, 2008). However, this gland is absent in the male small-toothed palm civet (*Arctogalidia trivirgata*) (Pocock, 1915c) and is questionable in the male Sulawesi palm civet (*Macrogalidia musschenbroekii*) (Wemmer *et al.*, 1983). Also, the presence of a perineal gland is debated in the African oyans, in which it may be absent in both sexes (Gaubert *et al.*, 2005c) or rudimentary (Pocock, 1933a, Rosevear, 1974; Wozencraft, 1984). The absence or reduction of this gland in these species may be a secondary loss, which could be related to their highly arboreal habits (Patou *et al.*, 2008).

The structure and shape of the perineal gland provides a phylogenetic character that is informative when investigating the relationships within the Viverridae (Pocock, 1915a,b,c,d; Gaubert *et al.*, 2005c; and see Jennings and Veron, 2009). The structure of the plantar pads (see Pocock, 1915a,b,c,d; Veron, 1999; Jennings and Veron, 2009) is non-homoplasic in the viverrids (Gaubert *et al.*, 2005c; Patou *et al.*, 2008) and also provides synapomorphies for the viverrid subfamilies or clades. For example, the African oyans (*Poiana*) have a foot structure close to that of *Genetta* (Mivart, 1882; Veron, 1999); in fact, Gray (1864) noticed that *Poiana* is 'very like *Prionodon* in external appearance, but with the feet of *Genetta*'. Within the Paradoxurinae, the structure and shape of the footpads are characteristic of this subfamily (Mivart, 1882; Pocock, 1915c;

Table 3.1 Classification of the Viverridae (Jennings and Veron, 2009; from Veron and Heard, 2000; Gaubert and Veron, 2003; Gaubert *et al.*, 2004a,b, 2005b; Wozencraft, 2005; Gaubert and Cordeiro-Estrela, 2006; Patou *et al.*, 2008).

Genettinae		
	Genetta	*abyssinica*
	Genetta	*angolensis*
	Genetta	*bourloni*
	Genetta	*cristata*
	Genetta	*felina*
	Genetta	*genetta*
	Genetta	*johnstoni*
	Genetta	*maculata*
	Genetta	*pardina*
	Genetta	*piscivora*
	Genetta	*poensis*
	Genetta	*servalina*
	Genetta	*thierryi*
	Genetta	*tigrina*
	Genetta	*victoriae*
	Poiana	*leightoni*
	Poiana	*richardsoni*
Hemigalinae		
	Chrotogale	*owstoni*
	Cynogale	*bennettii*
	Diplogale	*hosei*
	Hemigalus	*derbyanus*
Paradoxurinae		
	Arctictis	*binturong*
	Arctogalidia	*trivirgata*
	Macrogalidia	*musschenbroekii*
	Paguma	*larvata*
	Paradoxurus	*zeylonensis*
	Paradoxurus	*jerdoni*
	Paradoxurus	*hermaphroditus*
Viverrinae		
	Civettictis	*civetta*
	Viverra	*civettina*
	Viverra	*megaspila*
	Viverra	*tangalunga*
	Viverra	*zibetha*
	Viverricula	*indica*

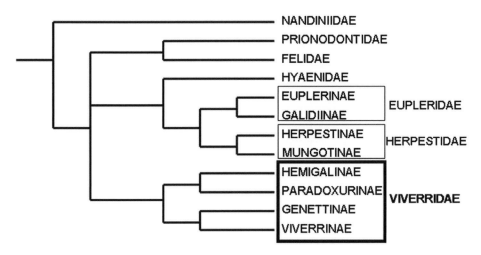

Figure 3.5 Synthetic tree of the Feliformia (from Gaubert and Veron, 2003; Yoder *et al.*, 2003, Gaubert and Cordeiro-Estrela, 2006; Veron *et al.*, 2004a; Koepfli *et al.*, 2006; Holliday, 2007; Patou *et al.*, 2008). The position of the Viverridae within the Feliformia is still debated. Their emergence is estimated at *ca.* 25–35 Mya (Gaubert and Cordeiro-Estrela, 2006; Koepfli *et al.*, 2006; Patou *et al.*, 2008).

Veron 1999; Jennings and Veron, 2009); moreover, the clade (*Arctictis, Paguma, Paradoxurus*), obtained in a recent molecular phylogeny (Patou *et al.*, 2008), is supported by a single morphological synapomorphy, the union of the third and fourth digit pads of the hind-foot (Pocock, 1915c; Veron, 1994). Although hypercarnivorous dentition has been shown to occur several times in the feliforms and to be homoplasic (Flower, 1869; Mivart, 1882; Petter, 1974; Veron, 1994, 1995; Gaubert *et al.*, 2005c), the hypocarnivorous dentition (the reduction of the size of the teeth and cusps) in the Paradoxurinae may provide reliable synapomorphies for this subfamily (Patou *et al.*, 2008).

Dating inferences from molecular studies suggest that the Viverridae appeared in Eurasia *ca.* 25–35 Mya (estimation at 34.29 Mya in Gaubert and Cordeiro-Estrela, 2006; 26.9 Mya in Patou *et al.*, 2008; and 25.2 Mya in Koepfli *et al.*, 2006). The oldest known viverrid fossils are *ca.* 23 Mya old (*Herpestides*, Hunt, 1991, 1996; *Semigenetta*, Helbing, 1927; Ginsburg, 1999). The two subfamilies of Asian palm civets (Hemigalinae and Paradoxurinae) have been shown to be sister groups and are estimated through molecular dating to have emerged *ca.* 20–29 Mya (Patou *et al.*, 2008); few fossils have yet been found for these viverrid subfamilies. According to the dating estimates by Gaubert and Cordeiro-Estrela (2006), the terrestrial civets (Viverrinae) originated in Eurasia during the middle Miocene (*ca.* 16 Mya), with an African civet ancestor emigrating to Africa *ca.* 12 Mya. A second event of migration leading to the

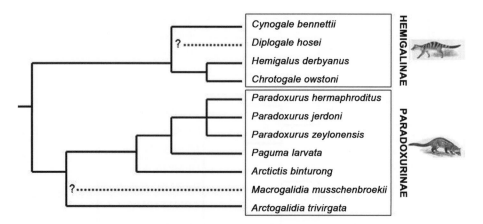

Figure 3.6 For colour version see Plate 2. Synthetic tree of the Asian palm civets (Hemigalinae and Paradoxurinae) (from Patou, 2008; Patou *et al.*, 2008). Dashed lines indicate the species not yet included in a molecular phylogeny. Images from Francis (2008).

Genettinae was estimated *ca.* 11.5 Mya (Gaubert and Cordeiro-Estrela, 2006). The genus *Viverra* is recorded in Asia *ca.* 8.5 Mya (and possibly earlier; Barry, 1995; Barry *et al.*, 2002) and the genus *Genetta* is acknowledged in Africa *ca.* 7.4 Mya (McDougall and Feibel, 2003; Werdelin, 2003).

The subfamily Hemigalinae is monophyletic and comprises at least four species that are found in Southeast Asia (Nowak, 2005; Wozencraft, 2005; Patou *et al.*, 2008; Patou, 2008; Jennings and Veron, 2009; see Figure 3.6). Two species of otter civet have been described: the otter civet (*Cynogale bennettii*) in the Sundaic region, and Lowe's otter civet (*Cynogale lowei*) in North Vietnam and southern China (Pocock, 1933a; and see Schreiber *et al.*, 1989), but the latter is often considered a synonym of *C. bennettii* (Wozencraft, 1993, 2005). In fact, the description of Lowe's otter civet is based on an immature animal that was bought in a village in North Vietnam (its actual origin is unknown). Although some skins have been reported from villages in China, no further specimens have been found to confirm its status or its presence in the Indochina region (see Veron *et al.*, 2006). The semi-aquatic otter civet (*C. bennettii*) has a peculiar and specialised morphology (Pocock, 1915b) that led some authors to place it in a separate subfamily, the Cynogalinae Pocock, 1933 (Pocock, 1933a; Gregory and Hellman, 1939) or in a separate tribe Cynogalini, in the subfamily Hemigalinae (Simpson, 1945). However, its basicranial region, plantar pads and perineal gland display features typical of the Hemigalinae (Pocock, 1915b; Veron, 1994, 1995, 1999; Jennings and Veron, 2009). It has now been shown by a recent molecular study (using two mitochondrial genes, cytochrome *b* and ND2, and two nuclear genes, beta-fibrinogen intron 7 and IRBP exon 1) to be the sister taxon of the rest of the hemigaline taxa (Patou *et al.*, 2008; see Figure 3.6).

The remaining taxa of the Hemigalinae are the banded palm civet (*Hemigalus derbyanus*), Owston's palm civet (*Chrotogale owstoni*), and Hose's palm civet (*Diplogale hosei*). The latter two have been placed in the genus *Hemigalus* by some authors (Payne *et al.*, 1985; Corbet and Hill, 1992) on the basis of their quite similar morphology, but other authors considered that they should be placed in separate genera (Thomas, 1912; Pocock, 1915d, Simpson, 1945; Wozencraft, 1993, 2005). Recent molecular studies have suggested that Owston's palm civet should remain in its separate genus *Chrotogale* (Veron *et al.*, 2004b); *D. hosei* has not been included in a molecular phylogeny. Although Owston's palm civet has a restricted range in Vietnam, Laos, and South China, recent genetic analyses have shown two distinct geographic clades within this species that might be considered subspecies (Veron *et al.*, 2004b).

The subfamily Paradoxurinae comprises at least seven species of Asian arboreal civets, and includes the masked palm civet (*Paguma larvata*) and the binturong (*Arctictis binturong*), the largest civet species (Nowak, 2005; Wozencraft, 2005; Patou *et al.*, 2008; Patou, 2008; Jennings and Veron, 2009; see Figure 3.6). The small-toothed palm civet (*Arctogalidia trivirgata*) has a peculiar dentition, is highly arboreal, and males do not possess a perineal gland. These features, together with an early fusion of the ectotympanic and ento-tympanic bones (tympanic bullae), led some authors to place the small-toothed palm civet in a monotypic subfamily, the Arctogalidiinae Pocock, 1933 (Pocock, 1933a; Gregory and Hellman, 1939) or in a monotypic tribe Arctogalidiini (Simpson, 1945). A recent molecular study suggested that *A. trivirgata* is the sister taxon of the other paradoxurine species (Patou *et al.*, 2008); the absence of a perineal gland in males may be a secondary loss rather than a primitive feature.

Paradoxurus is the only polytypic genus within the Paradoxurinae. The number of species in this genus is still uncertain (see Corbet and Hill, 1992); the Mentawai palm civet (*Paradoxurus hermaphroditus lignicolor*) is either a subspecies of the common palm civet (*Paradoxurus hermaphroditus*) (Chasen and Kloss 1927; Pocock 1934c; Wozencraft 1993, 2005) or a distinct species (Miller, 1903; Schreiber *et al.*, 1989; Corbet and Hill, 1992). The common palm civet has a large distribution throughout Asia and varies in size, coat pattern, and colour throughout its range; at least 30 subspecies have been described (Pocock 1934a,b,c, 1939; Corbet and Hill, 1992), but it appears that the delimitation of this species and its subspecies need to be reconsidered (Patou, 2008). The other *Paradoxurus* species, the brown palm civet (*Paradoxurus jerdoni*) and the golden palm civet (*Paradoxurus zeylonensis*) were also described mainly on the basis of coat colour and pattern, and could either be different morphotypes of *P. hermaphroditus* or valid species (Pocock 1939; Corbet and Hill, 1992; Wozencraft, 1984, 2005). Patou *et al.* (2008) revealed significant genetic distances between the brown palm civet and the common palm civet, which suggests that the brown palm civet is a valid taxon; this is also supported by its distinctive morphology (brown uniform colour, no spots, no facial pattern, the hairs on the neck are reversed, and there is a large parastyle on the upper carnassial). The golden palm civet, endemic to Sri Lanka, may also be a valid taxon (Patou, 2008).

The Sulawesi palm civet, endemic to Sulawesi, has yet to be investigated by molecular studies; its morphological features suggest that it may be a sister taxon to the clade (*Arctictis, Paguma,* and *Paradoxurus*) (see Jennings and Veron, 2009).

The subfamily Genettinae comprises 17 species of slender, semi-arboreal, semi-digitigrade genets and oyans, all found in Africa (except for the common genet, which is also found in Europe and on the Arabian peninsula) (see Jennings and Veron, 2009). Until recently, the genets and oyans were included in the subfamily Viverrinae (Wozencraft, 1993, 2005), but Gaubert and Cordeiro-Estrela (2006) have suggested that they should be separated from the terrestrial civets and placed in their own subfamily, the Genettinae. As already described, the Central African oyan and Leighton's oyan (genus *Poiana*) are the sister group of the genet clade (*Genetta*), and are not related to the Asian linsangs (Gaubert and Veron, 2003). An unusual genettine, the aquatic genet (*Genetta piscivora*), has a uniform brown pelage, naked feet, a piscivorous dentition, and a dorsal position of the nostrils (Van Rompaey, 1988). All these features are linked to its presumed adaptation to a semi-aquatic way of life, and were sufficiently unique that it was placed in its own genus, *Osbornictis* J. A. Allen, 1919 (see Van Rompaey, 1988; Wozencraft, 1993). Gaubert *et al.* (2004a,b) have now shown that it is in fact a derived genet and should be placed within the genus *Genetta*, as was also suggested by Verheyen (1962) and Stains (1983), and followed by Wozencraft (2005). This species thus shows fast morphological evolution compared to its relatively low genetic distance to other genets. It is interesting to compare this to the otter civet (*C. bennettii*), another semi-aquatic viverrid, which has both a derived morphology and a high genetic distance to the other hemigaline taxa, and thus does not have unbalanced genetic and morphological divergence rates (see Patou *et al.*, 2008).

The classification of the remaining genets (*Genetta*) has been fraught with complex taxonomic and species delimitation problems and has long been debated (see e.g. Wenzel and Haltenorth, 1972; Crawford-Cabral, 1981;

Schlawe, 1981; Crawford-Cabral and Pacheco, 1992; Crawford-Cabral and Fernandes, 2001; Gaubert, 2003a; Gaubert *et al.*, 2004b). Species boundaries were difficult to assess using traditional morphological characters (cranial measurements or characters, and coat patterns), with the similar species of the large-spotted genet complex being the most problematic. Morphological analyses suggested that this complex should be divided geographically into three groups that correspond to valid species: west of the Volta River (*G. pardina*), east of the Volta River (*G. maculata*), and the coastal area of South Africa (*G. tigrina*) (see Gaubert *et al.*, 2005b). *G. maculata* was also thought to comprise several valid species. Other debated genets were several 'forest forms' of the large-spotted genets, the servaline genets, and the feline genet (*G. felina*), which has been considered a subspecies of the common genet (*G. genetta*). Recent molecular and morphological studies have proposed that 17 *Genetta* species should be recognised (including the aquatic genet), but the validity of some species still needs to be confirmed (Gaubert *et al.*, 2004a,b, 2005a,b; Jennings and Veron, 2009).

The first species to branch off the genet lineage were the Hausa genet (*G. thierryi*) and Abyssinian genet (*G. abyssinica*), followed by the giant genet (*G. victoriae*) (Gaubert *et al.*, 2004b, 2005b; and see Figure 3.7). The servaline genets are now confirmed to be two species, the servaline genet (*G. servalina*) and crested genet (*G. cristata*). The aquatic genet (*G. piscivora*) and Johnston's genet (*G. johnstoni*) are closely related and are characterised by distinct morphology. The aquatic genet is adapted to a semi-aquatic way of life (Van Rompaey, 1988), and Johnston's genet has a derived morphology related to its adaptation to insectivory (Lamotte and Tranier, 1983). The small-spotted genet complex is a paraphyletic group and now consists of two species, the common genet (*G. genetta*) and the feline genet (*G. felina*), which have been for a long time considered conspecific; *G. felina* is in fact more closely related to the Angolan genet (*G. angolensis*) than to *G. genetta* (see Gaubert *et al.*, 2004b). The large-spotted genet complex is monophyletic and comprises at least five species: rusty-spotted genet (*G. maculata*), cape genet (*G. tigrina*), pardine genet (*G. pardina*), king genet (*G. poensis*), and the newly described Bourlon's genet (*G. bourloni*) (Gaubert, 2003b). Two other forms within this complex, *letabe* and *schoutedeni*, have been proposed as valid species (Gaubert *et al.*, 2004b, 2005b), but further research is required to confirm their status. Hybridisation between some species of the large-spotted genet complex has been detected, which has important evolutionary and conservation implications (Gaubert *et al.*, 2005a).

The subfamily Viverrinae is monophyletic and comprises six species of large, digitigrade terrestrial civets (Veron and Heard, 2000; Gaubert and

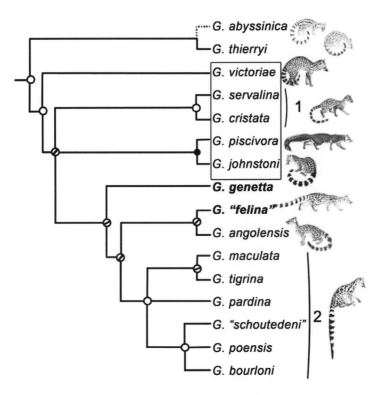

Figure 3.7 For colour version see Plate 3. Synthetic tree of the genets from Gaubert *et al.* (2004b) (1: true servaline genets, 2: large spotted genet complex; in bold: small spotted genet). Dashed lines indicate the species not yet included in a molecular phylogeny.

Cordeiro-Estrela, 2006). This subfamily previously included the Asian linsangs (see Honacki *et al.*, 1982; Wozencraft, 1993, 2005), which have now been placed in the family Prionodontidae (see Gaubert and Veron, 2003), and the genets and oyans, which are now in the subfamily Genettinae (see Gaubert and Cordeiro-Estrela, 2006). The African civet (*Civettictis civetta*) is found in Africa; the Malay civet (*Viverra tangalunga*), large Indian civet (*Viverra zibetha*), large-spotted civet (*Viverra megaspila*), Malabar civet (*Viverra civettina*), and the small Indian civet (*Viverricula indica*), all occur in Asia (Corbet and Hill, 1992; Nowak, 2005; Jennings and Veron, 2009). The small Indian civet (*V. indica*) is a sister taxon to the other terrestrial civets; it is the smallest viverrine species and shows some distinctive features (such as no alisphenoid canal, fusion of the third and fourth digital pads, scent gland pouch simple, and no erectile crest; Wozencraft, 1984). The rare Malabar civet (*V. civettina*), endemic to the Western Ghats in India, is very similar to the large-spotted civet (*V. megaspila*) and was believed to be conspecific with this species by some authors (Ellerman and Morrison-Scott, 1951), but is now recognised as a valid species (Wozencraft, 1989b; 2005); however, this has not yet been confirmed by molecular analyses. Corbet and Hill (1992) doubted the separation of these two species, as there are very few morphological features that differ between them. The Malabar civet (*V. civettina*) was listed as 'possibly extinct' by the IUCN in 1978, but it was rediscovered in 1987 (Kurup, 1987). However, since then no records have been obtained for this species. In 1997, a new terrestrial civet species, the Taynguyen civet (*Viverra tainguensis*), was described in Vietnam (Sokolov *et al.*, 1997). However, insufficient evidence has been presented to show that the Taynguyen civet is a distinct species: it shows many similarities with the large Indian civet (*V. zibetha*), and the morphological features considered as diagnostic for the Taynguyen civet can be observed in the large Indian civet in different parts of its range (Walston and Veron, 2001).

Conclusions

Recent molecular studies have shed some light on the phylogeny of the Viverridae and have clarified the systematic position of some debated taxa that were previously placed within this family. The African palm civet (*Nandinia binotata*), the Malagasy viverrids (*Cryptoprocta ferox, Eupleres goudotii,* and *Fossa fossana*), and the Asian linsangs (*Prionodon pardicolor* and *Prionodon linsang*) are no longer included in the Viverridae, which now consists of four subfamilies: the Hemigalinae (Asian palm civets and otter civet), the Paradoxurinae (Asian palm civets and binturong), the Genettinae (genets and oyans), and the Viverrinae (terrestrial civets).

Recent molecular studies have not only resulted in a new classification of the Viverridae, but they also highlighted the difficulties of using traditional morphological features to define systematic relationships within the feliform carnivores. Some dental and basicranial features were found to be convergent features or primitive generalised characters in this group. In contrast, some soft anatomy features were shown to be reliable characters to establish relationships.

Acknowledgements

I would like to thank the many collaborators and students who contributed to the study of the systematics of the Viverridae and viverrid-like taxa. I am particularly indebted to P. Gaubert and M.-L. Patou (Muséum National d'Histoire Naturelle, France). I would also like to thank A. Tillier,

J. Lambourdière and C. Bonillo (Service de Systématique Moléculaire, CNRS IFR 101, Muséum National d'Histoire Naturelle) for their help in the molecular laboratory, and C. Denys, L. Deharveng and E. Pasquet (Muséum National d'Histoire Naturelle, France) for their support. I also thank A. Friscia (University of California, Los Angeles, USA) and A. Goswami (University of Cambridge, UK) for inviting me to participate in the Carnivore Symposium at the 67th Annual Meeting of the Society of Vertebrate Paleontology and to contribute to this volume. Many thanks to A. P. Jennings (Muséum National d'Histoire Naturelle, France) for improving the early drafts of this paper. I also thank two anonymous reviewers for their fruitful comments.

REFERENCES

Albignac, R. (1973). *Faune de Madagascar. 36. Mammifères carnivores.* Paris: O.R.S.T.O.M., C.N.R.S.

Amigues, S. (1999). Les belettes de Tartessos. *Anthropozoologica*, **29**, 55–64.

Barry, J. C. (1995). Faunal turnover and diversity in the Terrestrial Neogene of Pakistan. In *Paleoclimate and Evolution with Emphasis on Human Origins*, ed. E. S. Vrba, G. H. Denton, T. C. Partridge and L. H. Burckle. New Haven, CT: Yale University Press, pp. 115–34.

Barry, J., Morgan, M., Flynn, L., *et al.* (2002). Faunal and environmental change in the Late Miocene Siwaliks of Northern Pakistan. *Paleobiology*, **28**, 1–71.

Beaumont, G. D. (1964). Remarques sur la classification des Felidae. *Eclogae Geologicae Helvetia*, **57**, 837–45.

Bennett, E. T. (1833). Notice of a new genus of viverridous Mammalia from Madagascar. *Proceedings of the Zoological Society of London*, **1833**, 46.

Bininda-Emonds, O. R., Gittleman, J. L. and Purvis, A. (1999). Building large trees by combining phylogenetic information: a complete phylogeny of the extant Carnivora (Mammalia). *Biological Review*, **74**, 143–75.

Bonaparte, C. (1845). *Catalogo metodico dei Mammiferei Europei.* Milano: Coi tipi di Luigi di Giacomo Pirola.

Bugge, J. (1978). The cephalic arterial system in carnivores with special reference to the systematic classification. *Acta Anatomica*, **101**, 45–61.

Carlsson, A. (1911). Uber *Cryptoprocta ferox. Zoologische Jahrbücher, Abt. Syst.*, **30**, 419–70.

Chasen, F. N. and Kloss, C. B. (1927). Spolia Mentawiensia. Mammals. *Proceedings of the Zoological Society of London*, **1927**, 797–840.

Chenu, J. (1851–1858). *Encyclopedie d'Histoire Naturelle – Carnassiers avec la collaboration avec M. E. Desmarest.* Paris.

Coetzee, C. G. (1977). Order Carnivora. Part 8. In *The Mammals of Africa: An Identification Manual*, ed. J. Meester and H. W. Setzer. Washington, DC: Smithsonian Institution Press, pp. 1–42.

Corbet, G. B. and Hill, J. E. (1992). *The Mammals of the Indomalayan Region: A Systematic Review.* Oxford: Oxford University Press.

Crawford-Cabral, J. (1981). The classification of the genets (Carnivora, Viverridae, genus *Genetta*). *Boletimm da Sociedade Portuguesa de Ciências Naturais,* **20,** 97–114.

Crawford-Cabral, J. (1993). A comment on the systematic position of *Poiana. Small Carnivore Conservation,* **9,** 8.

Crawford-Cabral, J. and Fernandes, C. A. (2001). The Rusty-spotted genets as a group with three species in Southern Africa (Carnivora: Viverridae). In *African Small Mammals,* ed. L. Granjon, A. Poulet and C. Denys. Paris: I.R.D., pp. 65–80.

Crawford-Cabral, J. and Pacheco, A. P. (1992). Are the Large-spotted and the Rusty-spotted genets separate species (Carnivora, Viverridae, genus *Genetta*). *Garcia de Orta, Sér. Zool.,* **16,** 7–17.

Dargel, B. (1990). A bibliography on Viverrids. *Mitteilungen aus dem Hamburgischen Zoologischen Museum und Institut,* **87,** 1–184.

Doyère, M. (1835). Notice sur un mammifère de Madagascar, formant le type d'un nouveau genre de la famille des carnassiers insectivores de M. Cuvier. *Annales des Sciences Naturelles, Zoologie,* **4,** 270–83.

Ellerman, J. R. and Morrison-Scott, T. C. S. (1951). *Checklist of Palearctic and Indian Mammals 1758 to 1946.* London: British Museum.

Ewer, R. F. (1973). *The Carnivores.* Ithaca and London: Comstock Publishing Associates, Cornell University, Press.

Filhol, H. (1879). Mammifères fossiles de l'Allier. *Bulletin de l'Ecole Pratique des Hautes Etudes, Sciences naturelles,* **19,** 192–201.

Filhol, H. (1894). Sur quelques points de l'anatomie du cryptoprocte de Madagascar. *Comptes-Rendus de l'Académie des Sciences,* **118,** 1060–62.

Flower, W. H. (1869). On the value of the characters of the base of the cranium in the classification of the order Carnivora and the systematic position of *Bassaris* and other disputed forms. *Proceedings of the Zoological Society of London,* **1869,** 4–37.

Flynn, J. J., Neff, N. A. and Tedford, R. H. (1988). Phylogeny of the carnivores. In *The Phylogeny and Classification of Tetrapods,* ed. M. J. Benton. Oxford: Systematics Association Special Volume, Clarendon Press, pp. 73–776.

Flynn, J. J. and Nedbal, M. A. (1998). Phylogeny of the Carnivora (Mammalia): congruence vs incompatibility among multiple data sets. *Molecular Phylogenetics and Evolution,* **9,** 414–26.

Flynn, J. J., Finarelli, J. A., Zehr, S., Hsu, J. and Nedbal, M. A. (2005). Molecular phylogeny of the Carnivora (Mammalia): assessing the impact of increased sampling on resolving enigmatic relationships. *Systematic Biology,* **54,** 317–37.

Fredga, K. (1972). Comparative chromosome studies in mongooses (Carnivora, Viverridae). 1. Idiograms of 12 species and karyotype evolution in Herpestinae. *Hereditas,* **71,** 1–74.

Francis, C. M. (2008). *A Field Guide to the Mammals of South-East Asia.* London: New Holland Publishers.

Gaubert, P. (2003a). *Systématique et phylogénie du genre Genetta et des énigmatiques "genet-like taxa"* Prionodon, Poiana *et* Osbornictis *(Carnivora, Viverridae): caractérisation de la sous-famille des Viverrinae et étude des patrons de diversification au sein du continent africain.* PhD Dissertation, Muséum National d'Histoire Naturelle, Paris.

Gaubert, P. (2003b). Description of a new species of genet (Carnivora; Viverridae; genus *Genetta*) and taxonomic revision of forest forms related to the Large-spotted Genet complex. *Mammalia*, **67**, 85–108.

Gaubert, P. and Cordeiro-Estrela, P. (2006). Phylogenetic systematics and tempo of evolution of the Viverrinae (Mammalia, Carnivora, Viverridae) within feliformians: implications for faunal exchanges between Asia and Africa. *Molecular Phylogenetics and Evolution*, **41**, 266–78.

Gaubert, P. and Veron, G. (2003). Exhaustive sample set among Viverridae reveals the sister-group of felids: the linsangs as a case of extreme morphological convergence within Feliformia. *Proceedings of the Royal Society of London, B*, **270**, 2523–30.

Gaubert, P., Veron, G. and Tranier, M. (2002). Genets and "genet-like" taxa (Carnivora, Viverrinae): phylogenetic analysis, systematics and biogeographic implications. *Zoological Journal of the Linnean Society*, **134**, 317–34.

Gaubert, P., Tranier, M., Delmas, A.-S., Colyn, M. and Veron, G. (2004a). First molecular evidence for reassessing phylogenetic affinities between genets (*Genetta*) and the enigmatic genet-like taxa *Osbornictis*, *Poiana* and *Prionodon* (Carnivora, Viverridae). *Zoologica Scripta*, **33**, 117–29.

Gaubert, P., Fernandes, C. A., Bruford, M. W. and Veron, G. (2004b). Genets in Africa: an evolutionary synthesis based on cytochrome *b* sequences and morphological characters. *Biological Journal of the Linnean Society*, **81**, 589–610.

Gaubert, P., Taylor, P. J., Fernandes, C. A., Bruford, M. W. and Veron, G. (2005a). Patterns of cryptic hybridization revealed using an integrative approach: a case study on genets (Carnivora, Viverridae, *Genetta* spp.) from the southern African subregion. *Biological Journal of the Linnean Society*, **86**, 11–33.

Gaubert, P., Taylor, P. J. and Veron, G. (2005b). Integrative taxonomy and phylogenetic systematics of the genets (Carnivora, Viverridae, genus *Genetta*): a new classification of the most speciose carnivoran genus in Africa. In *African Biodiversity: Molecules, Organisms, Ecosystems. Proceedings of the 5th International Symposium on Tropical Biology, Museum Koenig, Bonn*, ed. B. A. Huber, B. J. Sinclair and K. H. Lampe. Bonn: Springer Verlag, pp. 371–83.

Gaubert, P., Wozencraft, W. C., Cordeiro-Estrela, P. and Veron, G. (2005c). Mosaic of convergences, noise and misleading morphological phylogenies: what's in a viverrid-like carnivoran? *Systematic Biology*, **54**, 865–94.

Gilchrist, J. S., Jennings, A. P., Veron, G. and Cavallini, P. (2009). Family Herpestidae. In *Handbook of the Mammals of the World, Volume 1, Carnivores*, ed. D. Wilson and R. A. Mittermeier. Barcelona: Lynx edicions, pp. 262–328.

Ginsburg, L. (1999). Order Carnivora. In *The Miocene: Land Mammals of Europe*, ed. G. E. Rössner and K. Heissig. München: *Pfeil*, pp. 109–48.

Gray, J. E. (1821). On the natural arrangement of vertebrose animals. *London Med Repository*, **15**, 296–310.

Gray, J. E. (1864). A revision of the genera and species of viverrine animals (Viverridae), founded on the collection in the British Museum. *Proceedings of the Zoological Society of London*, **1864**, 502–79.

Gregory, W. K. and Hellman, H. (1939). On the evolution and major classification of the civets (Viverridae) and allied fossil and recent Carnivora; a phylogenetic study of the skull and dentition. *Proceedings of the American Philosophical Society*, **81**, 309–92.

Helbing, H. (1927). Une genette Miocène trouvée dans les argiles de Captieux (Gironde). *Verhandlungen der Naturforschenden Gesellschaft in Basel*, **38**, 305–15.

Holliday, J. A. (2007). *Phylogeny and character change in the feloid Carnivora*. PhD Dissertation. College of Arts and sciences, The Florida State University, USA.

Honacki, H. H., Kinman, K. E. and Koeppl, J. W. (1982). *Mammals Species of the World. A Taxonomic and Geographic Reference*. Lawrence, KS: Allen Press, Ass. Syst. Collections.

Horsfield, T. (1821). *Zoological Researches in Java, and the Neighbouring Islands*. London: Kingsbury, Parbury, & Allen.

Hunt, R. M. (1974). Auditory bullae in Carnivora: an anatomical basis for reappraisal of carnivore evolution. *Journal of Morphology*, **134**, 21–76.

Hunt, R. M. (1987). Evolution of aeluroid Carnivora: significance of auditory structure in the Nimravid cat *Dinictis*. *American Museum Novitates*, **2886**, 1–74.

Hunt, R. M. (1989). Evolution of the aeluroid Carnivora: significance of the ventral promontorial process of the petrosal and the origin of basicranial patterns in the living families. *American Museum Novitates*, **2930**, 1–32.

Hunt, R. M. (1991). Evolution of the aeluroid Carnivora: viverrid affinities of the Miocene carnivoran: *Herpestides*. *American Museum Novitates*, **3023**, 1–34.

Hunt, R. M. (1996). Biogeography of the Order Carnivora. In *Carnivore Behavior, Ecology and Evolution*, ed. J. L. Gittleman. Ithaca and London: Comstock Publishing Associates, Cornell University Press, pp. 485–541.

Hunt, R. M. Jr. (2001). Basicranial anatomy of the living linsangs *Prionodon* and *Poiana* (Mammalia, Carnivora, Viverridae), with comments on the early evolution of aeluroid Carnivora. *American Museum Novitates*, **3330**, 1–24.

Hunt, R. M. and Tedford, R. H. (1993). Phylogenetic relationships within the aeluroid Carnivora and implications of their temporal and geographic distribution. In *Mammal Phylogeny: Placentals*, ed. F. S. Szalay, M. J. Novacek and M. C. McKenna. New York, NY: Springer-Verlag, pp. 53–73.

Jennings, A. P. and Veron, G. (2009). Family Viverridae. In *Handbook of the Mammals of the World, Volume 1, Carnivores*, ed. D. Wilson and R. A. Mittermeier. Barcelona: Lynx edicions, pp. 174–233.

Johnson, W. E., Eizirik, E., Pecon-Slattery, J., *et al.* (2006). The Late Miocene radiation of modern Felidae: a genetic assessment. *Science*, **311**, 73–77.

Kingdon, J. (1977). *East African Mammals. An Atlas of Evolution in Africa*. London: Academic Press.

Koepfli, K. P., Jenks, S. M., Eizirik, E., Zahirpour, T., Van Valkenburgh, B. and Wayne, R. K. (2006). Molecular systematics of the Hyaenidae: relationships of a relictual lineage resolved by a molecular supermatrix. *Molecular Phylogenetics and Evolution*, **38**, 603–20.

Kurup, G. U. (1987). The rediscovery of the Malabar civet *Viverra megaspila civettina* Blyth in India. *Cheetal*, **28**, 1–4.

Lamotte, M. and Tranier, M. (1983). Un spécimen de *Genetta (Paragenetta) johnstoni* collecté dans la région du Nimba (Côte d'Ivoire). *Mammalia*, **47**, 430–32.

Lesson, R. P. (1842). *Nouveau tableau du règne animal*. Paris: A. Bertrand.

Linnaeus, C. (1758). *Systema Naturae per regna tria naturae, secundum classis, ordines, genera, species, cum characteribus, differentiis, synonymis, locis. Tenth ed. Vol. 1.* Stockholm: Laurentii Salvii.

McDougall, I. and Feibel, C. S. (2003). Numerical age control for the Miocene–Pliocene succession at Lothagam, a hominoid-bearing sequence in the northern Kenya rift. In *Lothagam, the Dawn of Humanity in Eastern Africa*, ed. M. G. Leakey and J. M. Harris. New York, NY: Columbia University Press, pp. 43–64.

McKenna, M. C. and Bell, S. K. (1997). *Classification of Mammals above the Species Level*. New York, NY: Columbia University Press.

Miller, G. S. (1903). Seventy new Malayan mammals. *Smithsonian Miscelleanous Collections*, **45**, 1–74.

Milne-Edwards, A. and Grandidier, A. (1867). On the organisation of *Cryptoprocta ferox*. *Annals of Natural History*, **20**, 382–84.

Mivart, St.-G. (1882). On the classification and distribution of the Ailuroidea. *Proceedings of the Zoological Society of London*, **1882**, 135–208.

Neff, N. A. (1983). *The basicranial anatomy of the Nimravidae (Mammalia: Carnivora): character analyses and phylogenetic inferences*. PhD Dissertation, City University of New York.

Novacek, M. J. (1977). Aspects of the problem of variation, origin and evolution of the eutherian auditory bulla. *Mammal Review*, **7**, 131–49.

Nowak, R. M. (2005). *Walker's Carnivores of the World*. Baltimore, MD: Johns Hopkins University Press.

Patou, M.-L. (2008). *Systématique et biogéography des Herpestidae et Viverridae (Mammalia, Carnivora) en Asie*. PhD Dissertation, Muséum National d'Histoire Naturelle, Paris.

Patou, M.-L., Debruyne, R., Jennings, A. P., Zubaid, A., Rovie-Ryan, J. J. and Veron, G. (2008). Phylogenetic relationships of the Asian palm civets (Hemigalinae & Paradoxurinae, Viverridae, Carnivora). *Molecular Phylogenetic and Evolution*, **47**, 883–92.

Payne, J., Francis, C. M. and Phillipps, K. (1985). *A Field Guide to the Mammals of Borneo*. Sabah: Sabah Society and WWF Malaysia.

Perez, M., Li, B., Tillier, A., Cruaud, A. and Veron, G. (2006). Systematic relationships of the bushy-tailed and black-footed mongooses (genus *Bdeogale*, Herpestidae, Carnivora) based on molecular, chromosomal and morphological evidence. *Journal of Zoological Systematics and Evolutionary Research*, **44**, 251–59.

Petter, G. (1974). Rapports phylétiques des Viverridés. Les formes de Madagascar. *Mammalia*, **38**, 605–36.

Pocock, R. I. (1915a). On some of the external characters of *Cynogale bennetti*, Gray. *Annals and Magazine of Natural History*, **15**, 351–60.

Pocock, R. I. (1915b). On some of the external characters of the genus *Linsang*, with notes upon the genera *Poiana* and *Eupleres*. *The Annals and Magazine of Natural History*, **16**, 341–51.

Pocock, R. I. (1915c). On the feet and glands and other external characters of the Paradoxurine genera *Paradoxurus*, *Arctictis*, *Arctogalidia* and *Nandinia*. *Proceedings of the Zoological Society of London*, **2**, 387–412.

Pocock, R. I. (1915d). On some of the external characters of the palm civet (*Hemigalus derbyanus*, Gray) and its allies. *The Annals and Magazine of Natural History*, **93**, 153–62.

Pocock, R. I. (1915e). On some external characters of *Galidia*, *Galidictis* and related genera. *The Annals and Magazine of Natural History*, **16**, 351–56.

Pocock, R. I. (1916a). On the external characters of the mongooses (Mungotidae). *Proceedings of the Zoological Society of London*, **1**, 349–74.

Pocock, R. I. (1916b). On some of the external characters of *Cryptoprocta*. *The Annals and Magazine of Natural History*, **27**, 413–25.

Pocock, R. I. (1916c). On the course of the internal carotid artery and the foramina connected therewith in the skulls of the Felidae and Viverridae. *The Annals and Magazine of Natural History*, **17**, 261–69.

Pocock, R. I. (1919). The classification of the mongooses (Mungotidae). *The Annals and Magazine of Natural History*, **23**, 515–24.

Pocock, R. I. (1929). *Carnivora. Encyclopaedia Britannica*, 14th ed., **IV**: 896–900.

Pocock, R. I. (1933a). The rarer genera of oriental Viverridae. *Proceedings of the Zoological Society of London*, **1933**, 969–1035.

Pocock, R. I. (1933b). The civet-cats of Asia. *Journal of the Bombay Natural History Society*, **36**, 423–49.

Pocock, R. I. (1933c). The civet-cats of Asia. Part II. *Journal of the Bombay Natural History Society*, **36**, 629–56.

Pocock, R. I. (1934a). The palm civet or 'toddy cats' and the genera *Paradoxurus* and *Paguma* inhabiting British India Part II. *Journal of the Bombay Natural History Society*, **37**, 172–92.

Pocock, R. I. (1934b). The palm civets or 'toddy cats' of the genera *Paradoxurus* and *Paguma* inhabiting British India. Part III. *Journal of the Bombay Natural History Society*, **37**, 314–46.

Pocock, R. I. (1934c). The geographical races of *Paradoxurus* and *Paguma* found to the east of the bay of Bengal. *Proceedings of the Zoological Society of London*, **1934**, 613–83.

Pocock, R. I. (1939). *The Fauna of British India, including Ceylon and Burma. Mammalia. Vol. 1*. London: Taylor and Francis.

Pocock, R. I. (1940). *Cryptoprocta* should be placed in Viverridae near *Paradoxurus* and not in Felidae. *The Annals and Magazine of Natural History*, **11**, 312.

Radinsky, L. (1975). Viverrid neuroanatomy: phylogenetic and behavioural implications. *Journal of Mammalogy*, **56**, 130–50.

Rosevear, D. R. (1974). *The Carnivores of West Africa*. London: British Museum (Natural History).

Schlawe, L. (1981). Material, Fundorte, Text- und Bildquellen als Grundlagen für eine Artenliste zur Revision der Gattung *Genetta* G. Cuvier, 1816. *Zoologische Abhandlungen Staatliches Museum für Tierkunde Dresden*, **37**, 85–182.

Schreiber, A., Wirth, R., Riffel, M. and Van Rompaey, H. (1989). *Weasels, Civets and Mongooses, and Their Relatives. An Action Plan for the Conservation of Mustelids and Viverrids*. Gland: IUCN/SSC Mustelid and Viverrid Specialist Group.

Simpson, G. G. (1945). The principles of classification and a classification of mammals. *Bulletin of the American Museum of Natural History*, **85**, 1–350.

Skinner, J. D. and Smithers, R. H. N. (1990). *The Mammals of the Southern African Subregion [new ed.]*. Pretoria: University of Pretoria.

Sokolov, V. E., Rozhnov, V. V. and Pham Tchong, A. (1997). New species of the genus *Viverra* (Mammalia, Carnivora) from Vietnam. *Zoologichesky Zhurnal*, **76**, 585–89.

Stains, H. J. (1983). Calcanea of members of the Viverridae. *Bulletin of the Southern California Academy of Science*, **82**, 17–38.

Stains, H. J. (1987). Carnivores and pinnipeds. In *Recent Mammals of the World. A Synopsis of Families*, ed. S. Anderson and J. K. Jones. New York, NY: The Ronald Press Company, pp. 325–54.

Taylor, M. E. (1988). Foot structure and phylogeny in the Viverridae (Carnivora). *Journal of Zoology*, **216**, 131–39.

Teilhard de Chardin, P. (1915). Les Carnassiers des Phosphorites du Quercy. *Annales de Paléontologie*, **9**, 103–91.

Thomas, O. (1912). Two new genera and a new species of viverrine Carnivora. *Proceedings of the Zoological Society of London*, **1912**, 498–503.

Van Rompaey, H. (1988). *Osbornictis piscivora. Mammalian Species*, **309**, 1–4.

Verheyen, W. (1962). Quelques notes sur la zoogéographie et la craniologie d'*Osbornictis piscivora* Allen 1919. *Revue de Zoologie et Botanique Africaine*, **65**, 121–28.

Veron, G. (1994). *Méthodes de recherche en biotaxonomie des mammifères carnivores. Confrontation des méthodes de phylogénie traditionnelle et moléculaire dans la recherche de la position systématique de* Cryptoprocta ferox *(Aeluroidea)*. PhD Dissertation, Muséum National d'Histoire Naturelle, Paris.

Veron, G. (1995). La position systématique de *Cryptoprocta ferox* (Carnivora). Analyse cladistique des caractères morphologiques de carnivores Aeluroidea actuels et fossiles. *Mammalia*, **59**, 551–82.

Veron, G. (1999). Pads morphology in the Viverridae (Carnivora). *Acta Theriologica*, **44**, 363–76.

Veron, G. and Catzeflis, F. (1993). Phylogenetic relationships of the endemic Malagasy carnivore *Cryptoprocta ferox* (Aeluroidea): DNA/DNA hybridization experiments. *Journal of Mammalian Evolution*, **1**, 169–85.

Veron, G. and Heard, S. (2000). Molecular systematics of the Asiatic Viverridae (Carnivora) inferred from mitochondrial Cytochrome *b* sequence analysis. *Journal of Zoological Systematics and Evolutionary Research*, **38**, 209–17.

Veron, G., Colyn, M., Dunham, A. E., Taylor, P. and Gaubert, P. (2004a). Molecular systematics and origin of sociality in mongooses (Herpestidae, Carnivora). *Molecular Phylogenetics and Evolution*, **30**, 582–98.

Veron, G., Heard Rosenthal, S., Long, B. and Roberton, S. (2004b). The molecular systematics and conservation of an endangered carnivore, the Owston's palm civet *Chrotogale owstoni* (Thomas, 1912) (Carnivora, Viverridae, Hemigalinae). *Animal Conservation*, **7**, 107–12.

Veron, G., Gaubert, P., Franklin, N., Jennings, A. P. and Grassman, L. (2006). A reassessment of the distribution and taxonomy of the endangered otter civet, *Cynogale bennettii* (Carnivora: Viverridae) of South-east Asia. *Oryx*, **40**, 42–49.

Walston, J. and Veron, G. (2001). Questionable status of the "Taynguyen civet", *Viverra tainguensis* Sokolov, Rozhnov and Pham Trong Anh, 1997 (Mammalia: Carnivora: Viverridae). *Zeitschrift für Säugetierkunde*, **66**, 181–84.

Wayne, R. K., Benveniste, R. E., Janczewski, D. N. and O'Brien, S. J. (1989). Molecular and biochemical evolution of Carnivora. In *Carnivore Behaviour, Ecology and Evolution*, ed. J. L. Gittleman. London: Chapman and Hall, pp. 465–95.

Wenzel, E. and Haltenorth, T. (1972). System der Schleichkatzen (Viverridae). *Säugetierkundliche Mitteilungen*, **20**, 110–27.

Wemmer, C., West, J., Watling, D., Collins, L. and Lang, K. (1983). External characters of the Sulawesi palm civet *Macrogalidia musschenbroekii* Schlegel, 1879. *Journal of Mammalogy*, **64**, 133–36.

Werdelin, L. (2003). Mio-Pliocene Carnivora from Lothagam, Kenya. In *Lothagam: The Dawn of Humanity in Africa*, ed. M. G. Leakey and J. M. Harris. New York, NY: Columbia University Press, pp. 261–330.

Wozencraft, W. C. (1984). *A phylogenetic reappraisal of the Viverridae and its relationship to other Carnivora*. PhD Dissertation. University of Kansas, Lawrence.

Wozencraft, W. C. (1989a). The phylogeny of the Recent Carnivora. In *Carnivore Behavior, Ecology, and Evolution*, ed. J. L. Gittleman. New York, NY: Cornell University Press, pp. 495–535.

Wozencraft, W. C. (1989b). Classification of the recent Carnivora. In *Carnivore Behavior, Ecology, and Evolution*, ed. J. L. Gittleman. New York, NY: Cornell University Press, pp. 569–93.

Wozencraft, W. C. (1993). Order Carnivora. In *Mammal Species of the World – A Taxonomic and Geographic Reference*, ed. D. E. Wilson and D. M. Reeder. Washington, London: Smithsonian Institution Press, pp. 279–348.

Wozencraft, W. C. (2005). Order Carnivora. In *Mammal Species of the World: A Taxonomic and Geographic Reference*, 3rd ed., ed. D. E. Wilson and D. M. Reeder. Baltimore, MD: The Johns Hopkins University Press, vol. **1**, pp. 532–628.

Wurster, D. H. (1969). Cytogenetic and phylogenetic studies in Carnivora. In *Comparative Mammalian Cytogenetics*, ed. K. Benirschke. Berlin, Heidelberg, New York: Springer-Verlag, pp. 310–29.

Wyss, A. R. and Flynn, J. J. (1993). A phylogenetic analysis and definition of the Carnivora. In *Mammal Phylogeny: Placentals*, ed. F. S. Szalay, M. J. Novacek and M. C. MacKenna. Berlin: Springer-Verlag, pp. 32–52.

Yoder, A. D., Burns, M. M., Zehr, S., *et al.* (2003). Single origin of Malagasy Carnivora from an African ancestor. *Nature*, **421**, 734–37.

4

Molecular and morphological evidence for Ailuridae and a review of its genera

MICHAEL MORLO AND STÉPHANE PEIGNÉ

Introduction

The red panda, *Ailurus fulgens*, is a peculiar recent carnivoran whose systematic relationships have been disputed since its first description in 1825 (Figure 4.1). While occurring with a single species today, Ailuridae was represented by several genera in the past. Flynn and Nedbal (1998) were the first to place *Ailurus* in its own family based on molecular evidence. While subsequent analyses of molecular data have confirmed this, paleontological findings of the last 10 years shed much more light on the natural history of Ailuridae and the morphology of its members. Before 1997, Ailuridae consisted of 5 genera with about 16 species. Only two genera belonged to Ailurinae (*Ailurus* and *Parailurus*). Today, the family contains 9 genera (5 of which are considered to be ailurines) with about 26 species. It now is agreed that the earliest ailurid is *Amphictis*, the earliest species of which are known from the Late Oligocene (MP 28, about 25 Mya) of Europe. Even though *Ailurus* today is restricted to Southeast Asia, it is apparent that ailurids were once present in all Northern continents and were most diverse in Europe throughout most of their history.

In this chapter, we examine the recent and fossil evidence supporting the assignment of *Ailurus fulgens* and its ancestors (including *Amphictis*) to a distinct family, Ailuridae. We find such an assignment fully justified and corroborated by morphological data. We review all molecular and morphological studies in which *Ailurus* has been included. While new molecular studies continue to support a family Ailuridae, a review of the morphology of *A. fulgens* led to the identification of a number of characters which could be traced within the fossil relatives of *Ailurus*. This survey led to the development of emended diagnoses of all included genera. We also include a list of all ailurid species and resolve some outstanding issues concerning the taxonomy or systematic relationships of several taxa. Finally, we briefly recapitulate the natural history of

Carnivoran Evolution: New Views on Phylogeny, Form, and Function, ed. A. Goswami and A. Friscia. Published by Cambridge University Press. © Cambridge University Press 2010.

Figure 4.1 For colour version see Plate 4. The only extant species of Ailuridae, the red panda, *Ailurus fulgens*.

Ailuridae. We follow Smith and Dodson (2003) for anatomical notation and orientation of dentitions.

Molecular evidence

Since biochemical studies (from the late 1960s) and molecular studies (since the mid 1980s) were first applied to carnivorans, the systematic position of the red panda has varied greatly within the Arctoidea. This variation is correlated with major parameters: the number of taxa included in the analyses (especially the number of arctoid species and families) and, especially for molecular cladistic analyses, the number, completeness, diversity (from both nuclear and mitochondrial genomes), and phylogenetic performance of genes. Its morphology led authors from the nineteenth and first half of the twentieth centuries to believe that the red panda was either a giant panda or raccoon relative (see discussion below). Moreover, due to its vernacular name and geographic occurrence, both of which seem to favour at least some relationship to the giant panda, the first molecular and biochemical studies mainly focused on this possibility. At that time, the systematic position of the great panda was also the subject of controversy, but Davis (1964) showed that *Ailuropoda melanoleuca* is an ursid with specialisations toward bamboo feeding (see Mayr 1986 for an historical perspective). Despite the work of Davis (1964), the relationships between the two pandas (with, generally, special attention to the giant panda) has continued to be debated among molecular biologists. Consequently, the only arctoids included in those studies, in addition to *Ailurus* and *Ailuropoda*, were other bears, some procyonids (e.g. Todd and Pressman, 1968; Sarich, 1973; O'Brien *et al.*, 1985; Goldman *et al.*, 1989), and occasionally, a few other arctoids (Tagle *et al.*, 1986). With such a limited number of taxa, the significance of the results of these studies was limited and, not surprisingly, the red panda was closely aligned either with bears (Sarich, 1973), procyonids (O'Brien *et al.*, 1985; Goldman *et al.*, 1989), or *Ailuropoda* (Todd and Pressman, 1968; Tagle *et al.*, 1986). The position of the giant panda was clarified by the end of the 1980s, so molecular biologists focused on other topics and somewhat neglected the red panda.

With some exceptions (e.g. Zhang and Ryder, 1993; Pecon Slattery and O'Brien, 1995), molecular studies that have included *Ailurus fulgens* aimed to resolve the relationships of either the whole of Carnivora (e.g. Ledje and Arnason, 1996b; Flynn *et al.*, 2005), or the Caniformia (e.g. Ledje and Arnason, 1996a; Delisle and Strobeck, 2005; Yu and Zhang, 2006), while some studies dealt more specifically with the lesser panda itself (e.g. Flynn *et al.*, 2000). Molecular biologists made an effort to improve both the quantity and quality of data by using a rich, diverse taxonomic sampling and the (various) combination of a greater number and longer molecular sequences from both nuclear and mitochondrial genes in a single analysis (Flynn *et al.*, 2000). These studies resulted in more robust phylogenetic hypotheses for the relationships within Carnivora in general, and within Caniformia in particular. Besides the long-known dichotomy of the order into Feliformia + Caniformia and, among the latter, the sister group relationships of Cynoidea (Canidae) to Arctoidea (Ursidae, Procyonidae, Mustelidae s.l. [i.e. including the skunks], Pinnipedia, and *Ailurus*), the relationships of the arctoid families also became much better supported. In particular, most of the molecular analyses since 1994 have identified three major monophyletic clades within the Arctoidea: Ursidae, Pinnipedia, and Musteloidea (comprising *Ailurus*, Mephitidae, Procyonidae, and Mustelidae s.s. [i.e. without skunks], but see Vrana *et al.*, 1994 for a different view). These studies supported the lack of a close relationship between *Ailurus* and the ursids (in particular *Ailuropoda melanoleuca*) or the procyonids, relationships that mainly had resulted from biases in taxonomic sampling in previous studies (see above).

This improvement in taxonomic sampling and molecular sequencing also resulted in a more stable systematic position for *Ailurus fulgens*. Table 4.1 summarises the origin and length of the analysed molecular sequences, and results obtained for the molecular studies that included *Ailurus fulgens* since Vrana *et al.* (1994). The works of Zhang and Ryder (1993) and Peccon Slattery and O'Brien (1995), which only included some ursids and procyonids in addition to the red panda for the Arctoidea, are not included. Table 4.1 shows that analyses using only one nuclear or mitochondrial gene most often resulted in weaker support or more poorly resolved relationships among arctoid families and *Ailurus fulgens* (e.g. Ledje and Arnason, 1996a,b). Completeness of the molecular sequences also has some effect. Thus, the placement of *Ailurus* as an early offshoot of an Ursidae + Pinnipedia clade in Vrana *et al.* (1994) probably partly results from the incompleteness and mitochondrial origin of the molecular sequences (for the problem of phylogeny based on a single gene system or origin, see, e.g. Degnan, 1993); in that study, the cytochrome *b* sequence is only 307 bp (base pairs) in length (the complete gene is 1140 bp), while that of the small ribosomal subunit (12S RNA) is 394 bp (the complete gene is *ca.* 964 bp). On the contrary, studies using a combination of data sets resulted in more robust phylogenetic hypotheses than any data set used alone, as already pointed out in many recent molecular studies (e.g. Flynn *et al.*, 2005). Since Ledje and Arnason (1996a), all analyses using multiple, diverse, and complementary (i.e. nuclear and mitochondrial sequences) data sets have rejected the placement of the red panda within or as sister taxon to either the Procyonidae or the Ursidae. Studies since Flynn and Nedbal (1998) have supported the recognition of a monotypic family Ailuridae within Musteloidea, but only some of those studies included species of mephitids (Delisle and Strobeck, 2005; Domingo-Roura *et al.*, 2005; Flynn *et al.*, 2000, 2005; Fulton and Strobeck, 2007). Other possible outcomes have suggested a sister group relationship between Ailuridae and Mephitidae (Delisle and Strobeck, 2005), or a branching of Ailuridae after Mephitidae and before the Procyonidae + Mustelidae s.s. clade (Fulton and Strobeck, 2006; Sato *et al.*, 2006). A recent analysis, based on the concatenated sequence of 5 nuclear genes (total of 5497 exon nucleotides) and including 42 species of Ursidae, Pinnipedia and Musteloidea (Wolsan and Sato, 2007), strongly supports the latter hypothesis. Yu *et al.* (2004) and Yu and Zhang (2006) obtained the Ailuridae as sister taxon to the Procyonidae + Mustelidae s.s. clade, but their data set included a limited number of taxa and/or no mephitids. Marmi *et al.* (2004) supported a monophyletic Ailuridae basal to the other musteloid families (Mephitidae, Procyonidae, and Mustelidae s.s.), but their data set did not include any ursids and canids, and their study mainly focused on the Mustelidae.

Table 4.1 Summary of molecular phylogenetic studies including the red panda, *Ailurus fulgens*. The origin, numbers of base pairs, new sequences, and Carnivora species, and the results of the phylogenetic analysis are indicated. nucl: nuclear; mt: mitochondrial; *: complete or partial sequences for multiple sample, the number of species concerned is 8 (see Marmi *et al.*, 2004: appendix 1); W and F: Wyss and Flynn (1993); Pi: Pinnipedia; U: Ursidae; Pi: Pinnipedia; Mu: Mustelidae; Me: Mephitidae; Pr: Procyonidae; A: Ailuridae or *Ailurus*; C: Canidae; '–': polytomy. In analyses using combined data, support for the clade including the red panda is indicated in bold (strong support), normal (moderate or variable support depending on the statistics used), and italic (weak support).

Sequence data	Genome	N° of bp	N° new seq	N Carnivora (N Caniformia: N species for each family)	Results
Fulton and Strobeck 2007					
Choline... CHRNA1 intron 8	nucl	370	1	15 (15: 1C, 2Pi, 1U, 1Me, 4Mu, 5Pr, A)	
GHR intron 9	nucl	632	1	idem	
IRBP exon 1	nucl	1187	1	17 (17: 1C, 2Pi, 1U, 1Me, 4Mu, 7Pr, A)	
COI	mt	1545	6	15 (15: 1C, 2Pi, 1U, 1Me, 4Mu, 5Pr, A)	
NADH subunit 2	mt	1044	6	idem	
Cytochrome b	mt	1140	3	idem	
All combined		5918		17 (17: 1C, 2Pi, 1U, 1Me, 4Mu, 7Pr, A)	**C(U(Pi((A+Me)(Pr+Mu))))**
Sato et al. 2006					
Apolipoprotein B	nucl	963	34	34 (29: 2U, 4Pi, 1 Me, 19Mu, 2Pr, A)	U(Pi((A+Me)(Mu+Pr)))
IRBP1 exon 1	nucl	1188	5	idem	U(Pi(Me(A(Mu+Pr))))
RAG1	nucl	1095	6	idem	U(Pi(Me(Mu(A+Pr))))
All combined		3228		idem	*U(Pi(Me(A(Mu+Pr))))*
Fulton and Strobeck 2006					
Feline sarcome FES intron 14	nucl	454	33	57 (56: 3C, 18Pi, 4U, 2Me, 24Mu, 4Pr, A)	
Choline... CHRNA1 intron 8	nucl	394	40	63 (57: 3C, 18Pi, 4U, 3Me, 24Mu, 4Pr, A)	
Growth GRH intron 9	nucl	652	41	65 (59: 4C, 19Pi, 4U, 2Me, 24Mu, 4Pr, A)	

RHOI intron 3	nucl	280	39	62 (56: 3C, 17Pi, 4U, 2Me, 24Mu, 4Pr, A)	C(U(Pi:(Me(A(Mu+Pr))))))
IRBP exon 1	nucl	1194	37	72 (66: 4C, 19Pi, 8U, 2Me, 25Mu, 6Pr, A)	
All combined		2974		85 (79: 4C, 21Pi, 8U, 3Me, 36Mu, 6Pr, A)	C(U(Pi:(Me(A(Mu+Pr))))))
Yu and Zhang 2006					
β-fibrinogen intron 4	nucl	580	19	20 (18: 3C, 3Pi, 4U, 4Mu, 3Pr, A)	C(Pi:(U(A(Mu+Pr))))
β-fibrinogen intron 7	nucl	650	13	20 (17: 3C, 3Pi, 4U, 4Mu, 3Pr, A)	C((A+U)(Pi(Mu+Pr)))
NADH subunit 2	mt	1044	11	19 (17: 2C, 3Pi, 4U, 4Mu, 3Pr, A)	C(U(Pi:(A(Mu+Pr))))
IRBP exon 1	nucl	1280	0	17 (15: 1C, 2Pi, 4U, 4Mu, 3Pr, A)	See Yu et al., 2004
Tranthyretin intron 1	nucl	857	0	17 (15: 1C, 2Pi, 4U, 4Mu, 3Pr, A)	See Yu et al., 2004
β-fibrinogen 4 and 7, IRBP, TTR combined		3373		18 (16: 2C, 2Pi, 4U, 4Mu, 3Pr, A)	C((U+Pi)(A(Pr+Mu)))
All five genes combined		4417		idem	**C(U(Pi:(A(Pr+Mu))))**
Domingo-Roura et al. 2005					
Melo8 (μsatellite)	nucl	219–232	NA	23 (23: 3Pi, 2Me, 15Mu, 2Pr, A)	*Pi(A(Me(Mu+Pr)))*
Delisle and Strobeck 2005					
Cytochrome c subunit 1–3	mt	Np	11	38 (35: 3C, 20Pi, 4U, 1Me, 5Mu, 1Pr, A)	Not shown
Cytochrome b	mt	Np	4	idem	Not shown
NADH subunit 1–5	mt	Np	11	idem	Not shown
ATPsynthase sub 6,8	mt	Np	11	idem	Not shown
All 12 genes combined		10842		idem	C(Pi:(U((Me+A)(Mu+Pr))))
Flynn et al. 2005					
Cytochrome b	mt	1149	0	59 (35: 3C, 6Pi, 4U, 2Me, 16Mu, 3Pr, A)	Not shown
subunit 12S rRNA	mt	1067	0	35 (28: 2C, 5Pi, 3U, 4Me, 10Mu, 3Pr, A)	Not shown
NADH subunit II	mt	1050	39	66 (37: 3C, 6Pi, 3U, 4Me, 18Mu, 2Pr, A)	Not shown
Transthyretin intron 1	nucl	1491	29	67 (37: 4C, 5Pi, 3U, 4Me, 18Mu, 2Pr, A)	Not shown

Table 4.1 (*cont.*)

Sequence data	Genome	N° of bp	N° new seq	N Carnivora (N Caniformia: N species for each family)	Results
IRBP	nucl	1043	5	39 (18: 1C, 2Pi, 1U, 2Me, 9Mu, 2Pr, A)	Not shown
TBG	nucl	443	30	30 (20: 3C, 5Pi, 3U, 3Me, 3Mu, 2Pr, A)	Not shown
Combined data		6243		76 (42: 4C, 7Pi, 4U, 4Me, 19Mu, 3Pr, A)	**C(U(Pi(A(Me(Pr+Mu))))**
Reduced-taxa combined data				58 (34: 3C, 6Pi, 3U, 4Me, 15Mu, 2Pr, A)	**C(U(Pi(A(Me(Mu+Pr))))**
Yu *et al.* 2004					
IRBP	nucl	ca 1300	21	37 (17: 2C, 2Pi, 4U, 5Mu, 3Pr, A)	U(Pi(C(A(Pr+Mu))))
Transthyretin intron 1	nucl	ca 1000	15	idem	C(U(Pi(A(Pr+Mu))))
All combined		2341		idem	C(U(Pi(A(Pr+Mu))))
Marmi *et al.* 2004					
Cytochrome b	mt	1140	14*	37 (37: 1Pi, 2Me, 2Pr, 31Mu, A)	Pi-A(Me(Pr-Mu))
Cytochrome b	mt	337		20 (20: 1Pi, 2Me, 1Pr, 15Mu, A)	Pi-A(Me(Pr-Mu))
Mel08 (µsatellite)	nucl		NA	idem	**Pi-A(Me(Pr-Mu))**
Flynn *et al.* 2000					
Cytochrome b	mt	1140	0	17 (15: 2C, 3Pi, 3U, 2Me, 2Mu, 2Pr, A)	
12S rRNA	mt	964	0	idem	
16S rRNA	mt	495	3	idem	
Transthyretin intron 1	nucl	851	2	idem	Polytomy: A+Mu+Pr
All combined		3450			C(U(Pi((A+Me)(Pr+Mu))))
Flynn and Nedbal 1998					
Transthyretin intron 1	nucl	851	22	22 (14: 2C, 5Pi, 3U, 2Mu, 2Pr, A)	C(U(Pi(A(Pr+Mu))))
Cytochrome b	mt	851	0	idem	Pi(C(U(A(Pr+Mu))))

					in conflict
12S rRNA	mt		0	idem	
All combined				idem	**C(U(Pi(A(Pr+Mu))))**
Ledje and Arnason 1996a					
Cytochrome b	mt	1140	15	30 (26: 3C, 9Pi, 4U, 5Mu, 2Me, 2Pr, A)	Monotypic Ailuridae, unresolved relationships within Caniformia
Ledje and Arnason 1996b					
12S rRNA	mt	954–966	29	32 (28: 3C, 11Pi, 4U, 5Mu, 2Me, 2Pr, A)	
Combined cyto and 12S rRNA					Monotypic Ailuridae, unresolved relationships within Caniformia
Vrana et al. 1994					
Cytochrome b	mt	307	11	27 (23: 4Pi, 5U, 7Mu, 1Me, 4Pr)	Not shown
Subunit 12S rRNA	mt	394	11	idem	Not shown
All combined				idem	Me(Mu(Pr(A(Pi+U))))

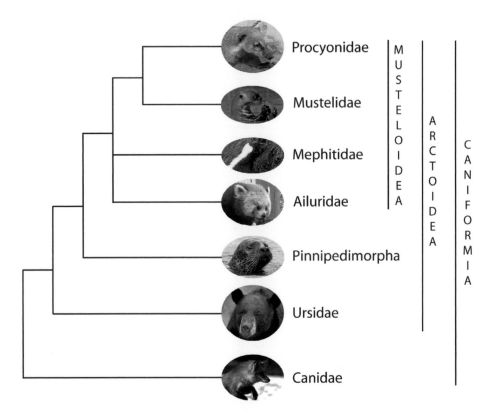

Figure 4.2 For colour version see Plate 5. Consensus phylogenetic tree of the extant arctoid families, based on molecular data.

It is obvious from the above discussion that the interfamilial relationships among the Musteloidea are still debated, with the exception of the strongly robust Procyonidae + Mustelidae clade. Nevertheless, whatever relationships these three clades have, molecular evidence for placing *Ailurus fulgens* in a monotypic family Ailuridae within the Musteloidea is well supported. Figure 4.2 presents a consensus molecular-based phylogeny of arctoid families.

Morphological evidence

Table 4.2 presents a summary of morphological studies dealing with the red panda, including the phylogenetic placement resulting from each study, the morphological characters used, and the taxonomic composition of the comparative data set. These parameters are relatively easy to identify for most of the studies. For some others, however, especially those studies with a general scope of proposing a classification of the order Carnivora (e.g. Simpson, 1945;

Table 4.2 Summary of morphological studies discussing the morphology and systematic position of the red panda *Ailurus fulgens* since 1825. D: dentition; Sk: skull; Pc: postcranial; genes: genetic; SA: soft anatomy; Bc: basicranium; Ecol evid: ecological evidence. *: classification of Baskin is unclear (see text). The study of Wesley-Hunt and Flynn (2005) is not included because, while they use *Ailurus*, the goal of their study was to resolve basal relationships of Carnivoramorpha. In addition, they did not discuss the position of extant clades, so their study cannot be regarded as really dealing with the red panda. In bold, studies including at least one fossil ailurid in the discussion in addition to *Ailurus fulgens*: *Simocyon, Parailurus, Pristinailurus, Amphictis,* or *Alopecocyon.*

Author	Classification	Characters	Comparative dataset
	Among the Procyonidae		
Milne-Edwards 1868–1874	Procyonidae	D, Sk	Procyonidae, Ursidae
Mivart 1885	Ailurinae (with *Ailuropoda*) in Procyonidae	Multi data set	Procyonidae
Flower and Lydekker 1901	Procyonidae	Multi data set	Procyonidae
Lankester in Lydekker 1901	Ailurinae (with *Ailurus*) in Procyonidae	D, Pc, Sk	Ursidae, Procyonidae
Weber 1904	Procyonidae	Not precise	undefined
Bardenfleth 1914	Procyonidae	Multi data set	Ursidae
Weber 1928	Procyonidae	Not precise	not precise
Gregory 1936	Ailurinae (with *Ailuropoda*) in Procyonidae	D, Sk, Bc	Ursidae, Procyonidae
Raven 1936	Close to *Ailuropoda* in Procyonidae	SA	Ursidae, Procyonidae, (Canidae)
Simpson 1945	Ailurinae (with *Ailuopoda*) in Procyonidae	Undefined	undefined
Davis 1964	Procyonidae	Multi data set	Arctoidea, mainly Ursidae and Procyonidae
Thenius 1979a	Procyonidae	D	Ursidae
Thenius 1979b	Procyonidae	D, Pc, genes	Ursidae, Procyonidae
Wang 1997	Ailurinae (sister to Simocyoninae) in Procyonidae	D, Sk, Bc	Musteloidea
Baskin 2004	Procyonidae (implicitly)	D, Sk	Procyonidae
Nowak 2005	Ailurinae in Procyonidae	Undefined	undefined

Table 4.2 (*cont.*)

Author	Classification	Characters	Comparative dataset
	Distinct from other families (= Ailuridae)		
Gray 1843	Ailuridae	Undefined	undefined
Turner 1848	Ailuridae, close to Procyonidae	Bc, SA, D	all
Flower 1869	Ailuridae	Bc, SA	
Flower 1870	Ailuridae or Ailurinae in Procyonidae	SA	Procyonidae, Ursidae
Gervais 1870	Ailuridae, close to Procyonidae	SA	undefined
Gill 1872	Ailuridae in Arctoidea procyoniformia	Multi data set	
Pocock 1921	Ailuridae, close to Procyonidae	SA, Sk, D	Procyonidae, *Ailuropoda*
Pocock 1928a	Distinct from Procyonidae and Ursidae	Bc	Procyonidae, Ursidae
Pocock 1928b	Ailuridae	SA	Ursidae
Segall 1943	Ailuridae (*Ailurus* + *Ailuropoda*)	Bc	Arctoidea
Hunt 1974	Distinct from other carnivores	Bc	Carnivora
Schmidt-Kittler 1981	Basal musteloid	Bc, D	Fossil arctoids
Ginsburg 1982	Ailuridae, sister to Ursidae+Otariidae in Ursoidea	D, Sk, Bc	Arctoidea
Roberts and Gittleman 1984	Ailuridae, close to Procyonidae	Ecol evid	undefined
Flynn et al. 1988	Close to Procyonidae	D, Sk, Bc	Carnivora
Wozencraft 1989	*Ailurus* sister to Ursidae	D, Sk, Bc	Carnivora
Wyss and Flynn 1993	Sistergroup to Ursida (Pinnipedia+Ursoidea (Ursidae+Amphicyonidae))	D, Sk, Bc	Carnivora
Wolsan 1993	*Ailurus* basal Musteloidea	D, Sk, Bc	Fossil musteloids
Ginsburg et al. 1997	Ailuridae	D	Procyonidae, *Simocyon*
Baskin 1998*	Ailuridae or unnamed taxon	D, Sk, Bc	Procyonidae

Reference	Classification	Data	Outgroup/Comparison
Ginsburg 1999	Ailuridae, sister to Procyonidae	D, Sk, Bc	Carnivora
Ginsburg *et al.* 2001	Ailurinae in Ailuridae	D	undefined
Wallace and Wang 2004	Ailuridae	D	Fossil ailurids
Wang *et al.* 2005	Ailuridae, sister to Procyonidae	D, Sk, Bc	Fossil musteloids
Wilson and Reeder 2005	Ailuridae	Multi data set	undefined
Salesa *et al.* 2006	Ailuridae	Pc	
Alternative Classifications			
Cuvier, in Geoffroy Saint-Hilaire and Cuvier 1825	Between Viverridae and Ursidae, near Procyonidae	D, Pc, SA	Viverridae, Ursidae, Mustelidae, Procyonidae
Blainville 1841	Subursi (with *Procyon, Potos, Meles, Myadus, Arctictis*)	D, Pc	undefined
Trouessart 1899, 1904	Ailurinae (with *Ailuropoda*) in Ursidae		undefined

Nowak, 2005; Wilson and Reeder, 2005), it is more difficult to distinguish the morphological characters used and the taxonomic composition of the comparative database, hence the label 'undefined' is used in these cases to address the parameters mentioned above. Below, we provide a summary of the classification of the red panda. For clarity, we propose to distinguish four major periods in the historical overview of morphological studies dealing with the red panda, with three key dates: the description of the giant panda (Milne-Edwards, 1868–1874), the treatise on the anatomy of the giant panda (Davis, 1964), and the first cladistic analysis including the red panda (Flynn *et al.*, 1988). This is followed by a summary of the morphological characters used (see Appendix 4.1) in the studies presented in Table 4.2, which mostly dealt with extant species, and a discussion of the contribution of the fossil data, for they are critical for understanding and supporting the position of the red panda in a monotypic Ailuridae. As a final step we extracted those characters from the Appendix 4.1 list which we believe to be apomorphic for *Ailurus*. The survey through all ailurids (Table 4.3) reveals whether these characters are autopomorphies of *Ailurus*, Ailurinae, or Ailuridae, respectively.

Historical perspective

Description of the red panda (1825) to description of the giant panda (1870)

The red panda was described by Cuvier, in Geoffroy Saint-Hilaire and Cuvier (1825). The last paragraph of the description reveals how the relationships of the red panda were rather confused in Cuvier's mind. In this brief description of the animal, the red panda is said to resemble the Mustelidae, the Procyonidae, or the Viverridae, depending on the features used (general body shape and colour, muzzle length, teeth and claw morphology). After Geoffroy Saint-Hilaire and Cuvier (1825), however, authors consistently placed the red panda within the Arctoidea, excluding the species from any close relationship to Viverridae. In these early works, the red panda was implicitly a member of its own family or group, even if this classification was not supported by morphological evidence. Blainville (1841) classified the red panda in a group named Subursus or 'small bears' that also included one mustelid (*Meles*), one mephitid (*Mydaus*), two procyonids (*Procyon* and *Potos*), and one viverrid (*Arctitis*). Gray (1843) was the first to erect the name Ailuridae (at subfamily rank). Although it was not based on evidence of any kind, this author proposed to divide Carnivora in two groups, one comprising the Ailuridae, the Ursidae, and the Procyonidae (the latter family was then divided in Procyonidae and Cercoleptidae) and the other group including the Felidae, Hyaenidae, Viverridae, Canidae, and Mustelidae.

Table 4.3 Distribution of 31 morphological characters regarded as diagnostic or derived for *Ailurus* among fossil ailurids (soft characters excluded). Data sources for *Parailurus*: Morlo and Kundrát (2001), Sasagawa *et al.* (2003); for *Pristinailurus*: Wallace and Wang (2004); for *Magerictis*: Ginsburg *et al.* (1997); for Ailurinae indet.: Ginsburg *et al.* (2001); for *Simocyon*: Wolsan (1993), Wang (1997), Peigné *et al.* (2005), Kullmer *et al.* (2008); for *Protursus*: Peigné *et al.* (2005); for *Alopecocyon*: Ginsburg (1961), Wolsan (1993), Baskin (1998, data of *Actiocyon*); for *Amphictis*: Dehm (1950), Cirot and Bonis (1993), Heizmann and Morlo (1994), Nagel (2003), and own unpublished observations. Character present: +; character variably present: ±; character absent: −; character not preserved in fossil specimens: ?

N° of character in Appendix 4.1	Character	*Ailurus*	*Parailurus*	*Pristinailurus*	*Magerictis*	Ailurinae indet.	*Simocyon*	*Protursus*	*Alopecocyon*	*Amphictis*
5	Palate prolonged behind last upper molar	+	?	?	?	?	−	?	?	−
7	Zygomatic arches sharply bowed outward and downward	+	?	?	?	?	+	?	?	?
16	Long auditory tube present	+	?	?	?	?	+	?	?	+
17	Presence of posterolateral process of promontorium	+	?	?	?	?	+	?	?	+
20	Posterior carotid foramen in front of and very close to posterior lacerate foramen	+	?	?	?	?	+	?	?	+
21	Crista tympanica ends anteriorly and posteriorly by a small spine	+	?	?	?	?	+	?	?	+
22	No epitympanic sinus	+	?	?	?	?	+	?	?	+
24	Small, but consistently present, processus muscularis for the insertion of the tensor tympani on the malleus	+	?	?	?	?	+	?	?	+
28	Ventrally ridged paroccipital process, that is mediolaterally compressed and blade-like	+	?	?	?	?	+	?	?	+
32	Mastoid process ventrally expanded	+	?	?	?	?	+	?	?	+
34	Hiatus subarcuatus present	+	?	?	?	?	+	?	?	+
35	Dorsal depression of middle ear not clearly divided into two depressions	+	?	?	?	?	+	?	?	+
42	Anterior edge of coronoid inclined forwards	+	?	?	?	?	+	?	?	−
43	Presence of lateral grooves on canines	+	+	+	?	?	+	?	+	+

Table 4.3 (*cont.*)

N° of character in Appendix 4.1	Character	*Ailurus*	*Parailurus*	*Pristinailurus*	*Magerictis*	Ailurinae indet.	*Simocyon*	*Protursus*	*Alopecocyon*	*Amphictis*
47	M2 enlarged	+	+	?	?	?	−	?	−	−
48	Enlargement of labial cusps on M1−2	+	+	+	?	+	−	?	−	−
50	M2 three-rooted and subequal in size to P4	+	+	?	?	?	−	?	−	−
51	Anterior and posterior cingula of M1 continuous around lingual base of protocone	+	±	+	?	?	+	?	+	+
52	Inner portion of P4 formed by two distinct cusps, the protocone and hypocone	+	+	?	?	?	−	?	−	−
53	P4 as large as M1, protocone and hypocone form more than half of the tooth, they are supported by a very strong root which is placed mainly under the mesial cusp	+	+	?	?	?	−	?	−	−
54	P4 protocone conical, not formed by cingulum entirely	+	+	?	?	?	−	?	−	−
56	P2 and P3 large and three-rooted	+	−	?	?	?	−	?	−	−
57	P3 with 5 cusps and closely resembling P4	+	?	?	?	?	−	?	−	−
62	p4 with 5 cusps	+	±	?	?	?	−	?	−	−
63	m1 entoconulid poorly differentiated (ridge-like or cuspule-like)	+	+	?	?	?	+	?	+	+
64	Elongation of m2	+	+	?	+	?	+	+	+	+
65	Elongation of the talonid of m2	+	+	?	+	?	+	+	+	+
66	Enlargement of m2 hypoconulid	+	+	?	+	?	−	−	−	−
67	Entoconid and entoconulid of m2 prominent, cusp-like	+	+	?	+	?	−	−	−	−
68	Protoconid and metaconid of m2 elongated and narrow, separated by a longitudinal trough that reaches the distal border of the tooth	+	?	?	+	?	+	+	+	±
80	Extra carpal bone or radial sesamoid moderate or small (relative to that of *Ailuropoda*)	+	?	?	?	?	+	?	?	?

This classification did evolve with the early studies of the auditory region by Turner (1848) and Flower (1869), who provided evidence to separate the terrestrial Carnivora into Aeluroidea (Felidae, Hyaenidae, Viverridae), Cynoidea (Canidae), and Arctoidea (Ursidae, Procyonidae, Ailuridae, Mustelidae). Based on basicranial features, Turner (1848) and Flower (1869) supported the position of *Ailurus* in Arctoidea, but they did not provide any morphological evidence for recognition of a family Ailuridae. The study of the external and soft anatomy of the red panda by Flower (1870) did not reveal any further evidence. The only difference with his previous work (Flower, 1869) was to propose that *Ailurus* should be classified within Procyonidae, or in its own family Ailuridae, with a close relationship to Procyonidae.

Description of the giant panda (see Milne-Edwards 1868–1874) to the study of Davis (1964)

Works subsequent to the description of the giant panda frequently included comparisons between the 'pandas'. In his study of the brain of mammals, Gervais (1870) revised the Subursus of de Blainville and excluded from this group *Meles*, *Arctonyx*, and *Mydaus*, which were placed in the Mustelidae, and *Arctictis*, which he placed in the Viverridae. Based on the cerebral characteristics, Gervais favoured a monotypic family Ailuridae closely related to the Procyonidae. This is the same position adopted by Gill (1872), who placed *Ailurus* in his Arctoidea procyoniformia (i.e. Aeluridae [*sic*], Cercoleptidae, Procyonidae, Bassaridae). The Ailuridae of Gill is distinguished in having an alisphenoid canal developed, a very small auditory bulla with the auditory meatus developed into a very long tube, a long and trigonal paroccipital process that stands backwards and outwards and that is unconnected with the bulla, and the presence of only three premolars and two molars in the upper and lower toothrow. During this period, most of the morphological studies that included the red panda mainly focused on the newly discovered giant panda, *Ailuropoda melanoleuca*. As a consequence, these studies primarily addressed the systematic position of the giant panda, with a particular interest in its position relative to the red panda, to the bears, and to procyonids, but they never addressed the position of the red panda directly. Authors simply followed previous hypotheses by placing *Ailurus fulgens* in the Procyonidae, in the Ursidae, in a monotypic family Ailuridae (e.g. Milne-Edwards, 1868–1874; Flower and Lydekker, 1891; Schlosser, 1899; Trouessart, 1899, 1904; Weber, 1904, 1928; Bardenfleth, 1914; Raven, 1936; Segall, 1943; Simpson, 1945), or suggested an ancestor-descendant relationship between the two pandas, the red panda being the ancestor of the giant panda (e.g. Gregory, 1936). Based on a thorough comparison between the two pandas, Gregory tried to demonstrate that every morphological feature of the giant panda derived from the condition observed in the red panda by arguing that: (i) the 'condition' of the giant panda was 'derivable' from that of the red panda; (ii) was not 'inconsistent' with the conclusion that the giant panda is a specialised ailurine (he did so for the characters of the masticatory apparatus and of the brain-case); or (iii) did 'not eliminate *Ailurus* from relatively close relationship to the direct ancestors of *Ailuropoda*', as for the absence of m3 in *Ailurus* (Gregory, 1936, p. 1314).

In other studies of morphological features of *Ailurus*, *Ailuropoda*, and other bears (e.g. Bardenfleth, 1914), it is difficult to distinguish which features are regarded as diagnostic by the respective authors. Furthermore, the support for a monotypic Ailuridae is almost never discussed. Based on soft tissues and external features of the animal and a review of its dental and osteological anatomy, Pocock (1921) argued for a monotypic Ailuridae distinct from the procyonids and *Ailuropoda*. The auditory region also received much attention during this period. Earlier studies on this anatomical region by Turner (1848) and Flower (1869) gave evidence for the placement of *Ailurus fulgens* in the Arctoidea, but these authors failed to provide strong evidence to support the validity of a family Ailuridae. For Pocock (1928b), 'there is no definite character by which the bulla and its accessory structures in *Ailurus* can be distinguished from those of the Procyonidae as a whole'. Based on the study of the auditory region and ossicles, Segall (1943) rather confirmed this view by classifying the red and great pandas in a group that also included the Procyonidae and the Ursidae.

As can be seen from Table 4.2, in a vast majority of the studies, the red panda was classified in the Procyonidae or regarded as a sister taxon (hence a mono-typic Ailuridae) of another family or clade. Only Cuvier, in Geoffroy Saint-Hilaire and Cuvier (1825), Blainville (1841), and Trouessart (1899, 1904) propose a different (i.e. inside another family) or unclear classification (i.e., among groups including phylogenetically diverse taxa).

Davis (1964) to first cladistic analysis including the red panda (Flynn *et al.* 1988)

The taxonomic position of the giant panda became clear thanks to the work of Davis (1964). Based on an extensive comparison of the giant panda with other carnivorans, Davis (1964) demonstrated that the giant panda was a member of the Ursidae and was specialised towards herbivory. A second consequence of Davis' work was that he clearly demonstrated that the red panda was not related to the giant panda, and hence was not an ursid, although its systematic position remained unresolved at that time.

Additional studies of the auditory region failed to provide support for the existence of a family Ailuridae. This led Hunt (1974) to conclude that *Ailurus* retained a primitive auditory structure, similar in evolutionary grade to that of the ursids, mephitids, lutrine mustelids, otariids, and odobenids. Schmidt-Kittler (1981) confirmed the plesiomorphic character of the bulla of *Ailurus fulgens* by placing the species at the base of the Mustelida, at an evolutionary grade similar to that of a series of primitive fossil arctoids (*Mustelictis, Amphictis, Amphicynodon,* and *Cephalogale*). Despite the work of Davis (1964), Ginsburg (1982) still argued that the red panda was a close relative of Ursidae based on dental characters (loss of M3, hypocone on P4, quadrangular shape of M1 and M2 resulting from the lengthening of the lingual border). Other authors continued to follow previous classifications (in particular that of Simpson, 1945) that placed the red panda among Procyonidae (e.g. Bugge, 1978).

Flynn *et al.* (1988) to today

From Cuvier (in Geoffroy Saint-Hilaire and Cuvier, 1825) to Flynn *et al.* (1988), morphological studies used the general principle of resemblance to classify species, but such methods do not distinguish primitive from derived character states. Furthermore, with a few exceptions (e.g. Davis, 1964), these morphological studies suffered from the same taxonomic biases as early molecular studies (see discussion above), that is: comparisons with the red panda only included a limited number of taxa. Furthermore, comparisons never included fossil taxa, with a few exceptions (e.g. Schmidt-Kittler, 1981). After Flynn *et al.* (1988), studies including a wider taxonomic data set of extinct Carnivora became more frequent. Flynn *et al.* (1988) provided the first cladistic analysis for the order Carnivora. The scope of their study was the whole of Carnivora, not the position of the red panda specifically. Hence, support is rather weak for the position of the red panda as closely related to Procyonidae, with only two synapomorphic characters (processus muscularis for the insertion of the tensor tympani on the malleus present and only two upper and two lower molars). After Flynn *et al.* (1988), most phylogenetic studies that include the red panda have also included some fossil taxa. However, even in these analyses, the position of the red panda has remained unresolved. A few morphological studies have considered the species as a member of the Procyonidae (subfamily Ailurinae) (Wang, 1997; Baskin, 1998, 2004), while others considered the red panda as a distinct lineage (formalised as a monotypic Ailuridae or not) that was either closely related to the Ursidae (Wozencraft, 1989; Wyss and Flynn, 1993), to the Procyonidae (Ginsburg, 1999; Ginsburg *et al.*, 1997; Wang *et al.*, 2005), or that had unresolved relationships at the base of the Musteloidea (Wolsan, 1993).

Overview of morphological characters

In Appendix 4.1 we present those characters that have been regarded as diagnostic for *Ailurus*, or its monotypic family or subfamily. Characters from cladistic analyses (i.e. studies since Flynn *et al.*, 1988) were easily selected. For earlier studies, we retain characters listed by the various authors to distinguish *Ailurus fulgens* (or Ailuridae or Ailurinae) from the other groups (mostly arctoids). Appendix 4.1 includes 83 characters: 36 for the skull (including 28 for the basicranium), 6 for the mandible, 26 for the dentition, and 15 for the postcranium. Some dental and osteological features are particularly redundant in morphological studies, such as the retention of the alisphenoid canal (Turner, 1848; Gill, 1872, Mivart, 1885; Lydekker in Lankester, 1901; Pocock, 1921; Wyss and Flynn, 1993), the presence of only two molars in the upper and lower dentition (Turner, 1848; Gill, 1872, Mivart, 1885; Lydekker in Lankester, 1901; Bardenfleth, 1914; Ginsburg, 1982; Flynn *et al.*, 1988), the presence of a long auditory tube (Flower, 1869; Gill, 1872; Bardenfleth, 1914; Wolsan, 1993; Wang, 1997; Baskin, 1998), and the prolongation of the palate beyond the last upper molar (Lydekker in Lankester, 1901; Wozencraft, 1989; Wolsan, 1993; Wang, 1997; Wang *et al.*, 2005). The first two characters do not provide any support for the placement of the red panda. The alisphenoid canal is independently lost in many groups among Carnivora. It is known in all the primitive musteloids (Schmidt-Kittler, 1981). Furthermore, earlier members of a clade (e.g. Felidae, Hyaenidae, Mustelidae) may have an alisphenoid canal, while more derived and/or extant relatives may not (see, e.g. Veron, 1995; Wolsan, 1993). The presence of two upper and lower molars, a character state that may also be defined as the loss of m3, is primitive for Carnivora. Extant Carnivora lack an M3; only some extinct taxa (Amphicyonidae, 'Miacidae') retained it. The presence of 'only' two upper molars is therefore primitive for nearly all families of the order. Finally, the loss of m3 may have occurred independently at least three times during the evolutionary history of Carnivora (Viverravidae, Feliformia, Arctoidea). The absence of m3 is therefore not significant to support any placement of *Ailurus*. The two other characters are discussed below (i.e. auditory tube and posterior extension of the palate).

Some anatomical structures received more attention than others, such as the paroccipital process (characters 25–28), the upper molars (characters 44–51, 55), and the m2 (characters 64–68). Finally, the definitions of some characters are sometimes in contradiction with others, like characters 31 (well-developed mastoid process; Flower, 1869) and 32 (mastoid process small; Bardenfleth, 1914 and others; see Table 4.2), and characters 25 and 28. The latter is retained here.

Critical discussion of every character listed in Appendix 4.1 lies beyond the scope of the present overview. Nevertheless, some general comments that may help to clarify the pertinence of the characters is warranted. As pointed out earlier, the goal of many earlier studies was to compare the two pandas or to discuss the position of the giant panda relative to bears, and the red panda, relative to Procyonidae (see Table 4.2). One major consequence is that morphological characters were surveyed in a limited, select number of taxa, and that most of them are of limited use for all of Carnivora, or even all of Arctoidea. Many characters observed in a limited number of taxa proved to be plesiomorphic at a larger taxonomic scale. This is true for the majority of the numerous characters mentioned by Bardenfleth (1914) and Gregory (1936). Moreover, a majority of the characters listed in Appendix 4.1 are not significant for the resolution of the position of the red panda for various reasons. As pointed out above for the loss of M3/m3, some characters are clearly primitive or, if derived, have independently appeared in many groups (e.g. 1, 6, 14, 29). Other characters display intraspecific variation (e.g. 6, 39–41); and, finally, others are not applicable because of a too general or imprecise description (e.g. characters 1, 6, 45). For these reasons, characters 1–4, 6, 8, 12–15, 27, 29–31, 33, 37–41, 45, 46, 55, 58–61, 69, 71–79, and 81–83 are probably not pertinent for the resolution of the red panda.

Despite the significant contribution of Davis (1964), *Ailurus* has been recognised as a sister taxon to Ursidae by several recent cladistic analyses (e.g. Wozencraft 1989; Wyss and Flynn, 1993). Characters used by Wozencraft (1989) to support the placement of *Ailurus fulgens* among the Ursidae (Appendix 4.1: characters 5, 9–11, 26, 36, 49, 70) have been critically revised by Wolsan (1993) and Wang (1997). The latter author demonstrated that characters 26, 36, 70 (Appendix 4.1) are primitive for the Arctoidea, that two others (corresponding to character 49 in Appendix 4.1) are not homologous in ursids and ailurids and are independently derived for the Ailurinae (i.e. *Parailurus* and *Ailurus* in Wang, 1997), and that three others (characters 9, 10, and 11) display great intraspecific or intrafamilial variation: the lacrimal varies in size, in occurrence, and in shape in the Ursidae, enclosing or not the lacrimal foramen (character 9; see Wolsan, 1993) and the fossa for the inferior oblique muscle is in fact rarely adjacent to the nasolacrimal foramen (character 10; see Wang, 1997). Moreover, the postscapular fossa of *Ailurus* (character 70 in Appendix 4.1) is not strictly comparable to that of ursids (see Salesa *et al.*, 2008). The five synapomorphies supporting the sister-group relationships of *Ailurus* to a Pinnipedia + Ursoidea (Ursidae + Amphicyonidae) clade in Wyss and Flynn (1993) (characters 2, 18, 19, 23, and 49 in Appendix 4.1) were also discussed by Wang (1997) and Wolsan (1993). These characters are clearly primitive for Arctoidea (characters 2, 18), not homologous in ursids and ailurids (character 49), or have a condition in *Ailurus* that is not different from that observed in fossil mustelidans (characters 19, 23). The hypothesis that *Ailurus* is closely related to ursids is therefore not supported. Finally, the remaining characters in Appendix 4.1 include 12 cranial, 1 mandibular, 17 dental, and 1 postcranial, for a total of 31 characters that may prove to be diagnostically useful for phylogenetic relationships of and within ailurids. To evaluate phylogenetic signals provided by these characters, their occurrence in fossil ailurids is detailed below and summarised in Table 4.3.

Considering fossil taxa

Considering fossils is critical for the support of the family Ailuridae. Phylogenetic study of the extant red panda frequently supports a close relationship with Procyonidae (Table 4.2; and see historical perspective above). This relationship appears to be not strongly supported because its specialisations, especially in diet and arboreal lifestyle, often obscure the phylogenetic position of the red panda. Recent discoveries of fossil ailurids reveal a much more complex picture of the evolution of this family. Several issues regarding the fossil record of Ailuridae are important to distinguish here: (i) the existence of close fossil relatives of the red panda, i.e. those presenting the same dental specialisation toward herbivory; (ii) the relationships between *Ailurus* and *Simocyon*; and (iii) the relationships of *Simocyon* with earlier genera such as *Alopecocyon* and *Amphictis*.

Specialised genera very close to *Ailurus*: *Parailurus*, *Pristinailurus*, and other related forms

The fossil record of the red panda is relatively poor, but the existence of extinct close relatives has been long known. *Parailurus anglicus* was described from Pliocene sediments of England and central Europe as early as the end of the nineteenth century (Dawkins, 1888; Schlosser, 1899). Several recent discoveries revealed an earlier ancestor for the red panda, *Pristinailurus bristoli* (Wallace and Wang, 2004, 2007). Dental similarity leaves little doubt about the close relationships between *Ailurus*, *Parailurus*, and *Pristinailurus* (Wallace and Wang, 2004). However, as pointed out previously (e.g. Wang, 1997; Sotnikova, 2008), the dentitions of these species are already very specialised (as is that of the red panda) and are therefore not very useful for the resolution of the systematic relationships of the lineage. At the least, these fossils indicate that the lineage extends at least to the late Miocene (Wallace and Wang, 2004).

Two additional taxa probably represent ailurids and share a similar herbivorous diet: *Magerictis imperialis* from the middle Miocene (*ca.* 17 Mya) of Spain

(Ginsburg *et al.*, 1997) and Ailurinae indet. from the middle Miocene (*ca.* 12 Mya) of Four, France (Ginsburg *et al.*, 2001). Both taxa are known by the most diagnostic teeth of the family, m2 (for *Magerictis*) and M1 (for Ailurinae indet.). The morphology of the m2 of *Magerictis imperialis* strongly supports a close relationship to the red panda lineage (Ginsburg *et al.*, 1997). Recent discoveries of more complete dental remains of *M. imperialis* from the middle Miocene of Madrid Basin, Spain (J. Morales, personal communication, 2007) will be critical for assessing the relationships of this genus to the red panda lineage. The relationship of Ailurinae indet. with the red panda lineage remains to be determined due to the fragmentary nature of the material, but the M1 already shows the molarisation characteristic of the *Ailurus–Parailurus–Pristinailurus* clade (Ginsburg *et al.*, 2001; Wallace and Wang, 2004).

Relationships of *Ailurus* to *Simocyon*

The genus *Simocyon* includes musteloid species that present a mixture of primitive and derived characters. In previous studies, its taxonomic position fluctuates within Caniformia (see Wang, 1997), partly due to the lack of adequate fossil material. The adaptations of *Simocyon* are very different from those of the red panda. The species of the genus represent another evolutionary trend toward carnivory and durophagy. The dentition of *Simocyon* spp. is indeed so different from that of *Parailurus* or *Ailurus* that previous authors failed to recognise *Simocyon* as a close relative of these herbivorous genera. However, cranial and basicranial synapomorphies undoubtedly relate these genera. As pointed out by Schmidt-Kittler (1981) and Baskin (1998), the discovery of well-preserved basicranial material is necessary to resolve the relationships of *Simocyon*. Recently, thanks to such basicranial material, Wang (1997) was the first to provide evidence to support a close relationship of *Simocyon* and *Ailurus*. The revision of Wang (1997) is also critical in assessing morphological characters previously regarded as supporting an ursoid–*Ailurus* relationship (e.g. Wozencraft, 1989; Wyss and Flynn, 1993; Vrana *et al.*, 1994) and demonstrated the weakness of such a hypothesis. Particularly well-preserved basicrania provide good support for a sister relationship of the Simocyoninae (*Simocyon*, *Protursus*, and *Alopecocyon*) to the Ailurinae (*Ailurus*, its closest relatives *Parailurus*, *Pristinailurus*, *Magerictis*, and Ailuridae indet.). The clade Simocyoninae + Ailurinae is the sister taxon to the Procyoninae in the phylogeny proposed by Wang (1997). According to Wang (1997, p. 196), synapomorphies of the whole clade include: 'highly arched zygomatic arch, posteriorly extended posterior palatine border, long external auditory meatus, presence of a postero-lateral process of promontorium, ventrally ridged paroccipital process, anteriorly inclined coronoid crest, and presence of lateral grooves on canines' (but see below for a re-assessment of these characters).

An independent study by Baskin (1998), which was in press at the time of the publication of Wang (1997), resulted in a similar conclusion, i.e. that the red panda lineage is closely related to procyonids. The conclusions of Baskin (1998) are, however, somewhat confusing, since he placed the clade *Parailurus* + *Simocyon* inside the Procyonidae under the name 'unnamed taxon or Ailuridae', while in the text he placed 'Ailuridae or unnamed group' under the same hierarchical heading as 'Procyonidae', which may implicitly suggest that this author regards Ailuridae as a separate family; furthermore, in his figure 8.4 (Baskin, 1998), the genera discussed are clearly separated into Procyonidae (Procyoninae) and Ailuridae (Ailurinae and Simocyoninae). Nevertheless, Baskin (2004) considered that his 1998 paper supported the division of Procyonidae into Procyoninae, Ailurinae, and Simocyoninae.

A close relationship between *Simocyon* and *Ailurus* became even more strongly supported in recent years by discoveries from the late Miocene Spanish locality of Batallones-1 (MN 10), Province of Madrid. This locality yielded a large number of fossil carnivorans (Morales *et al.*, 2004), including *Simocyon batalleri*. Species of *Simocyon* were previously known only from cranial and dental remains, but the postcranial material remained extremely rare (e.g. Gaudry, 1862–1867; Tedrow *et al.*, 1999). *S. batalleri* from Batallones-1 is represented by nearly its entire skeleton (Salesa *et al.*, 2008).

This material demonstrates the presence of an enlarged radial sesamoid or 'false-thumb' in *S. batalleri* as in the pandas (Salesa *et al.*, 2006). This bone exists in carnivorans, but such an enlargement is unique to the pandas and *Simocyon batalleri*. The presence of a relatively large radial sesamoid in the carpus is long known in the giant panda (Gervais, 1875) and in the red panda (Lankester, 1901). The functional morphology and myology of this false-thumb was described much later for *Ailuropoda melanoleuca* (Wood-Jones, 1939; Davis, 1964; Antón *et al.*, 2006 and references therein) and only recently for *Ailurus fulgens* (see Antón *et al.*, 2006 and references therein). The false-thumb of these bamboo feeders was interpreted as an adaptation for grasping and as a perfect example of 'contraption', i.e. it originally did not evolve for that purpose but was co-opted from existing structures to become an inelegant, yet useful device (Gould, 1978). The discovery of a false-thumb in *S. batalleri* from Batallones-1, a non-herbivore relative of *Ailurus fulgens*, supported the idea that this structure for grasping bamboos is indeed an exaptation (Salesa *et al.*, 2006). Anatomical and myological features of the radial sesamoid of *S. batalleri* are extremely similar to those in *Ailurus fulgens*, but they are distinct from those displayed by the radial sesamoid of the giant panda (Antón *et al.*, 2006). This singular feature is an additional, strong support for the clade Simocyoninae + Ailurinae.

Relationships of the clade Simocyoninae + Ailurinae with the Procyonidae

Baskin (1998, figure 8.3) proposed a single character (M2 enlarged relative to [other] procyonids), which was also proposed by Wang (1997), to support the placement of the Ailurinae + Simocyoninae clade inside the Procyonidae. In Wang (1997), the only additional character supporting this clade is the elongation of the m2 talonid. However, the validity of these synapomorphies (the latter previously used in Flynn *et al.*, 1988 to assign *Simocyon* to the Procyonidae) has been challenged with the discovery of *Magerictis imperialis* by Ginsburg *et al.* (1997). In the m2 of the Procyonidae, the trigonid is well separated from the talonid by a transverse crest connecting the protoconid and the metaconid, the trigonid fossa is closed and transversely elongated. In *Ailurus*, *Parailurus*, *Magerictis*, and some *Amphictis* the protoconid and the metaconid are elongated and narrow, separated by a longitudinal trough that developed over the tooth posteriorly to its distal border. The distinct structure of the m2 in the red panda and its closest relatives suggests that the support for placing them among the Procyonidae is weak. The corresponding enlargement of M2 is also debatable as morphology and size of the M2 of *Simocyon* is extremely dissimilar to M2 of the Procyonidae and *Ailurus*. This tooth in *Simocyon* is considerably smaller than in the *Ailurus* + *Parailurus* + *Pristinailurus* clade and displays no molarisation. Based on the current molecular phylogeny (Figure 4.2), a small M2 is probably plesiomorphic for the clade Musteloidea. An enlarged M2 probably appeared convergently in hypocarnivorous taxa like procyonids and ailurids. While the elongation of m2 relative to m1 (and especially of the talonid) is convergent in Procyonidae and Ailuridae, the structural evolution of this tooth distinguishes the two families.

Relationships of *Simocyon* with earlier genera such as *Alopecocyon* and *Amphictis* (Amphictinae)

A close relationship of *Simocyon* with other fossil taxa such as *Alopecocyon* (?=*Actiocyon*) and *Amphictis* has been suggested from dental anatomy (Thenius, 1949; Viret, 1951; Beaumont, 1964, 1976), but none of the characters involved have been regarded as synapomorphies by successive authors (e.g. Schmidt-Kittler, 1981; Wolsan, 1993). No authors have suggested any relationships between any one of these genera (*Simocyon*, *Alopecocyon* or *Amphictis*) and *Ailurus* until Wang (1997). Here, we classify *Alopecocyon* together with *Simocyon* and separate *Amphictis* as 'basal paraphyletic ailurid stock' from all other Ailuridae (see below).

In conclusion, cranial, dental, and postcranial evidence revealed by the 31 morphological characters (Table 4.3) extracted from Appendix 4.1 list strongly supports a close relationship between *Ailurus* and *Simocyon*, then a sister-group relationship of the Ailurinae (*Ailurus, Parailurus, Pristinailurus, Magerictis*) to the Simocyoninae (*Simocyon, Protursus, Alopecocyon*). Given this morphological distinction, these two subfamilies may be classified into a distinct family Ailuridae. However, so far, the species of *Amphictis*, although possibly representing early ailurids, do not present unquestionable diagnostic features that would allow them to be firmly classified in one of the subfamilies. Regarding this topic, the future description of the lower dentition of *Magerictis imperialis* will be critical, as this may allow the rooting of Ailurinae within *Amphictis* (see below).

Systematic paleontology

Order CARNIVORA BOWDICH, 1821

Family AILURIDAE GRAY, 1843

Diagnosis. All Ailuridae share a plesiomorphic musteloid basicranium (='amphictid middle ear type'; Schmidt-Kittler, 1981) in combination with lateral grooves on the canines, poorly developed entoconulid in m1, relative to other musteloids enlarged, double-rooted m2 with protoconid and metaconid completely separated (except in some species of *Amphictis*), and (except in *Parailurus baikalicus*) M1 with the anterior and posterior cingula continuous around the lingual base of the protocone (and hypocone if present).

Remarks. As an anteriorly inclined anterior coronoid edge is known in *Ailurus* and *Simocyon* but not *Amphictis*, we tentatively regard this character as independently evolved within Ailurinae and Simocyoninae. The character is directly linked to the functional morphology of the jaw and also known from other carnivores (i.e. dasyurids, proviverrine creodonts): placing the masseter muscle more anterior by enlarging the coronid anteriorly also moves forward the point where the vector of highest muscle force meets the lower tooth row (Greaves, 1983, 1985). The presence of an extra-carpal bone or radial sesamoid also may be another autapomorphy of Ailuridae, but as no postcranial remains of the hand of *Amphictis* have been described yet, this remains questionable.

Subfamily AILURINAE GRAY, 1843

Type genus *Ailurus* F. Cuvier, 1825

Other genera *Parailurus* Schlosser, 1899; *Magerictis* Ginsburg, Morales, Soria, and Herraez, 1997; *Pristinailurus* Wallace and Wang, 2004; Ailurinae indet. Ginsburg, Maridet, and Mein, 2001.

Diagnosis. Ailurinae differs from Simocyoninae and *Amphictis* in containing strictly hypocarnivorous taxa. Within Ailurinae, the palate is prolonged behind the last upper molar, the M2 is enlarged to about the size of P4, P4 contains a hypocone and is about as large as M1, m2 has the hypoconulid elongated, and entoconid and entoconulid are cusp-like. Where known, the shearing function of m1 and P4 are reduced or absent.

Remarks. The oldest records of this subfamily are *Magerictis* from the middle Miocene of Spain (17 Mya) and Ailurinae indet. from the middle Miocene of France (12 Mya). Another middle Miocene member of the subfamily was briefly reported from China (Qi in Sasagawa *et al.*, 2003), but no more information has been given. As Ailurinae may be rooted in the paraphyletic genus *Amphictis* (see below), the related species of *Amphictis* should be placed into Ailurinae as well.

Genus *AILURUS* F. Cuvier, **1825**

Type species *A. fulgens* F. Cuvier, **1825**

Other species none

Differential diagnosis. *Ailurus* differs from all other ailurines in having p4 always with four cusps and P2–3 large and three-rooted; it differs additionally from *Parailurus* in being 50% smaller. It may have four lower premolars. M1 is clearly more broad than long and has the mesostylar cusp higher, and m1 lacks a metaconulid and the hypoconid is not higher than the entoconid and entoconulid.

Distribution. *Ailurus* is endemic to the Himalayas in Bhutan, southern China, India, Laos, Nepal, and Myanmar. The global population of the red panda ranges from 10,000 to 20,000 individuals and is decreasing due to the massive habitat loss and fragmentation, increased human activity and poaching (Choudhury, 2001; Wang *et al.*, 2008). Fossil specimens are only known from the Chinese Pleistocene (e.g. Bien and Chia, 1938; Chen and Qi, 1978) and these M1s clearly differ from *Parailurus* in being smaller in overall size and in relative length (Sasagawa *et al.*, 2003).

Remarks. *Ailurus fulgens* is separated today into an eastern and western subspecies, but fossil specimens have never been assigned at subspecies level. Recent information on the conservation status, genetic diversity and ecology of this last ailurid can be found in Choudhury (2001), Pradhan *et al.* (2001), Su *et al.* (2001), Li *et al.* (2005) and Zhang *et al.* (2006a,b). The red panda is currently regarded as a vulnerable species by the IUCN; more information is available from Wang *et al.* (2008).

Genus *PARAILURUS* Schlosser, **1899**

Type species *Parailurus anglicus* (Dawkins, **1888**)

Other species *Parailurus hungaricus* Kormos, 1935, *Parailurus baikalicus* Sotnikova, 2008 from the early late Pliocene (MN 16a) of Udunga, Russia, *Parailurus* sp. from North America (Tedford and Gustafson, 1977), *Parailurus* sp.

from the late Pliocene of Včeláre (Morlo and Kundrát, 2001), *Parailurus* sp. from the late Pliocene of Japan (Sasagawa *et al.*, 2003).

Differential diagnosis. *Parailurus* differs from *Ailurus* in being 50% larger, always has only three lower premolars, M1 about as long as broad and its mesostylar cusp generally lower, m1 with a metaconulid present and the entoconulid and entoconid being lower than the hypoconid. *Parailurus* differs from all other ailurines except *Ailurus* in M1 having metaconule and protocone connected and a mesostylar cusp present.

Distribution. *Parailurus* is now known to be present in the Pliocene of all northern continents, even if more than 90% of the specimens were found in Europe (see Sotnikova, 2008, for the last review). The North American and Japanese records consist of a single tooth in each case. Recently, the first continental Asian material from MN 16a of the Transbaikalian site Udunga has been published in detail (Sotnikova, 2008, for brief notes on the new species see references therein).

Remarks. After the generic separation of *Parailurus* from *Ailurus* had been established by Schlosser (1899), nearly all subsequent authors have followed this division. European *Parailurus* was long separated into two species, *P. anglicus* and *P. hungaricus*. Based on the large sample of *P. anglicus* from MN 15 of Wölfersheim, Germany, Morlo and Kundrát (2001) demonstrated that the late Pliocene specimens attributed to *P. hungaricus* fall within the range of variation of the type species. This was not fully accepted, however, as Fejfar and Sabol (2004) reinstated *P. hungaricus* based on differences in the upper and lower first molars compared to *P. anglicus*. Because the view of Fejfar and Sabol (2004) has been followed by Sotnikova (2008), we tentatively leave *P. hungaricus* as a separate species pending further comparisons. Morlo and Kundrát (2001) also reported a possible separate species from Včeláre, Slovakia, but this material has not yet been described. Recently, Sotnikova (2008) erected *P. baikalicus* on the first *Parailurus* specimens from continental Asia. The few known specimens of *Parailurus* from outside of Europe are insufficient to determine their exact affinities (Tedford and Gustafson, 1977; Sasagawa *et al.*, 2003), even if their respective morphology separates them from the exisiting species. Sotnikova (2008) speculated that the P4 from Japan does not represent an ailurid at all, but this is hard to judge because we have not seen the specimen.

European *Parailurus* seems to be too apomorphic to be ancestral to *Ailurus* due to consistent absence of p1 (Schlosser, 1899; Wang, 1997) and the same is true for the apomorphic Asian *Parailurus baikalicus* (Sotnikova, 2008). It therefore remains unclear from which lineage *Ailurus* arose. Sotnikova (2008) interprets *Parailurus* and *Ailurus* as sister-taxa, but this creates a ghost lineage of *Ailurus* of at least 2 Mya, as the earliest undisputed *Ailurus* is middle Pleistocene in age (Chen and Qi, 1978; Sotnikova, 2008). More material of Pliocene ailurines, especially from subtropical Asia, is necessary to address this problem.

Genus *PRISTINAILURUS* Wallace and Wang, **2004**

Type species *P. bristoli* Wallace and Wang, 2004

Other species none

Differential diagnosis. The type specimen, an isolated M_1, differs from *Ailurus* and *Parailurus* mainly in being longer than broad, having an enlarged metaconule which is separated from the protocone by a notch, and lacking a mesostylar cusp. It differs from Ailurinae indet. from locality Four in France in having the paraconule crest not connected to the protocone.

Distribution. Late Miocene to early Pliocene, Tennessee, USA.

Remarks. While in the publication of Wallace and Wang (2004) only the M_1 was discussed, the discovery of an upper canine was noted in the comments of this publication. It was placed into *Pristinailurus* due to the presence of a lateral groove. The rather plesiomorphic character of the genus, especially in comparison with *Parailurus* sp. from North America (Tedford and Gustafson, 1977), indicates a multiple immigration of Ailurinae from Eurasia into North America (Wallace and Wang, 2004). Recently, a brief note was published on the discovery of a lower jaw, containing c, p4, m_{1-2} of *Pristinailurus* (Wallace and Wang, 2007). Based on this specimen, the authors noted that the molars are similar to those of *Ailurus* and the p4 is plesiomorphic and similar to *Simocyon* in containing several cusps, thereby verifying the plesiomorphic character of *Pristinailurus*.

Genus *MAGERICTIS* Ginsburg, Morales, Soria and Herraez, **1997**

Type species *M. imperialis* Ginsburg, Morales, Soria and Herraez, 1997

Other species none

Differential diagnosis. *Magerictis* is smaller than *Parailurus* and about the size of *Ailurus*. It differs from *Ailurus* and *Parailurus* in having m2 completely basined with all cuspids, including that of the trigonid, being located at the tooth margin, which implies that protoconid and metaconid are not connected. This m2 is not directly comparable to *Pristinailurus* and Ailurinae indet. of which only the M_1 are described. However, the M_1 of *Pristinailurus* is structurally similar to that of *Parailurus* while Ailuridae indet. has a metacone and metaconulid much too low to occlude with the strong trigonid cuspids of *Magerictis* (Ginsburg *et al.*, 2001).

Distribution. Madrid, early middle Miocene of Spain.

Remarks. Originally based on a single m2, a couple of new specimens were recently found from the middle Miocene of Madrid (J. Morales, personal communication, 2007). These specimens still await description. Until such time, *Magerictis* remains an enigma and its relationships, especially to Ailurinae indet., are unresolved. However, an unconnected protoconid and metaconid is also known from the Oligocene *Amphictis ambigua* and *A. milloquensis* and lower Miocene *A. borbonica* (Heizmann and Morlo, 1994; Morlo, 1996), even if the gap is much smaller than in *Magerictis*. From these, only *Amphictis borbonica* has a relatively large m2 (but clearly smaller than *Magerictis*). *Magerictis* – and consequently Ailurinae – possibly represents a member of a lineage rooted in early Miocene *Amphictis*.

Ailurinae indet. Ginsburg, Maridet, and Mein, 2001

Differential diagnosis. This taxon is solely based on an isolated and fragmented M1. In contrast to *Ailurus*, *Parailurus*, and *Pristinailurus*, a single crest runs from the paraconule to the protocone, separating the cusps only by low notches. In all other taxa the crest of the paraconule is not connected to that of the protocone.

Distribution. Late middle Miocene of Four, France.

Remarks. As this fragmented M1 is the only specimen of the taxon, nothing can be said concerning its relationships except establishing its ailurine nature.

Subfamily SIMOCYONINAE DAWKINS, 1868

Type genus *Simocyon* Wagner, 1858

Other genera *Alopecocyon* Camp and Vanderhoof, 1940 (=*Viretius* Kretzoi, 1947); *Protursus* (see Peigné *et al.*, 2005).

Diagnosis. Within Ailuridae, Simocyoninae are characterised by having carnivorous to hypercarnivorous dentitions. Simocyoninae differs from Ailurinae in having the palate not prolonged beyond the last upper molar, P3 simple, P4 much longer than broad, because only a protocone is developed, M2 and m2 smaller, and m1 with smaller and less differentiated talonid. All of these characters are typical of a carnivorous to hypercarnivorous dietary pattern. Simocyoninae differs from *Amphictis*, the paraphyletic basal ailurid, in having M2 and m2 more enlarged.

Remarks. We follow Beaumont (1964, 1976, 1982, 1988), Wang (1997), Ginsburg *et al.* (1997), Baskin (1998), and Tedrow *et al.* (1999) in placing *Alopecocyon* within this subfamily and not Ginsburg (1999) and Ginsburg *et al.* (2001) in placing it together with *Amphictis* into a paraphyletic Amphictinae. Technically, even some species of *Amphictis* probably should be placed within Simocyoninae (Wang, 1997; Baskin, 1998; Tedrow *et al.* 1999), but at the moment it is unclear which species these are, even if *A. wintershofensis* and *A. prolongata* are the most likely candidates (see below).

Genus SIMOCYON WAGNER, 1858

Type species *S. primigenius* Roth and Wagner, 1854 (=*S. zdanskyi*, *S. marshi*)

Other species *S. diaphorus* Kaup, 1832, *S. batalleri* Viret, 1929a (?=*S. diaphorus*), *S. hungaricus* Kadic and Kretzoi, 1927, *Simocyon* small sp. (Wang *et al.*, 1998).

Differential diagnosis. *Simocyon* differs from all other ailurids in being the largest and the most hypercarnivorous taxon. This is especially evident in m1, which lacks a metaconid and has its talonid relatively reduced.

Distribution. Late middle Miocene to early Pliocene of Eurasia and North America.

Remarks. *Simocyon* is a widespread genus with a complex taxonomic history beginning with its first description in 1832 as *Gulo diaphorus* from Eppelsheim (Kaup, 1832). Today, the genus is placed in Ailuridae (Wang, 1997; Ginsburg *et al.*, 2004; Peigné *et al.*, 2005; Salesa *et al.*, 2006; Kullmer *et al.*, 2008). Several different species have been erected since *Gulo diaphorus*, but besides this species only two others were considered to be valid in the last contribution to the systematics of the genus (Kullmer *et al.*, 2008): the type species *S. primigenius* and *S. hungaricus*, with *S. batalleri* being regarded as a presumable junior synonym of *S. diaphorus*. As one of us (SP) is inclined to doubt whether *S. batalleri* is indeed a junior synonym of *S. diaphorus*, we keep *S. batalleri* as a valid species here. Most of the record comes from Europe (*S. diaphorus*, *S. batalleri*, *S. hungaricus*, *S. primigenius*), but *S. primigenius* is also known from Asia (as '*S. zdanskyi*') (Wang, 1997) and as '*S. marshi*', from the North American Hemphilian of Oregon, Utah, and possibly Nevada (Tedrow *et al.*, 1999). The geologically oldest record of the genus comes from Asia. Two small specimens (IVPP V7732, IVPP V11505) from the late middle Miocene of the Chinese Junggar Basin were assigned to *Simocyon* sp. by Wang *et al.* (1998). Interestingly, they are the only *Simocyon* co-occurring with *Alopecocyon*.

Genus *PROTURSUS* Crusafont and Kurtén, **1976**

Type species *Protursus simpsoni* Crusafont and Kurtén, 1976

Other species none

Differential diagnosis. The only known specimen of *Protursus*, an isolated m2, is similar to *Simocyon* in the relative proportions of the talonid and trigonid, but differs in detail from that genus in being smaller, 'more elongated and in having no paraconid, a smaller and more posteriorly located metaconid, and a talonid less structured' (Peigné *et al.*, 2005, p. 229).

Distribution. Early late Miocene (MN 9) from Can Llobateras, Spain.

Remarks. Originally placed in ursids by its first authors, *Protursus* was moved into *Simocyon* by Thenius (1977). This decision was followed by all subsequent authors until Peigné *et al.* (2005) demonstrated the differences between *Protursus* and *Simocyon* and reinstated the genus within ailurid simocyonines.

Genus *ALOPECOCYON* Camp and Vanderhoof, **1940**

(=*Viretius* Kretzoi, **1947**; ?=*Actiocyon* Stock, **1947**, see Webb, **1969**, and Baskin, **1998**; =*Ichneugale* Jourdan, **1862**, nomen oblitum)

Type species *A. goeriachensis* (Toula, 1884)

Questionable other species. *A. getti* Mein, 1958; *A. leardi* Stock, 1947 (*=Actiocyon*)

Differential diagnosis. *Alopecocyon* differs from *Simocyon* in being smaller and having mı with a metaconid.

Distribution. Middle Miocene of Eurasia.

Remarks. In his description of the new species *Viverra leptorhyncha*, Filhol (1883) mentions *Ichneugale* Jourdan as a synonym as this author had given this uninominal name to the holotype of *V. leptorhyncha*. In disagreement with Wolsan (1993), we regard *Ichneugale* as nomen oblitum, because it is uninominal, not sufficiently described, and, in concordance to article 23.9 of the International Code of Zoological Nomenclature (1999), has not been regarded as valid after 1899 (except by Wolsan, 1993, but even there by a citation of the name, only). Moreover, the name *Alopecocyon* has been regarded as valid in at least 25 studies, by at least 10 authors during the last 50 years (Mein, 1958, 1989; Ginsburg, 1961, 1963, 1972, 1974, 1980, 1990a,b, 1999, 2000, 2001, 2002; Beaumont, 1968, 1982; Webb, 1969; Ginsburg *et al.*, 1981; Schmidt-Kittler, 1981; de Bruijn *et al.*, 1992; Baskin, 1998, 2003; Wang *et al.*, 1998; Wu *et al.*, 1998; Nagel, 2003; Wallace and Wang, 2004; Peigné *et al.*, 2005). We also include those studies here, which regard *Viretius* Kretzoi, 1947 as valid, because none of those authors mentioned *Ichneugale*, but discussed the validity of *Viretius* only in reference to *Alopecocyon* (Ginsburg, 1999, 2000, 2001, 2002; Ginsburg *et al.*, 2001; Nagel, 2003), with *Alopecocyon* clearly having priority as it was named in 1940, seven years before *Viretius*. Baskin (1998) synonymised *Actiocyon* with *Alopecocyon*, as had Webb (1969) before him. The only species ever assigned to *Alopecocyon* besides the type species is *A. getti*, which is based on an M2 and a fragmented Mı. Recently, however, Ginsburg (2002) assigned the material – and thus the species – to a new genus, *Meiniogale*, and placed it into the family Amphicyonidae, based on the great resemblance of these teeth with those of amphicyonids such as *Ysengrinia tolosana*, *Amphicyon laugnacensis*, and *A. giganteus*. Given the fragmentary nature of the material, we prefer to retain the species in *Alopecocyon*.

The origin of *Alopecocyon* is rooted in Miocene *Amphictis*, presumably *A. wintershofensis* which has the relatively longest m2 of all *Amphictis* (Heizmann and Morlo, 1994: figure 3). Wolsan (1993) even synonymised the genus with *Amphictis*, a decision not followed by subsequent authors, due to the clearly more enlarged m2 in *Alopecocyon* (e.g. Wang, 1997; Nagel, 2003). Contrastingly, the close relationship of *Alopecocyon* to *Simocyon* was established early (Viret, 1951; Beaumont, 1964, 1976) and was never questioned later, even if Schmidt-Kittler (1981) and Wolsan (1993) regarded the characters used by Beaumont as plesiomorphic (see above). A direct ancestor–descendant relationship between

Simocyon and *Alopecocyon* is, however, unlikely, as the presumed more plesiomorphic *Simocyon* co-occurs with *Alopecocyon* in the latest middle Miocene of China (Wang *et al.*, 1998). As the material is fragmentary, however, these taxonomic assignments need to be confirmed.

Paraphyletic basal Ailuridae (=AMPHICTINAE Winge, 1895)

Genus *Amphictis* Pomel, 1853

Other genera none

Remarks. *Amphictis* is the basal taxon of Ailuridae and as such is paraphyletic with respect to Ailurinae and Simocyoninae. However, as long as the intrageneric relationships of *Amphictis* remain unresolved, the exact relationships of its species to the ailurid subfamilies will not be clear.

We do not follow Ginsburg (1999) and Ginsburg *et al.* (2001) in placing *Alopecocyon* in the same subfamily as the paraphyletic *Amphictis*. As its connection to *Simocyon* has confidently been established (Beaumont, 1964, 1976), a placement of *Alopecocyon* in Simocyoninae (Beaumont, 1982, 1988; Wang, 1997; Ginsburg *et al.*, 1997; Baskin, 1998) seems to be much more appropriate.

Other Oligocene taxa regarded as close to *Amphictis* based on their shared basicranial morphology are *Bavarictis* and *Mustelictis* (Schmidt-Kittler, 1981; Mödden, 1991; Wolsan, 1993). For this reason, *Bavarictis* has been included in ailurids by Nagel (2003). However, *Bavarictis* and *Mustelictis* have a reduced M2 and no prolonged m2 (Cirot and Bonis, 1993; Wolsan, 1993) and therefore lack one of the key features of Ailuridae (see above and Table 4.3). Moreover, *Bavarictis* has an isolated p4 protocone lobe, an apomorphic character after Wolsan (1993). We therefore place neither *Bavarictis* nor *Mustelictis* in ailurids.

Genus *AMPHICTIS* Pomel, 1853

Type species *A. antiqua* (de Blainville, 1842)

Other species *A. ambigua* (Gervais, 1872); *A. borbonica* Viret, 1929b; *A. cuspida* Nagel, 2003; *A. milloquensis* (Helbing, 1928); *A. prolongata* Morlo, 1996; *A. schlosseri* Heizmann and Morlo, 1994; *A. wintershofensis* Roth in Heizmann and Morlo, 1994.

Differential diagnosis. *Amphictis* differs from all other ailurids in having the relatively smallest m2 and M2, and a plesiomorphic mustelid dentition.

Distribution. The genus is known from the late Oligocene (MP 28) to the late early Miocene (MN 4) of Europe.

Remarks. *Amphictis* is by far the best documented fossil ailurid. However, due to the plesiomorphic characters of the numerous described species, its intrageneric relationships are unclear and disputed. While Heizmann and Morlo (1994) and Morlo (1996) concentrated on the Miocene species, Cirot and Wolsan (1995) focused solely on the Oligocene taxa. A systematic revision of the whole genus has never been provided, even if it is clear now that an understanding of ailurid origins is only possible after understanding *Amphictis* in detail. Heizmann and Morlo (1994) recognised four different groups.

1. The Oligocene *A. ambigua* and *A. milloquensis* have the relatively shortest m2 among *Amphictis* species, with protoconid and metaconid being widely separated (Heizmann and Morlo, 1994; Cirot and Wolsan, 1995). This last character can also be found in *A. borbonica*, but this species has the m2 talonid relatively much larger than the (other) Oligocene taxa (Morlo 1996: figure 8). The combination of both characters in *A. borbonica* is unique among *Amphictis* and foreshadows the m2 morphology of the ailurine *Magerictis*. *A. borbonica* is known from MP 29 to MN 2a of France and Germany.

2. Early Miocene *A. antiqua* and *A. schlosseri* are separated from the other species in having m2 protoconid and metaconid connected by a continuous crest and the m2 relatively long. This character resembles the situation in early procyonids such as *Broiliana* (Morlo, 1996), but clearly represents the plesiomorphic condition in musteloids.

3. *A. prolongata*, *A. wintershofensis*, and *A. cuspida*, as the remaining species, are interpreted to represent a single evolutionary lineage (Morlo, 1996; Nagel, 2003). In these species the m2 protoconid and metaconid are connected by a notched crest. *A. cuspida* is the latest and largest *Amphictis*, but probably represents an endemic form from the middle Miocene of South-Eastern Europe (Nagel, 2003). *A. wintershofensis*, on the other hand, probably gave rise to *Alopecocyon* (Beaumont, 1964, 1976; Wolsan, 1993) and eventually *Simocyon* (Beaumont, 1964).

4. Some other species were included in the genus, but Heizmann and Morlo (1994) removed all of them, including *A. major* (Teilhard de Chardin, 1915) and *A. nana* (Teilhard de Chardin, 1915), a position followed by Morlo (1996) and Ginsburg (1999).

Amphictis is obviously paraphyletic: *A. borbonica* may well be the oldest representative of ailurines, and *A. wintershofensis* may have given rise to either *Alopecocyon*/Simocyoninae and *A. cuspida*. To avoid paraphyly, Wolsan (1993) included *Alopecocyon* in *Amphictis*, but that has not been accepted (see above) and, moreover, has become moot after rooting ailurines in *Amphictis* as well. Because a change in the intrageneric taxonomy of *Amphictis* would require a complete revision of the genus, which is outside of the scope of this contribution, we consider it to be paraphyletic.

The origin of *Amphictis*, and thus the origin of ailurids, remains unclear. Bonis (1976) had rooted the genus in *Cephalogale* or *Amphicynodon*, but both of these early Oligocene taxa are now interpreted as early ursoids (Wang *et al.*,

2005). *Amphicticeps*, another early Oligocene arctoid, may in fact represent one of the earliest musteloids, but shows in the lingually shifted position of M2 an apomorphic character which places it at the beginning of semantorids and, consequently, pinnipeds (Wolsan, 1993; Wang *et al.*, 2005; Morlo and Nagel, 2007).

Historical summary of Ailuridae

The evolution of Ailuridae mainly occurred in Europe. Its earliest representative, *Amphictis*, is known from this continent only (Cirot and Wolsan, 1995; Morlo, 1996; Ginsburg, 1999), and its latest and only Middle Miocene representative, *A. cuspida*, is known from the southeastern corner of the continent (Nagel, 2003). *Alopecocyon*, the earliest simocyonine, probably evolved in Europe as well, as its morphology is very close to the latest Early Miocene *Amphictis wintershofensis*. Unlike *Amphictis*, however, *Alopecocyon* is also known from the late middle Miocene of Asia (Wang *et al.*, 1998) and, under the name *Actiocyon*, in the late Miocene of North America (Webb, 1969; Baskin, 1998). These specimens also represent the earliest migrations of Ailuridae into Asia and North America, respectively. However, most of the fossils of *Alopecocyon* are known from Europe, which corroborates the hypothesis that the genus originated here. Of the two later simocyonines, *Simocyon* and *Protursus*, the latter is known by a single tooth from Spain (Peigné *et al.*, 2005). *Simocyon*, on the other hand, occurred in all northern continents, and is known by a few specimens from outside of Europe, among them the oldest and most primitive taxon from the latest middle Miocene of China (Wang *et al.*, 1998). In the late Miocene, *Simocyon* reappears in Asia again, but not before the latest Miocene (Wang *et al.*, 1998), contemporary with the first North American record (Baskin, 1998; Tedrow *et al.*, 1999). In the meantime, especially in the early late Miocene (MN 9–10) and the earliest latest Miocene (MN 11), *Simocyon* underwent a fairly well documented evolution in Europe (Morlo, 1997; Roussiakis, 2002; Peigné *et al.*, 2005; Kullmer *et al.*, 2008). It seems likely that *Simocyon* arrived in Europe with the faunal transition at the end of the middle Miocene and re-migrated to Asia and, eventually, North America in the latest Miocene. *Simocyon*, and with it Simocyoninae, vanished in Europe and North America at the end of the Miocene, but may have persisted in China until the early Pliocene (Wang, 1997). Its disappearance may be correlated with a worldwide change towards wetter climates.

As with simocyonines, the highest diversity of Ailurinae is documented from Europe. In the middle Miocene, two ailurines were present, *Magerictis* and Ailuridae indet. (Ginsburg *et al.*, 1997, 2001), even if a brief note on the presence of a middle Miocene ailurine in China exists (Sasagawa *et al.*, 2003). Interestingly, Ailurinae are unknown in the late Miocene of Europe (ghost lineage), possibly because of the development of drier climates globally. Surprisingly, the next member of the family, *Pristinailurus*, shows up in the latest Miocene/early Pliocene of North America, but as part of an immigrating Eurasian fauna (Wallace and Wang, 2004). It therefore seems likely that Ailurinae survived the dry late Miocene somewhere in South Asia in wetter climates. In Eurasia, the next known taxon is *Parailurus* which emerged in the early Pliocene (MN 14) of Europe (Morlo and Kundrát, 2001; Sotnikova, 2008), while all Asian records are late Pliocene (MN 16) in age (Sasagawa *et al.*, 2003; Sotnikova, 2008). Contemporary to the late Eurasian record, *Parailurus* is also represented in North America by a single tooth (Tedford and Gustafson, 1977; Sotnikova, 2008). Nevertheless, *Parailurus* had its main distribution in Europe, from where the most specimens, the greatest morphological variation, and the longest temporal distribution (MN 14–16) are documented. *Parailurus* went extinct before the earliest remains of *Ailurus* from the middle Pleistocene of China appear (Bien and Chia, 1938). Since that time, Ailuridae are restricted to Asia. Fossil remains of *Ailurus fulgens* are known only from Yunnan Province.

The range of the single extant species, *Ailurus fulgens*, is restricted and is decreasing (Wang *et al.*, 2008). The species is endemic to southern China (Yunnan and Sichuan provinces) and several areas of the Himalayas, but had a much larger distribution during recent historical time in China (Qinghai, Shaanxi, Gansu, northern Guizhou provinces; Li *et al.*, 2005). Two subspecies are now recognised, a western subspecies, *Ailurus fulgens fulgens*, known from southwestern China (Tibet, western Yunnan), central Nepal, Bhutan, northeastern India (Darjeeling, Sikkim and Arunachal Pradesh states), and northern Burma, and an eastern subspecies, *Ailurus fulgens styani*, known from eastern and southern Sichuan (Qionglai, Minshan, Xiangling, Liangshan mountains) and western Yunnan (Daxueshan, Shalulishan, Gaoligongshan mountains), China (Wei *et al.*, 2000; Pradhan *et al.*, 2001; Li *et al.*, 2005). Old records mention the species in Laos, but there is no recent evidence confirming the presence of the species (Wang *et al.*, 2008). The present distribution and genetic diversity of populations of the red panda are the result of habitat fragmentation and expansion from glacial refugia (Li *et al.*, 2005).

Acknowledgements

We thank the editors for enabling us to include this chapter in the book. We are also grateful to Dr Shintaro Ogino (Primate Research Institute,

Inuyama, Japan) who allowed one of us (MM) to have a closer look at the *Parailurus* of Udungu. We thank P. Gaubert (MNHN) for his comments on the part of the manuscript dealing with the molecular evidence, Gregg F. Gunnell (University of Michigan) for correcting the English, and the two anonymous reviewers for their remarks that have clearly improved the manuscript.

REFERENCES

Antón, M., Salesa, M. J., Pastor, J. F., Peigné, S. and Morales, J. (2006). Implications of the functional anatomy of the hand and forearm of *Ailurus fulgens* (Carnivora, Ailuridae) for the evolution of the 'false-thumb' in pandas. *Journal of Anatomy*, **209**, 757–64.

Bardenfleth, K. S. (1914). On the systematic position of *Aeluropus melanoleucus*. *Mindeskrift i anledning af hundredaaret for Japetus Steenstrups Fødsel udgivet af en kreds af naturforskere*, **1**(17), 1–15.

Baskin, J. A. (1998). Procyonidae. In *Evolution of Tertiary Mammals of North America. Volume 1: Terrestrial Carnivores, Ungulates, and Ungulatelike Mammals*, ed. C. M. Janis, K. M. Scott and L. L. Jacobs. Cambridge: Cambridge University Press, pp. 144–51.

Baskin, J. A. (2003). New Procyonines from the Hemingfordian and Barstovian of the Gulf Coast and Nevada, including the first fossil record of the Potosini. *Bulletin of the American Museum of Natural History*, **279**, 125–46.

Baskin, J. A. (2004). *Bassariscus* and *Probassariscus* (Mammalia, Carnivora, Procyonidae) from the early Barstovian (middle Miocene). *Journal of Vertebrate Paleontology*, **24**, 709–20.

Beaumont, G. de (1964). Essai sur la position taxonomique des genres *Alopecocyon* Viret et *Simocyon* Wagner (Carnivora). *Eclogae geologicae Helvetiae*, **57**, 829–36.

Beaumont, G. de (1968). Note sur l'ostéologie crânienne de *Plesiogale* Pomel (Mustelidae, Carnivora). *Archives des Sciences*, **21**, 27–34.

Beaumont, G. de (1976). Remarques préliminaires sur le genre *Amphictis* Pomel (Carnivore). *Bulletin de la Société vaudoise de Sciences naturelles*, **73**, 171–80.

Beaumont, G. de (1982). Qu'est-ce que le *Plesictis leobensis* Redlich (mammifère, carnivore)? *Archives des Sciences*, **35**, 143–52.

Beaumont, G. de (1988). Contributions à l'étude du gisement Miocène supérieur de Montredon (Hérault) – les grands mammifères, 2 – Les carnivores. *Palaeovertebrata, Mémoire extraordinaire*, **1988**, 15–42.

Bien, M. N. and Chia, L. P. (1938). Cave and rock-shelter deposits in Yunnan. *Bulletin of the Geological Society of China*, **18**, 325–47.

Blainville, H. M. D. de (1841). *Ostéographie ou description iconographique comparée du squelette et du système dentaire des cinq classes d'animaux vertebras récents et fossils pour servir de base à la zoologie et à la géologie. Mammifères. Carnassiers: Vespertilio. Talpa. Sorex. Erinaceus. Phoca. Ursus. Subursus. – Des petits-ours (G. Subursus)*. Paris: Arthus Bertrand.

Bonis, L. de (1976). Découverte d'un crane d'*Amphictis* (Mammalia, Carnivora) dans l'Oligocène supérieur des Phosphorites du Quercy (Lot). *Comptes rendus de l'Académie des Sciences, série D*, **283**, 327–30.

Bowdich, T. E. (1821). *An Analysis of the Natural Classifications of Mammalia, for the use of Students and Travellers*. Paris: J. Smith.

Bruijn, H. de, Daams, R., Daxner-Höck, G., *et al.* (1992). Report of the RCMNS working group on fossil mammals, Reisensburg 1990. *Newsletters on Stratigraphy*, **26**, 65–118.

Bugge, J. (1978). The cephalic arterial system in carnivores, with special reference to the systematic classification. *Acta Anatomica*, **101**, 45–61.

Camp, C. L. and Vanderhoof, V. L. (1940). Bibliography of fossil vertebrates 1928–1933. *Geological Society of America Special Papers*, **27**, 1–503.

Chen, D. Z. and Qi, G. (1978). Human remains and mammals accompanied in Xi Chuo, Yunnan. *Vertebrata PalAsiatica*, **16**, 35–46.

Choudhury, A. (2001). An overview of the status and conservation of the red panda *Ailurus fulgens* in India, with reference to its global status. *Oryx*, **35**, 250–59.

Cirot, E. and Bonis, L. de (1993). Le crâne d'*Amphictis ambiguus* (Carnivora, Mammalia): son importance pour la compréhension de la phylogénie des mustéloïdes. *Comptes rendus de l'Académie des Sciences, Série II*, **316**, 1327–33.

Cirot, E. and Wolsan, M. (1995). Late Oligocene amphictids (Mammalia: Carnivora) from La Milloque, Aquitaine Basin, France. *Geobios*, **28**, 757–67.

Crusafont-Pairó, M. and Kurtén, B. (1976). Bears and bear-dogs from the Vallesian of the Vallés-Penedés Basin, Spain. *Acta Zoologica Fennica*, **144**, 1–29.

Davis, D. D. (1964). The giant panda. A morphological study of evolutionary mechanisms. *Fieldiana: Zoology Memoirs*, **3**, 1–339.

Dawkins, W. B. (1868). Fossil animals and geology of Attica, by Albert Gaudry. (Critical-summary). *Quarterly Journal of the Geological Society of London*, **24**, 1–7.

Dawkins, W. B. (1888). On *Ailurus anglicus*, a new carnivore from Red Crag. *Quarterly Journal of the Geological Society of London*, **44**, 228–31.

Degnan S. M. (1993). The perils of single gene trees – mitochondrial versus single-copy nuclear DNA variation in white-eyes (Aves: Zosteropidae). *Molecular Ecology*, **2**, 219–25.

Dehm, R. (1950). Die Raubtiere aus dem Mittel-Miocän (Burdigalium) von Wintershof-West bei Eichstätt in Bayern. *Abhandlungen der Bayerischen Akademie der Wissenschaften, Mathematisch-naturwissenschaftliche Klasse, Neue Folge*, **58**, 1–141.

Delisle, I. and Strobeck, C. (2005). A phylogeny of the Caniformia (order Carnivora) based on 12 complete protein-coding mitochondrial genes. *Molecular Phylogenetics and Evolution*, **37**, 192–201.

Domingo-Roura, X., López-Giráldez, F., Saeki, M. and Marmi, J. (2005). Phylogenetic inference and comparative evolution of a complex microsatellite and its flanking region in carnivores. *Genetical Research*, **85**, 223–33.

Fejfar, O. and Sabol, M. (2004). Pliocene carnivores (Carnivora, Mammalia) from Ivanovce and Hajnáča (Slovakia). *Courier Forschungsinstitut Senckenberg*, **246**, 15–53.

Filhol, H. (1883). Notes sure quelques mammifères fossiles de l'époque Miocène. *Archives du Muséum d'Histoire Naturelle de Lyon*, **3**, 1–97.

Flower, W. H. (1869). On the value of the characters of the base of the cranium in the classification of the Order Carnivora, and on the systematic position of *Bassaris* and other disputed forms. *Proceedings of the Scientific Meetings of the Zoological Society of London*, **1869**, 4–37.

Flower, W. H. (1870). On the anatomy of *Aelurus fulgens*, Fr. Cuv. *Proceedings of the Scientific Meetings of the Zoological Society of London*, **1870**, 752–69.

Flower, W. H. and Lydekker, R. (1891). *An Introduction to the Study of Mammals Living and Extinct*. London: Adam and Charles Black.

Flynn, J. J. and Nedbal, M. A. (1998). Phylogeny of the Carnivora (Mammalia): congruence vs incompatibility among multiple data sets. *Molecular Phylogenetics and Evolution*, **9**, 414–26.

Flynn, J. J., Finarelli, J. A., Zehr, S., Hsu, J. and Nedbal, M. A. (2005). Molecular phylogeny of the Carnivora (Mammalia): assessing the impact of increased sampling on resolving enigmatic relationships. *Systematic Biology*, **54**, 317–37.

Flynn, J. J., Nedbal, M. A., Dragoo, J. W. and Honeycutt, R. L. (2000). Whence the red panda? *Molecular Phylogenetics and Evolution*, **17**, 190–99.

Flynn, J. J., Neff, N. A. and Tedford, R. H. (1988). Phylogeny of the Carnivora. In *The Phylogeny and Classification of the Tetrapods, Volume 2: Mammals*, ed. M. J. Benton. Oxford: Clarendon Press, Systematics Association Special Volume 35B, pp. 73–116.

Fulton, T. L. and Strobeck, C. (2006). Molecular phylogeny of the Arctoidea (Carnivora): effect of missing data on supertree and supermatrix analyses of multiple gene data sets. *Molecular Phylogenetics and Evolution*, **41**, 165–81.

Fulton, T. L. and Strobeck, C. (2007). Novel phylogeny of the raccoon family (Procyonidae: Carnivora) based on nuclear and mitochondrial DNA evidence. *Molecular Phylogenetics and Evolution*, **43**, 1171–77.

Gaudry, A. (1862–1867). *Animaux fossiles et géologie de l'Attique*. Paris: F. Savy Editeur.

Geoffroy Saint-Hilaire, E. and Cuvier, F. (1825). *Histoire naturelle des Mammifères, avec des figures originales, coloriées, dessinées d'après des animaux vivans. Tome troisième. Livraison 50e*. Paris: A. Belin.

Gervais, P. (1870). Mémoire sur les formes cérébrales propres aux carnivores vivants et fossiles suivi de remarques sur la classification de ces animaux. *Nouvelles Archives du Muséum d'Histoire naturelle de Paris*, **6**, 103–62.

Gervais, P. (1872). Sur les Mammifères dont les ossements accompagnent les dépôts de chaux phosphatée de départements du Tarn-et-Garonne et du Lot. *Journal de Zoologie*, **1**, 261–68.

Gervais, P. (1875). De l'*Ursus melanoleucus* de l'Abbé Arman David. *Journal de Zoologie*, **4**, 79–87.

Gill, T. (1872). Arrangement of the families of mammals. With analytical tables. Prepared for the Smithsonian Institution. *Smithsonian Miscellaneous Collections*, **11**, 1–98.

Ginsburg, L. (1961). La faune des carnivores miocènes de Sansan (Gers). *Mémoires du Muséum National d'Histoire Naturelle, Séries C*, **9**, 1–190.

Ginsburg, L. (1963). Les mammifères fossiles récoltés à Sansan au cours du XIXe siècle. *Bulletin de la Société géologique de France, 7e série,* **5,** 3–15.

Ginsburg, L. (1972). Sur l'âge des Mammifères des faluns miocènes du Nord de la Loire. *Comptes-rendus de l'Académie des Sciences de Paris, Série D,* **274,** 3345–47.

Ginsburg, L. (1974). Les faunes de Mammifères burdigaliens et vindoboniens des bassins de la Loire et de la Garonne. *Mémoires du B.R.G.M.,* **78,** 153–67.

Ginsburg, L. (1980). Paléogéographie et âge de la mer des faluns d'après les Mammifères. *Mémoires de la Société d'Etudes scientifiques de l'Anjou,* **4,** 69–77.

Ginsburg, L. (1982). Sur la position systématique du petit panda, *Ailurus fulgens* (Carnivora, Mammalia). *Geobios, mémoire spécial* **6,** 247–58.

Ginsburg, L. (1990a). Les quatre faunes de Mammifères miocènes des faluns du synclinal d'Esvres (Val-de-Loire, France). *Comptes-rendus de l'Académie des Sciences de Paris, Série II,* **310,** 89–93.

Ginsburg, L. (1990b). The faunas and stratigraphical subdivisions of the Orleanian in the Loire Basin (France). In *European Neogene Mammal Chronology,* ed. E. H. Lindsay, V. Fahlbusch and P. Mein. New York, NY: Plenum Press, pp. 157–78.

Ginsburg, L. (1999). Order Carnivora. In *The Miocene Land Mammals of Europe,* ed. G. Rössner and K. Heissig. Munich: Verlag Dr F. Pfeil, pp. 109–48.

Ginsburg, L. (2000). Chronologie des dépôts miocènes du Blésois à la Bretagne. *Symbioses, nouvelle série,* **2,** 3–16.

Ginsburg, L. (2001). Les faunes de mammifères terrestres du Miocène moyen des Faluns du bassin de Savigné-sur-Lathan (France). *Geodiversitas,* **23,** 381–94.

Ginsburg, L. (2002). Un Amphicyonidae (Carnivora, Mammalia) nouveau du Miocène moyen de Vieux-Collonges (Rhône). *Symbioses, nouvelle série,* **7,** 55–57.

Ginsburg, L., Huin, J. and Locher, J.-P. (1981). Les Carnivores du Miocène inférieur des Beilleaux à Savigné-sur-Lathan (Indre-et-Loire). *Bulletin du Muséum national d'Histoire naturelle, Paris, 4e série,* **3** C, 183–94.

Ginsburg, L., Morales, J., Soria, D. and Herraez, E. (1997). Découverte d'une forme ancestrale du Petit Panda dans le Miocène moyen de Madrid (Espagne). *Comptes-Rendus de l'Académie des Sciences de Paris, Sciences de la Terre et des planètes,* **325,** 447–51.

Ginsburg, L., Maridet, O. and Mein, P. (2001). Un Ailurinae (Mammalia, Carnivora, Ailuridae) dans le Miocène moyen de Four (Isère, France). *Geodiversitas,* **23,** 81–85.

Goldman, D., Giri, P. R. and O'Brien, S. J. (1989). Molecular genetic-distance estimates among the Ursidae as indicated by one- and two-dimensional protein electrophoresis. *Evolution,* **43,** 282–95.

Gould, S. J. (1978). The Panda's peculiar thumb. *Natural History,* **87,** 20–30.

Gray, J. E. (1843). *List of the Specimens of Mammalia in the Collection of the British Museum.* London: British Museum (Natural History) Publications.

Greaves, W. S. (1983). A functional analysis of carnassial biting. *Biological Journal of the Linnean Society,* **2,** 353–63.

Greaves, W. S. (1985). The generalized carnivore jaw. *Zoological Journal of the Linnean Society,* **85,** 267–74.

Gregory, W. K. (1936). On the phylogenetic relationships of the giant panda (*Ailuropoda*) to other arctoid Carnivora. *American Museum Novitates*, **878**, 1–29.

Helbing, H. (1928). Carnivoren des oberen Stampien. *Abhandlungen der Schweizerischen Palaeontologischen Gesellschaft*, **50**, 1–36.

Heizmann, E. P. J. and Morlo, M. (1994). *Amphictis schlosseri* n. sp. – eine neue Carnivoren-Art (Mammalia) aus dem Unter-Miozän von Südwestdeutschland. *Stuttgarter Beiträge zur Naturkunde B*, **216**, 1–25.

Hunt, R. H. Jr. (1974). The auditory bulla in Carnivora: an anatomical basis for reappraisal of Carnivore evolution. *Journal of Morphology*, **143**, 21–76.

Jourdan, C. (1862). La description de restes fossiles de grands mammifères. *Revue des sociétés savantes. Sciences mathématiques, physiques et naturelles*, **1**, 126–30.

Kadic, O. and Kretzoi, N. (1927). Vorläufiger Bericht über die Ausgrabungen in der Csákvárer Höhlung. *Barlankutatás*, **14–15**, 1–21.

Kaup, J. J. (1832). Vier neue Arten urweltlicher Raubthiere, welche im zoologischen Museum zu Darmstadt aufbewahrt werden. *Archiv für Mineralogie, Geognosie, Bergbau– und Hüttenkunde*, **5**, 150–58.

Kormos, T. (1935). Beiträge zur Kenntnis der Gattung *Parailurus*. *Mitteilungen aus dem Jahrbuch der Königlich-Ungarischen Geologischen Anstalt*, **30**, 1–39.

Kretzoi, M. (1947). New names for mammals. *Annales historico-naturales musei nationalis Hungarici*, **40**, 285–87.

Kullmer, O., Morlo, M., Sommer, J., *et al.* (2008). The second specimen of *Simocyon diaphorus* (Kaup, 1832) (Mammalia, Carnivora, Ailuridae) from the type-locality Eppelsheim (early Late Miocene, Germay). *Journal of Vertebrate Paleontology*, **28**, 928–32.

Lankester, E. R. (1901). On the affinities of *Aeluropus melanoleucus*, A. Milne-Edwards. *Transactions of the Linnean Society of London, 2nd series, 8, Zoology*, part 6, 163–72.

Ledje, C. and Arnason, U. (1996a). Phylogenetic analyses of complete cytochrome *b* genes of the Order Carnivora with particular emphasis on the Caniformia. *Journal of Molecular Evolution*, **42**, 135–44.

Ledje, C. and Arnason, U. (1996b). Phylogenetic relationships within caniform Carnivores based on analyses of the mitochondrial 12S rRNA. *Journal of Molecular Evolution*, **43**, 641–49.

Li, M., Wei, F., Goossens, B., *et al.* (2005). Mitochondrial phylogeography and subspecific variation in the red panda (*Ailurus fulgens*): implications for conservation. *Molecular Phylogenetics and Evolution*, **36**, 78–89.

Marmi, J., López-Giráldez, J. F. and Domingo-Roura, X. (2004). Phylogeny, evolutionary history and taxonomy of the Mustelidae on sequences of the cytochrome *b* gene and a complex repetitive flanking region. *Zoologica Scripta*, **33**, 481–99.

Mayr, E. (1986). Uncertainty in science: is the giant bear a bear or a raccoon? *Nature*, **323**, 769–71.

Mein, P. (1958). Les mammifères de la faune sidérolitique de Vieux-Collonges. *Nouvelles Archives du Muséum d'Histoire naturelle de Lyon*, **5**, 1–118.

Mein, P. (1989). Updating of MN zones. In *European Neogene Mammal Chronology*, ed. E. H. Lindsay, V. Fahlbusch, and P. Mein. New York, NY: Plenum Press, pp. 73–90.

Milne-Edwards, H. (1868–1874). *Recherches pour servir à l'histoire naturelle des Mammifères comprenant des considerations sur la classification de ces animaux*. Paris: G. Masson.

Mivart, St. George (1885). On the anatomy, classification, and distribution of the Arctoidea. *Proceedings of the Scientific Meetings of the Zoological Society of London*, **1885**, 340–404.

Mödden, C. (1991). *Bavarictis gaimersheimensis* n. gen. n. sp., ein früher Mustelide aus der oberoligozänen Spaltenfüllung Gaimersheim bei Ingolstadt. *Mitteilungen aus der Bayerischen Staatssammlung für Paläontologie und Historische Geologie*, **31**, 125–47.

Morales J., Alcalá, L., Álvarez-Sierra, M. A., *et al.* (2004). Paleontología del sistema de yacimientos de mamíferos miocenos del Cerro de los Batallones, Cuenca de Madrid. *Geogaceta*, **35**, 139–42.

Morlo, M. (1996). Carnivoren aus dem Unter-Miozän des Mainzer Beckens. *Senckenbergiana lethaea*, **76**, 193–249.

Morlo, M. (1997). Die Raubtiere (Mammalia, Carnivora) aus dem Turolium von Dorn–Dürkheim 1 (Rheinhessen). Teil 1: Mustelida, Hyaenidae, Percrocutidae, Felidae. *Courier Forschungsinstitut Senckenberg*, **197**, 11–47.

Morlo, M. and Kundrát, M. (2001). The first carnivoran fauna from the Ruscinium (MN 15) of Germany. *Paläontologische Zeitschrift*, **75**, 163–87.

Morlo, M. and Nagel, D. (2007). The carnivore guild of the Taatsiin Gol area: Hyaenodontidae (Creodonta), Carnivora, and Didymoconida from the Oligocene of Central Mongolia. *Annalen des Naturhistorischen Museums Wien*, **108A**, 217–31.

Nagel, D. (2003). Carnivora from the middle Miocene hominoid locality of Çandir (Turkey). *Courier Forschungsinstitut Senckenberg*, **240**, 113–31.

Nowak, R. M. (2005). *Walker's Carnivores of the World*. Baltimore, MD: The Johns Hopkins University Press.

O'Brien, S. J., Nash, W. G., Wildt, D. E., Bush, M. E. and Benveniste, R. E. (1985). A molecular solution to the riddle of the giant panda's phylogeny. *Nature*, **317**, 140–44.

Pecon Slattery, J. and O'Brien, S. J. (1995). Molecular phylogeny of the red panda (*Ailurus fulgens*). *Journal of Heredity*, **86**, 413–22.

Peigné, S., Salesa, M. J., Antón, M. and Morales, J. (2005). Ailurid carnivoran mammal *Simocyon* from the late Miocene of Spain and the systematics of the genus. *Acta Palaeontologica Polonica*, **50**, 219–38.

Pocock, R. I. (1921). The external characters and classification of the Procyonidae. *Proceedings of the General Meetings for Scientific Business of the Zoological Society of London*, **1921**, 389–422.

Pocock, R. I. (1928a). The structure of the auditory bulla in the Procyonidae and the Ursidae, with a note on the bulla of *Hyaena*. *Proceedings of the General Meetings for Scientific Business of the Zoological Society of London*, **1928**, 963–74.

Pocock, R. I. (1928b). Some external characters of the giant panda (*Ailuropoda melanoleuca*). *Proceedings of the General Meetings for Scientific Business of the Zoological Society of London*, **1928**, 975–81.

Pomel, A. (1853). Catalogue des vertébrés fossils (suite). *Annales scientifiques, littéraires et industrielles de l'Auvergne*, **26**, 81–229.

Pradhan, S., Saha, G. K. and Khan, J. A. (2001). Ecology of the red panda *Ailurus fulgens* in the Singhalila National Park, Darjeeling, India. *Biological Conservation*, **98**, 11–18.

Raven, H. C. (1936). Notes on the anatomy of the viscera of the giant panda (*Ailuropoda melanoleuca*). *American Museum Novitates*, **877**, 1–23.

Roberts, M. S. and Gittleman, J. L. (1984). *Ailurus fulgens*. *Mammalian Species*, **222**, 1–8.

Roth, J. and Wagner, A. (1854). Die fossilen Knochenüberreste von Pikermi in Griechenland. *Abhandlungen der mathematisch–physikalischen Classe der Königlich Bayerischen Akademie der Wissenschaften*, **7**, 371–464.

Roussiakis, S. J. (2002). Musteloids and feloids (Mammalia, Carnivora) from the late Miocene locality of Pikermi (Attica, Greece). *Geobios*, **35**, 699–719.

Salesa M. J., Antón, M., Peigné, S. and Morales, J. (2006). Evidence of a false thumb in a fossil carnivore clarifies the evolution of pandas. *Proceedings of the National Academy of Sciences of the USA*, **103**, 379–82.

Salesa M. J., Antón, M., Peigné, S. and Morales, J. (2008). Functional anatomy and biomechanics of the postcranial skeleton of *Simocyon batalleri* (Viret, 1929) (Carnivora, Ailuridae) from the late Miocene of Spain. *Zoological Journal of the Linnean Society*, **152**, 593–621.

Sarich, V. M. (1973). The giant panda is a bear. *Nature*, **245**, 218–20.

Sasagawa, I., Takahashi, K., Sakumoto, T., Nagamori, H., Yabe, H. and Katoh, S. (2003). Discovery of the extinct red panda *Parailurus* (Mammalia, Carnivora) in Japan. *Journal of Vertebrate Paleontology*, **23**, 895–900.

Sato, J. J., Wolsan, M., Suzuki, H., *et al.* (2006). Evidence from nuclear DNA sequences sheds light on the phylogenetic relationships of Pinnipedia: single origin with affinity to Musteloidea. *Zoological Science*, **23**, 125–46.

Schlosser, M. (1899). *Parailurus anglicus* and *Ursus böckhi* aus den Ligniten von Baróth-Köpecz, Comitat Háromszék in Ungard. *Mittheilungen aus dem Jahrbuche der Königlich Ungarischen Geologischen Anstalt*, **13**, 66–95.

Schmidt-Kittler, N. (1981). Zur Stammesgeschichte der marderverwandten Raubtiergruppen (Musteloidea, Carnivora). *Ecologae geologicae Helvetiae*, **74**, 753–801.

Segall, W. (1943). The auditory region of the arctoid carnivores. *Zoological Series of Field Museum of Natural History*, **29**, 33–59.

Simpson, G. G. (1945). The principles of classification and a classification of mammals. *Bulletin of the American Museum of Natural History*, **85**, 1–350.

Smith, J. B. and Dodson, P. (2003). A proposal for a standard terminology of anatomical notation and orientation in fossil vertebrate dentitions. *Journal of Vertebrate Paleontology*, **23**, 1–12.

Sotnikova, M. V. (2008). A new species of lesser panda *Parailurus* (Mammalia, Carnivora) from the Pliocene of Transbaikalia (Russia) and some aspects of ailurine phylogeny. *Paleontological Journal*, **42**(1), 90–99.

Stock, C. (1947). A peculiar new carnivore from the Cuyama Miocene, California. *Bulletin of the Southern California Academy of Sciences*, **46**, 84–89.

Su, B., Fu, Y., Wang, Y., Jin, L. and Chakraborty, R. (2001). Genetic diversity and population history of the red panda (*Ailurus fulgens*) as inferred from mitochrondrial DNA sequence variations. *Molecular Biology and Evolution*, **18**, 1070–76.

Tagle, D. A., Miyamoto, M. M. and Goodman, M. (1986). Hemoglobin of pandas: phylogenetic relationships of Carnivores as ascertained with protein sequence data. *Naturwissenschaften*, **73**, 512–14.

Tedford, R. H. and Gustafson, E. P. (1977). First North American record of the extinct panda *Parailurus*. *Nature*, **265**, 621–23.

Tedrow, A. R., Baskin, J. A. and Robison, S. F. (1999). An additional occurrence of *Simocyon* (Mammalia, Carnivora, Procyonidae) in North America. In *Vertebrate Paleontology in Utah*, ed. D. Gillette. Salt Lake City, UT: Utah Geological Survey Miscellaneous Publications **99**-1, pp. 487–93.

Teilhard de Chardin, P. (1915). Les Carnassiers des Phosphorites du Quercy. *Annales de Paléontologie*, **9**, 89–191.

Thenius, E. (1949). Zur Herkunft der Simocyoniden (Canidae, Mammalia). *Sitzungsberichten der Österreichischen Akademie der Wissenschaften, Mathematisch–Naturwissenschaftliche Klasse, Abteilung 1*, **158**, 799–810.

Thenius, E. (1977). Zur systematischen Stellung von *Protursus* (Carnivora, Mammalia). *Anzeiger der Österreichischen Akademie der Wissenschaften, Mathematisch–Naturwissenschaftliche Klasse*, **3**, 37–41.

Thenius, E. (1979a). Die taxonomische und stammesgeschichtliche Position des Bambousbären (Carnivora, Mammalia). *Anzeiger der Österreichischen Akademie der Wissenschaften, Mathematisch-naturwissenschaftliche Klasse*, **1979**, 67–78.

Thenius, E. (1979b). Zur systematischen und phylogenetischen Stellung des Bambusbären: *Ailuropoda melanoleuca* David (Carnivora, Mammalia). *Zeitschrift für Säugetierkunde*, **44**, 286–305.

Todd, N. B. and Pressman, S. R. (1968). The karyotype of the lesser panda (*Ailurus fulgens*) and general remarks on the phylogeny and affinities of the panda. *Carnivore Genetics Newsletter*, **5**, 105–08.

Toula, F. (1884). Ueber einige Säugethierreste von Göriach bei Turnau (Bruck a.M.), Nord Steiermark. *Jahrbuch der kaiserlich-königlichen geologischen Reichsanstalt*, **34**, 385–401.

Trouessart, E.-L. (1899). *Catalogus mammalium tam viventium quam fossilium. Nova editio (Prima completa). Tomus I. Primates, Prosimiae, Chiroptera, Insectivora, Carnivora, Rodentia, Pinnipedia*. Berolini: R. Friedländer and Sohn.

Trouessart, E.-L. (1904). *Catalogus mammalium tam viventium quam fossilium. Nova editio (Prima completa). Quinquennale supplementum*. Berolini: R. Friedländer and Sohn.

Turner, H. N. (1848). Observations relating to some of the foramina at the base of the skull in Mammalia, and on the classification of the order Carnivora. *Proceedings of the Zoological Society of London*, **1848**, 63–88.

Veron, G. (1995). La position systématique de *Cryptoprocta ferox* (Carnivora). Analyse cladistique des caractères morphologiques de carnivores Aeluroidea actuels et fossiles. *Mammalia*, **59**, 551–82.

Viret, J. (1929a). *Cephalogale batalleri* carnassier du Pontien de Catalogne. *Bulletin de la Société d'Histoire Naturelle de Toulouse*, **58**, 567–68.

Viret, J. (1929b). Les faunes de mammifères de l'Oligocène supérieur de la Limagne Bourbonnaise. *Annales de l'Université de Lyon n.s., Sciences, Médecine*, **47**, 1–328.

Viret, J. (1951). Catalogue critique de la faune des mammifères miocènes de La Grive Saint-Alban (Isère). 1. Chiroptères, Carnivores, Édentés, Pholidotes. *Nouvelles Archives du Muséum d'Histoire naturelle de Lyon*, **4**, 1–197.

Vrana, P. B., Milinkovitch, M. C., Powell, J. R. and Wheeler, W. C. (1994). Higher level relationships of the arctoid Carnivora based on sequence data and 'total evidence'. *Molecular Phylogenetics and Evolution*, **3**, 47–58.

Wagner, A. (1858). *Geschichte der Urwelt, mit besonderer Berücksichtigung der Menschenrassen und des mosaischen Schöpfungsberichtes*, 2nd ed. Leipzig: Leopold Voss.

Wallace, S. C. and Wang, X. (2004). Two new carnivores from an unusual late Tertiary forest biota in eastern North America. *Nature*, **431**, 556–59.

Wallace, S. and Wang, X. (2007). First mandible and lower dentition of *Pristinailurus bristoli* with comments on life history and phylogeny. *Journal of Vertebrate Paleontology*, **27**, supplement to 3, 162A.

Wang, X. (1997). New cranial material of *Simocyon* from China, and its implications for phylogenetic relationships to the red panda (*Ailurus*). *Journal of Vertebrate Paleontology*, **17**, 184–98.

Wang, X., Ye, J., Meng, J., Wu, W., Liu, L. and Bi, S. (1998). Carnivora from middle Miocene of Northern Junggar Basin, Xinjiang autonomous region, China. *Vertebrata PalAsiatica*, **36**, 218–43.

Wang, X., McKenna, M. C. and Dashzeveg, D. (2005). *Amphicticeps* and *Amphicynodon* (Arctoidea, Carnivora) from Hsanda Gol Formation, Central Mongolia and phylogeny of basal arctoids with comments on zoogeography. *American Museum Novitates*, **3483**, 1–57.

Wang, X., Choudhry, A., Yonzon, P., Wozencraft, C. and Than Zaw (2008). *Ailurus fulgens. In 2008 IUCN Red List of Threatened Species*, ed. IUCN 2008. Cambridge: IUCN (www.iucnredlist.org).

Webb, S. D. (1969). The Pliocene Canidae of Florida. *Bulletin of the Florida State Museum*, **14**(4), 273–308.

Weber, M. (1904). *Die Säugetiere. Einführung in die Anatomie und Systematik der recenten und fossilen Mammalia*. Jena: Verlag von Gustav Fischer.

Weber, M. (1928). *Die Säugetiere. Einführung in die Anatomie und Systematik der recenten und fossilen Mammalia. Band II. Systematische teil*. Jena: Verlag von Gustav Fischer.

Wei, F., Feng, Z., Wang, Z. and Hu, J. (2000). Habitat use and separation between the giant panda and the red panda. *Journal of Mammalogy*, **80**, 448–55.

Wilson, D. E. and Reeder, D. M. (2005). *Mammal Species of the World*. Baltimore, MD: The Johns Hopkins University Press.

Winge, H. (1895). Jordfundne og nulevende Rovdyr (Carnivora) fra Lagoa Santa, Minas Geraes, Brasilien. Med Udsigt over Rovdyrenes indbyrdes Staegtskab. In *E Museo Lundii. En Samling af Afhandlinger om de i det indre Brasiliens Kalkstenshuler af Professor*

Dr. Peter Vilhelm Lund udgravede og i den Lundske palaeontologiske Afdeling af Kjben-havns Universitets zoologiske Museum opbevarede Dyre-og Menneskeknogler, Bind 2 (2, 4), ed C. F. Lotken. Copenhagen: H. Hagerups Boghandel.

Wolsan, M. (1993). Phylogeny and classification of early European Mustelida (Mammalia, Carnivora). *Acta Theriologica*, **38**, 345–84.

Wolsan, M. and Sato, J. (2007). Pinniped and red panda affinities elucidated using exon nucleotide sequences of five nuclear genes. *Journal of Vertebrate Paleontology*, **27**, supplement to 3, 168A.

Wood-Jones, F. (1939). The forearm and manus of the giant panda, *Ailuropoda melanoleuca*, M.-Edw. with an account of the mechanism of its grasp. *Proceedings of the Zoological Society, Series B*, **1939**, 113–29.

Wozencraft, W. C. (1989). The phylogeny of the recent Carnivora. In *Carnivore Behavior, Ecology, and Evolution*, ed. J. L. Gittleman. Ithaca, NY: Cornell University Press, pp. 495–535.

Wu, W., Ye, J., Meng, J., *et al.* (1998). Progress of the study of Tertiary biostratigraphy in North Junggar Basin. *Vertebrata PalAsiatica*, **36**, 24–31.

Wyss, A. R. and Flynn, J. J. (1993). A phylogenetic analysis and definition of the Carnivora. In *Mammal Phylogeny – Placentals*, ed. F. S. Szalay, M. J. Novacek and M. C. McKenna. New York, NY: Springer Verlag, pp. 32–52.

Yu, L. and Zhang, Y.-P. (2006). Phylogeny of the caniform Carnivora: evidence from multiple genes. *Genetica*, **127**, 65–79.

Yu, L., Li, Q.-W., Ryder, O. A. and Zhang, Y.-P. (2004). Phylogenetic relationships within mammalian order Carnivora indicated by sequences of two nuclear DNA genes. *Molecular Phylogenetics and Evolution*, **33**, 694–705.

Zhang, Y.-P. and Ryder, O. A. (1993). Mitochondrial DNA sequence evolution in the Arctoidea. *Proceedings of the National Academy of Sciences of the USA*, **90**, 9557–61.

Zhang, Z., Wei, F., Li, M. and Hu, J. (2006a). Winter microhabitat separation between giant and red pandas in *Bashania faberi* bamboo forest in Fengtongzhai Nature Reserve. *Journal of Wildlife Management*, **70**, 231–35.

Zhang, Z., Wei, F., Li, M., Zhang, B., Liu, X. and Hu, J. (2006b). Microhabitat separation during winter among sympatric pandas, red pandas, and tufted deer: the effect of diet, body size, and energy metabolism. *Canadian Journal of Zoology*, **82**, 1451–58.

Appendix 4.1 Morphological characters regarded as diagnostic or derived for *Ailurus* (or Ailuridae, or Ailurinae) in the literature. Only prominent differences with Ailuropoda and bears cited by Bardenfleth (1914) are mentioned. Cladistic analyses are given in bold.

Skull

1   Bony forehead of moderate width; Gregory (1936)
2   Alisphenoid canal present; Turner (1848), Flower (1869), Gill (1872), Mivart (1885), Lydekker in Lankester (1901), Pocock (1921), **Wyss and Flynn (1993)**
3   Foramen rotundum minute, lying beneath anterior lacerate foramen, the two separated by a thin plate of bone and sunk in a common pit; Pocock (1921)
4   Foramen oval elongate; Pocock (1921)
5   Palate prolonged behind last upper molar; Lydekker in Lankester (1901), **Wozencraft (1989), Wolsan (1993), Wang (1997), Wang *et al.* (2005)**
6   Sagittal crest moderate; Gregory (1936)
7   Zygomatic arches sharply bowed outward and downward; Gregory (1936), modified in **Wang (1997**: highly arched zygomatic arch) and **Wang *et al.* (2005**: zygomatic arch dorsally arched)
8   Postorbital process present (on the zygomatic only for Bardenfleth); Lydekker in Lankester (1901), Bardenfleth (1914)
9   Lacrimal vestigial and restricted to the area round the lacrimal foramen; **Wozencraft (1989)**
10   Inferior oblique muscle fossa closely adjacent to nasolacrimal foramen; **Wozencraft (1989)**
11   Orbital wing of the palatine reaches lacrimal, narrow contacts; **Wozencraft (1989)**
12   Space between posterior ends of palate and hamular processes of pterygoids most narrow between the palates; Bardenfleth (1914)
13   Processus postglenoideus very high, separated from the bulla by a narrow space; Bardenfleth (1914)
14   Postglenoid foramen large; Turner (1848), Flower (1869), Segall (1943)

Auditory region and surrounding processes

15   Shape of the bulla: very small, simple, nottle-shaped, inflated in its inner part; Turner (1848), Flower (1869), Gill (1872), Bardenfleth (1914), Segall (1943)
16   Long auditory tube present; Flower (1869), Gill (1872), Bardenfleth (1914), **Wolsan (1993), Wang (1997), Baskin (1998)**
17   Presence of posterolateral process of promontorium; **Wang (1997), Wang *et al.* (2005)**

18   Carotid canal large and distinct, behind the middle of the inner edge of the bulla; Flower (1869), modified in **Wyss and Flynn** (**1993**: posterior entrance of carotid posterior)

19   Carotid canal enclosed in the medial wall of the tympanic cavity, medial to or below the promontorium; Segall (1943), simplified in **Wyss and Flynn** (**1993**: carotid enclosed in a tube)

20   Posterior carotid foramen in front of and very close to posterior lacerate foramen; Segall (1943), **Wolsan (1993)**

21   Crista tympanica ends anteriorly and posteriorly by a small spine; Segall (1943)

22   No epitympanic sinus, in contrast to the Procyonidae; Segall (1943)

23   Inferior petrosal sinus in excavation in basioccipital; **Wyss and Flynn (1993)**

24   Small, but consistently present, processus muscularis for the insertion of the tensor tympani on the malleus; **Flynn et al. (1988)**

25   Paroccipital process prominent, neither flattened on the surfaces of the bulla nor laterally compressed; Turner (1848)

26   Paroccipital process long and trigonal, standing backwards and outwards, quite unconnected with the bulla, curved inwards at the extremity in old animals; Flower (1869), Gill (1872), Segall (1943), simplified in **Wozencraft** (**1989**: paroccipital process long)

27   Paroccipital process longer than mastoid process; Gregory (1936)

28   Ventrally ridged paroccipital process that is mediolaterally compressed and blade-like; **Wang (1997), Wang et al. (2005)**

29   Foramen condyloid exposed or distinct; Turner (1848), Flower (1869)

30   Well-developed mastoid process, distinct from paroccipital process; Flower (1869)

31   Mastoid processus small (relative to Ailuropoda and bears); Bardenfleth (1914), Gregory (1936), Segall (1943)

32   Mastoid process ventrally expanded; **Wang et al. (2005)**

33   Posterior base of postglenoid process not overlapping tympanic bone; Gregory (1936)

34   Hiatus subarcuatus present; Segall (1943)

35   Dorsal depression of middle ear not or not clearly divided into two depressions; Schmidt-Kittler (1981)

36   Suprameatal fossa of the external auditory meatus shallow; Schmidt-Kittler (1981), **Wozencraft (1989)**

Mandible

37   Upper face of the mandibular condyle concave; Bardenfleth (1914)

38   Level of condyle far above plane of cheek teeth; Gregory (1936)

39   Mandibular body equally thick in both ends, its inferior border convex; Bardenfleth (1914)

40   Mandibular inferior border convex; Bardenfleth (1914), Gregory (1936)

41 Symphysis rather long, not anchylosed; Bardenfleth (1914)

42 Anterior edge of coronoid inclined forwards; Pocock (1921), **Wang (1997), Baskin (1998), Wang et al. (2005)**

Dentition

43 Presence of lateral grooves on canines; **Wang (1997)**; mentioned but not used as diagnostic in earlier studies.

44 2M/2m or loss of m3; Turner (1848), Gill (1872), Mivart (1885), Lydekker in Lankester (1901), Bardenfleth (1914), Ginsburg (1982), **Flynn et al. (1988)**

45 Molar very complexly tuberculate; Mivart (1885)

46 M2 broader than long, shorter than M1; Bardenfleth (1914)

47 M2 enlarged; **Wang et al. (2005)**

48 Enlargement of labial cusps on M1–2; Ginsburg (1982)

49 Enlargement of the upper molar conules (especially the metaconule); **Flynn et al. (1988)**, modified in **Wozencraft (1989**, presence of hypocone in M1 and M2), modified in **Wyss and Flynn (1993**, hypocone in M2); modified in Ginsburg (1999, metaconule of M1–2 very important); **Wallace and Wang (2004**, at least in M1)

50 M2 three-rooted and subequal in size to P4; **Wolsan (1993)**

51 Anterior and posterior cingula of M1 continuous around lingual base of protocone; **Wolsan (1993)**

52 Inner portion of P4 formed by two distinct cusps, the protocone and hypocone; Lydekker in Lankester (1901), modified in Ginsburg (1999: hypocone of P4 very important)

53 P4 as large as M1, protocone and hypocone form more than half of the tooth, they are supported by a very strong root which is placed mainly under the mesial cusp; Bardenfleth (1914), modified in **Wolsan (1993**: protocone and hypocone prominent and subequal in size)

54 P4 protocone conical, not formed by cingulum entirely; **Wolsan (1993)**

55 Widening of P3–M2; Ginsburg (1999)

56 P2 and P3 large and three-rooted; Bardenfleth (1914), Pocock (1921)

57 P3 with 5 cusps and closely resembling P4; Pocock (1921)

58 Molarisation of P3 and P4; Ginsburg (1982); modified in Ginsburg (1999: hypocone of P3–4 very important) 59 – Lost of P1; Bardenfleth (1914), **Wolsan (1993)**

60 p1 single-rooted to absent; **Wolsan (1993)**

61 p3 with one rather blunt cusp; Bardenfleth (1914)

62 p4 with 5 cusps; Bardenfleth (1914)

63 m1 entoconulid poorly differentiated (ridge-like or cuspule-like); **Wolsan (1993)**

64 Elongation of m2; Ginsburg (1982), **Wang et al. (2005)**

65 Elongation of the talonid of m2, enlargement of the hypoconulid; **Flynn *et al.* (1988), Baskin (1998), Wang *et al.* (2005)**

66 Enlargement of m2 hypoconulid; **Flynn *et al.* (1988), Baskin (1998)**

67 Entoconid and entoconulid of m2 prominent, cusp-like; **Wolsan (1993)**

68 Protoconid and metaconid of m2 elongated and narrow, separated by a longitudinal trough that reaches the distal border of the tooth; Ginsburg *et al.* (1997)

Postcranial

69 Scapula small; the suprascapular border almost not extending beyond the ridge running along the glenoid border; Bardenfleth (1914)

70 Postscapular fossa present; **Wozencraft (1989)**

71 Upper face greater tubercle of humerus somewhat oblique; Bardenfleth (1914)

72 Supinator ridge not very prominent; Bardenfleth (1914)

73 Entepicondyle flat, somewhat expanded; Bardenfleth (1914)

74 Entepicondylar foramen on humerus; De Blainville (1841), Lankester (1901), Lydekker in Lankester (1901)

75 Strongly marked entepicondylar ridge; Lankester (1901)

76 Inner crest of trochlea (surface of ulna) very little prominent; Bardenfleth (1914)

77 Preaxial malleolus of radius forms a short point; Bardenfleth (1914)

78 Baculum more developed than bears; De Blainville (1841)

79 Baculum short; Pocock (1921)

80 Extra-carpal bone or radial sesamoid moderate or small (relative to that of the giant panda); Lankester (1901), Bardenfleth (1914)

81 Femur rather long and slender; Bardenfleth (1914)

82 Feet plantigrade; Lydekker in Lankester (1901)

83 Tail long; Lydekker in Lankester (1901)

Soft characters

No caecum; Turner (1848)

No Cowper's gland; Turner (1848)

Rhomboid area visible in the front region of the brain; Bardenfleth (1914)

General form of the brain mostly procyonoid; Bardenfleth (1914)

Cerebrellum mostly overlapped by cerebrum; Bardenfleth (1914)

Prepuce close to scrotum; Pocock (1921)

Pads of feet reduced and functionless, completely concealed by woolly hair; Pocock (1921)

Carpal pad remote from planter pad; Pocock (1921)

Anus in centre of glandular depressed area; Pocock (1921)

Absence of major a4 arterial shunt; **Wyss and Flynn (1993)**

5

The influence of character correlations on phylogenetic analyses: a case study of the carnivoran cranium

ANJALI GOSWAMI AND P. DAVID POLLY

Introduction

Character independence is a major assumption in many morphology-based phylogenetic analyses (Felsenstein, 1973; Emerson and Hastings, 1998). However, the fact that most studies of modularity and morphological integration have found significant correlations among many phenotypic traits worryingly calls into question the validity of this assumption. Because gathering data on character correlations for every character in every taxon of interest is unrealistic, studies of modularity are more tractable for assessing the impact of character non-independence on phylogenetic analyses in a real system because modules summarise broad patterns of trait correlations. In this study, we use empirically derived data on cranial modularity and morphological integration in the carnivoran skull to assess the impact of trait correlations on phylogenetic analyses of Carnivora.

Carnivorans are a speciose clade of over 270 living species, with an extremely broad range of morphological and dietary diversity, from social insectivores to folivores to hypercarnivores (Nowak, 1999; Myers, 2000). This diversity offers many opportunities to isolate various potential influences on morphology, and, in this case, to study the effects of trait correlations on cranial morphology. Carnivorans also have an excellent fossil record, providing the opportunity to examine morphologies not represented in extant species, such as in the sabre-toothed cat *Smilodon*. Perhaps most importantly, several recent molecular and morphological studies of carnivoran phylogeny (Hunt and Tedford, 1993; Wyss and Flynn, 1993; Tedford *et al.*, 1995; Flynn and Nedbal, 1998; Flynn *et al.*, 2000, 2005; Flynn and Wesley-Hunt, 2005; Wesley-Hunt and Flynn, 2005; Flynn *et al.*, this volume) provide the necessary resolution to assess the influence of character correlations on morphology-based phylogenetic analyses.

Carnivoran Evolution: New Views on Phylogeny, Form, and Function, ed. A. Goswami and A. Friscia. Published by Cambridge University Press. © Cambridge University Press 2010.

Here, we present morphometric analyses of 47 species (38 extant and 9 fossil), representing 44% of extant genera, and 15% of extant species, and including all extant terrestrial families and the extinct families Nimravidae and Amphicyonidae. Using both simulations and empirically derived data, we test the following specific questions: (1) Do individual modules differ in the relationship between shape and phylogenetic relatedness? (2) Do individual modules differ in the relationship between similarity of pattern of integration and phylogenetic relatedness? (3) Do highly correlated characters show significantly more coordinated shifts in discrete character states than do uncorrelated characters? (4) Have correlated characters significantly misled previous phylogenetic analyses of Carnivora based on morphology?

Integration and modularity

The idea that the skull is composed of a series of autonomous 'functional components' dates to van der Klaauw (1948–1952) and has since become an important framework for examining the evolution of cranial morphology in mammals (Moss and Young, 1960; Schwenk, 2001). The concept of independent evolutionary units, however, has appeared in many forms before and since then. Developmental studies in the early twentieth century focused on morphogenetic fields and their evolutionary importance as 'discrete units of embryonic development', an idea contested at the time by geneticists who argued that the gene is the primary unit of evolutionary significance (Gilbert *et al.*, 1996). Decades later, and with the emergence of evolutionary developmental biology, it is clear that aspects of both positions may be valid. Structures and processes as diverse as signalling pathways and colonial individuals have been reasonably described as independent units of evolutionary change (Schlosser and Wagner, 2004). Yet, despite the early recognition of evolutionary 'parts' in genetic, developmental, and morphological systems, it is only in recent years that these fields have begun exploring the relationships among these different scales. The study of modules, autonomous subsets of highly correlated traits within larger systems of any type, and its application to understanding diverse biological systems (Schlosser and Wagner, 2004), thus may herald a new, more inclusive synthesis of evolutionary theory.

For morphologists and paleontologists, this emergence of modularity is particularly important, because the quantitative methods used to identify modularity can be applied equally to living, extinct, or rare taxa. Perhaps the first quantitative examination of phenotypic trait relationships can be attributed to Olson and Miller (1951), expounded in their book *Morphological Integration* (1958). Their argument was a simple one: many trait changes that occur during the course of evolution do not occur independently of each other. More specifically, traits that are related by proximity in development or function have greater influence on each other than on more distant traits.

Trait associations potentially influence evolutionary paths in many ways, from constraining the variability of individual traits to facilitating transformations of functional sets (Olson and Miller, 1958; Vermeij, 1973; Atchley and Hall, 1991; Cheverud, 1996b; Wagner, 1996; Wagner and Altenberg, 1996; Emerson and Hastings, 1998; Bolker, 2000; Polly, 2005; Goswami and Polly, 2010). Thus, integration and modularity have been tied to some of the most fundamental and interesting questions in morphological evolution, including evolvability and constraints on morphological variation, the generation of novelties, and the production of morphological diversity (Vermeij, 1973; Wagner, 1995; Cheverud, 1996b; Wagner, 1996; Wagner and Altenberg, 1996; Chernoff and Magwene, 1999; Polly et al., 2001; Eble, 2004; Shubin and Davis, 2004).

Integration involves linked interactions among traits, whereas modularity emphasises the autonomy of units. In a sense, integration and modularity can be taken as antagonistic forces, because, when applied to the same structure or process, they describe the opposite relationships among characters. However, both integration and modularity are structured in a hierarchical framework. Modules are autonomous from other modules, but the elements that compose them are highly integrated within themselves. Likewise, integration of genetically, developmentally, or functionally related traits implies autonomy from unrelated traits. Units that are modular or autonomous may, and in most cases must, interact with other units within the larger system. This implicit inverse relationship between the effects of integration and modularity is central to their potential importance to the evolutionary process.

Total independence among traits would allow each trait to vary independently and to respond to selection pressures in an optimal way. Correlations among traits may limit the variation of any individual trait by necessitating a coordinated response from several traits, perhaps preventing any one trait from responding optimally to selection. Conversely, functional or developmental units that require coordination among traits would suffer from complete independence among traits (in a sense, all traits independently have the same selective optimum).

Wagner and Altenberg (1996) proposed an evolutionary mechanism for modifying the relationships among traits: new modes of integration arise to link traits involved in new functional or developmental interactions, while new modularity (parcellation or fragmentation) decouples previously restrictive relationships (Figure 5.1). Some researchers have also hypothesised that modularity has generally increased during the course of evolution to circumvent

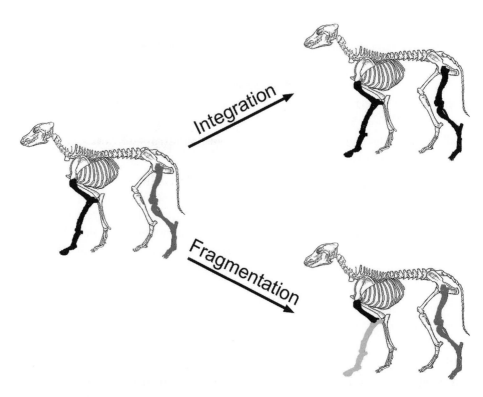

Figure 5.1 For colour version, see Plate 6. The two paths by which modules may evolve: integration, where ancestrally independent modules evolve strong correlations; or fragmentation, where ancestrally correlated traits become independent of each other, shown in the postcranial skeleton of a dog. Elements shaded with the same colour are integrated.

canalisation, the evolution of developmental constraints as systems become more complex, and its genetic counterpart, pleiotropy (Vermeij, 1973; Wagner and Altenberg, 1996). Because fragmentation of parts increases the scope for each part to vary and respond to selection, many have considered fragmentation to increase the 'adaptability' or 'evolvability' of organisms.

The breadth of studies of morphological integration has extended greatly since the publication of Olson and Miller's (1958) book, and the diversity of research in integration is apparent from contents of recent published collections (Pigliucci and Preston, 2004; Schlosser and Wagner, 2004). In recent years, morphological integration has been empirically or theoretically tied to quantitative genetics, molecular pathways, novelty, life-history strategies, and macroevolutionary trends (for recent reviews, see Pigliucci and Preston, 2004; Schlosser and Wagner, 2004).

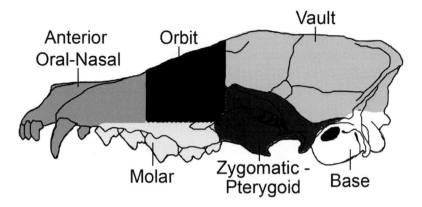

Figure 5.2 For colour version, see Plate 7. The six morphometrically derived cranial modules (Goswami, 2006a) upon which analyses of discrete character evolution are based.

The skull is a particularly good system to test for morphological integration and modularity, as it is a unified structure, yet is also both functionally and developmentally complex. The skull serves several functions (Moss and Young, 1960; Schwenk, 2001), from feeding and respiration, to housing the sensory organs and protecting the brain. Developmentally, in mammals it arises from two major tissues, the neural crest and the paraxial mesoderm, and is composed of both dermal and endochondral bones (Thorogood, 1993). The complexity of the skull thus provides many potential functionally or developmentally integrated units for assessing morphological integration, modularity, and their evolutionary significance (Atchley *et al.*, 1982; Cheverud, 1982, 1988, 1989, 1995, 1996a,b; Zelditch, 1988; Zelditch and Carmichael, 1989a, 1989b; Steppan, 1997; Ackermann and Cheverud, 2000, 2004; Badyaev and Foresman, 2000, 2004; Marroig and Cheverud, 2001; Zelditch *et al.*, 2001; Marroig *et al.*, 2004; Goswami, 2006a, 2006b, 2007a, 2007b).

Several recent studies have focused on modularity and integration in the carnivoran cranium (Goswami, 2006a, 2006b; Goswami and Polly, 2010). One study demonstrated that patterns of phenotypic modularity are strongly conserved in the cranium of carnivorans (Goswami, 2006a). Morphometric analyses of 3D cranial landmarks identified six sets of traits that were consistently recovered in the examined species (Figure 5.2): anterior oral–nasal; molar; orbit; zygomatic–pterygoid; vault; and basicranium. Correlations among traits that were not in the same cluster were consistently zero or not significantly different from zero. While all of the six groups of traits fulfilled the practical definition of phenotypic modularity, having significantly stronger correlations within the module than across modules in at least some taxa. However, only three modules (anterior oral–nasal, molar, and basicranial) were significantly integrated in most taxa. In contrast, the orbit, zygomatic–pterygoid region, and cranial vault were not integrated in most taxa.

Correlated characters and phylogeny analysis

Modularity and integration have important consequences. They describe the correlated evolution of characters, and character independence is a well-known requirement of phylogenetic analysis (Kluge and Farris, 1969; Felsenstein, 1973, 1985; Kluge, 1989; Kluge and Wolf, 1993; Kangas et al., 2004). Correlated characters cheat the parsimony algorithm by causing the same underlying evolutionary change to be counted more than once, spuriously increasing the signal-to-noise ratio. If character correlations are pervasive, treating characters as independent may mislead interpretations of phylogenetic relationships among taxa. However, determining when two discrete characters are correlated can be difficult because the limited number of character states combined with the fairly small number of taxon observations in most data sets leave very little statistical power to detect a correlation.

Because of the great potential of correlated character evolution to skew phylogenetic analyses, many studies have focused on estimating the effects of correlated characters on tree topologies, tree lengths, and tree support (Wagner, 1998; Huelsenbeck and Nielsen, 1999; Sadleir and Makovicky, 2008) and on identifying correlated characters from character distributions or character matrices (Read and Nee, 1995; Maddison, 2000; O'Keefe and Wagner, 2001). One of the most conservative methods considers characters that have identical state distributions (Harris et al., 2003). Perfectly correlated characters are qualitatively evaluated for anatomical, developmental, or functional links suggesting that the correlation is due to biological interaction, in which case one of the characters is dropped or the two are recoded as a single composite character. While this method is unlikely to mistakenly conflate two uncorrelated characters, it will miss characters with an underlying and more subtle biological correlation, as can be ascertained qualitatively or with statistical analysis of continuous quantitative data, but whose discrete character states are not identical. A less conservative method uses principal coordinates analysis (PCO) to confirm correlations between characters that do not have identical state distributions (Naylor and Adams, 2001). Like the method of Harris et al. (2003), potentially correlated characters are first identified on the basis of anatomical, developmental, or functional criteria and then quantitatively assessed for whether they group in PCO space. The multivariate PCO space is derived from a pairwise character distance matrix such that characters whose states are distributed similarly across taxa will cluster together. A close clustering is interpreted as supporting the hypothesis that the characters are correlated, whereas a significantly more distant clustering is interpreted as falsifying that hypothesis.

A consistent drawback in most existing studies examining the effect of correlated characters on phylogenetic analyses is that they do not use an independent measure of character correlations, or rigorously identify correlated characters a priori. Here, we used the observed differences in the cranial modules of the carnivoran skull and the quantitatively derived correlations among cranial traits, described above, to address whether correlated characters influence phylogenetic analyses of Carnivora. First, we examined whether there are differences among the modules in the relationship between phylogeny and module shape, thereby testing whether some cranial modules better reflect phylogenetic relationships among carnivorans. We also expanded the previous studies of modularity and integration in the carnivoran skull, combining the topics discussed above to establish whether the six cranial modules differ in the relationship between phylogenetic relatedness and within-module similarity in morphological integration. We used both methods described above to assess the effects of empirically derived trait correlations on the distribution of discrete character states in Carnivora, first assessing the power of the two methods using Monte Carlo simulations. Lastly, we examined previous morphology-based phylogenetic analyses of Carnivora to assess whether the focus on basicranial and molar traits is justified or has consistently misled interpretations of the relationships among carnivorans.

Methods

Phylogenetic signal in module shape and integration

Specimens

Three-dimensional landmark data were gathered with an Immersion Microscribe G2X 3-D digitiser. Fifty-one landmarks were gathered from across the skull (Figure 5.3) from a total of 744 specimens, representing 47 species (9 extinct, 38 extant; Table 5.1). Landmarks were distributed across the skull and are assigned to one of the six modules based on previous study of correlations (Goswami, 2006a).

Module shape

To test if modules differ in their relationship to phylogeny, each module was oriented across all 47 taxa with Generalised Procrustes Analysis, and partial

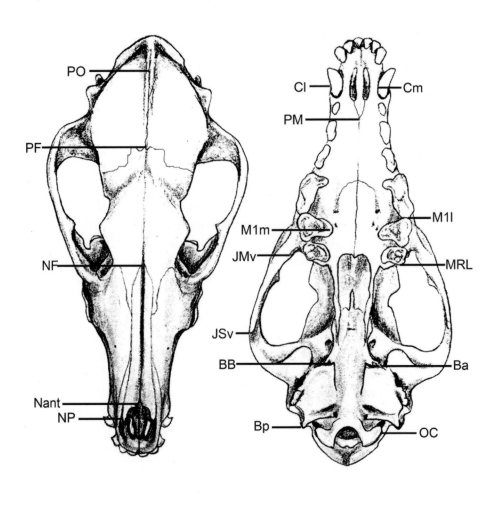

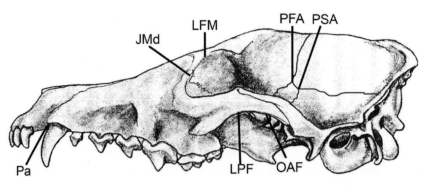

Figure 5.3 The 51 3D landmarks used in the analyses of shape disparity and integration. Symmetrical landmarks are represented by two numbers and shown on one side only.

Table 5.1 List of species and numbers of specimens used in analyses.

Suborder	Family	Species
Caniformia	Amphicyonidae	*Daphoenus* sp.* 11
	Canidae	*Hesperocyon* sp.* 13
		Mesocyon sp.* 12
		Canis lupus 18
		*Canis dirus** 20
		Cerdocyon thous 18
		Otocyon megalotis 16
		Vulpes vulpes 22
	Ursidae	*Ursus americanus* 14
		Melursus ursinus 15
		Tremarctos ornatus 15
		Ailuropoda melanoleuca 15
	Ailuridae	*Ailurus fulgens* 16
	Mephitidae	*Mephitis mephitis* 15
		Spilogale putorius 17
	Procyonidae	*Procyon lotor* 18
		Procyon cancrivorous 18
		Potos flavus 20
		Nasua nasua 15
	Mustelidae	*Melogale personata* 15
		Meles meles 15
		Enhydra lutris 15
		Martes pennanti 15
		Taxidea taxus 15
		Gulo gulo 16
Feliformia	Nimravidae	*Hoplophoneus* sp.* 19
		Dinictis sp.* 19
	Nandiniidae	*Nandinia binotata* 16
	Felidae	*Acinonyx jubatus* 15
		Lynx rufus 16
		Felis viverrina 15
		Felis bengalensis 18
		*Panthera atrox** 11
		*Smilodon fatalis** 20
	Viverridae	*Paradoxurus hermaphroditus* 19
		Civettictis civetta 15
		Genetta genetta 20
	Eupleridae	*Eupleres goudotii* 12
		Cryptoprocta ferox 13
		Fossa fossana 15
		Galidia elegans 15

Table 5.1 (*cont.*)

Suborder	Family	Species
	Herpestidae	*Cynictis penicillinatus* 15
		Herpestes ichneumon 21
		Ichneumia albicauda 15
	Hyaenidae	*Proteles cristatus* 15
		Crocuta crocuta 18
		Thalassictis sp.* 13

Note: *Indicates extinct species.

Procrustes distance was calculated for each pair of species. This quantification was repeated for each of the six cranial modules (Figure 5.2), resulting in six matrices of module distance across all 47 taxa. A patristic distance matrix was constructed using recent phylogenetic analyses (Figure 5.4), primarily based on molecular data for Recent taxa (Flynn *et al.*, 2005; Wesley-Hunt and Flynn, 2005). The six matrices of module distance were each compared to the patristic distance matrix using matrix correlation analysis with Mantel's test (10,000 repetitions) for significance.

Module integration
To test if the patterns of integration within modules differ in their relationship to phylogeny, correlation matrices were generated for each of the six modules (Figure 5.2) for each species. A matrix of similarity of integration (MSI) for each module was generated by pairwise matrix correlation analysis of species-specific correlation matrices. The six module MSIs were then compared to the patristic distance matrix using matrix correlation analysis with Mantel's test for significance.

Monte Carlo simulations

We assessed the power of existing methods for identifying correlation in character matrices using Monte Carlo simulations. We simulated character state evolution using a threshold model in which the state would change depending on the change in an underlying continuous variable (Otto and Day, 2007). A state change was triggered when the underlying continuous change was greater than a threshold value. Continuous changes were drawn from normal distributions, each with a mean of 0.0 and standard deviation of 1.0. The probability of state changes per step was controlled by setting the threshold to the appropriate number of standard deviations above or below 0.0.

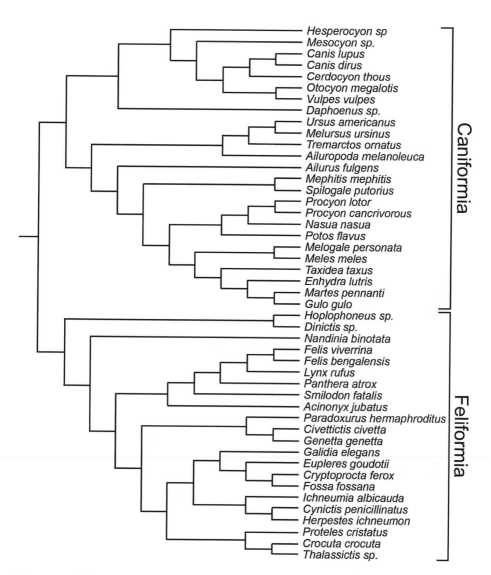

Figure 5.4 The phylogenetic tree for Carnivora (Flynn *et al.*, 2005; Flynn and Wesley-Hunt, 2005) that provided the model for the Monte Carlo simulations of discrete character evolution.

One random number was selected per character per step, yielding a *k* length vector **r** of random changes at each step, where *k* is the number of characters.

Correlations were introduced by dividing characters into blocks associated with the six cranial modules described above and imposing the corresponding module correlation onto the underlying continuous random variables for each block. The module correlations were empirically derived from the same carnivoran taxa as used in this study (Goswami, 2006a). The following mean correlations were used for each module: Anterior Oral–Nasal, 6 characters, $r = 0.73$; Molar, 5 characters, $r = 0.47$; Orbit, 5 characters, $r = 0.37$; Zygomatic–Pterygoid, 8 characters, $r = 0.40$; Cranial Vault, 4 characters, $r = 0.40$; Basicranium, 4 characters, $r = 0.64$ (Figure 5.2). Correlations between traits from different modules were all set at 0.

To impose the empirical character correlations onto the continuous random variables, the Cholesky decomposition $\mathbf{G}$ of a $k \times k$ matrix of pairwise correlation coefficients (where k is the number of characters being simulated) was multiplied by the k length vector $\mathbf{r}$ of random changes in the continuous traits as follows to give the k length vector $\mathbf{r}^*$ of correlated random changes: $\mathbf{r}^* = \mathbf{r} \cdot \mathbf{G}$. Character state changes were assessed by applying the threshold criterion to $\mathbf{r}^*$. Note that even strong correlation in the underlying continuous variables does not necessarily result in perfect correlation among discrete character state changes (Figure 5.5a).

Character evolution was simulated on a tree with 47 tips (Figure 5.4), corresponding to taxa in which character correlations were studied in previous analyses (Goswami, 2006a), and the same topology as recent phylogenetic analyses of Carnivora (Flynn et al., 2005; Flynn and Wesley-Hunt, 2005). Each simulation started at the base of the tree with all characters in the ancestral state 0 (Figure 5.5b). The simulation proceeded along each branch of the tree with character states changing randomly as determined by the threshold and character correlations. The simulations were run using a punctuational and anagenetic model of evolution. In the punctuational model, there was only one chance for character state change along each branch; in the anagenetic model, there were 100 chances for change. Two consequences of the anagenetic model are that reversals can erase character transformations that occur along a single branch and there is a higher probability of independent changes in characters that are correlated.

In addition to varying the number of opportunities for characters to change, we varied the probability of change, from equal (branching probability $b = 0.5$), high ($b = 0.9$), and low ($p = 0.1$), for a total of six simulations. Each simulation was repeated 200 times.

The effect of the underlying correlations on the character state matrix in the simulations was measured with three statistics. The first statistic was the proportion of correlated characters with identical character state distributions across the tip taxa. This metric is related to the Harris et al. (2003) method, which used identical distributions of states as confirmation of underlying character correlation.

The second statistic was the mean pairwise distance between correlated and uncorrelated characters. Even without a perfect correlation in character state

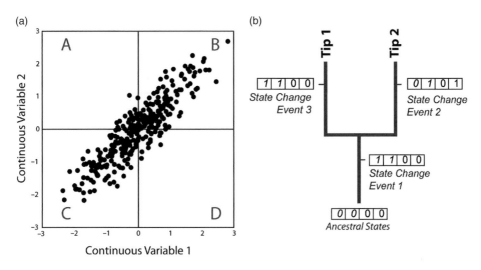

Figure 5.5 (a) Diagram showing relationship between the underlying continuous change and state change for a pair of characters when the underlying correlation is 0.9 and the threshold for state change is 50%. One hundred random changes in two correlated continuous variables are shown as points on the graph. Perfectly correlated character state distributions occur when the continuous points lie in quadrant C (no change in either character) or B (change in both characters). Seemingly uncorrelated character state changes occur when the continuous points lie in quadrant A (change in character 2, but not character 1) or D (change in character 1, but not character 2). (b) Diagram of how character state evolution was simulated, shown here with four characters and two tip taxa. The first pair of characters in this diagram has an underlying correlation and the second pair does not. Each simulation starts with all character states in the ancestral condition of 0. At each state change event (one per branch for punctuated simulations, 100 per branch for anagenetic simulations) a random change in the underlying triggers changes in the character states if the continuous change exceeds the threshold.

changes, it can be expected that correlated characters will be more similar to one another than are uncorrelated characters.

The third statistic was the mean distance of correlated and uncorrelated characters in PCO space. This metric was used by Naylor and Adams (2001) to assess whether potentially correlated characters are truly correlated. To calculate the mean PCO distances, characters were projected into PCO space by calculating a $k \times k$ pairwise squared distance matrix, where k is the number of characters, converting it to a similarity matrix by multiplying the squared distances by -0.5, double-centring it by subtracting the mean column and mean row values and adding the matrix mean, and calculating eigenvectors from the double-centred matrix using singular value decomposition (Gower,

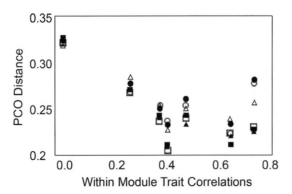

Figure 5.6 PCO distance plotted against mean within-module character correlations for the six Monte Carlo simulations: open symbols, anagenetic; closed symbols, punctuated; circles, equal probability of change; squares, high probability; triangles, low probability.

1966). Because PCO is a Q-mode analysis, the elements of the eigenvectors are the scores of the characters in PCO space. Mean pairwise distances on the first two axes between correlated and uncorrelated characters in the space were calculated. Mean and standard deviations from the 200 repetitions of each of the 6 simulations are reported for each measure.

We used these statistics to determine whether module correlations are likely to have an adverse effect on the character state matrix: when the proportion of identical state distributions was, on balance, higher in correlated than uncorrelated characters, when the mean pairwise character distance was, on balance, smaller in correlated than uncorrelated characters, and when the mean PCO distance was, on balance, smaller in correlated than uncorrelated characters.

Character distributions in previous phylogenetic analyses

In order to assess whether character correlations may actively have had an effect on previous phylogenetic analyses of Carnivora, we classified clades as valid or invalid and determined whether characters supporting invalid clades are predominantly correlated in modules. We first tabulated the characters supporting each clade for the most extensive recent phylogenetic analysis based solely on morphological characters (Wyss and Flynn, 1993). We compared the clades identified in that study with more recent studies (Flynn and Nedbal, 1998; Yoder et al., 2003; Flynn et al., 2005; Wesley-Hunt and Flynn, 2005), including several molecular studies using up to six mitochondrial and nuclear genes. We presumed that the later studies are more correct than the earlier morphological study, and the Wyss and Flynn (1993) clades (hereafter WF) were classified as valid if

Table 5.2 Correlation between similarity in shape, similarity in integration, and phylogenetic relatedness. $^*p < 0.05$, $^{**}p < 0.01$.

Module	Carnivora		Caniformia		Feliformia	
	Shape	Integration	Shape	Integration	Shape	Integration
Ant. Oral–Nasal	0.28**	0.16	0.54*	0.17	0.40	0.24
Molar	0.43**	0.10	0.57*	0.18	0.42	0.16
Orbit	0.38**	0.13	0.40	0.17	0.57**	0.19
Zyg–Pter	0.24**	0.23*	0.56**	0.29	0.52*	0.35
Vault	0.37**	0.11	0.46*	0.22	0.45	0.14
Base	0.53**	0.24**	0.69**	0.36**	0.53*	0.37
All	0.46**	0.17**	0.69**	0.36	0.55*	0.33**

upheld by more recent studies, or invalid if no longer considered to be a monophyletic group. We tabulated the characters supporting each clade and binned them into one of the six cranial modules described above. If character correlations have significantly misled this morphological phylogenetic analysis, then characters supporting an invalid clade are expected to represent fewer modules than those supporting valid clades. We calculated total character support and module range (the number of modules represented by characters) for each clade and compared these measures between valid and invalid clades.

Results

Phylogenetic signal in module shape

When module shape distance and patristic distance were compared across all carnivorans, all six modules showed significant correlations at the $p = 0.01$ level (Table 5.2). Total cranial shape (incorporating all 6 modules) was also significantly correlated with phylogenetic relatedness ($p < 0.01$). When analyses were conducted within Caniformia, all but the orbit module were significantly correlated with patristic distance at the $p = 0.05$ level, but only the zygomatic–pterygoid and basicranium were significant at the $p = 0.01$ level. Conversely, within Feliformia only the orbit was significantly correlated with patristic distance at the $p = 0.01$ level, while the zygomatic–pterygoid and basicranium were significantly correlated at the $p = 0.05$ level. Overall, caniforms showed stronger correspondence between cranial shape and phylogenetic relationship than do feliforms, and the zygomatic–pterygoid and basicranium were the only modules significantly correlated with patristic distance in both Feliformia and Caniformia.

Phylogenetic signal in module integration

It has previously been demonstrated that similarity in the pattern of integration across the whole skull is significantly correlated with phylogenetic relatedness in Carnivora ($p < 0.05$) and Feliformia ($p < 0.01$), but not in Caniformia (Table 5.2). When individual modules are considered separately, only the basicranium showed a significant correlation between similarity in pattern of integration and phylogenetic relatedness across Carnivora ($p < 0.01$) and within Caniformia ($p < 0.01$), but not within Feliformia.

Monte Carlo simulations

Identical character state scores

In only 3 of the 36 analyses were there any cases of identical character state scores for highly correlated characters. All of these cases were in the highly integrated anterior oral–nasal module, with a punctuational model and high probability of change (3.33% of 200 runs), a punctuational model and low probability of change (10%), and an anagenetic model and high probability of change (6.67%). All of the other modules failed to produce even a single instance of identical character state scores.

Mean pairwise character distance

In none of the 36 simulations was mean pairwise character distance significantly different between correlated characters within the six modules than between uncorrelated characters spanning the modules.

PCO distance

In all 36 cases, the mean PCO distance between uncorrelated characters was significantly greater than between correlated characters (Table 5.3). However, there were not significant differences among the modules, despite a large range of magnitude of mean within-module correlations (Figure 5.2).

Character distributions in previous phylogenetic analyses

Out of 31 clades within Carnivora that were identified in Wyss and Flynn (1993), 9 are not supported in more recent analyses (Miacinae, Viverravinae + Carnivora, Felidae + Hyaenidae, Viverride, Procyonidae + Ursdia + Ailurus, Ailurus + Ursida, Ursida, Mustelidae, Mephitidae + Lutrinae). Fifty-one characters used in Wyss and Flynn (1993) can be assigned to one of the 6 cranial modules. Because of homoplasy and multistate characters, the 31 clades were supported by a total of 129 apparent synapomorphies. Forty-four of these

Table 5.3. Mean PCO distances and standard deviations (s.d.) for each module and for uncorrelated traits for the six Monte Carlo simulations. *b*, probability of character changing, ranging from low (0.1) to high (0.9).

Module	Punctuation					
	$b = 0.5$	s.d.	$b = 0.9$	s.d.	$b = 0.1$	s.d.
Ant. Oral–Nasal	0.279	0.052	0.227	0.065	0.225	0.067
Molar	0.257	0.055	0.240	0.059	0.233	0.062
Orbit	0.250	0.049	0.242	0.056	0.240	0.061
Zyg.–Pterygoid	0.279	0.040	0.270	0.046	0.271	0.047
Vault	0.231	0.057	0.210	0.063	0.208	0.072
Basicranium	0.234	0.058	0.210	0.064	0.221	0.066
Uncorrelated	0.325	0.013	0.326	0.014	0.326	0.013

	Anagenesis					
	$b = 0.5$	s.d.	$b = 0.9$	s.d.	$b = 0.1$	s.d.
Ant. Oral–Nasal	0.277	0.053	0.230	0.063	0.257	0.054
Molar	0.254	0.050	0.239	0.054	0.252	0.057
Orbit	0.254	0.052	0.238	0.064	0.253	0.058
Zyg.–Pterygoid	0.279	0.039	0.266	0.043	0.284	0.039
Vault	0.236	0.059	0.202	0.070	0.229	0.066
Basicranium	0.238	0.059	0.224	0.062	0.239	0.061
Uncorrelated	0.323	0.011	0.326	0.015	0.324	0.013

supported invalid clades, and 105 supported valid clades. Characters were very unevenly distributed across the skull. Of the character support, 59.7% was derived from the molar region, and 28.7% was basicranial. The remaining characters were divided between the orbit (4.7%), zygomatic–pterygoid (5.4%), and anterior oral–nasal (1.6%). There were no characters from the cranial vault. Of the 31 clades, 2 were not supported by any cranial characters (only postcranial), 9 were supported by characters from a single module (ranging from 1 to 5 total character support), 13 were supported by 2 modules (2 to 8 total character support), 8 were supported by 3 modules (4 to 11 total character support), and 1 clade, Phocoidea, was supported by 7 characters from 5 modules. Module representation for invalid clades ranged from one to three, with character support ranging from one to seven. Module representation for valid clades ranged from 1 to 5, with total character support ranging from 1 to 11. Although the clades with the most character support and broadest module representation were supported by more recent molecular phylogenetic analyses, there were no significant differences between invalid and valid clades in module representation or total character support.

Discussion

It appears that character correlations may well have affected morphological phylogenetic analyses of Carnivora. Our simulations of correlated character evolution, using empirically derived character correlations (Goswami, 2006a) suggest that simply identifying characters with identical character state scores across taxa will underestimate the number of correlated characters. However, PCO distances were significantly greater among uncorrelated than correlated characters, demonstrating that character correlations are affecting character state changes across complex phylogenies and a range of evolutionary models. Even the most weakly integrated modules, with relatively low, but non-zero, correlations among traits, were significantly closer than were uncorrelated characters.

Tabulating character distributions in a large-scale morphological analysis of carnivoran phylogeny demonstrated that cranial characters are overwhelmingly concentrated in the molar and basicranial regions. Of course, this is not a new observation, and has been well appreciated in many previous studies (Flynn and Wesley-Hunt, 2005). However, from the analyses presented here, there is no evidence that this concentration on only a few regions of highly correlated characters has significantly misled phylogenetic analyses. Clades supported by characters from several modules were found to be invalidated by recent molecular analyses as often as clades only supported by characters from a single module. Furthermore, our analysis of phylogenetic signal in cranial module shape demonstrated that the basicranium and the zygomatic–pterygoid, which includes some of the anterior basicranium, have the strongest phylogenetic signal. As the phylogenies used for these analyses are predominantly based on molecular data, this does not simply reflect the fact that the major divisions within Carnivora are based on basicranial morphology. Feliformia and Caniformia are identified by their distinct bullar morphologies, which is included in the basicranial module. Understandably, recent and ongoing studies of the stem carnivoran groups Viverravidae and 'Miacoidea' also focus on basicranial morphology to untangle the relationships of these enigmatic taxa. The strong phylogenetic signals of basicranial and zygomatic–pterygoid shape shown in this study support the reliance on basicranial characters in morphological analyses of carnivoran phylogeny. However, the potential for correlated characters to display coordinated character state changes urges caution in basing phylogenetic analyses on characters from only a single module.

A relevant debate on the selection and atomisation of character has been occurring among phylogenetic systematists for decades (Rieppel and Kearney, 2002), alongside related debates on the selection pressures and lability of certain types of characters and levels of homoplasy in different systems (Sanchez-Villagra and Williams, 1998; Williams, 2007). As Rieppel and Kearney (2002) note, many morphological phylogenetic analyses focus on increasing the quantity of characters, rather than on increasing the quality of characters. In fact, more complex suites of characters may serve as better representatives of discretely evolving traits (Strait, 2001; Gonzáles-José et al., 2008), but it is difficult to determine the boundaries of such biological units. Because modules may well be stable across large clades (Goswami, 2006a), they can provide a more practical way to assess whether over-emphasis of a cranial region or atomisation of a module does negatively influence phylogenetic analyses, particularly in studies involving large amounts of fragmentary fossil material or in clades, such as Carnivora, where great emphasis is placed on a few anatomical regions.

While shape is the most obvious aspect of a module to consider, the relationships among traits within a module are flexible and may well change over evolutionary time, even if the actual composition of the module is stable (Goswami, 2006a, 2006b, 2007a). The basicranium was the only module to show any phylogenetic signal in its patterns of integration, and only when compared across all Carnivora and within Caniformia. Feliformia, which showed the strongest phylogenetic signal in whole-cranium integration (Goswami, 2006b) did not show significant phylogenetic signal in the patterns of integration for any individual module. This result again justifies the attention paid to the basicranium in phylogenetic analyses of Carnivora.

It is difficult to make a conclusive statement on the effect of character correlations on phylogenetic analyses of Carnivora. On the one hand, simulated character evolution shows unquestionably that correlated characters do shift in a coordinated matter on evolutionary time scales, reflected in their significantly lesser distances in PCO analyses. Perhaps even more surprisingly, these coordinated shifts are apparent even in the most weakly integrated of modules, and little difference is seen among any modules in PCO distance. This suggests that any correlation, however weak, has the potential to affect character state changes and, in turn, phylogenetic analyses based on morphological characters. This result on its own would suggest that workers should use extreme caution when focusing on a single cranial region, such as molars or the basicranium, when building a character matrix, or when interpreting the results of such an analysis.

On the other hand, the region that dominates our understanding of carnivoran phylogeny and provides the morphological support for the most fundamental divisions within Carnivora, the basicranium, shows the strongest phylogenetic signal in its shape when compared to molecular phylogenies. It also shows the strongest phylogenetic signal in its pattern of morphological integration. Furthermore, there is no evidence from examination of the broadest morphological analysis of carnivoran phylogeny that the reliance on the molar and basicranial regions has in fact consistently and significantly misled analyses. There are no significant differences between valid and invalid clades in the modular distribution of their character support. Quite possibly, this result simply reflects the paucity of characters from other regions – only ~11% of characters come from modules other than the molar and basicranium. None the less, five of the eight clades supported by characters from only a single module are still considered monophyletic in recent molecular studies, and two of the nine clades supported by characters from three or more modules have been invalidated, leading to the conclusion that sampling across multiple modules does not necessarily translate in better phylogenetic analyses.

Thus, for the workhorse of carnivoran morphological phylogenetics, the basicranium, there is good support that its morphology strongly tracks phylogenetic relationships, as determined by molecular analysis. Perhaps more interestingly, the concordance between basicranial integration and phylogeny suggests that the changing relationships among basicranial traits retains a strong signal of their evolutionary history. However, in an ideal world, characters would be better distributed across the organism, and our simulations of character evolution do suggest that even the more weakly correlated characters display some coordination of state changes, which has the often discussed but little acted upon potential to mislead phylogenetic analyses based on morphological characters from a single anatomical region.

Acknowledgements

We thank J. J. Flynn, S. Harris, L. Van Valen, P. J. Wagner, J. Marcot, J. Finarelli, M. R. Sánchez-Villagra, V. Weisbecker, M. Wilkinson, K. Sears, and G. Wesley-Hunt for many helpful discussions of the concepts in this study. This study is based upon research supported by the US National Science Foundation International Research Fellowship OISE #0502186 (to AG). Morphometric data were collected during AG's doctoral research, supported by US National Science Foundation DDIG# 0308765, the Field Museum's Women-in-Science Fellowship, the Society of Vertebrate Paleontology Predoctoral Fellowship, the American Museum of Natural History collections study grant, the University of California Samuel P. and Doris Welles Fund, and the University of Chicago Hinds Fund. We thank W. Simpson (FMNH), W. Stanley (FMNH), D. Diveley (AMNH), J. Spence (AMNH), C. Shaw (Page Museum), P. Holroyd (UCMP), X. Wang (LACM), S. McLeod (LACM), D. Brinkman (YPM), L. Gordon (SI-NMNH), P. Jenkins (NHM), K. Krohmann (Senckenberg), and O. Roehrer-Ertl (SAPM) for access to specimens.

REFERENCES

Ackermann, R. R. and Cheverud, J. M. (2000). Phenotypic covariance structure in tamarins (genus *Saguinus*): a comparison of variation patterns using matrix correlation and common principal components analysis. *American Journal of Physical Anthropology*, **III**, 489–501.

Ackermann, R. R. and Cheverud, J. M. (2004). Morphological integration in primate evolution. In *Phenotypic Integration*, ed. M. Pigliucci and K. Preston. Oxford: Oxford University Press, pp. 302–19.

Atchley, W. R. and Hall, B. K. (1991). A model for development and evolution of complex morphological structures. *Biological Reviews*, **66**, 101–57.

Atchley, W. R., Rutledge, J. J. and Cowley, D. E. (1982). A multivariate statistical analysis of direct and correlated response to selection in the rat. *Evolution*, **36**, 677–98.

Badyaev, A. V. and Foresman, K. R. (2000). Extreme environmental change and evolution: stress-induced morphological variation is strongly concordant with patterns of evolutionary divergence in shrew mandibles. *Proceedings of the Royal Society of London Biological Sciences Series B*, **267**, 371–77.

Badyaev, A. V. and Foresman, K. R. (2004). Evolution of morphological integration. I. Functional units channel stress-induced variation in shrew mandibles. *American Naturalist*, **163**, 868–79.

Bolker, J. A. (2000). Modularity in development and why it matters to evo-devo. *American Zoologist*, **40**, 770–76.

Chernoff, B. and Magwene, P. M. (1999). Afterword. In *Morphological Integration*, ed. E. C. Olson and R. L. Miller. Chicago, IL: University of Chicago Press, pp. 319–48.

Cheverud, J. M. (1982). Phenotypic, genetic, and environmental morphological integration in the cranium. *Evolution*, **36**, 499–516.

Cheverud, J. M. (1988). Spatial-analysis in morphology illustrated by rhesus macaque cranial growth and integration. *American Journal of Physical Anthropology*, **75**, 195–96.

Cheverud, J. M. (1989). A comparative analysis of morphological variation patterns in the papionines. *Evolution*, **43**, 1737–47.

Cheverud, J. M. (1995). Morphological integration in the saddle-back tamarin (*Saguinus fuscicollis*) cranium. *American Naturalist*, **145**, 63–89.

Cheverud, J. M. (1996a). Quantitative genetic analysis of cranial morphology in the cotton-top (*Saguinus oedipus*) and saddle-back (*S. fuscicollis*) tamarins. *Journal of Evolutionary Biology*, **9**, 5–42.

Cheverud, J. M. (1996b). Developmental integration and the evolution of pleiotropy. *American Zoologist*, **36**, 44–50.

Eble, G. (2004). The macroevolution of phenotypic integration. In *Phenotypic Integration*, ed. M. Pigliucci and K. Preston. Oxford: Oxford University Press, pp. 253–73.

Emerson, S. B. and Hastings, P. A. (1998). Morphological correlations in evolution: Consequences for phylogenetic analysis. *The Quarterly Review of Biology*, **73**, 141–62.

Felsenstein, J. (1973). Maximum likelihood and minimum-steps methods for estimating evolutionary trees from data on discrete characters. *Systematic Zoology*, **22**, 240–49.

Felsenstein, J. (1985). Phylogenies and the comparative method. *American Naturalist*, **125**, 1–15.

Flynn, J. J. and Nedbal, M. A. (1998). Phylogeny of the carnivora (mammalia): Congruence versus incompatability among multiple data sets. *Molecular Phylogenetics and Evolution*, **9**, 414–26.

Flynn, J. J. and Wesley-Hunt, G. D. (2005). Carnivora. In *The Rise of Placental Mammals: Origins and Relationships of the Major Extant Clades*, ed. D. Archibald and K. Rose. Baltimore, MD: Johns Hopkins University Press, pp. 175–98.

Flynn, J. J., Nedbal, M. A., Dragoo, J. W. and Honeycutt, R. L. (2000). Whence the red panda? *Molecular Phylogenetics and Evolution*, **17**, 190–99.

Flynn, J. J., Finarelli, J. A., Zehr, S., Hsu, J. and Nedbal, M. A. (2005). Molecular phylogeny of the Carnivora (Mammalia): assessing the impact of increased sampling on resolving enigmatic relationships. *Systematic Biology*, **54**, 317–37.

Gilbert, S. F., Opitz, J. M. and Raff, R. A. (1996). Resynthesizing evolutionary and developmental biology. *Developmental Biology*, **173**, 357–72.

Gonzáles-José, R., Escapa, I., Neves, W. A., Cúneo, R. and Pucciarelli, H. M. (2008). Cladistical analysis of continuous modularized traits provides phylogenetic signal in *Homo* evolution. *Nature*, **453**, 775–78.

Goswami, A. (2006a). Cranial modularity shifts during mammalian evolution. *American Naturalist*, **168**, 270–80.

Goswami, A. (2006b). Morphological integration in the carnivoran skull. *Evolution*, **60**, 169–83.

Goswami, A. (2007a). Cranial integration, phylogeny, and diet in australodelphian marsupials. *PLoS One*, **2**, e995.

Goswami, A. (2007b). Cranial modularity and sequence heterochrony in mammals. *Evolution & Development*, **9**, 290–98.

Goswami, A. and Polly, P. D. (2010). The influence of modularity on cranial morphological disparity in carnivora and primates (Mammalia). *PLoS One*, **5**, e9517.

Gower, J. C. (1966). Some distance properties of latent root and vector methods used in multivariate analysis. *Biometrika*, **53**, 325–38.

Harris, S. R., Gower, D. J. and Wilkinson, M. (2003). Intraorganismal homology, character construction, and the phylogeny of aetosaurian archosaurs (Reptilia, Diapsida). *Systematic Biology*, **52**, 239–52.

Huelsenbeck, J. P. and Nielsen, R. (1999). Effect of nonindependence substitution on phylogenetic accuracy. *Systematic Biology*, **48**, 317–28.

Hunt, R. M. J. and Tedford, R. H. (1993). Phylogenetic relationships within aeluroid Carnivora and implications of their temporal and geographic distribution. In *Mammal Phylogeny* ed. F. S. Szalay, M. J. Novacek and M. C. McKenna. New York, NY: Springer, pp. 53–73.

Kangas, A. T., Evans, A. R., Thesleff, I. and Jernvall, J. (2004). Non-independence of mammalian dental characters. *Nature*, **432**, 211–14.

Kluge, A. G. (1989). A concern for evidence and a phylogenetic hypothesis of relationships among Epicrates (Boidae, Serpentes). *Systematic Zoology*, **38**, 7–25.

Kluge, A. G. and Farris, J. S. (1969). Quantitative phyletics and evolution of anurans. *Systematic Zoology*, **18**, 1–32.

Kluge, A. G. and Wolf, A. J. (1993). Cladistics: what's in a word? *Cladistics – The International Journal of the Willi Hennig Society*, **9**, 183–99.

Maddison, W. P. (2000). Testing character correlation using pairwise comparisons on a phylogeny. *Journal of Theoretical Biology*, **202**, 195–204.

Marroig, G. and Cheverud, J. M. (2001). A comparison of phenotypic variation and covariation patterns and the role of phylogeny, ecology, and ontogeny during cranial evolution of New World monkeys. *Evolution*, **55**, 2576–600.

Marroig, G., Vivo, M. and Cheverud, J. M. (2004). Cranial evolution in Sakis (Pithecia, Platyrrhini) ii: evolutionary processes and morphological integration. *Journal of Evolutionary Biology*, **17**, 144–55.

Moss, M. O. and Young, R. W. (1960). A functional approach to craniology. *American Journal of Physical Anthropology*, **18**, 281–91.

Myers, P. (2000). Carnivora (on-line), animal diversity web. http://animaldiversity.ummz. umich.edu/site/accounts/information/carnivora.html.

Naylor, G. J. P. and Adams, D. C. (2001). Are the fossil data really at odds with the molecular data? Morphological evidence for Cetartiodactyla phylogeny reexamined. *Systematic Biology*, **50**, 444–53.

Nowak, R. M. (1999). *Walker's Mammals of the World*, 6th ed. Baltimore, MD: Johns Hopkins University Press.

O'Keefe, F. R. and Wagner, P. J. (2001). Inferring and testing hypotheses of cladistic character dependence by using character compatibility. *Systematic Biology*, **50**, 657–75.

Olson, E. C. and Miller, R. L. (1951). A mathematical model applied to the evolution of species. *Evolution*, **5**, 325–38.

Olson, E. C. and Miller, R. L. (1958). *Morphological Integration*. Chicago, IL: University of Chicago Press.

Otto, S. P. and Day, T. (2007). *A Biologist's Guide to Mathematical Modeling in Ecology and Evolution*. Princeton, NJ: Princeton University Press.

Pigliucci, M. and Preston, K. (2004). *Phenotypic Integration*. Oxford: Oxford University Press.

Polly, P. D. (2005). Development and phenotypic correlations: the evolution of tooth shape in Sorex araneus. *Evolution & Development*, **7**, 29–41.

Polly, P. D., Head, J. J. and Cohn, M. J. (2001). Testing modularity and dissociation: the evolution of regional proportions in snakes. In *Beyond Heterochrony: The Evolution of Development*, ed. M. L. Zelditch. New York, NY: Wiley-Liss, pp. 305–35.

Read, A. F. and Nee, S. (1995). Inference from binary comparative data. *Journal of Theoretical Biology*, **173**, 99–108.

Rieppel, O. and Kearney, M. (2002). Similarity. *Biological Journal of the Linnean Society*, **75**, 59–82.

Sadleir, R. W. and Makovicky, P. J. (2008). Cranial shape and correlated characters in crocodile evolution. *Journal of Evolutionary Biology*, **21**, 1578–96.

Sanchez-Villagra, M. R. and Williams, B. (1998). Levels of homoplasy in the evolution of the mammalian skeleton. *Journal of Mammalian Evolution*, **5**, 113–26.

Schlosser, G. and Wagner, G. P. (2004). *Modularity in Development and Evolution*. Chicago, IL: University of Chicago Press.

Schwenk, K. (2001). Functional units and their evolution. In *The Character Concept in Evolutionary Biology*, ed. G. P. Wagner. San Diego, CA: Academic Press, pp. 165–98.

Shubin, N. and Davis, M. C. (2004). Modularity in the evolution of vertebrate appendages. In *Modularity in Development and Evolution*, ed. G. Schlosser and G. P. Wagner. Chicago, IL: University of Chicago Press, pp. 429–40.

Steppan, S. J. (1997). Phylogenetic analysis of phenotypic covariance structure. II. Reconstructing matrix evolution. *Evolution*, **51**, 587–94.

Strait, D. S. (2001). Integration, phylogeny, and the hominid cranial base. *American Journal of Physical Anthropology*, **114**, 273–97.

Tedford, R. H., Taylor, B. E. and Wang, X. (1995). Phylogeny of the Canidae (Carnivora: Canidae): the living taxa. *American Museum Novitates*, **3146**, 1–37.

Thorogood, P. (1993). Differentiation and morphogenesis of cranial skeletal tissues. In *The Skull*, ed. J. Hanken and B. K. Hall. Chicago, IL: University of Chicago Press, pp. 112–52.

Van Der Klaauw, C. J. (1948–1952). Size and position of the functional components of the skull. *Archives Neerlandaises de Zoologie*, **9**, 1–559.

Vermeij, G. J. (1973). Adaptation, versatility, and evolution. *Systematic Zoology*, **22**, 466–77.

Wagner, G. P. (1995). Adaptation and the modular design of organisms. *Advances in Artificial Life*, **929**, 317–28.

Wagner, G. P. (1996). Homologues, natural kinds and the evolution of modularity. *American Zoologist*, **36**, 36–43.

Wagner, G. P. and Altenberg, L. (1996). Perspective: complex adaptations and the evolution of evolvability. *Evolution*, **50**, 967–76.

Wagner, P. J. (1998). A likelihood approach for evaluating estimates of phylogenetic relationships among fossil taxa. *Palaeobiology*, **24**, 430–49.

Wesley-Hunt, G. D. and Flynn, J. J. (2005). Phylogeny of the Carnivora: basal relationships among the carnivoramorphans, and assessment of the position of 'Miacoidea' relative to crown-clade Carnivora. *Journal of Systematic Palaeontology*, **3**, 1–28.

Williams, B. (2007). Comparing levels of homoplasy in the primate skeleton. *Journal of Human Evolution*, **52**, 480–89.

Wyss, A. R. and Flynn, J. J. (1993). A phylogenetic analysis and definition of the carnivora. In *Mammal Phylogeny*, ed. F. S. Szalay, M. J. Novacek and M. C. McKenna. New York, NY: Springer, pp. 32–52.

Yoder, A. D., Burns, M. M., Zehr, S., *et al.* (2003). Single origin of Malagasy Carnivora from an African ancestor. *Nature*, **421**, 734–37.

Zelditch, M. L. (1988). Ontogenetic variation in patterns of phenotypic integration in the laboratory rat. *Evolution*, **42**, 28–41.

Zelditch, M. L. and Carmichael, A. C. (1989a). Growth and intensity of integration through postnatal growth in the skull of Sigmodon fulviventer. *Journal of Mammalogy*, **70**, 477–84.

Zelditch, M. L. and Carmichael, A. C. (1989b). Ontogenetic variation in patterns of developmental and functional integration in skulls of Sigmodon fuliviventer. *Evolution*, **43**, 814–24.

Zelditch, M. L., Sheets, H. D. and Fink, W. L. (2001). The spatial complexity and evolutionary dynamics of growth. In *Beyond Heterochrony: The Evolution of Development*, ed. M. L. Zelditch. New York, NY: Wiley-Liss, pp. 145–94.

6

What's the difference? A multiphasic allometric analysis of fossil and living lions

MATTHEW H. BENOIT

Introduction

Differentiating between various species in the fossil record is one of the most vital tasks in paleontology. As such, evaluating the morphological features that we use to make these taxonomic distinctions is critical. Without any confirmation from molecular lines of evidence, morphological analyses are the only option for such studies. Determining the validity and independence of character changes is a major part of that evaluation. Compounding this limitation to morphological analyses is the fact that assembling a significant sample size of fossil specimens for a single taxon is frequently very difficult, if not impossible. Often, paleontologists compare a single fossil specimen with a single specimen of a closely related extant taxon or representatives of several such taxa. Analyses of this nature, while valuable first glimpses, do not account for variation within populations (of either the fossil or the extant groups), and therefore may result in inaccurate conclusions regarding the relationships of the organisms in question. In this chapter, I present an example of a species–status conflict within the pantherine felids and use allometric analyses to evaluate some of the morphological characteristics that have been used as evidence to support arguments in this conflict.

Since its first official use by Pocock (1930), the generic designation of *Panthera* for the clade consisting of the lion (*P. leo*), tiger (*P. tigris*), leopard (*P. pardus*), jaguar (*P. onca*), and now the snow leopard (*P. uncia*) has reached standard usage. However, the attribution of species or subspecies status below the rank of genus has not been so readily settled, especially for fossil groups that seem to show a relationship to one of the extant pantherine cats. One of these fossil groups is the 'American lion' (*Panthera leo* cf. *atrox*). There has been some argument regarding the nature of the relationship of *P. atrox* and *P. spelea* (the 'cave lion') within *Panthera*, and several authors have maintained a *P. tigris* or

Carnivoran Evolution: New Views on Phylogeny, Form, and Function, ed. A. Goswami and A. Friscia. Published by Cambridge University Press. © Cambridge University Press 2010.

P. onca affinity for *P. atrox* (e.g. Simpson, 1941; Groiss, 1996). However, as the majority of authors have discussed the affinities of *P. atrox* in relation to modern *P. leo*, those will be the comparisons that I will address in this chapter. When using the term 'lion', I am referring to any member of the extant species *Panthera leo* and any known fossil specimen that is more closely related to that species than they are to any other extant species. This definition includes, but is not limited to, specimens attributed to the species *P. atrox* (the 'American Lion') and *P. spelea*.

Palaeogeography of lions

The current geographic distribution of lions is limited to sub-Saharan Africa, with a relic population in northwestern India. This limited range is a fairly recent development as lions are known throughout North Africa and the Middle East as late as the twentieth century (Sunquist and Sunquist, 2002). Prehistorically, however, lions were very widespread indeed. The first lion-like (and lion-sized) species of the genus *Panthera* appears in eastern Africa almost 3.5 Mya ago (Barry, 1987). The earliest fossils that have been attributed to the species *P. leo* come from sediments in the Olduvai Gorge in Tanzania dated at around 1.87–1.7 Mya (Petter, 1973). During the Pleistocene, the lion spread out of Africa and across Eurasia. The earliest lion remains known from North America are found in Alaska and date to roughly 300 Kya ago (Kurtén and Anderson, 1980; Herrington, 1987; Yamaguchi *et al.*, 2004). From there, lions spread south into western North America and South America. Lion fossils have been found as far south in the Americas as the Talara region in northwestern Peru (Lemon and Churcher, 1961). The presence of lions in the Americas persisted until about 10,000 years ago (Harington, 1977; Yamaguchi *et al.*, 2004).

Species-status arguments

Considering the taxonomic arguments that have plagued study of the entire family Felidae (Haas *et al.*, 2005; Bona, 2006), it is not surprising that the species status of fossil lions has been a subject of some debate throughout the years. The American Lion was first described by Leidy (1853) as a separate species (within the genus *Felis*, which, at that time, included all cats). A fuller and more complete description came from Merriam and Stock (1932), who were working with a larger sample of specimens from the La Brea Tar Pits in Los Angeles, California. They, too, chose to designate the American lion as its own species, and proposed that it might be the ancestral stock from which the modern lion and tiger descended. Their conclusions, however, may have resulted from the information available to them at that time. No lion fossils were known from Siberia or Beringia at that time, and the geographically closest extant *Panthera* species was the tiger (Harington, 1977). Their morphological arguments supported a close relationship with the lion, so it is likely that this paleogeographic consideration prompted their argument for an affinity with *P. tigris*, and perhaps their conclusion that the American lion may be ancestral to both *P. tigris* and *P. leo*.

During much of this early work, species-status was given to these fossils partly because the paleontologists were not entirely certain to which modern *Panthera* species these fossils were most closely related. Most authors agree that the cave lion and the American lion most closely resemble each other (Harington, 1969; Vereshchagin, 1971; Hemmer, 1974; Sotnikova and Nikolskiy, 2006). Simpson (1941) proposed that the American lion was actually an oversized jaguar, although distinct from the extant jaguar (*P. onca*). However, most authors have noted that the American and cave lions shared affinities with modern lions and modern tigers (Leidy, 1853; Pocock, 1930; Merriam and Stock, 1932; Vereshchagin, 1971; Kurtén, 1985; Groiss, 1996; Sotnikova and Nikolskiy, 2006). Aside from Groiss (1996), who felt that braincase similarities were enough to place the cave and American lions within *P. tigris*, most researchers have concluded that the fossil specimens more closely resemble the lion, and that the several tiger-like features are plesiomorphic (Sotnikova and Nikolskiy, 2006). The most promising recent evidence for this conclusion may be the fact that fossil molecular work done on the cave lion placed it as a sister taxon to all modern lions (represented by several subspecies) in an analysis that included both *P. pardus* and *P. tigris* (Burger *et al.*, 2004).

Since the general consensus (although by no means the only possibility) is that the fossil specimens are most closely related to *P. leo*, the main phylogenetic contention has become the status of these groups as separate species (*P. atrox* and *P. spelea*) versus a subspecific designation within *P. leo* (*P. l. atrox* and *P. l. spelea*). This latter assignment has been used by many authors and seems to represent the majority opinion in most of the current literature (e.g. Harington, 1971; Hemmer, 1974, 1979; Kurtén and Anderson, 1980; Haas *et al.*, 2005). There are those who disagree with this assessment, claiming that the fossil lions show synapomorphies separate from modern *P. leo* (Sotnikova and Nikolskiy, 2006). Despite the general agreement that the cave and American lions most closely resemble each other, several authors have pointed out that the American lion is more derived and may be distinct from even the Siberian *P. l. spelea* specimens examined (Kurtén, 1985; Sotnikova and Nikolskiy, 2006). As such, an analysis addressing the differences between fossil and living lions should focus on this most disconnected of the available groups.

While most work discerning the various subspecies in extant *P. leo* uses soft tissue and molecular data (Hemmer, 1974; Dubach *et al.*, 2005; Haas *et al.*, 2005), fossil specimens generally provide only skeletal morphology for examination. Therefore, the designation of species status with regard to these fossil groups has relied heavily on this morphology in conjunction with the geographic distribution of the specimens. However, given that the fossil record is somewhat capricious and arbitrary in the amount and types of information it provides, statistical samples of fossil features can be difficult to attain. In this chapter, I will present several skeletal features of the skull that have been proposed by various authors as distinguishing between the extinct *P. l. atrox* and the extant *P. leo*. These characteristics will be quantitatively examined using a multiphasic allometric methodology that allows for a population-level analysis of these features as distinguishing taxonomic characters.

Methods

Proposed phenotypic differences between the crania of *P. l. leo* and *P. l. atrox*

The species-status arguments regarding *P. l. atrox* are based upon numerous phenotypic differences with relation to the modern *P. leo* that have been proposed. The larger size of *P. l. atrox* has been noted by most authors and remains uncontested. In order to address the question of species-status in the American lion, I analysed measurements obtained from *P. l. atrox* specimens and compared them to similar measurements from extant lion specimens. These measurements are solely from the exterior of the skull, so analysis is necessarily restricted to cranial features that are externally visible. While some arguments regarding separation of these groups based on hide pattern (Harington, 1977), mane presence/absence (Yamaguchi *et al.*, 2004), and brain endocast morphology (Groiss, 1996) have been made, these are not features that can be statistically analysed from the samples available. Fortunately, there are multiple external cranial features that are different between *P. l. atrox* and modern *P. leo* and have been used as diagnostic characters. The features addressed in this chapter can be seen in Figure 6.1.

The size of the braincase in *P. l. atrox* relative to modern lions has been described in several studies, some of which say it is larger (Kurtén and Anderson, 1980), while others claim that it is smaller (Merriam and Stock, 1932; Harington, 1969; Martin and Gilbert, 1978; Sotnikova and Nikolskiy, 2006). Groiss (1996) argued that brain size was irrelevant for taxonomic assignment, apparently in an effort to focus on brain morphology. Most of these studies refer to endocasts of the braincase, as opposed to external measurements of the skull. Due to this,

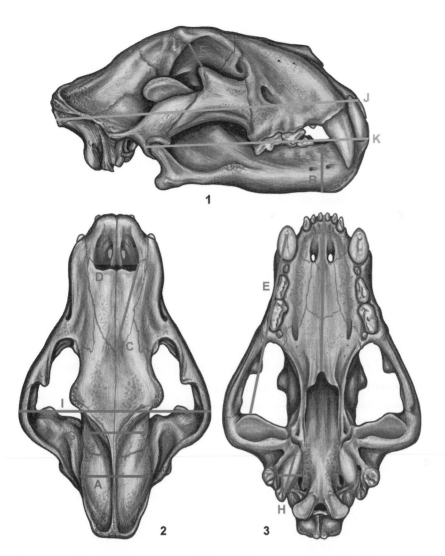

Figure 6.1 For colour version, see Plate 8. Measurements analysed for multiphasic analyses. A, Braincase width (BCW); B, mandibular flange (i.e. symphyseal) depth (MFD); C, nasal bone length (NBL); D, external narial area (NRA); E, facial length measured by length of the palate (PLL) and the length from the glenoid to the tip of the canine (GCL); F, orbit size measured by the distance between the left or right postorbital processes of the frontal and zygoma (LRPP); G, postorbital constriction width (POC); H, auditory bulla size measured as anteroposterior length (TBL) and mesiolateral width (TBW); I, zygomatic arch width measured as the greatest distance across the skull at the zygomatic arches (ZAW); J, skull length (SKL); K, mandible length (GML). The measurements are colour-coded based on whether that measurement should be larger (red), smaller (blue), or ambiguous (purple) in the American lion, according to descriptions in the literature as discussed above. The baseline measurements are shown in green. GML is the allometric baseline for MFD analyses. Skull of *P. l. atrox* illustrated by Emma Schachner, redrawn from Merriam and Stock (1932). Skulls are drawn in 1. lateral, 2. dorsal, and 3. ventral views.

they generally only describe one or two specimens. Single specimen analyses between closely related groups with natural levels of morphological variation may account for this discrepancy in the literature regarding braincase size. For this study, external braincase width (BCW) is the measurement used to address this feature.

While discussing the affinity of *P. l. atrox* to modern *P. leo*, Martin and Gilbert (1978) stated that the orbits in *P. l. atrox* were large and forward-facing. However, they made no mention as to whether they considered this to be in excess of the already large and forward-facing orbits of most members of the genus. In general, their description seems to be in line with their desire to portray *P. l. atrox* as a cursorial cat convergent on *Acinonyx* (or more likely *Miracinonyx*). The distance between the tips of the postorbital processes (LRPP) on either the left or right side, depending on the availability of measurement, was used to examine orbit size variations. While this measurement is expressly chosen to test the variation hypothesised by Martin and Gilbert (1978), it should also be indicative of any other variation in orbit structure.

In one of the earliest descriptions from the La Brea tar pits, Merriam and Stock (1932) report that mandible of *P. l. atrox* had a sharp chin. While in extant lions the ventral surface slopes back from the lower incisors, they reported that American lion mandibles dropped sharply down from the incisors to a more abrupt corner before the ventral surface swept back under the rest of the toothrow. The measurement of the dorsoventral depth of the mandibular ramus at the postcanine diastema (MFD) is used here to analyse this feature.

The external nares of *P. l. atrox* open somewhat dorsally, as in *P. leo*, although not quite to the same extent. The rostral tip of the nasal suture is closer to the rostral tip of the premaxillae in *P. l. atrox* (Merriam and Stock, 1932). However, this difference contrasts slightly with a report that the nasal bones of *P. l. atrox* are shorter (which would imply that the rostral tip of their suture might be farther away from that of the premaxilla) (Martin and Gilbert, 1978). However, Martin and Gilbert (1978) also described a shorter face and larger nareal area in *P. l. atrox*, which may have resulted from their observations regarding the nasal bones. Several measures were chosen to analyse these two features (craniofacial length and narial orientation). The distance from the glenoid to the canine (GCL), the palate length (PLL), and the nasal bone length (NBL) are informative with regard to the shortness of the preorbital face. The external nareal area (NRA) and the length of the nasal bone (NBL) are informative with regard to the external nareal opening.

The postorbital constriction of *P. l. atrox* is reportedly 'less pronounced', as a result of the posterior skull (including the braincase) being fuller and more robust (Merriam and Stock, 1932). More recent work has also reported that the cave lion is also less constricted than the modern lion (Sotnikova and Nikolskiy, 2006). These same researchers, however, note that the La Brea specimens have still greater breadth across the constriction, which is important when considering this analysis (see 'Sampling effects' below). The width of the skull at the postorbital constriction (POC) is informative with regard to this feature. A 'less pronounced' constriction shows up as a wider measurement.

Several reports have claimed that the auditory bullae in *P. l. atrox* are relatively small, although most authors have not designated whether they are small for a lion of that size or whether they are absolutely smaller than those of extant *P. leo* (Martin and Gilbert, 1978). Interestingly, Merriam and Stock (1932) did not note these small bullae in their first description of the skull, possibly because they did not feel that this feature was helpful in determining whether *P. l. atrox* had a closer affinity to *P. leo* or *P. tigris*. Two bullar measurements were used in this study: total bullar length (TBL) and total bullar width (TBW).

Finally, Merriam and Stock (1932) reported that the American lion had two phenotypes, one of which had very wide zygomatic arches, the other having more narrow arches. A specimen from northern Alaska was also described as having a large zygomatic arch breadth (Harington, 1969). However, Sotnikova and Nikolskiy (2006) claimed that the modern lion is derived in having 'strongly arched zygomata'. Unfortunately, they gave no indication whether they mean broader zygomata or simply zygomata that are arched in a different shape (rounder arching in the dorsal view). In order to examine this feature, the width across the zygomatic arches (ZAW) was examined.

Measurement acquisition

To obtain the measurements for this study, I visited the paleontological and osteological collections of several museums. The majority of extant *P. leo* measurements in this study were obtained from specimens in the Mammalogy Department at the American Museum of Natural History (AMNH) in New York City. Their extensive collection provided over 90 *P. leo* skulls that were complete enough for measurement. *P. atrox* measurements were obtained from specimens at the Canadian Museum of Nature (CMN) in Ottawa, ON and the Page Museum at the La Brea Tar Pits, which is a branch of the Los Angeles County Museum of Natural History (LACMNH) in Los Angeles, CA.

The specimens were measured using a Microscribe G2X digitiser. The digitiser provided a quick, accurate, and efficient method for taking multiple measurements from a large quantity of specimens quickly with little handling of the specimens themselves. The range of the arm is 127 cm (50 in), which was easily large enough to obtain measurements from every felid skull encountered.

The precision of the digitizer was 0.23 mm (0.009 in), which is less than most calipers would provide. However, the measurements taken were large enough that the digitiser precision was acceptable.

When taking the measurements for this study, each skull was mounted using clay so that it would not be able to move during the measurement session. In order to obtain all of the measurements, most skulls had to be mounted in three separate positions: skull (without mandibles) upright, skull upside-down, and mandibles upright. After taking the locations of each of the measurement endpoints available, the skull was examined for damage or deformation that could skew or invalidate the measurements.

Allometric analysis

Nearly all authors have commented upon the fact that *P. l. atrox* is larger than the modern *P. leo*. Despite this broad agreement, very little research has aimed to statistically distinguish which of the cranial features described above correspond with allometric trends in *P. leo*. In order to do this, I used multiphasic allometric analyses of various measurements of the skulls of *P. leo* and *P. l. atrox* (see below for the procedural details). Multiphasic allometric analysis involves analysing shape and size proxies with an eye on the possibility that their relationships are different at different size ranges. Multiphasic regression analysis fits different regression lines to different sections of the data, finding the best statistical fit. When applied intraspecifically, regression analyses of this type allow for the recognition of ontogenetic allometries consisting of more than one growth phase, illuminating periods of ontogenetic development where the trait is developing faster or slower. In this way, multiphasic allometric analysis allows a researcher to discover more complex allometric relationships without relying on age determination (see Figure 6.2 for a comparison of multiphasic and monophasic allometric analyses). This feature of multiphasic analyses is particularly important for the comparison of modern *P. leo* with fossil *P. l. atrox*. In most wild felids, tooth eruption is the most reliable form of age determination. However, the teeth of most felids finish erupting before the organism has completed its developmental growth. As mentioned below, most museum specimens of large pantherine felids have little to no age data. In the absence of these data, multiphasic allometry allows a researcher to find developmental growth phases based on the size of the organism.

Having designated these periods of different allometric relationships, one can compare these patterns with other groups (subspecies or closely related species) and see if interspecific size differences account for perceived shape differences. For the comparison of postorbital constriction widths of *P. leo*

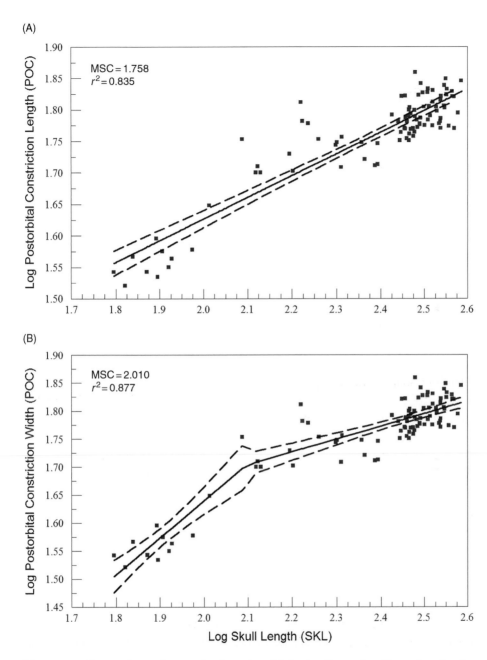

Figure 6.2 Comparison of monophasic and multiphasic allometries. Data plots for the logged postorbital constriction width measurements of the *P. leo* specimens plotted against logged skull length. All scale units are log$_{10}$ millimetres. Fit statistics r^2 and MSC are given for each regression. A, Monophasic (simple) allometric regression through the data showing 95% confidence intervals. B, Multiphasic allometric regression through the same data showing 95% confidence intervals. This regression is a two-phase regression with a phase change at 2.1 (the value of X_{pi} in Equation 1) on the x-axis.

(see Figure 6.2) and *P. l. atrox*, the use of a monophasic (simple) regression would not reflect the biphasic growth pattern seen in the data. As such, comparison of that allometry with the *P. l. atrox* allometry would not be representative of this more complex nature. Without many juvenile *P. l. atrox* specimens available (see below), the determination of the final phase of ontogenetic development for each trait in *P. leo* became even more important, as that was the only phase available for comparison in *P. l. atrox*.

I compared the cranial measurements of statistically viable numbers of extant *P. leo* ($n = 93$) and *P. l. atrox* ($n = 46$) in order to examine whether the differences in the features outlined above are the result of simple extension of allometries (ontogenetic scaling) resulting from the American lion's greater size or truly different characters derived separately within that taxon. This analysis allows for an evaluation of independence of these characters from the size difference noted between the taxa. I compared the two taxa using two different types of allometric analysis: monophasic and multiphasic. In order to analyse these allometric relationships, each cranial measurement was logged and plotted against the log skull length (SKL).

The use of skull length as the allometric baseline deserves some explanation. The skull of vertebrates has been identified as a developmental module for quite some time. The pattern of skull growth from early in ontogeny separates the development of the skull from that of the rest of the body (Jacobson, 1993), apparently as a result of differentiation of the neural crest (Langille and Hall, 1993). Therefore, treatment of the skull as a distinct developmental module is justified. Using skull length as the proxy for the size of that module is appropriate, as it is a measurement that encompasses the entire module. Furthermore, a principal components analysis (PCA) was performed on all of the skull measurements. In such analyses of biological entities, the first principal component generally reflects the overall size of the body or body part being measured (McKinney and McNamara, 1991), provided the first principal component explains the vast majority of the variance (Hammer and Harper, 2006). In the PCA performed, the first principal component accounted for 98% of the variance, and all loadings were positive, strongly indicating that this component represents overall size. SKL had the highest loading (0.56) on the first component, indicating that its variance is more closely correlated with the overall size of the skull than any other measurement. Based on these arguments, skull length was deemed appropriate for use as the allometric baseline of this study.

I performed linear best-fit analyses on these bivariate plots to determine the allometric parameters for each measurement. Three different types of best-fit line were employed: original least squares (OLS or linear regression), major axis (MA), and standardised major axis (SMA, or reduced major axis). The best fits for each taxon were then compared (discussed below). Monophasic allometric analyses produced and compared single best-fit lines for the entire specimen sample of each taxon, including skulls of the youngest cubs available (some only a week or two old). Multiphasic allometric analyses, however, required the determination of the final growth phase of development in the taxa for comparison.

To compare multiphasic allometric relationships, I first performed a multiphasic allometric analysis on the *P. leo* sample, determining the optimum number of regression segments and their location. To perform these regressions, I used the following equation from Vrba (1998):

$$Y = \beta_1 X + \alpha_1 + \sum_{i=2}^{n} I_i \beta_i (X - X_{P(i-1)}) \tag{1}$$

where Y is the log of the feature measurement; X is log skull length; β_i is the regression coefficient (line slope) for the ith phase; α_1 is the Y-intercept of the first phase (P_1) line; I_i is an indicator variable such that $I_i = 0$ if $X < X_{P(i-1)}$ and $I_i = 1$ if $X > X_{P(i-1)}$; X_{Pi} is log skull length value at the end of growth phase i. In order to determine which model (number of phases) fit the data best, I used the Model Selection Criterion (MSC) in PSI-Plot (2002), which is an adaptation of the Akaike Information Criterion (AIC) (Akaike, 1974). The MSC evaluates the correlation between the best-fit line and the data, but penalises best-fit formulae for each parameter they contain. Therefore, while the correlation will likely go up with the number of phases allowed in the regression formula, the MSC will be lower for models that use too many phases for a relatively small increase in correlation.

Ontogenetic age designations were almost never available on the museum tag for the *P. leo* specimens. Therefore, for this group, age determinations were performed using the calendar of Smuts *et al.* (1978) for tooth eruption. Because these determinations used only tooth eruption schedules, all individuals with fully erupted adult dentition are referred to as adults. However, one should note that fully erupted adult dentition occurs well before sexual adulthood and the end of ontogenetic development in lions. Error in age estimation based on tooth eruption increases with age. Based on this methodology, age class estimates were established for the *P. leo* population. The distribution of specimens in the various age classes can be seen in Table 6.1. As one would expect, the majority of specimens are classified as 'adult' (meaning that their adult dentition have fully erupted); however, there is a relatively good representation of each age class through development. There were no prenatal specimens in these analyses.

For *P. l. atrox*, very few young juvenile specimens were preserved in the fossil record, reducing the overall spread of ages sampled. There are juveniles in the sample, however, the use of the Smuts *et al.* (1978) calendar would not be

Table 6.1 Distribution of specimens of *P. leo* and *P. l. atrox* by age class as determined using the Smuts *et al.* (1978) tooth eruption calendar. As discussed in the text, this calendar is not appropriate for *P. l. atrox* and is used here only to show the dearth of very young specimens of this group.

Smuts *et al.* (1978) age class	No. of *P. leo* specimens	No. of *P. l. atrox* specimens
0–7 days	7	0
7–21 days	4	0
1–2 months	3	0
2–3 months	3	0
4–8 months	4	0
9–11 months	3	1
12–14 months	4	1
14–17 months	3	1
18–24 months	1	0
2–3 years	2	0
>2–3 years ('Adult')	59	43

appropriate for designation of ontogenetic age, because it is based on modern *P. leo* specimens and cannot be extrapolated to other *Panthera* species (and presumably not to extinct subspecies either). One should note that the *P. l. atrox* juveniles are not the very young juveniles that were available for the extant sample (see Table 6.1). Analyses of the *P. l. atrox* specimens showed that there were no multiphasic allometries discernible from the sample. Because of this, all *P. l. atrox* allometries in this study are monophasic, encompassing the final phase of ontogenetic allometry. Therefore, comparison of *P. l. atrox* allometries were performed only with the final phase of the multiphasic *P. leo* allometries. Using monophasic allometric analyses of *P. leo* does not account for the complexity of development, and lumps all variation into the single regression line provided. However, the use of multiphasic allometric analysis to determine the final developmental phase allowed for the most inclusive, late-developmental growth phases to be accurately compared between the taxa, allowing a comparison that is more appropriate and informative.

Comparing best-fit lines

Until recently, best-fit lines have only been statistically comparable if they were OLS regressions compared using an analysis of covariance (ANCOVA). OLS regression (and therefore ANCOVA) is not always appropriate for use with allometric data, although it is commonly used. For a review of these conditions, please refer to Sokal and Rohlf (1995) or Warton et al. (2006). Recently, however, software has been developed to allow for statistical comparison of both MA and SMA regression lines (Warton et al., 2006). I used this software (Standardised Major Axis Tests & Routines, SMATR) to determine and compare the best-fit lines for the various allometric relationships using all three best-fit techniques (OLS, MA, and SMA). The software utilises an algorithm that is comparable to a likelihood analysis, because it tests statistical compatibility through iterative testing (Warton and Weber, 2002; Warton et al., 2006). SMATR allows for comparison of allometric slopes, elevational shifts in allometries (same slope, but significantly different y-intercepts), and shifts along allometries (an extension of an allometry into a larger size range).

There is often argument about the applicability or usefulness of MA vs. SMA for various data sets, so results from both analyses are provided to accommodate such discussion. OLS comparisons are also presented so that the regressions discussed here may be compared with previous studies that may have used that methodology to determine allometric relationships.

Culling measurements

Measurements were culled from analyses based on the sample size, the coefficient of determination (COD or r^2), or the feasibility of the allometric slope obtained. Sample sizes of less than five were considered useless for analyses, because their confidence intervals were too broad to allow for any meaningful interpretation. If the COD was too low (accounting for less than 30% of the variance), then the results for that measurement were rendered ambiguous and removed from comparison. If the slope of the allometry was extremely unlikely to represent the natural condition (i.e. if it was a negative value), then the measurement was considered untenable for analysis.

Sampling effects

The vast majority of American lion specimens which I measured were from the La Brea Tar Pits in Los Angeles. These specimens are housed at the Page Museum at the La Brea Tar Pits and were collected under the auspices of the Los Angeles County Museum of Natural History (LACMNH). The mode of entrapment in the tar pits was one that favoured the preservation of carnivores. A prey animal, such as a mammoth, would get stuck in the tar. Its distress calls or general scent would attract carnivores, often represented in the pits by dire wolves (Canis dirus), sabre-toothed cats (in this case Smilodon *fatalis*), and the American lion. These carnivores would also get trapped in the tar. Attesting to this scenario, the Page Museum at the La Brea Tar Pits is full of the remains of carnivores and scavengers, yet contains relatively few herbivorous taxa.

While *P. l. atrox* remains are not nearly as plentiful in the tar pits as those of *S. fatalis* and *C. dirus*, they are substantially more abundant there than at any other Pleistocene site in the Americas. The American lion is rare at most sites, and, if found, is usually represented by a few teeth or a mandibular fragment, such as the type specimen described by Leidy (1853). Due to this preservational constraint, in order to examine the properties of the crania of these animals in a statistical manner, I was forced to rely on a large percentage of samples (100% for many measurements) from a single fossil locality. While previous morphometric analyses on the mandibles and dentition of the American lion have determined that there is no significant difference between specimens from La Brea and those from other localities across the Americas (Kurtén, 1985), the danger of local ecological or preservational bias is still present.

The sampling for extant lions should produce no bias. The skulls measured are housed at the American Museum of Natural History (AMNH) in New York City. These specimens were collected from all over Africa as well as from several zoos. Due to this diverse sampling, the variation found in extant lions should prove to be higher than would be expected for any subspecific group that may be within extant *P. leo*. As such, there should be no subspecific bias in the measurements for the extant lions used in this study.

Results and discussion

Previous comparative descriptions have resulted mainly from the analysis of one or a few skulls of each group (e.g. Martin and Gilbert, 1978; Groiss, 1996; Sotnikova and Nikolskiy, 2006). There has been some statistical analysis with regard to fossil and extant lions. Kurtén (1985) examined metric characters of various fossil lions from across Eurasia and the Americas. However, his analysis, while recognising some difficulties with including young specimens, did not analyse them allometrically, but merely compared the averages of dental and mandibular measurements. Hemmer (1974) performed an allometric analysis on extinct lions, although the measurements he analysed did not address the features outlined above. Hemmer's analysis also examined only adult specimens with a simple linear regression, which cannot detect multiphasic ontogenetic allometric relationships that might exist between the groups.

Every one of the measurements examined for this study exhibited multiphasic allometry in extant *P. leo*, as determined by MSC values. All of the measurements showed biphasic allometric relationships in *P. leo*, except for braincase width (BCW), which had a triphasic allometric regression. The slopes and r^2 values of all analyses (mono- and multiphasic) for both taxa can be seen in Table 6.2. The intertaxon comparisons of these parameters are presented in Tables 6.3. and 6.4, respectively. Glenoid–canine length (GCL) and left/right postorbital process gap (LRPP) measurements did not produce sample sizes or correlations (r^2) that allowed for allometric analyses. As such, statements about species status based on these features should be considered tentatively.

The multiphasic comparisons showed different significance patterns for several of the measurements taken. In general, the use of multiphasic allometric analysis found lower *F*-statistics and higher *p*-values for the slope comparisons. For the analysis of palate length (PLL), the monophasic OLS regression indicated significantly different slopes between the two taxa. However, when the phasic nature of the development of this feature was taken into account, all three best-fit analyses indicated no significant difference in either the slope or the elevation (*y*-intercept). In the multiphasic analyses, the only measurement to show a significant difference in slope was postorbital constriction width (POC). This measurement demonstrated poor correlation with skull length in both taxa ($r^2 = 0.54$ and 0.46 for *P. leo* and *P. l. atrox*, respectively), which may account for this result. However, it should be noted that the POC values for *P. l. atrox* were consistently larger than for extant *P. leo*.

Elevation shifts (which can be detected if the slopes are not significantly different) were statistically detected in mandibular flange depth (MFD), nareal area (NRA), bullar length (TBL), and zygomatic arch width (ZAW). MFD and NRA both showed an elevational increase in *P. l. atrox*, while TBL and ZAW showed drops in elevation in this taxon relative to extant lions. POC also displayed an increase in elevation in *P. l. atrox* for the OLS analysis (the only one to find no significant difference in slope). Bullar width (TBW) showed a significant increase in elevation in only the MA and SMA analyses, with OLS showing no significant difference. The reverse is found in the braincase width analyses, with OLS showing a barely significant increase elevation in *P. l. atrox*.

Almost every measurement exhibited a shift in *P. l. atrox* that extended the allometry of modern lions into larger ranges. This result is unsurprising, as the increased size of *P. l. atrox* has already been noted across the board. This shift was detected for all but one of the measurements which also exhibited an elevational change. TBW showed no significant shift for the MA and SMA analyses. These analyses showed a significantly lower elevation for the *P. l. atrox* allometry, which may have affected their ability to detect a shift.

Arguments regarding the species status of *P. l. atrox* have been morphological in nature, but most fail to account for variation in the population and allometric

Table 6.2 Results of allometric regressions for *P. leo* and *P. l. atrox*. The beginning of the final phase of multiphasic regressions (Final Phase Start) is given as the value of the logged baseline measurement (SKL or GML) at the start of that phase.

Meas. analysis	Panthera leo							Panthera leo atrox		
	Monophasic			Multiphasic						
	r^2	Slope	y-intercept	r^2	# of phases	Final phase start	Final phase slope	r^2	Slope	y-intercept
BCW	0.97			0.98	3	2.43		0.71		
OLS		0.540	0.644				0.591		0.442	0.905
MA		0.544	0.634				0.688		0.473	0.824
SMA		0.550	0.621				0.741		0.525	0.692
MFD	0.95			0.95	2	2.14		0.45		
OLS		0.904	−0.540				0.669		0.670	0.091
MA		0.927	−0.593				0.898		1.00	−0.703
SMA		0.929	−0.597				0.925		1.00	−0.701
NBL	0.98			0.99	2	2.20		0.65		
OLS		1.123	−0.827				1.006		0.964	−0.450
MA		1.133	−0.852				1.051		1.254	−1.191
SMA		1.132	−0.849				1.049		1.200	−1.053
NRA	0.99			0.99	2	2.46		0.90		
OLS		1.069	−0.987				0.761		1.159	−1.196
MA		1.077	−1.005				0.989		1.236	−1.392
SMA		1.076	−1.004				0.992		1.223	−1.358
PLL	0.98			0.99	2	1.94		0.88		
OLS		1.077	−0.533				1.016		0.850	0.042
MA		1.091	−0.567				1.023		0.898	−0.081
SMA		1.090	−0.564				1.022		0.904	−0.096

POC	0.835			0.88	2	2.10		0.53		
OLS		0.344	0.939				0.226		0.579	0.402
MA		0.351	0.922				0.239		0.727	0.025
SMA		0.376	0.862				0.332		0.790	−0.136
TBL	0.95			0.96	2	2.32		0.49		
OLS		0.922	−0.601				0.627		0.600	0.158
MA		0.946	−0.660				0.839		0.806	−0.370
SMA		0.948	−0.664				0.882		0.860	−0.508
TBW	0.88			0.93	2	2.25		0.53		
OLS		0.787	−0.513				0.225		0.715	−0.367
MA		0.830	−0.616				1.207		0.971	−1.024
SMA		0.839	−0.639				1.042		0.979	−1.044
ZAW	0.99			0.99	2	1.94		0.70		
OLS		1.033	−0.255				1.062		0.658	0.682
MA		1.037	−0.265				1.071		0.751	0.444
SMA		1.037	−0.265				1.070		0.786	0.354

Table 6.3 Comparison of monophasic allometric analyses of *P. l. atrox* and *P. leo.*

Measurement analysis	Slope F-test	Slope P-value	Elevation WALD	Elevation P-value	Shift WALD	Shift P-value
BCW						
OLS	0.952	0.330	3.571	0.059	47.181	0.000*
MA	0.423	0.518	2.884	0.089	50.002	0.000*
SMA	0.060	0.806	2.211	0.137	52.611	0.000*
MFD						
OLS	2.452	0.121	54.127	0.000*	47.933	0.000*
MA	0.127	0.722	44.745	0.000*	69.535	0.000*
SMA	0.261	0.609	44.405	0.000*	71.714	0.000*
NBL						
OLS	0.266	0.609	5.204	0.023*	38.958	0.000*
MA	0.103	0.751	5.874	0.015*	32.712	0.000*
SMA	0.052	0.823	5.808	0.016*	33.541	0.000*
NRA						
OLS	0.238	0.624	7.043	0.008*	32.830	0.000*
MA	0.716	0.394	6.243	0.012*	35.264	0.000*
SMA	0.677	0.409	6.299	0.012*	35.152	0.000*
PLL						
OLS	4.077	0.039*	N/A	N/A	N/A	N/A
MA	2.373	0.122	1.255	0.263	46.886	0.000*
SMA	2.499	0.114	1.186	0.276	46.921	0.000*
POC						
OLS	1.492	0.222	51.488	0.000*	47.848	0.000*
MA	2.780	0.094	49.629	0.000*	55.967	0.000*
SMA	7.420	0.006*	N/A	N/A	N/A	N/A
TBL						
OLS	2.284	0.132	37.430	0.000*	42.083	0.000*
MA	0.233	0.629	41.507	0.000*	27.500	0.000*
SMA	0.173	0.679	41.997	0.000*	26.497	0.000*
TBW						
OLS	0.109	0.745	11.494	0.001*	42.083	0.000*
MA	0.256	0.615	15.737	0.000*	31.166	0.000*
SMA	0.461	0.499	17.231	0.000*	28.429	0.000*
ZAW						
OLS	1.982	0.157	7.746	0.005*	46.073	0.000*
MA	0.781	0.372	8.199	0.004*	43.829	0.000*
SMA	0.818	0.365	8.179	0.004*	43.938	0.000*

Note: *Significant at $p < 0.05$.

Table 6.4 Comparison of multiphasic allometric analyses of *P. l. atrox* and *P. leo.*

Final phase comparison	Slope F-test	Slope P-value	Elevation WALD	Elevation P-value	Shift WALD	Shift P-value
BCW						
OLS	1.622	0.203	4.946	0.026*	15.665	0.000*
MA	2.469	0.114	1.411	0.235	19.863	0.000*
SMA	2.663	0.101	0.323	0.57	22.657	0.000*
MFD						
OLS	0.000	0.995	87.315	0.000*	23.773	0.000*
MA	0.197	0.653	47.915	0.000*	53.047	0.000*
SMA	0.222	0.639	52.001	0.000*	71.864	0.000*
NBL						
OLS	0.019	0.891	2.359	0.125	21.108	0.000*
MA	0.306	0.582	3.484	0.062	15.507	0.000*
SMA	0.272	0.607	3.456	0.063	15.800	0.000*
NRA						
OLS	2.986	0.081	10.765	0.001*	4.614	0.032*
MA	1.258	0.260	6.669	0.010*	6.818	0.009*
SMA	1.364	0.245	7.972	0.005*	6.810	0.009*
PLL						
OLS	2.413	0.119	0.005	0.944	40.705	0.000*
MA	1.142	0.287	0.014	0.904	42.878	0.000*
SMA	1.165	0.279	0.013	0.909	42.846	0.000*
POC						
OLS	3.039	0.081	60.903	0.000*	34.6000	0.000*
MA	4.205	0.041*	N/A	N/A	N/A	N/A
SMA	9.141	0.003*	N/A	N/A	N/A	N/A
TBL						
OLS	0.017	0.899	19.100	0.000*	22.345	0.000*
MA	0.013	0.908	26.015	0.000*	8.431	0.004*
SMA	0.011	0.915	32.231	0.000*	6.000	0.014*
TBW						
OLS	3.096	0.078	0.515	0.473	8.606	0.003*
MA	0.083	0.762	4.823	0.028*	3.419	0.064
SMA	0.054	0.815	8.042	0.005*	3.348	0.067
ZAW						
OLS	2.216	0.135	8.968	0.003*	37.715	0.000*
MA	0.919	0.334	9.650	0.002*	35.406	0.000*
SMA	0.995	0.314	9.604	0.002*	35.631	0.000*

Note: * Significant at $p < 0.05$.

relationships. The results presented here indicate that such considerations are warranted. Of the eight major features examined here that have been argued as distinguishing of *P. l. atrox*, only five showed unambiguous differences when multiphasic allometric analyses were applied. Braincase width showed very little difference between the taxa, although OLS did show a weakly significant difference between the taxa. As mentioned in the methods section, however, OLS, while the most commonly used form of regression, is rarely applicable to allometric data. In the analyses done here, MA and SMA were consistent with regard to significant differences, with any inconsistency (which was rare) coming from OLS. At best, BCW seems to be ambiguous with regard to its independence of size.

The depth of the mandibular symphysis did show significant allometric differences. The elevated allometry of MFD for *P. l. atrox* indicates that the character noted by Merriam and Stock (1932) is indeed valid and allows for some level of distinction between the groups independent of size. The sharper chin resulting from this feature is not surprising in a felid coming from the Hollywood area and may indicate that this region of the country was a celebrity haven some tens of thousands of years ago just as it is today.

The shorter face of *P. l. atrox* does not seem to be a feature that is independent of the size change. The nasals show no allometric difference, indicating that *P. l. atrox* nasals are exactly the length that one should expect from a lion of that size. The length of the palate, another indicator of facial length, also showed no allometric deviation. The only other indicator of this feature (GCL) provided no information due to low sample size. There is, therefore, little evidence that the shorter face observed by some researchers in *P. l. atrox* is independent of the size difference of this organism.

The more dorsal opening of the external nares may have some validity as a feature distinct from size that distinguishes *P. l. atrox* from modern *P. leo*. While the nasals showed no difference, as mentioned above, the overall measure of nareal area did show significant allometric elevation in *P. l. atrox*. The fact that this feature, which is very much a product of the shape of the front of the face, shows deviation when those associated with the face 'shortness' showed no such difference may indicate that the shape of the nareal opening is altered such that it opens more dorsally.

The postorbital constriction of *P. l. atrox* from the La Brea region is definitely more robust than that found in extant lions. Even without strong allometries discernible in either taxon, the POC measurements of *P. l. atrox* were clearly higher, despite their poor correlation with size.

The auditory bullae's lower allometric elevation in *P. l. atrox* is both striking and intriguing. The bullae were absolutely about the same size as one would expect in modern lions (although in the lower end of that spectrum). It is possible that the sound waves that were important in the life of an American lion were shorter than those that are important in modern lions. However, this speculation requires a great deal of further investigation. Nonetheless, the auditory bullae measurements show independence from the size difference between modern and American lions.

The zygomatic arches of *P. l. atrox* are narrower than those of *P. leo*; a feature that, according to the allometric analyses, is independent of the size difference. This result runs counter to the previous literature, which has indicated more strongly arching or wider zygomatic arches. The nature of this feature would not have been discernible without an allometric perspective. The significant difference of zygomatic arch width independent of the size difference supports the argument for species status for *P. l. atrox*. Contrary to previous argument, however, the width across the zygomas is narrower in *P. l. atrox* than it is in *P. leo*.

Conclusions

Assigning fossil specimens to extant taxa is a tricky business. The size difference between *P. l. atrox* and modern *P. leo* has led to several misidentifications of distinguishing features, because the differences between the taxa were not considered allometrically. The features examined in this study revealed that while many of the characteristics used to argue for/against species status are allometrically distinct in the two taxa, several are not. Braincase width and the facial shortness should not be considered differences between the taxa that are independent of size. When measured indirectly through the gap between postorbital processes, orbit size is too variable to be used as a distinguishing character, though other orbital measurements may be of value. Arguments based on such traits only serve to confuse and confound paleontological inquiry. The research presented here does not disprove the hypothesis that the American lion and modern lions are separate species. However, the argument for this hypothesis is weakened by broadly sampled, multiphasic, allometric analyses of these character traits. With the extensive museum collections we have accumulated and the ever-increasing computer power available, investigations into the allometry and variability of diagnostic characters will help refine our understanding of past species and their relationships to extant relatives.

Acknowledgements

I would like to thank the Yale University Department of Geology and Geophysics for providing equipment and funding for this study. The American Museum of Natural History, Page Museum at the La Brea Tar Pits, and the

Canadian Museum of Nature granted me access to their collections and allowed me to take measurements. People at all three institutions were very friendly and helpful, and to them I extend my thanks, especially Eileen Westwig, Michel Gosselin, Margaret Feuerstack, Shelley Cox, and Christopher Shaw. I am deeply grateful to Dr Elisabeth Vrba for introducing me to multiphasic analysis and the types of evolutionary questions that it can address and for her guidance throughout this research. I would also like to thank Brian Andres, Faysal Bibi, Madalyn Blondes, Una Farrell, Emma Schachner, Krister Smith, and two anonymous reviewers for help putting together this chapter and the original presentation that went into it. I would also like to thank Melissa Cohen, Sterling Nesbitt, Stephanie Schollenberger, and Alan Turner for practical support. Finally, I would like to thank Anjali Goswami and Anthony Friscia for their patience and hard work in putting the symposium and this accompanying volume together.

REFERENCES

Akaike, H. (1974). New look at statistical-model identification. *Institute of Electrical and Electronic Engineers Transactions on Automatic Control*, **AC19**, 716–23.

Barry, J. C. (1987). Large carnivores (Canidae, Hyaenidae, Felidae) from Laetoli. In *Laetoli: A Pliocene Site in Northern Tanzania*, ed. M. D. Leakey and J. M. Harris. Oxford: Clarendon Press, pp. 235–59.

Bona, F. (2006). Systematic position of a complete lion-like cat skull from the Eemian ossiferous rubble near Zandobbio (Bergamo, North Italy). *Rivista Italiana Di Paleontologia E Stratigrafia*, **112**, 157–66.

Burger, J., Rosendahl, W., Loreille, O., *et al.* (2004). Molecular phylogeny of the extinct cave lion *Panthera leo spelaea*. *Molecular Phylogenetics and Evolution*, **30**, 841–49.

Dubach, J., Patterson, B. D., Briggs, M. B., *et al.* (2005). Molecular genetic variation across the southern and eastern geographic ranges of the African lion, *Panthera leo*. *Conservation Genetics*, **6**, 15–24.

Groiss, J. T. (1996). Der höhlentiger panthera *Tigris spelaea* (*Goldfuss*). *Neues Jahrbuch Fur Geologie Und Palaontologie-Monatshefte*, **7**, 399–414.

Haas, S. K., Hayssen, V. and Krausman, P. R. (2005). *Panthera leo*. *Mammalian Species*, **762**, 1–11.

Hammer, Ø. and Harper, D. A. T. (2006). *Paleontological Data Analysis*, 1st ed. Malden, MA: Blackwell Publishing.

Harington, C. R. (1969). Pleistocene remains of lion-like cat (*Panthera atrox*) from Yukon Territory and Northern Alaska. *Canadian Journal of Earth Sciences*, **6**, 1277–88.

Harington, C. R. (1971). Pleistocene lion-like cat (*Panthera atrox*) from Alberta. *Canadian Journal of Earth Sciences*, **8**, 170–74.

Harington, C. R. (1977). *Pleistocene Mammals of the Yukon Territory*. Edmonton, Alberta: Department of Zoology, University of Alberta, p. 1060.

Hemmer, H. (1974). Untersuchungen zur Stammesgeschichte der Pantherkatzen (Pantherinae) Teil iii: zur Artgeschichte des Löwen *Panthera (Panthera) leo* (Linnaeus 1758). *Veröffentlichungen der Zoologischen Staatssammlung München*, **17**, 167–280.

Hemmer, H. (1979). Fossil history of living felidae. *Carnivore*, **2**, 58–61.

Herrington, S. J. (1987). Subspecies and the conservation of *Panthera tigris*: preserving genetic heterogeneity. In *Tigers of the World: The Biology, Biopolitics, Management, and Conservation of an Endangered Species*, ed. R. L. Tilson and U. S. Seal. Park. Ridge, NJ: Noyes Publications, pp. 51–62.

Jacobson, A. G. (1993). Somitomeres: mesodermal segments of the head and trunk. In *The Skull*, ed. B. K. Hall and J. Hanken. Chicago, IL: University of Chicago Press, pp. 42–76.

Kurtén, B. (1985). The Pleistocene lion of Beringia. *Annales Zoologici Fennici*, **22**, 117–21.

Kurtén, B. and Anderson, E. (1980). *Pleistocene Mammals of North America, 1*. New York, NY: Columbia University Press.

Langille, R. M. and Hall, B. K. (1993). Pattern formation and the neural crest. In *The Skull*, ed. J. Hanken and B. K. Hall. Chicago, IL: Chicago University Press, pp. 77–111.

Leidy, J. (1853). Description of an extinct species of American lion: *Felis atrox*. *Transactions of the American Philosophical Society*, **10**, 319–21.

Lemon, R. R. H. and Churcher, C. S. (1961). Pleistocene geology and paleontology of the Talara region, northwest Peru. *American Journal of Science*, **259**, 410–29.

Martin, L. D. and Gilbert, B. M. (1978). An American lion, *Panthera atrox*, from natural trap cave, north central Wyoming. *Contributions to Geology*, **16**, 95–101.

McKinney, M. L. and McNamara, K. J. (1991). *Heterochrony: The Evolution of Ontogeny*, New York, NY: Plenum Press.

Merriam, J. C. and Stock, C. (1932). *The Felidae of Rancho La Brea*, 1st ed. Washington, DC: Carnegie Institution of Washington.

Petter, G. (1973). Carnivores pleistocène du ravin d'olduvai. In *Fossil Vertebrates of Africa*, ed. L. S. B. Leakey, R. J. G. Savage and S. C. Coryndon. London: Academic Press, pp. 43–100.

Pocock, R. I. (1930). The lions of Asia. *Journal of the Bombay Natural Historical Society*, **34**, 638–65.

PSI-Plot (2002). Scientific Spreadsheet and Technical Plotting. Pearl River, NY: PolySoftware International.

Simpson, G. G. (1941). Large Pleistocene felines of North America. *American Museum Novitates*, **1136**, 1–27.

Smuts, G. L., Anderson, J. L. and Austin, J. C. (1978). Age-determination of African lion (*Panthera leo*). *Journal of Zoology*, **185**, 115–46.

Sokal, R. R. and Rohlf, F. J. (1995). *Biometry: The Principles and Practice of Statistics in Biological Research*, 3rd ed. New York, NY: W.H. Freeman and Company.

Sotnikova, M. and Nikolskiy, P. (2006). Systematic position of the cave lion *Panthera spelaea* (*Goldfuss*) based on cranial and dental characters. *Quaternary International*, **142**, 218–28.

Sunquist, M. and Sunquist, F. (2002). *Wild Cats of the World*. Chicago, IL: The University of Chicago Press.

Vereshchagin, N.K. (1971). The cave lion and its history in the Holarctic and on the territory of the U.S.S.R. *Trudy of Zoological Institute*, **49**, 123–99.

Vrba, E.S. (1998). Multiphasic growth models and the evolution of prolonged growth exemplified by human brain evolution. *Journal of Theoretical Biology*, **190**, 227–39.

Warton, D.I. and Weber, N.C. (2002). Common slope tests for bivariate errors-in-variables models. *Biometrical Journal*, **44**, 161–74.

Warton, D.I., Wright, I.J., Falster, D.S. and Westoby, M. (2006). Bivariate line-fitting methods for allometry. *Biological Reviews*, **81**, 259–91.

Yamaguchi, N., Cooper, A., Werdelin, L. and Macdonald, D.W. (2004). Evolution of the mane and group-living in the lion (*Panthera leo*): a review. *Journal of Zoology*, **263**, 329–42.

7

Evolution in Carnivora: identifying a morphological bias

JILL A. HOLLIDAY

Introduction

To understand the role of adaptation in generating macroevolutionary patterns, it is necessary to understand whether and in what ways specific features of the phenotype affect subsequent phenotypic diversification. This area has been much debated by both past and present workers, some of who considered whether certain morphologies might be 'channelled' (e.g. Gould, 1984; Emerson, 1988; Wagner, 1996) to appear once a specific starting morphology was attained. Less radically, a number of workers have suggested that possession of certain morphological character states may reduce the ability to attain certain other character states (Lauder, 1981; Maynard-Smith *et al.*, 1985; Emerson, 1988; Futuyma and Moreno, 1988; Wagner, 1996; Werdelin, 1996; Alroy, 2000; Donoghue and Ree, 2000; Wagner and Schwenk, 2000; Wagner, 2001; Wagner and Mueller, 2002; Porter and Crandal, 2003; Van Valkenburgh *et al.*, 2004; Polly, 2008), implying that, in some cases, taxa may be limited in their subsequent evolutionary trajectories. Both morphological channelling and a limitation on specific character states fall into the realm of a character change bias, where certain states are more likely to appear than others (Sanderson, 1993; Wagner, 1996; Donoghue and Ree, 2000; Wagner, 2001; Goldberg and Igic, 2008; Polly, 2008). Despite ongoing theoretical debate, however, there has been relatively little empirical exploration of the possibility of bias or directionality in morphological character change, and this area remains poorly understood (Arthur, 2001, 2004; Schluter *et al.*, 2004).

Much of the available empirical work evaluating questions of bias in character evolution has been performed in a functional context, assessing whether certain starting morphologies act to limit the specific kinds of phenotypes that are subsequently attained (Emerson, 1988; Richardson and Chipman, 2003). Such functional limitations, if they are shown to exist, will necessarily result in

Carnivoran Evolution: New Views on Phylogeny, Form, and Function, ed. A. Goswami and A. Friscia. Published by Cambridge University Press. © Cambridge University Press 2010.

a bias in the appearance of certain phenotypes. In recent years, there has also been increased interest in the study of morphological integration (Olson and Miller, 1958; Pigliucci and Preston, 2004; Goswami, 2006; Polly, 2005) and, in particular, Evolutionarily Stable Systems (ESS) (Wagner and Schwenk, 2000), where sets of characters function (and evolve) as a single, integrated unit. Morphological integration studies are predicated on the presence of strong biases in (or against) character state transitions, but are more frequently discussed in the context of constraint (usually functional). Recent work in the area of integration has also begun to tackle the role that embryological development plays in phenotypic evolution, particularly in the sense of non-independence of character sets and potential biases in character transformations (Donoghue and Ree, 2000; Goswami, 2006). Finally, irreversible evolution, often discussed in the context of 'Dollo's Law', is nothing more than an extreme bias against reversals: a character, once lost, cannot be regained (Igic et al., 2006; Goldberg and Igic, 2008).

Character evolution and specialisation theory

For many years, there has been a general sense among workers in ecology and evolution that ecologically or phenotypically specialised species are very unlikely to revert to a more generalised condition. A significant amount of theoretical discussion addresses the possibility that specialists may be subject to strong stabilising selection to maintain their particular niche, either through habitat tracking or simply due to lower fitness for variants that fall outside the basic (fundamental) niche (Harvey and Pagel, 1991; Holt and Gaines 1992; Losos and Irschick, 1994; Wagner, 1996). Other workers suggest that specialists may in fact be influenced by biased (or directional) selection toward an increasingly specialised morphotype as a result of increasingly fine niche partitioning among specialised forms (Van Valkenburgh, 1991; Losos and Irschick, 1994; Nosil, 2002; Van Valkenburgh et al., 2004).

Because this theoretical framework is already in place, the evolution of specialisation is an obvious choice for detailed testing of questions of bias or directional trends. Unfortunately, to date, empirical studies that evaluate specialisation and how it evolves have produced highly equivocal results (see the review by Futuyma and Moreno, 1988). Furthermore, tests of bias are not only difficult to interpret, but detecting even general patterns is problematic due to varying research scales and methodological approaches (Richardson and Chipman, 2003; see also Cunningham et al., 1998).

Typically, research into character evolution (and specialisation in particular) can be divided into two main areas: the question of total irreversibility (Bull and Charnov, 1985; Emerson, 1988; Moran, 1988; Igic et al., 2006; Goldberg and Igic, 2008) and the ease with which certain character states may be gained or lost (Sanderson, 1993; Wagner, 1995, 1996, Wiens, 1999; Wagner, 2001; McShea and Venit, 2002; Polly, 2008). Clearly, these two areas are not necessarily mutually exclusive: an extreme bias in favour of character state gain over loss would be interpreted as irreversibility. Regardless, studies of either type are often qualitative in nature and their power is accordingly weak: researchers frequently seek to identify a change or reversal in order to test an 'always or never' (or mostly versus seldom) hypothesis (Siddall et al., 1993; Rouse, 1999; Omland and Lanyon, 2000) and few studies provide quantified transition frequencies despite their potential importance (but see McShea, 2001 for metazoans; Igic et al., 2006; see also McShea and Venit, 2002 for the importance of considering frequencies). The need for quantified transition rates is especially evident in irreversibility studies where, without a frequency value, a single instance of a reversal supports a null, despite the recognition that a strong bias would certainly be biologically relevant and in many cases a more interesting finding.

At present, attempts to identify and quantify biases in transition rates remain uncommon (but see Jensen, 1992; Rouse, 1999; McShea, 2001; Bokma, 2002; McShea and Venit, 2002; Geeta, 2003; Igic et al., 2006), although an increasing number of methods allow for quantification of character transitions in both a statistical and non-statistical framework (Harvey and Pagel, 1991; Sanderson, 1993; Hansen and Martins, 1996). Furthermore, much of the available empirical research to date that explicitly assesses differences in transition rates (also known as gain:loss bias) has focused on broad-scale transitions, such as increasing complexity or changes in a macroevolutionary hierarchy (e.g. McShea, 2001; McShea and Venit, 2002; Marcot and McShea, 2007), ecological niche changes (Geeta, 2003) or patterns of sexual dimorphism (Omland, 1997). Only a handful of studies consider specific morphologies or the role particular characters or character complexes play with respect to subsequent adaptive change (Emerson, 1988; Wagner, 1996, 2001; Wagner and Schwenk, 2000, Polly, 2001; Richardson and Chipman, 2003; Holliday and Steppan, 2004).

Understanding potential biases in character evolution is of fundamental importance to workers in fields as diverse as ecology, conservation, systematics, and evolutionary biology, since a bias or limitation in character change can – and will – have a significant effect on the ability of a taxon to survive and adapt. If character change is biased in a particular direction, this bias has implications for patterns of ecological interactions, including competitive ability and guild or community composition. The goal of this contribution is to present a general approach to testing for a gain:loss bias using a variation of the method presented by Sanderson (1993). I provide an example from Carnivora, where

I use replicated sister group comparisons to assess bias in character change relative to a specific dental morphology, that of the hypercarnivore.

The order Carnivora

The order Carnivora is composed of 13 extant and 2 extinct families descended from a mid-Eocene radiation of primitively meat-eating mammals (Wesley-Hunt and Flynn, 2005). The diagnostic character for Carnivora is the carnassial pair, the fourth upper premolar and first lower molar, which in this group have been modified as shearing blades for effective slicing of meat. Although the shearing carnassials are a synapomorphy for Carnivora, member taxa have diversified to occupy a wide range of ecological niches and include highly carnivorous clades such as the cats and weasels as well as generalists, insectivores, omnivores, and even strict herbivores such as the giant panda. Variation in ecology is strongly reflected in the dentition (Van Valkenburgh, 1988, 1989), so a more omnivorous diet is accompanied by a relative increase in grinding surfaces while a more highly carnivorous diet is reflected in a relative decrease in grinding surfaces and an increase in shearing edges (Van Valkenburgh 1988, 1989).

In a series of papers that explored dental morphospace in carnivorans, Van Valkenburgh showed that variables including relative blade length, canine-tooth shape, premolar size and shape, and grinding area of the lower molars distinguished between dietary types in extant carnivores (Van Valkenburgh, 1988, 1989). Using these variables, Van Valkenburgh compared guild compositions of living and extinct carnivoran communities, concluding that each guild comprised a broadly similar set of morphotypes occupying a limited number of ecological niches (Van Valkenburgh, 1988, 1989). There is thus a substantial overlap in morphospace (Crusafont-Pairo and Truyols-Santonja, 1956; Radinsky, 1982; Van Valkenburgh, 1988, 1989) resulting from convergence on similar ecomorphological types, including meat specialists, bone-crackers/scavengers, omnivores, and generalists (Van Valkenburgh, 1988, 1989; Werdelin, 1996).

Of the recognised carnivoran ecomorphs, the niche of the meat specialist, or hypercarnivore, is associated with a diet comprising more than 70% meat, in contrast to the generalist (Van Valkenburgh, 1988, 1989), which may eat 50–60% meat with vegetable matter and invertebrates making up the remainder of the diet. Ecological specialisation to hypercarnivory is associated morphologically with changes in the skull and dentition that include a relative lengthening of the shearing edges (the trigon of the upper fourth premolar and the trigonid of the lower first molar), and reduction or loss of the postcarnassial dentition (the second and third lower molars and first and second upper molars, teeth used for

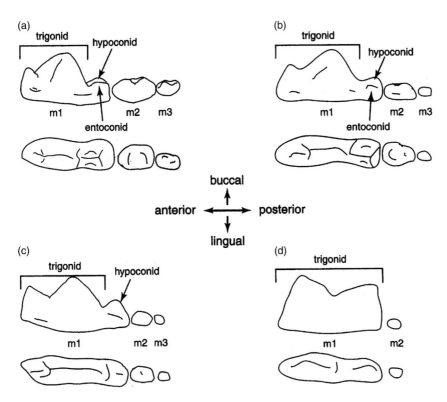

Figure 7.1 'Typical' dentitions indicating increasing specialisation. a, A generalist dentition. The talonid is basined, with the hypoconid and entoconid cusps roughly equal in size. Note that the m2 and m3 are unreduced. b, Dentition trending toward hypercarnivory. The shearing blade is slightly elongate, while the hypoconid and entoconid are unequal in size (hypoconid is larger). The m2 and m3 are somewhat reduced in size. c, Trenchant talonid. The shearing blade is elongate, and the hypoconid is enlarged and medial, while the entoconid is completely reduced. The m2 and m3 are reduced. d, Note the loss of the hypoconid and entoconid. The shearing blade extends the entire length of the m1, the m2 is reduced or absent, and the m3 is absent.

grinding food; Van Valkenburgh, 1989; Hunt, 1998). The facial portion of the skull frequently shortens as well, an alteration thought to be related to maintaining high bite force (Van Valkenburgh and Ruff, 1987; Radinsky, 1981a, 1981b; Biknevicius and Van Valkenburgh, 1996). Figure 7.1a illustrates a generalised carnivoran with a 'typical' tooth formula; individual cusps are labelled. Figure 7.1b–d illustrates hypercarnivorous modifications in order of increasing specialisation. Certain extant or extinct members of such diverse lineages as mustelids, viverrids, canids, hyaenids, amphicyonids, and ursids have all evolved phenotypes characteristic of hypercarnivory (Van Valkenburgh, 1991;

Werdelin and Solounias 1991; Wang 1994; Biknevicius and Van Valkenburgh, 1996; Werdelin, 1996, Wang *et al.*, 1999), although the most extreme cases appear to be in the families Felidae and Nimravidae (Van Valkenburgh, 1991; Holliday and Steppan, 2004).

In a study of evolution of hypercarnivory in the family Canidae, Van Valkenburgh (1991) commented on the low apparent variability in the cranial and dental morphologies of felids and nimravids relative to canids, and suggested that this was possibly due to the extreme specialisation to hypercarnivory in the former two groups. Lack of variation in felid craniodental characteristics has been noted qualitatively by many authors (e.g. Radinsky, 1981a, 1981b; Flynn *et al.*, 1988), and low morphological diversity was quantified by Holliday and Steppan (2004). Cause and effect, however, have yet to be ascertained. There are, of course, a variety of reasons why any particular group might not exhibit certain morphologies, including lack of genetic variation, functional constraint, stabilising selection, or competition (Maynard-Smith *et al.*, 1985; Brooks and McLennan, 1991; Polly, 2008). Additional causes may include intrinsically low rates of evolution or a recent rapid radiation (Schluter, 2000), either of which might suggest a pattern of constraint but actually reflect a lack of time. Because of these possibilities, it is recognised that any study of phenotypic evolution will benefit from the inclusion of as many different groups as possible that exhibit the relevant characteristic (Emerson, 1988; Harvey and Pagel, 1991; Schluter, 2000). Since morphological specialisation to hypercarnivory has occurred repeatedly within Carnivora (Werdelin, 1996; Van Valkenburgh, 1999), these natural replicates allow workers to control or eliminate confounding variables and focus instead on the trait of interest. Likewise, use of sister-group comparisons allows interpretations of character change without the effects of phylogeny. In Holliday and Steppan (2004), six distinct clades of hypercarnivorous taxa were compared to their sister groups in order to determine the effects of specialisation to hypercarnivory on subsequent morphological diversity (disparity and frequency of character change). This current study extends that work, and I utilise the same six pairs of sister groups to evaluate bias in character change on a finer scale.

Previous findings

Holliday and Steppan (2004) quantified the broad-scale effects of specialisation to hypercarnivory on subsequent character change, and showed that not only felids but hypercarnivores as a group are reduced in their morphological diversity relative to sister taxa. Holliday and Steppan (2004) compared the variance of factor scores (Foote, 1993; Wills *et al.*, 1994) from

Table 7.1 Disparity values obtained for each set of sister groups. Disparity was calculated as the sum of the scaled variance of the first five factor scores obtained from PCA. Originally published in Holliday and Steppan (2004).

Hypercarnivore	Disparity	Sister group	Disparity
Felidae	46.20	Hyaenidae	116.66

Hyaenidae:
  Chasmaporthetes
  Lycyaena/Hyaenictis
	44.91	Hyaenidae:
		  Crocuta/Hyaena
			109.95

Nimravidae	90.65	Felidae/Hyaenidae/ Viverridae/Herpestidae	79.01

Canidae:
  Hesperocyoninae
  Enhydrocyon/Philotrox/
  Sunkahetanka
	84.53	Canidae: Hesperocyoninae
		  Cynodesmus
			180.56

Viverridae:
  Cryptoprocta
	21.47	Viverridae:
		  Eupleres/Fossa
			107.64

Mustelidae: *Mustela*	36.87	Mustelidae:
		  Galictis/Ictonyx/
		  Pteronura/Lontra/
		  Enhydra/ Lutra/
		  Amblonyx/Aonyx
			117.60

principal components analysis of six sets of hypercarnivore clades and their sister groups and found that hypercarnivores occupy relatively less morphospace relative to their sister taxa (Table 7.1). They also mapped morphological characters onto phylogenies and calculated average rates of character change for hypercarnivores and their sister groups (Table 7.2). Based on a method proposed by Sanderson (1993), these 'frequency of change' measures indicated that hypercarnivores change character state less often on a given phylogeny (Holliday and Steppan, 2004). Together with the finding of lower variance in hypercarnivores, the lower number of character state transitions suggests that some cause may be acting to limit change within hypercarnivores. Holliday and Steppan (2004) evaluated only average rates of change, however, and did not address whether there might be a difference in rates of forward vs. reverse change (bias) in clades of interest. Consequently, the question remains: do hypercarnivores exhibit less change overall (due to, perhaps, reduced natural variation or stabilising selection [Harvey and Pagel, 1991; Holt and Gaines, 1992; Losos and Irschick, 1994; Wagner, 1995, 1996, 2001; Polly, 2008]) or is the

Table 7.2 Average frequency of change obtained for each set of sister groups, calculated as the number of independent derivations of a character state/number of nodes on the phylogeny. Originally published in Holliday and Steppan (2004).

Hypercarnivore	Average frequency of change	Sister group	Average frequency of change
Felidae	0.1370	Hyaenidae	0.1841
Hyaenidae: Chasmaporthetes/ Lycyaena/Hyaenictis	0.1206	Hyaenidae: Crocuta/ Hyaena	0.1637
Nimravidae	0.2289	Felidae/Hyaenidae/ Viverridae/Herpestidae	0.1838
Canidae: Hesperocyoninae Enhydrocyon/Philotrox/ Sunkahetanka	0.0714	Canidae: Hesperocyoninae *Cynodesmus*	0.1786
Viverridae: Cryptoprocta	0	Viverridae: Eupleres/Fossa	0.3182
Mustelidae: *Mustela*	0.1607	Mustelidae: Galictis/Ictonyx/ Pteronura/Lontra/ Enhydra/ Lutra/ Amblonyx/Aonyx	0.2195

rate difference between specialist hypercarnivores vs. their sister groups due to higher rates of change in one direction (directional evolution due to strong selection [Van Valkenburgh, 1991; Losos and Irschick, 1994; Nosil, 2002; Van Valkenburgh *et al.*, 2004]) and/or reduced rates of change in another (constraint [Wagner, 1995, 1996, 2001; Polly, 2008])?

Quantifying gain:loss bias

In carnivorans, increasing amounts of meat in the diet have been correlated with increasing length of the carnivoran shearing blade on the carnassial tooth (Van Valkenburgh, 1988, 1991). This character, relative blade length (RBL), is defined as the length of the carnassial shearing blade, or trigonid, relative to the length of the entire lower first molar, and is a key feature in understanding dental evolution within Carnivora (Van Valkenburgh, 1988, 1989, 1991; Holliday and Steppan, 2004; Dayan and Simberloff, 2005). In describing possible selective forces that would shape the evolution of the shearing blade in carnivorans, Van Valkenburgh (1991) proposed that strong competition for resources (e.g. meat), even among littermates, should result in directional selection for the evolution of a longer, more efficient, slicing blade. Thus, the first question to be tested is whether there is a difference in rates of change towards a longer shearing blade (higher RBL, described as forward change) relative to rates of change in the reverse direction (shorter RBL, described as reverse change). However, because hypercarnivory is not defined solely by RBL, but is made up of a combination of several characteristics, a second, more complete measure of bias is also required: are there differences in rates of forward change relative to reverse change for characters relevant to the hypercarnivore phenotype? In other words, are there more changes leading towards the hypercarnivorous specialisation rather than away from that specialisation? To answer these questions, I evaluate the competing hypotheses of stasis (stabilising selection, e.g. Holt and Gaines, 1992; Losos and Irschick, 1994; Wagner, 1995, 1996; Polly, 2008), directional selection toward increasing specialisation (e.g. Van Valkenburgh, 1991; Losos and Irschick, 1994; Nosil, 2002; Van Valkenburgh *et al.*, 2004) and the possibility of a limitation on reversals to a more generalised condition (constraint [Wagner, 1995, 1996, 2001; Polly, 2008]). These hypotheses are compared by mapping morphological characters onto phylogenies for each group of interest and calculating relative rates of forward to reverse change. When morphological character states are polarised so that the extreme 'hypercarnivore' phenotype represents an end state reached via multiple gains (and the hypocarnivorous phenotype as an end state reached by multiple losses), then stasis will be observed as no difference in rates of forward (towards hypercarnivory) to reverse change (away from hypercarnivory). Directionality, if present, should be observed as a higher rate of forward change (relatively more gains than losses), and constraint would be indicated by a reduced rate of reverse change (relatively fewer losses than gains).

As with any comparative study, use of sister groups or some other closely comparable taxon is a preferred approach (Lauder, 1981; Harvey and Pagel, 1991; Warheit *et al.*, 1999; Nosil, 2002; Holliday and Steppan, 2004). There are presently several methodological approaches that may be used to quantify transition rates in morphological characters, including the parsimony-based method of Sanderson (1993) and a Markov-based model (Harvey and Pagel, 1991; Pagel, 1994). Because the data used in this study have recognised limitations (missing data), I chose to apply Sanderson's (1993) method. Sanderson's approach has the advantage of being based on information directly available from the phylogeny (number of branches) and characters (number of changes) and mitigates additional complicating assumptions and the risk of over-analysis; the principle of parsimony and its assumptions are well-established.

Using Sanderson's method, a value for frequency of character change (forward changes, reverse changes) can be obtained using only the number of changes in a given character relative to the number of times that character could have possibly changed (Sanderson, 1993; McShea and Venit, 2002). While the accuracy of this method depends on both the underlying topology and on accurate ancestral state reconstructions, either (or both) of which may be subject to significant uncertainty, issues surrounding phylogeny estimation and ancestral state reconstruction have been discussed at length by other workers (e.g. Pagel, 1994; Schluter *et al.*, 1997; Cunningham *et al.*, 1998; Mooers and Schluter, 1999; Omland, 1999; Pagel, 1999; Martins, 2000; Oakley and Cunningham, 2000; Polly, 2001; Finarelli and Flynn, 2006; Igic *et al.*, 2006; Goldberg and Igic, 2008) and are not the focus here. Rather, I present a variation and extension of Sanderson's (1993) approach for testing gain:loss bias using hypercarnivores as a study group.

Methods

Sanderson (1993) presented a method of testing for the presence of a bias in rates of forward change versus rates of reverse change for binary characters. In its simplest form, this approach entails merely counting the number of character state changes on a given phylogenetic tree and then dividing those changes by the number of times the character could have possibly changed, as represented by the branches on the tree (Figure 7.2). The calculation (# changes/# opportunities for change) provides a frequency of change metric (i.e. rate of gain of a character versus rate of loss of a character) that can then be compared between groups to evaluate whether there is a bias. Here, I extend Sanderson's approach, which utilised only binary characters, to include multistate characters. I also use replicated sister-group comparisons to consider rates of forward and reverse change for multiple clades of hypercarnivore ecomorphs, comparing rates of forward change to rates of reverse change within clades and comparing rates of forward change between hypercarnivore clades and their non-hypercarnivorous sister groups. Rates of reversal are compared in the same way. Characters were mapped onto phylogenies using both ACCTRAN and DELTRAN optimisations; results from both optimisations are reported. Use of replicated sister-group comparisons allows testing for patterns that may be applicable across categorical designations (e.g. hypercarnivores as a group), and enables consideration of possible mechanisms that may be responsible for differences in patterns of change on a macroevolutionary scale. Statistical analyses of the paired data were performed using Wilcoxon signed-ranks test (Sokal and Rohlf, 1998). The modified

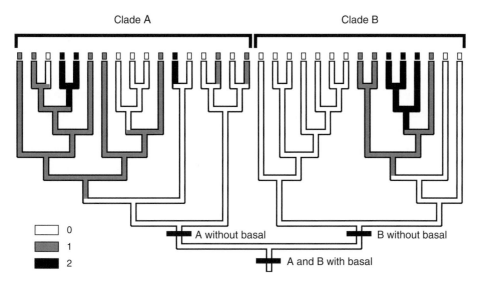

Figure 7.2 Method for calculating frequency of change for multistate characters for a hypothetical sister group pair. Characters are polarised and ordered, with 0 being the least specialised and increasing specialisation indicated by increasing numbered states. Any forward change counts as one step, even if a state is skipped (0–1 or 0–2 both = 1 forward step). Reversals are calculated in the same way (1–0 or 2–0 both = 1 reverse step). The total number of forward or reverse changes is divided by the number of branches that could possibly change (= opportunities for change). Branches already in state 0 cannot reverse further and are excluded when counting branches for reversals. Likewise, branches already in a highest terminal state cannot experience additional forward change and are excluded when counting branches for forward changes. The value obtained from these calculations is the frequency of change, and these values can be averaged over multiple characters for an average frequency of change. Calculations: Clade A: Without basal. There are 5 forward changes and 1 reverse change. There are 26 branches that can possibly move forward (branches in state 2 are excluded). There are 18 branches that can possibly reverse (branches in state 0 are excluded). Clade A: With basal. This adds one more branch to the calculations of forward change and no branches to the calculations of reverse change (this branch is already at state 0). Clade B: Without basal. There are 2 forward changes and no reverse changes. There are 22 branches that could possibly move forward and 4 branches that could possibly reverse. Clade B: With basal. This adds one more branch that could move forward and no additional branches that could reverse. Frequency of forward change: A without basal =.20; A with basal =.19; B without basal =.09; B with basal =.09. Frequency of reverse change: A without basal =.08; A with basal =.08; B without basal = 0; B with basal = 0.

approach presented here represents a novel application of Sanderson's (1993) method, and is also one of the first studies to use sister-group comparisons to quantify and explicitly test for gain:loss bias in a vertebrate group (see also Richardson and Chipman, 2003).

Previously, Holliday and Steppan (2004) used Sanderson's method to calculate average frequency of change for 23 morphological characters in hypercarnivores relative to their sister clades (Table 7.2). However, because the purpose of that study was to quantify overall patterns of morphological diversity rather than to assess bias, characters in that study were chosen to describe general skull and tooth proportions. Consequently, rates of change (=frequency of change) were calculated as total amounts of change on the phylogeny, and any character state change in any direction was considered a change. Unlike that study, the work presented here evaluates a specific subset of 'hypercarnivore' characters that, as recommended by Sanderson (1993), may be expected *a priori* to show bias in their evolution. Further, this study calculates forward changes and reverse changes separately, and then compares forward:reverse change both within hypercarnivore clades (testing for a gain:loss bias within hypercarnivores) and between hypercarnivores and their sister groups (testing rates of forward change in hypercarnivores relative to their sister groups, testing rates of reverse change in hypercarnivores and their sister groups).

Multistate characters

As presented, Sanderson's (1993) method considers differences in rates of gain versus rates of loss for a single binary character. Here, I have extended the method to include multistate characters. The primary difference in calculating frequency of change for binary versus multistate characters involves terminal character codes. For example, under Sanderson's method, a character with state zero cannot experience a reversal because it cannot go to a lower state. Likewise, a character in state one cannot experience a new forward change because, in a binary system, it is already at the highest state (Sanderson, 1993). When counting change, then, forward changes (0–1) are possible only for branches at state 0. Likewise, reverse changes (1–0) are possible only for branches at state one.

Incorporating multistate characters is fairly straightforward: regardless of the number of character states, if characters are ordered and polarised so that 0 represents the character value most distant from that observed in taxa with the hypercarnivorous specialisation (i.e. hypocarnivory), change is calculated in essentially the same way as for binary characters. Thus, a character at state 0 cannot reverse – but all other character states may – and its branch is not counted when calculating reversals. Likewise, a character in the terminal (highest) state cannot move forward – all other states can – and that branch is not counted when calculating forward change (Figure 7.2). End states are based on the entire range of values obtained from all carnivoran taxa in the study.

Data

Sanderson (1993) suggested that patterns of bias might be easier to detect if groups of characters could be selected *a priori* for some functional or biological reason. In these analyses, morphological characters were chosen based on their functional significance for hypercarnivores, and are a specific subset of characters previously described in Holliday and Steppan (2004). As stated previously, relative blade length (RBL) has been correlated with the amount of meat in the diet (Van Valkenburgh, 1988) and is considered a key character. Additional characters used in this study include length of the trigonid (the carnassial shearing blade) relative to the remaining tooth surfaces (blade/GSL), shape of the lower fourth premolar (p4 shape, width of the lower fourth premolar divided by its length), ratio of the length of the upper fourth premolar relative to the length of the upper first molar (P4/M1), length of the upper first molar relative to tooth row length (M1/TRL), and total grinding surfaces relative to the length of the lower carnassial tooth (GSL/m1). Each of these characters represents features that are known to modify as lineages move toward the hypercarnivore phenotype.

To ensure that character codings were consistent across all families, quantitative data from all taxa were pooled prior to coding each character. Values were then ordered from smallest to largest, segment coded, and discretised (Simon, 1983; see also Chappill, 1989). Segment coding was used instead of gap coding because the pooled data were continuously distributed (i.e. there were no gaps). Characters were polarised so that lower states indicate lower levels of carnivory; higher states represent changes toward hypercarnivory. Most characters were divided into between 4 and 6 states. Segment sizes were chosen to reflect the biological relevance of particular phenotypes as much as possible. For example, while a two-state character based on arbitrary segment sizes of 0–0.50 and 0.51–1.0 could very likely mask information about intermediate phenotypes, three segments of 0–0.35, 0.35–0.70, and 0.71–1.0 are much more informative. It is acknowledged that opinions may differ regarding the most appropriate method for coding quantitative data; however, no method has yet been shown to be 'best' (Wiens, 2000; Zelditch *et al.*, 2000). Missing data were not replaced; individual species or taxa for which data were not available were excluded for that character. For each clade, character changes were counted for each individual character and then divided by available branches. Any change was counted as a single step, regardless of the number of states in between (e.g. a change in state from 1 to 2 or from 1 to 3 was still a single step).

After frequencies of forward and reverse change were calculated for each character and for each clade under study, individual frequencies over all six characters (per clade) were combined to obtain an average rate of forward change and an average rate of reverse change. These characters together comprise the 'hypercarnivore complex', i.e. the characters that are expected to change as a unit as taxa become hypercarnivorous. In limited cases (*Mustela*, *Enhydrocyon*, *Lycyaena*), missing character data or incomplete taxon sampling prevented inclusion of one or two individual characters for that clade. In those cases, the average was calculated based on the remaining four or five characters. Thus, for every group, a set of metrics was produced that provided information on forward and reverse change both on an individual character basis and as an average rate of change overall. Within and between group comparisons were performed for the character RBL and for the average of all characters (the hypercarnivore complex). RBL was the only character fully assessed on an individual basis because this character is considered the most indicative of transformations toward a high meat diet (see Van Valkenburgh, 1988, 1989) and therefore the most likely to show an individual bias.

Phylogenies: ingroups

In their study of morphological diversity in hypercarnivorous carnivorans, Holliday and Steppan (2004) used molecular or morphological phylogenies drawn from the literature. In some cases, complete and/or robust phylogenies were not available and it was necessary to build composite trees based on partial phylogenies. Relevant details of taxa, phylogenetic hypotheses, and justifications for inclusion are presented in Holliday and Steppan (2004). At the time of the original study, they noted that the phylogenies were inadequate, and likewise recognised that new hypotheses could have an effect on their findings. Since that time, several new phylogenies have indeed been published and some clarification has been achieved regarding sister-group relationships between groups of taxa (see, e.g. Yoder *et al.*, 2003; Veron *et al.*, 2004; Flynn *et al.*, 2005; Gaubert and Cordeiro-Estrela, 2006; Johnson *et al.*, 2006; Koepfli *et al.*, 2006; Perez *et al.*, 2006, Holliday, 2007; see also review by Barycka, 2007). Despite this, for the present analyses, the earlier topologies were retained in order to maintain consistency with the original rate of change studies performed by Holliday and Steppan (2004).

Following Holliday and Steppan (2004), hypercarnivorous taxa included in this study are as follows: Felidae (cats), Nimravidae (sabre-toothed non-cats), the mustelid genus *Mustela* (weasels), the herpestid *Cryptoprocta ferox* (the Malagasy fossa), a clade of hypercarnivorous hyaenids (*Chasmaporthetes–Lycyaena–Hyaenictis*), and a clade of early hesperocyonine canids (*Enhydrocyon–Philotrox–Sunkahetanka*). Sister-group comparisons (hypercarnivore/sister group) included Felidae/Hyaenidae (hyenas), Nimravidae/Aeluroidea (felids, hyaenids, viverrids and herpestids), *Mustela/Enhydra–Lutra* (the otters), *Cryptoprocta/Eupleres–Fossa* (the falanouc and the Malagasy civet), *Chasmaporthetes–Lycyaena–Hyaenictis/Hyena–Crocuta* (bone-cracking hyenas, including fossil forms), and *Enhydrocyon–Philotrox–Sunkahetanka/Cynodesmus* (a small to mid-sized, generalist hesperocyonine canid).

Within-group comparisons

To understand the transition rates of the characters and character complexes under study, outgroup and sister-group comparisons were used to establish an 'expected' pattern of forward versus reverse change (Schwenk and Wagner, 2004). In these analyses, I calculated rates of forward and reverse change for clades of broadly generalised (non-specialised) carnivorans and for the sister groups of the hypercarnivorous taxa. I then compared the two metrics (forward: reverse) *within* each clade to determine whether unspecialised groups exhibit any detectable difference in their rate of forward change relative to their rate of reverse change. These analyses provided a baseline for subsequent comparisons of rates of forward and reverse change within hypercarnivore clades. Non-specialist taxa treated as outgroups for baseline values include Caninae (dogs and foxes), Herpestidae (mongooses), Hesperocyoninae (a basal canid subfamily), and Mustelidae (weasels, otters and stoats). It should be noted that some of these groups (Mustelidae, Hesperocyoninae) do have hypercarnivorous taxa (and their sister groups) nested within them. However, since on the whole these larger clades can be considered generalist or non-specialist, the more inclusive clade was evaluated as a 'generalist' taxon under the justification that any individual biases in rates of change due to presence of a hypercarnivore/sister clade would be balanced by the inclusion of the remainder of the family or subfamily as long as that group is broadly unspecialised.

Between-group comparisons

One limitation of within group comparisons is that, even when a rate difference exists, one cannot determine whether the difference is due to a relatively higher rate of forward change (directional selection) or a relatively lower rate of reverse change (constraint) without reference to an outgroup or a sister group. To address this difficulty, I used sister-group comparisons to compare relative rates of forward change between hypercarnivores and their sister groups. I also compared relative rates of reverse change between hyper-carnivores and their sister groups.

Methodological issues

Because the sister-group method introduces several variations of Sanderson's original approach, certain methodological issues must be addressed. The most striking of these is intrinsic to the use of paired sister groups, and involves the way in which branches and changes are counted. Sanderson's (1993) description of his method states that change should be counted 'from the root' – but he did not address the question of paired clades. Depending on how the root is defined (as the basal portion of an individual clade or the root of the entire sister-group pair; see Figure 7.2), a branch may be added or lost during rate calculations. In practice, inclusion or exclusion of this single branch should not significantly affect the frequency of change values unless a clade is extremely small. However, precisely this circumstance arose during rate calculations for the sister-group set composed of *Cryptoprocta/ Eupleres–Fossa*. In these analyses, the hypercarnivorous herpestid genus *Cryptoprocta* is represented by a single species, and inclusion of this individual taxon without including the shared basal branch leaves the frequency of change metric undefined ($(2 * 1-2)/1 = 0$), effectively eliminating it as a data point. For this reason, in comparisons labelled 'without basal branch', *Cryptoprocta* and its sister group are excluded. At the same time, exclusion of a valid hypercarnivorous taxon or comparison could bias the results, as well as reduce the already low statistical power by permitting only five sets of comparisons. All analyses were therefore performed both with and without the shared root of the sister group pairs (with and without the basal branch), and I was thus able to include *Cryptoprocta* and its putative sister group (*Eupleres/Fossa*) in some computations (all computations labelled w/basal). For all analyses that include this basal branch, the root state is established using outgroup comparison.

Models of evolution

When branch lengths are not included in a frequency of change analysis, a punctuated mode of evolution is implicit. Such a model assumes that all change occurs only at speciation, and branch lengths are not taken into account. Under this assumption, it is not possible to differentiate between a group that has recently undergone explosive radiation (e.g. felids) and an older

Table 7.3 Calculated rates of forward and reverse change for generalist groups for relative blade length and the hypercarnivore complex under both ACCTRAN and DELTRAN optimisations. There is no significant difference between rates of forward change and rates of reversal under either optimisation.

Relative blade length	ACCTRAN		DELTRAN	
	Forward	Reverse	Forward	Reverse
Mustelidae	0.15	0.11	0.13	0.09
Herpestidae	0.08	–	0.08	–
Hesperocyoninae	0.14	0.02	0.14	0.02
Caninae	0.11	0.06	0.11	0.06
	No significant difference		No significant difference	
Hypercarnivore Complex	Forward	Reverse	Forward	Reverse
Mustelidae	0.06	0.09	0.07	0.09
Herpestidae	0.10	0.08	0.08	0.07
Hesperocyoninae	0.08	0.02	0.08	0.02
Caninae	0.09	0.06	0.10	0.05
	No significant difference		No significant difference	

group that has had significant time to evolve (hyaenids or viverrids), despite the likely impact of time on patterns of change. Lack of branch length data is a recognised shortcoming of the current analysis (e.g. Finarelli and Flynn, 2006; Polly, 2008). It is anticipated that as new phylogenies and more complete data become available, this situation will be rectified.

Results

All analyses were performed using Wilcoxon Ranked Pairs, a non-parametric test for comparison of paired data that performs especially well when sample sizes are small (Sokal and Rohlf, 1998). Frequency of change values from both ACCTRAN and DELTRAN optimisations are provided in Tables 7.3–7.7. Tables 7.8 and 7.9 are summaries of p values (two-tailed, sig. at 0.05) for different comparisons.

Relative blade length: without basal branches (five comparisons)

For within-group comparisons, there was no significant difference between rates of forward and reverse change for generalised clades (Table 7.3). Differences in rates of forward and reverse change for sister groups showed only

Table 7.4 Frequency of change values for relative blade length for hypercarnivores and their sister groups under both ACCTRAN and DELTRAN optimisations. Calculations performed without basal branches.

ACCTRAN

Clade	Hypercarnivores			Sister groups	
	Forward	Reverse		Forward	Reverse
Enhydrocyon	0.10	0	Cynodesmus	0.50	0.50
Felidae	0.60	0.08	Viverridae	0.08	0.12
Mustela	0.17	0	Enhydra/Lutra	0	0.33
Chasmaporthetes	0.19	0.10	Hyena/Crocuta	0.08	0.18
Nimravidae	0.18	0.08	Aeluroidea	0.10	0.11

DELTRAN

Clade	Hypercarnivores			Sister groups	
	Forward	Reverse		Forward	Reverse
Enhydrocyon	0.10	0	Cynodesmus	0.50	0.50
Felidae	0.40	0.06	Viverridae	0.03	0.32
Mustela	0.17	0	Enhydra/Lutra	0	0.33
Chasmaporthetes	0.04	0	Hyena/Crocuta	0.14	0.11
Nimravidae	0.19	0	Aeluroidea	0.10	0.11

Table 7.5 Frequency of change values for relative blade length for hypercarnivores and their sister groups under both ACCTRAN and DELTRAN optimisations. Calculations performed with basal branches.

ACCTRAN

Clade	Hypercarnivores			Sister groups	
	Forward	Reverse		Forward	Reverse
Enhydrocyon	0.09	0	Cynodesmus	0.11	0.06
Felidae	0.50	0.08	Viverridae	0.07	0.08
Mustela	0.16	0	Enhydra/Lutra	0.09	0.19
Chasmaporthetes	0.18	0.09	Hyena/Crocuta	0.12	0.07
Nimravidae	0.17	0.07	Aeluroidea	0.06	0.06
Cryptoprocta	0	0	Eupleres/Fossa	0	0.28

Table 7.5 (*cont.*)

DELTRAN

Clade	Hypercarnivores			Sister groups	
	Forward	Reverse		Forward	Reverse
Enhydrocyon	0.09	0	Cynodesmus	0.33	0.33
Felidae	0.40	0.06	Viverridae	0.09	0.17
Mustela	0.16	0	Enhydra/Lutra	0	0.29
Chasmaporthetes	0.33	0	Hyena/Crocuta	0.13	0.10
Nimravidae	0.19	0	Aeluroidea	0.10	0.11
Cryptoprocta	0	0	Eupleres/Fossa	0	0.33

Table 7.6 Frequency of change values for the hypercarnivore complex for hypercarnivores and their sister groups under both ACCTRAN and DELTRAN optimisations. Calculations performed without basal branches.

ACCTRAN

Clade	Hypercarnivores			Sister groups	
	Forward	Reverse		Forward	Reverse
Enhydrocyon	0.09	0.03	Cynodesmus	0.25	0.13
Felidae	0.13	0.04	Viverridae	0.08	0.12
Mustela	0.09	0.03	Enhydra/Lutra	0.10	0.22
Chasmaporthetes	0.11	0.02	Hyena/Crocuta	0.14	0.08
Nimravidae	0.14	0.10	Aeluroidea	0.06	0.06

DELTRAN

Clade	Hypercarnivores			Sister groups	
	Forward	Reverse		Forward	Reverse
Enhydrocyon	0.09	0	Cynodesmus	0.25	0.13
Felidae	0.11	0.02	Viverridae	0.07	0.11
Mustela	0.05	0.06	Enhydra/Lutra	0.10	0.16
Chasmaporthetes	0.15	0	Hyena/Crocuta	0.18	0.04
Nimravidae	0.14	0.04	Aeluroidea	0.06	0.06

Table 7.7 Frequency of change values for the hypercarnivore complex for hypercarnivores and their sister groups under both ACCTRAN and DELTRAN optimisations. Calculations performed with basal branches.

ACCTRAN

Clade	Hypercarnivores			Sister groups	
	Forward	Reverse		Forward	Reverse
Enhydrocyon	0.12	0.02	Cynodesmus	0.11	0.06
Felidae	0.11	0.04	Viverridae	0.07	0.08
Mustela	0.08	0.03	Enhydra/Lutra	0.09	0.19
Chasmaporthetes	0.08	0.02	Hyena/Crocuta	0.12	0.07
Nimravidae	0.13	0.09	Aeluroidea	0.06	0.06
Cryptoprocta	0.33	0	Eupleres/Fossa	0	0.28

DELTRAN

Clade	Hypercarnivores			Sister groups	
	Forward	Reverse		Forward	Reverse
Enhydrocyon	0.12	0	Cynodesmus	0.11	0.06
Felidae	0.11	0.02	Viverridae	0.09	0.06
Mustela	0.04	0.05	Enhydra/Lutra	0.08	0.14
Chasmaporthetes	0.10	0	Hyena/Crocuta	0.14	0.03
Nimravidae	0.14	0.03	Aeluroidea	0.06	0.06
Cryptoprocta	0.25	0	Eupleres/Fossa	0	0.33

marginal significance under ACCTRAN ($p < 0.07$, Table 7.8) and no significant difference under DELTRAN ($p = 0.27$, Table 7.8). There was a significant difference between rates of forward and reverse change for hypercarnivore clades under either optimisation ($p < 0.04$, Table 7.8). It is worth noting that, for within-group comparisons, the differences in rates of forward and reverse change for the *sister* groups of hypercarnivores and for hypercarnivores were in opposite directions. More specifically, within-group comparisons showed that hypercarnivores exhibit relatively lower rates of reversal away from hypercarnivory (or higher rates of forward change towards hypercarnivory) while their sister groups exhibit relatively higher rates of reversal away from hypercarnivory (or lower rates of change in the direction of hypercarnivory) (Table 7.4). This suggests that different and potentially opposing selective forces may be influencing the evolution of RBL of both sets of taxa.

Table 7.8 Summary of p values for relative blade length. The table shows p values obtained from comparisons of ACCTRAN and DELTRAN optimisations as well as results obtained with and without the basal branches. The first column indicates the type of comparison; the top row whether comparisons were within taxon or between sister groups. The column labelled Hypercarnivores FOR:REV shows p values obtained from comparisons of relative rates of forward change to relative rates of reversal within hypercarnivore clades; the column labelled Sister FOR:REV indicates p values obtained from comparisons of relative rates of forward change to relative rates of reversal within the sister groups of hypercarnivores; the column labelled FORWARD indicates the p values obtained from sister group comparisons (hypercarnivores versus sister groups) for rates of forward change; the column labelled REVERSE shows p values from sister group comparisons for rates of reverse change.

| | WITHIN TAXON | | SISTER-GROUP COMPARISONS | |
	FOR:REV Hypercarnivores	FOR:REV Sister	FORWARD	REVERSE
ACCTRAN with basal	0.042	0.043	0.345	0.027
ACCTRAN without basal	0.042	0.068	0.345	0.043
DELTRAN with basal	0.043	0.138	0.345	0.027
DELTRAN without basal	0.043	0.273	0.500	0.043

For sister-group comparisons, relative rates of forward change for hypercarnivores versus their sister groups were not significantly different (Table 7.8). However, relative rates of reverse change for hypercarnivores versus their sister groups indicated significantly lower relative rates of reversal for hypercarnivores under either ACCTRAN or DELTRAN optimisations ($p < 0.04$, Table 7.8).

Relative blade length: with basal branches (six comparisons)

When basal branches were included in within-group comparisons, relative rates of forward:reverse change for hypercarnivores were all significantly different ($p < 0.05$, Table 7.8). The sister groups of hypercarnivores were significantly different under ACCTRAN ($p < 0.05$, Table 7.8) but not under DELTRAN ($p = 0.138$, Table 7.8) and, as in the analyses that did not include

Table 7.9 Summary of p values for the hypercarnivore complex, which is the average of values for the following characters: RBL, p4 shape, blade/GSL, P4/M1, M1/TRL, GSL/m1. Characters are explained in the text. Table format follows that of Table 7.8.

	WITHIN TAXON		SISTER-GROUP COMPARISONS	
	FOR:REV Hypercarnivores	FOR:REV Sister	FORWARD	REVERSE
ACCTRAN with basal	0.028	0.463	0.173	0.046
ACCTRAN without basal	0.043	0.893	0.893	0.080
DELTRAN with basal	0.046	0.917	0.463	0.028
DELTRAN without basal	0.080	0.686	0.686	0.043

basal branches, the directions of the difference for hypercarnivores and their sister taxa were in opposite directions.

For sister-group comparisons that included basal branches, rates of forward change for hypercarnivores and their sister groups and rates of reversal for hypercarnivores and their sister groups showed no significant difference in relative rates of forward change (Table 7.8), but a significant decrease in the relative rate of reversal for hypercarnivores as compared to their sister groups under either optimisation ($p < 0.05$, Table 7.8).

Hypercarnivore morphotype without basal branches (five comparisons)

Because the evolution of the hypercarnivore morphotype involves changes in more than a single character, it was important to test the evolution of the entire character complex. Results of within-group comparisons for analyses of the hypercarnivore complex were very similar to those obtained for RBL. There was no significant difference between rates of forward and reverse change for generalised (non-specialist) clades (Table 7.3). There was also no significant difference between forward and reverse rates for sister groups of hypercarnivore clades (Table 7.9). There was a significant difference between rates of forward and reverse change for hypercarnivore clades under ACCTRAN ($p < 0.03$) and DELTRAN ($p < 0.05$, Table 7.9), and rates of forward change were higher than rates of reverse change (Table 7.6).

Relative rates of forward change for hypercarnivores versus their sister groups and relative rates of reverse change for hypercarnivores versus their sister groups showed no significant difference in the rate of forward change for the two groups (Table 7.9), but a decreased rate of reversal for hypercarnivores relative to their sister taxa. For the five comparisons, this rate was non-significant ($p < 0.08$) under ACCTRAN and barely significant $p < 0.05$ under DELTRAN (Table 7.9).

Hypercarnivore morphotype with basal branches (six comparisons)

For the character complex, within-group comparisons indicated no significant difference between rates of forward and reverse change for sister groups of hypercarnivore clades (Table 7.9). There was a significant difference ($p < 0.03$, $p < 0.05$, Table 7.9) between rates of forward and reverse change for hypercarnivore clades under ACCTRAN and DELTRAN, respectively, with forward changes outnumbering reverse changes (Table 7.7).

Comparison of relative rates of forward change for hypercarnivores and their sister groups and rates of reverse change for hypercarnivores and their sister groups showed no significant difference in the rate of forward change for the two groups (Table 7.9), but a significantly decreased rate of reversal for hypercarnivores relative to their sister groups under either ACCTRAN or DELTRAN ($p < 0.05$, $p < 0.03$, respectively, Table 7.9).

Discussion

Given the availability of methods for evaluating character evolution and quantifying transition frequencies, identification of broad patterns in character evolution (e.g. trends or unusual levels of disparity in a particular clade) is only a first, necessary step toward addressing more specific questions of character bias and causality. Clearly, the presence of a constraint or limitation on character change, or strong selection for some feature or phenotype, can have significant effects on the directions and rates of morphological evolution and/or species diversification. It follows that these effects, when strong enough, will produce recognisable macroevolutionary patterns. Once these patterns are identified, however, the focus should shift to questions about mechanisms (e.g. Alroy, 2000; Wagner, 1995, 1996; Polly, 2008). For example, given an observed difference in morphological diversity for a specific taxon, is that difference a result of a higher rate of morphological change in one clade (strong selection for a particular kind of change), or a reduced rate in another (a constraint or limitation against a specific change)? How large (quantitatively) is the difference in rates? How strong is the pattern?

Previously, Holliday and Steppan (2004) showed that hypercarnivores occupy less morphospace and exhibit lower rates of character change relative to their sister taxa. To better understand the underlying causes of this disparity, I used sister group and outgroup comparisons to evaluate rates of forward and reverse change explicitly. As the results show, hypercarnivores have a significant difference in their rates of forward change relative to rates of reverse change for both RBL and for the hypercarnivore complex (Tables 7.8 and 7.9). Further, comparison with sister groups indicates that this is due to a decreased rate of reversal, not an increased rate of forward change (Table 7.9). Potential explanations for the lower overall frequency of change in hypercarnivores relative to their sister taxa included stabilising selection, directional selection, or constraint. By breaking down overall frequency of change into rates of forward change and reverse change, however, it becomes clear that neither stabilising selection nor strong directional selection can fully explain the observed patterns.

Stabilising selection, which would maintain the phenotype 'as is' without significant shifts towards or away from the optimum, is rejected because there is no apparent limitation on forward change for hypercarnivorous taxa. Strong directional selection is also not supported: although forward changes do occur at a higher rate than reversals within hypercarnivores, in no analyses are rates of forward change significantly different between hypercarnivores and their sister taxa. Instead, it appears that hypercarnivores are 'ratcheting' forward (see also Van Valkenburgh et al., 2004) into increasingly specialised morphospace, with no limitation on change in a forward direction but with significant limitations on reversals. This contrasts with the situation in generalists and sister taxa, where the rate of forward and reverse change is not significantly different and forward shifts are roughly balanced by reversals.

Based on current interpretations of macroevolutionary patterns, theory also contradicts directional selection as a causal mechanism for the evolution of hypercarnivory. Whether directional selection can produce a measurable trend is not at question in this study (see, e.g. Alroy, 2000; Knoll and Bambach, 2000). However, whether directional selection can persist over extremely long time scales (e.g. hundreds of thousands or millions of years) is a less likely possibility (McShea, 1994, 2001; Sheldon, 1996; Polly, 2004), and it is unlikely that hypercarnivores would have been continuously exposed to the same selective factors through their entire history. Over the history of any clade, it is reasonable to anticipate that there will be periods where various perturbing influences, including competition, climate change, or other ecological and environmental factors, will alter selection regimes and, hence, the direction of phenotypic evolution (Sheldon, 1996). During such times, when the shape of the adaptive landscape itself is varying, some changes in the observed phenotype should also result (Sheldon, 1996). This is a key point, since Van Valkenburgh et al. (2004) made an important observation regarding hyper-carnivorous canids: these taxa do not exhibit adaptive change when their environments change (i.e. they do not exhibit shifts in phenotype when selection is relaxed). Instead, they have a tendency to go extinct (Van Valkenburgh et al., 2004). While these workers' results can only be applied to the taxa studied (canids), both their findings and those presented here support the possibility that hypercarnivores as a group are strongly limited in their ability to respond to environmental/ecological change, at least in the sense of reversing to a more generalised condition. It appears that the lack of reversals to a more generalised condition has a greater effect on the evolution of hypercarnivory than does directional selection towards the specialisation.

Now that the 'how' has been established (low rate of reversal), the question that follows is, of course, 'Why?' Why can hypercarnivorous taxa not reverse or shift their phenotype as selective pressures change? What processes might be acting to bring about this constraint? Given the present data, several possibilities remain, including genetic or developmental limitations, and functional constraint.

A generally accepted definition of developmental constraint is 'a bias or limitation on the production of variant phenotypes' (Maynard-Smith et al., 1985). Such a constraint is typically viewed as the result of selection on gene interactions and gene products during embryological development (Arthur, 2001; Fusco, 2001; Salazar-Ciudad, 2006). Since developmental constraint is directly predicated on the viability of specific gene combinations (Maynard-Smith et al., 1985; Fusco, 2001; Schwenk and Wagner, 2001; Salazar-Ciudad, 2006), a selective bias in the production of certain phenotypes could feasibly occur for any number of reasons (Arthur, 2001, 2004; Fusco, 2001; Polly, 2008). Having said that, it is recognised that developmental constraint will be difficult to identify, much less reject, without significant detailed study (Arthur, 2001; Fusco, 2001; Schluter et al., 2004). In this case, the use of replicated sister-group comparisons is the best available evidence against such a constraint: it is unlikely (although by no means impossible) that the same developmental bias against reversals would appear repeatedly in separate taxa (but see, e.g. Schluter et al., 2004).

In contrast to developmental constraint, functional constraint affects variants of the phenotype from the standpoint of phenotypic construction or integration, and maintains the ability of a character complex to function effectively as a unit by eliminating less well-suited variants of that whole (Wagner and Schwenk, 2000; Schwenk and Wagner, 2001). Further, theoretical predictions involving morphological integration, arguably the most extreme version of functional constraint, suggest that the functionality of a particular suite of characters is affected not just by the existence of a complex of correlated (and functionally related characters), but by the specific interactions among the characters (Olson and Miller, 1958; Schwenk and Wagner, 2001, 2004; see also Eble, 2004). In fact, an integrated unit may lose its ability to function at all if an individual character varies outside a particular range (Wagner and Schwenk, 2000), and such integration would allow for only limited phenotypic change in any character at any time (Wagner and Schwenk, 2000). At the same time, any change that acts to improve the fit of the characters to the functional unit (i.e. forward change toward increasing integration, or in this context, increasing specialisation) would be selected for, while reverse changes in any one characteristic, which would decrease the integration of the characters and hence negatively affect functionality, would be strongly selected against (Bull and Charnov, 1985; Wagner and Schwenk, 2000). In the case of hypercarnivores, then, change in individual characters in the direction of increased adaptation to meat-eating would be reinforced – or at least not selected against – while reverse change, particularly for individual characters, would exhibit a very low rate of occurrence. This is precisely the pattern exhibited by both data sets.

Conclusions

Studies to date have evaluated the evolution of hypercarnivory at steadily increasing levels of detail (Holliday and Steppan, 2004; Van Valkenburgh et al., 2004). The patterns described here are clearly the result of some constraint – either developmental or functional or both. Determining which of these possibilities is the most likely – and whether there is a 'point of no return' in hypercarnivore morphospace, after which reversals cannot occur – will require even more fine-scale study, including a better understanding of character correlations and development (both evolutionary and ontogenetic) within Carnivora. Studies with the specific intent of quantifying types and amounts (degrees) of character correlations and levels of morphological integration would be ideal, and some work has already begun in this area (Meiri et al., 2005; Goswami, 2006; see also Polly, 2008). Improvements in quantitative methods should allow more specific testing of pattern/process hypotheses by enabling the comparison of developmental data for various species to the evolution of character complexes over time and across phylogenies (Steppan, 1997b; Marroig and Cheverud, 2001). Furthermore, recent methodological advances in evaluating and comparing the structure of phenotypic variance-covariance matrices at multiple phylogenetic levels (Steppan, 1997a,b; Marroig and Cheverud, 2001; Baker and Wilkinson, 2003; Polly, 2008) may offer significant insights into the myriad factors that affect evolutionary patterns.

Hypercarnivores have a significant difference in their rates of forward change to reverse change for both relative blade length and for the hypercarnivore complex (Tables 7.8 and 7.9). Comparison with sister groups indicates that this is primarily due to a decreased rate of reversal, not a strongly increased rate of forward change (Tables 7.4–7.7). These findings are little affected by character optimisation (ACCTRAN or DELTRAN) or minor modifications of the approach (inclusion or exclusion of basal branches), although inclusion of basal branches (which allows inclusion of *Cryptoprocta ferox* and its sister group) generally strengthens the findings, almost certainly because of the concomitant increase in statistical power.

Instead of support for the prediction of directional evolution toward a longer shearing blade or a more specialised hypercarnivore phenotype, I found a pattern of reduced reversals to a more primitive condition. These results suggest that the evolution of the hypercarnivore specialist morphotype proceeds very much as an evolutionary ratchet (see also Van Valkenburgh *et al.*, 2004), allowing taxa to passively move forward into an increasingly specialised morphospace but inhibiting reversals to a more generalised condition. The finding that hypercarnivorous taxa are able to move only in a forward direction leads to the inescapable conclusion that these taxa are under the influence of a strong constraint.

The approach described here, which combines the frequency of change method of Sanderson (1993) with the comparative tool of replicated sister-group comparisons, is a promising new way to handle old macroevolutionary questions – in this case, to test for directionality in character evolution, or gain: loss bias. It should be emphasised that the results presented in this paper must be interpreted cautiously; as with any character evolution study, new phylogenies can potentially alter these results (Pagel and Harvey, 1988). Despite these issues, many additional applications and extensions of this methodology are readily apparent: sister-group comparisons, character comparisons, comparison of ecomorphs or time series data. Further, as noted by Sanderson (1993), power may be increased by including sets of characters that, as with hypercarnivores, are all expected to exhibit the same behaviour, while rigour may be increased by incorporating branch length information or maximum likelihood estimates. In its most basic form, however, the method requires only a phylogeny, making it accessible to workers in a variety of fields.

Acknowledgements

I would like to express appreciation for discussion and support throughout the course of this study to S. Steppan, G. Erickson, J. Travis, D. Swofford, and B. Parker. I thank J. Albright, J. Burns, M. Reno, K. Rowe, and L. VandeVrede for technical discussion and assistance, and J. J. Flynn, L. Werdelin, G. Wesley-Hunt, and A. Goswami for their unselfish encouragement and helpful suggestions. I gratefully acknowledge all of those institutions and individuals who made specimens and samples available, including the American Museum of Natural History; The Brookfield Zoo; Carnegie Museum of Natural History; Louisiana State Museum; Texas Technical University; L. Heaney, B. Stanley, and S. Goodman, the Field Museum; Candace McCaffery and David Reed, Florida Museum of Natural History; J. Dragoo, Museum of Southwestern Biology; C. Conroy, Museum of Comparative Zoology, UC Berkeley; C. Matthee, Ellerman Museum, South Africa; Jerry Hooker and Daphne Hills, the British Museum of Natural History; The National Museum of Natural History, Paris; Harvard Museum of Comparative Zoology, Boston; The National Museum of Natural History, Washington, D.C.; the Page Museum, Los Angeles; the Los Angeles County Museum; Museum of Vertebrate Zoology, Berkeley; and the Swedish Museum of Natural History, Stockholm. A variety of funding sources contributed to this research, and support was received from the following: Florida State University Dissertation Research Grant, Florida State University Bennison Memorial Scholarship, Sigma Xi Grants in Aid of Research, The Society of Systematic Biologists, The American Society of Mammalogists, the American Museum of Natural History Collections Study Grant, and the American Museum of Natural History Theodore Roosevelt Scholarship, and NSF DDIG # 050–8848 and NSF DEB # 0108450 to Scott Steppan.

REFERENCES

Alroy, J. (2000). Understanding the dynamics of trends within evolving lineages. *Paleobiology*, **26**, 319–29.

Arthur, W. (2001). Developmental drive: an important determinant of the direction of phenotypic evolution. *Evolution and Development*, **3**, 271–78.

Arthur, W. (2004). The effect of development on the direction of evolution: toward a twenty-first century consensus. *Evolution and Development*, **6**, 282–88.

Baker, R. H. and Wilkinson, G. S. (2003). Phylogenetic analysis of correlation structure in stalk-eyed flies (*Diasemopsis*, Diopsidae). *Evolution*, **57**, 87–103.

Barycka, E. (2007). Evolution and systematics of the feliform Carnivora. *Mammalian Biology*, **72**, 257–82.

Biknevicius, A. R. and Van Valkenburgh, B. (1996). Design for killing: craniodental adaptations of predators. In *Carnivore Behavior, Ecology and Evolution*, ed. J. L. Gittleman. Ithaca, NY: Cornell University Press, pp. 393–428.

Bokma, F. (2002). A statistical test of unbiased evolution of body size in birds. *Evolution*, **56**, 2499–504.

Brooks, D. R. and McLennan, D. A. (1991). *Phylogeny, Ecology and Behavior*. Chicago, IL: University of Chicago Press.

Bull, J. J. and Charnov, E. L. (1985). On irreversible evolution. *Evolution*, **39**, 1149–55.

Chappill, J. A. (1989). Quantitative characters in phylogenetic analysis. *Cladistics*, **5**, 217–34.

Crusafont-Pairo, M. and Truyols-Santonja, J. (1956). A biometric study of the evolution of fissiped carnivores. *Evolution*, **10**, 314–32.

Cunningham, C. W., Omland, K. E. and Oakley, T. H. (1998). Reconstructing ancestral character states: a critical reappraisal. *Trends in Ecology and Evolution*, **13**, 361–66.

Dayan, T. and Simberloff, D. (2005). Ecological and community-wide character displacement: the next generation. *Ecology Letters*, **8**, 875–94.

Donoghue, M. J. and Ree, R. H. (2000). Homoplasy and developmental constraint: a model and an example from plants. *American Zoologist*, **40**, 759–69.

Eble, G. J. (2004). The macroevolution of phenotypic integration. In *Phenotypic Integration*, ed. M. Pigliucci and K. Preston. Oxford: Oxford University Press, pp. 253–73.

Emerson, S. B. (1988). Testing for historical patterns of change – a case-study with frog pectoral girdles. *Paleobiology*, **14**, 174–86.

Flynn, J. J., Neff, N. A. and Tedford, R. H. (1988). Phylogeny of the Carnivora. In *The Phylogeny and Classification of the Tetrapods, Vol 2: Mammals*, ed. M. J. Benton. Oxford: Clarendon Press, pp. 73–116.

Flynn, J. J., Finarelli, J. A., Zehr, S., Hsu, J. and Nedbal, M. A. (2005). Molecular phylogeny of the Carnivora (Mammalia): assessing the impact of increased sampling on resolving enigmatic relationships. *Systematic Biology*, **54**, 317–37.

Finarelli, J. S. and Flynn, J. J. (2006). Ancestral state reconstruction of body size in the Caniformia (Carnivora, Mammalia): the effects of incorporating data from the fossil record. *Systematic Biology*, **55**, 301–13.

Foote, M. (1993). Discordance and concordance between morphological and taxonomic diversity. *Paleobiology*, **19**, 185–204.

Fusco, G. (2001). How many processes are responsible for phenotypic evolution? *Evolution and Development*, **3**, 279–86.

Futuyma, D. J. and Moreno, G. (1988). The evolution of ecological specialization. *Annual Review of Ecology, Evolution and Systematics*, **19**, 207–33.

Gaubert, P. and Cordeiro-Estrela, P. (2006). Phylogenetic systematics and tempo of evolution of the viverrinae (Mammalia, Carnivora, Viverridae) within feliformians: implications for faunal exchanges between Asia and Africa. *Molecular Phylogenetics and Evolution*, **41**, 266–89.

Geeta, R. (2003). The origin and maintenance of nuclear endosperms: viewing development through a phylogenetic lens. *Proceedings of the Royal Society of London B*, **270**, 29–35.

Goldberg, E. E. and Igic, B. (2008). On phylogenetic tests of irreversible evolution. *Evolution*, **62**, 2727–41.

Goswami, A. (2006). Morphological integration in the carnivoran skull. *Evolution*, **60**, 169–83.

Gould, S. J. (1984). Morphological channeling by structural constraint; convergence in styles of dwarfing and gigantism in *Cerion*, with a description of two new fossil species and a report on the discovery of the largest *Cerion*. *Paleobiology*, **10**, 172–94.

Hansen, T. F. and Martins, E. P. (1996). Translating between microevolutionary process and macroevolutionary patterns: the correlation structure of interspecific data. *Evolution*, **50**, 1404–17.

Harvey, P. H. and Pagel, M. (1991). *The Comparative Method in Evolutionary Biology*. New York, NY: Oxford University Press.

Holliday, J. A. and Steppan, S. J. (2004). Evolution of hypercarnivory: the effect of specialization on morphological and taxonomic diversity. *Paleobiology*, **30**(1), 108–28.

Holliday, J. A. (2007). *Phylogeny and character change in the feloid Carnivora*. PhD dissertation, Florida State University, Tallahassee.

Holt, R. D. and Gaines, M. S. (1992). Analysis of adaptation in heterogeneous landscapes – implications for the evolution of fundamental niches. *Evolution and Ecology*, **6**, 433–47.

Hunt, R. M., Jr. (1998). Evolution of the aeluroid Carnivora: diversity of the earliest aeluroids from Eurasia (Quercy, Hsanda-Gol) and the origin of felids. *American Museum Novitates*, **3252**, 1–65.

Igic, B., Bohs, L. and Kohn, J. R. (2006). Ancient polymorphism reveals unidirectional breeding system shifts. *Proceedings of the National Academy of Science*, **103**, 1359–63.

Jensen, C. H. (1992). Measuring evolutionary constraints: a Markov model for phylogenetic transitions among seed dispersal syndromes. *Evolution*, **46**, 136–58.

Johnson, W. E., Eizirik, E., Pecon-Slattery, J., *et al.* (2006). The Late Miocene radiation of modern Felidae: a genetic assessment. *Science*, **311**, 73–77.

Knoll, A. H. and Bambach, R. K. (2000). Directionality in the history of life: diffusion from the left wall or repeated scaling of the right? In *Deep Time: Paleobiology's Perspective*, ed. D. H. Erwin and S. L. Wing. *Paleobiology*, **26**(Suppl. to no. 4), 1–14.

Koepfli, K., Jenks, S. M., Edizirik, E., Zihirpour, T., Van Valkenburgh, B. and Wayne, R. K. (2006). Molecular systematics of the Hyaenidae, relationships of a relictual lineage resolved by a molecular supermatrix. *Molecular Phylogenetics and Evolution*, **38**, 603–20.

Lauder, G. V. (1981). Form and function – structural analysis in evolutionary morphology. *Paleobiology*, **7**, 430–42.

Losos, J. B. and Irschick, D. J. (1994). Adaptation and constraint in the evolution of specialization of Bahamian *Anolis* lizards. *Evolution*, **48**, 1786–98.

Marcot, J. D. and McShea, D. W. (2007). Increasing hierarchical complexity throughout the history of life: phylogenetic tests of trend mechanisms. *Paleobiology*, **33**, 182–200.

Marroig, G. and Cheverud, J. M. (2001). A comparison of phenotypic variation and covariation patterns and the role of phylogeny, ecology, and ontogeny during cranial evolution of new world monkeys. *Evolution*, **55**, 2576–600.

Martins, E. P. (2000). Adaptation and the comparative method. *Trends in Ecology and Evolution*, **15**, 296–99.

Maynard-Smith, J., Burian, R., Kauffman, S., *et al.* (1985). Developmental constraints and evolution. *Quarterly Review of Biology*, **60**, 265–87.

McShea, D. W. (1994). Mechanisms of large-scale evolutionary trends. *Evolution*, **48**, 1747–63.

McShea, D. W. (2001). The minor transitions in hierarchical evolution and the question of a directional bias. *Journal of Evolutionary Biology*, **14**, 502–18.

McShea, D. W. and Venit, E. P. (2002). Testing for bias in the evolution of coloniality: a demonstration in cyclostome bryozoans. *Paleobiology*, **28**, 308–27.

Meiri, S., Dayan, T. and Simberloff, D. (2005). Variability and correlations in carnivore crania and dentition. *Functional Ecology*, **19**, 337–43.

Mooers, A. O. and Schluter, D. (1999). Reconstructing ancestor states with maximum likelihood: support for one- and two-rate models. *Systematic Biology*, **48**, 623–33.

Moran, N. A. (1988). The evolution of host-plant alternation in aphids – evidence for specialization as a dead end. *American Naturalist*, **132**, 681–706.

Nosil, P. (2002). Transition rates between specialization and generalization in phytophagous insects. *Evolution*, **56**, 1701–06.

Oakley, T. H. and Cunningham, C. W. (2000). Independent contrasts succeed where ancestor reconstruction fails in a known bacteriophage phylogeny. *Evolution*, **54**, 397–405.

Olson, E. C. and Miller, R. L. (1958). *Morphological Integration*. Chicago, IL: University of Chicago Press.

Omland, K. E. (1997). Examining two standard assumptions of ancestral reconstructions: repeated loss of dichromatism in dabbling ducks (*Anatini*). *Evolution*, **51**, 1636–46.

Omland, K. E. (1999). The assumptions and challenges of ancestral state reconstructions. *Systematic Biology*, **48**, 604–11.

Omland, K. E. and Lanyon, S. M. (2000). Reconstructing plumage evolution in orioles (*Icterus*): repeated convergence and reversal in patterns. *Evolution*, **54**, 2119–33.

Pagel, M. (1994). Detecting correlated evolution on phylogenies: a general method for the comparative analysis of discrete characters. *Proceedings of the Royal Society of London B*, **255**, 37–45.

Pagel, M. (1999). The maximum likelihood approach to reconstructing ancestral character states of discrete characters on phylogenies. *Systematic Biology*, **48**, 612–22.

Pagel, M. D. and Harvey, P. H. (1988). Recent developments in the analysis of comparative data. *Quarterly Review of Biology*, **63**, 413–40.

Perez, M., Li, B., Tillier, A., Craud, A. and Veron, G. (2006). Systematic relationships of the bushy-tailed and black-footed mongooses (genus *Bdeogale*, Herpestidae, Carnivora) based on molecular, chromosomal and morphological evidence. *Journal of Zoological Systematics and Evolutionary Research*, **44**, 251–59.

Pigliucci, M. and Preston, K. (2004). *Phenotypic Integration*. Oxford: Oxford University Press.

Polly, P. D. (2001). Paleontology and the comparative method: ancestral node reconstructions versus observed node values. *American Naturalist*, **157**, 596–609.

Polly, P. D. (2004). On the simulation of the evolution of morphological shape: multivariate shape under selection and drift. *Palaeontologia Electronica*, **7**(7A), 28. http://palaeo-electronica.org/2004_2/evo/issue2_04.htm.

Polly, P. D. (2005). Development and phenotypic correlations: the evolution of tooth shape in *Sorex araneus*. *Evolution and Development*, **7**, 29–41.

Polly, P. D. (2008). Developmental dynamics and G-Matrices: can morphometric spaces be used to model phenotypic evolution? *Evolutionary Biology*, **35**, 83–96.

Porter, M. L. and Crandall, K. A. (2003). Lost along the way: the significance of evolution in reverse. *Trends in Ecology and Evolution*, **18**, 541–47.

Radinsky, L. B. (1981a). Evolution of skull shape in carnivores. 1. Representative modern carnivores. *Biological Journal of the Linnean Society*, **15**, 69–388.

Radinsky, L. B.(1981b). Evolution of skull shape in carnivores. 2. Additional modern carnivores. *Biological Journal of the Linnean Society*, **16**, 337–55.

Radinsky, L. B. (1982). Evolution of skull shape in carnivores. 3. The origin and early radiation of the modern carnivore families. *Paleobiology*, **8**(3), 177–95.

Richardson, M. K. and Chipman, A. D. (2003). Developmental constraints in a comparative framework: a test case using variations in phalanx number during amniote evolution. *Journal of Experimental Zoology*, **296B**, 8–22.

Rouse, G. W. (2000). Bias? What bias? The evolution of downstream larval-feeding in animals. *Zoologica Scripta*, **29**, 213–36.

Salazar-Ciudad, I. (2006). Developmental constraints vs. variational properties: how pattern formation can help to understand evolution and development. *Journal of Experimental Zoology*, **306B**, 107–25.

Sanderson, M. J. (1993). Reversibility in evolution – a maximum-likelihood approach to character gain loss bias in phylogenies. *Evolution*, **47**, 236–52.

Schluter, D. (2000). *The Ecology of Adaptive Radiation*. Oxford: Oxford University Press.

Schluter, D., Price, T., Mooers, A. O. and Ludwig, D. (1997). Likelihood of ancestor states in adaptive radiation. *Evolution*, **51**, 1699–711.

Schluter, D., Clifford, E. A., Nemethy, M. and McKinnon, J. S. (2004). Parallel evolution and inheritance of quantitative traits. *American Naturalist*, **163**, 809–22.

Schwenk, K. and Wagner, G. P. (2001). Function and the evolution of phenotypic stability: connecting pattern to process. *American Zoologist*, **41**, 552–63.

Schwenk, K. and Wagner, G. P. (2004). The relativism of constraints on phenotypic evolution. In *Phenotypic Integration*, ed. M. Pigliucci and K. Preston. Oxford: Oxford University Press, pp. 390–408.

Sheldon, P. R. (1996). Plus ça change – a model for stasis and evolution in different environments. *Paleogeography, Palaeoclimatology, Palaeoecology*, **127**, 209–27.

Siddall, M. E., Brooks, D. R. and Desser, S. S. (1993). Phylogeny and the reversibility of parasitism. *Evolution*, **47**, 308–13.

Simon, C. (1983). A new coding procedure for morphometric data with an example from periodical cicada wing veins. In *Numerical Taxonomy*, ed. J. Felsenstein. Berlin: Springer, pp. 378–82.

Sokal, R. R. and Rohlf, F. J. (1998). *Biometry*. New York, NY: W.H. Freeman and Company.

Steppan, S. J. (1997a). Phylogenetic analysis of phenotypic covariance structure 1. Contrasting results from matrix correlation and common principal component analyses. *Evolution*, **51**, 571–86.

Steppan, S. J. (1997b). Phylogenetic analysis of phenotypic covariance structure 2. Reconstructing matrix evolution. *Evolution*, **51**, 587–94.

Van Valkenburgh, B. (1988). Trophic diversity in past and present guilds of large predatory mammals. *Paleobiology*, **14**, 155–73.

Van Valkenburgh, B. (1989). Carnivore dental adaptations and diet: a study of trophic diversity within guilds. In *Carnivore Behavior, Ecology and Evolution*, ed. J. L. Gittleman. Ithaca, NY: Cornell University Press, pp. 410–36.

Van Valkenburgh, B. (1991). Iterative evolution of hypercarnivory in canids (Mammalia, Carnivora) – evolutionary interactions among sympatric predators. *Paleobiology*, **17**, 340–62.

Van Valkenburgh, B. (1999). Major patterns in the history of carnivorous mammals. *Annual Review of Earth and Planetary Science*, **27**, 463–93.

Van Valkenburgh, B. and Ruff, C. B. (1987). Canine tooth strength and killing behavior in large carnivores. *Journal of Zoology*, **212**, 379–97.

Van Valkenburgh, B., Wang, X. M. and Damuth, J. (2004). Cope's rule, hypercarnivory, and extinction in North American canids. *Science*, **306**, 101–04.

Veron, G., Colyn, M., Dunham, A. E., Taylor, P. and Gaubert, P. (2004). Molecular systematics and origin of sociality in mongooses (Herpestidae, Carnivora). *Molecular Phylogenetics and Evolution*, **30**, 582–98.

Wagner, G. P. and Mueller, G. B. (2002). Evolutionary innovations overcome ancestral constraints: a re-examination of character evolution in male sepsid flies (Diptera: Sepsidae). *Evolution and Development*, **4**, 1–6.

Wagner, G. P. and Schwenk, K. (2000). Evolutionarily stable configurations: functional integration and the evolution of phenotypic stability. *Evolutionary Biology*, **31**, 155–217.

Wagner, P. J. (1995). Testing evolutionary constraint hypotheses with early Paleozoic gastropods. *Paleobiology*, **21**, 248–72.

Wagner, P. J. (1996). Contrasting the underlying patterns of active trends in morphologic evolution. *Evolution*, **50**, 990–1007.

Wagner, P. J. (2001). Rate heterogeneity in shell character evolution among lophospiroid gastropods. *Paleobiology*, **27**, 290–310.

Wang, X. (1994). Phylogenetic systematics of the Hesperocyoninae (Carnivora: Canidae). *Bulletin of the American Museum of Natural History*, **221**, 1–207.

Wang, X., Tedford, R. H. and Taylor, B. E. (1999). Phylogenetic systematics of the Borophaginae (Carnivora: Canidae). *Bulletin of the American Museum of Natural History*, **243**, 1–391.

Warheit, K. I., Forman, J. D., Losos, J. B. and Miles, D. B. (1999). Morphological diversification and adaptive radiation: a comparison of two diverse lizard clades. *Evolution*, **53**, 1226–34.

Werdelin, L. (1996). Carnivoran ecomorphology: a phylogenetic perspective. In *Carnivore Behavior, Ecology and Evolution, Volume II*, ed. J. L. Gittleman. Ithaca, NY: Cornell University Press.

Werdelin, L. and Solounias, N. (1991). The Hyaenidae: taxonomy, systematics and evolution. *Fossils and Strata*, **30**, 1–104.

Wesley-Hunt, G. D. and Flynn, J. J. (2005). Phylogeny of the Carnivora: basal relationships among the carnivoramorphans, and assessment of the position of 'Miacoidea' relative to Carnivora. *Journal of Systematic Palaeontology*, **3**(1), 1–28.

Wiens, J. J. (1999). Phylogenetic evidence for multiple losses of a sexually selected character in phrynosomatid lizards. *Proceedings of the Royal Society of London B*, **266**, 1529–35.

Wiens, J. J. (2000). Coding morphological variation within species and higher taxa for phylogenetic analysis. In *Phylogenetic Analysis of Morphological Data*, ed. J. J. Wiens. Washington: Smithsonian Institute Press, pp. 115–45.

Wills, M. A., Briggs, D. E. G. and Fortey, R. A. (1994). Disparity as an evolutionary index – a comparison of Cambrian and Recent arthropods. *Paleobiology*, **20**, 93–130.

Yoder, A. D., Burns, M. M., Zehr, S., *et al.* (2003). Single origin of Malagasy Carnivora from an African ancestor. *Nature*, **421**, 734–37.

Zelditch, M., Swiderski, D. and Fink, W. L. (2000). Discovery of phylogenetic characters in morphological data. In *Phylogenetic Analysis of Morphological Data*, ed. J. J. Wiens. Washington: Smithsonian Institute Press, pp. 115–45.

Appendix 7.1

Specimens used to calculate morphological diversity. Museum abbreviations are as follows: The American Museum of Natural History (AMNH); The Frick Collection, The American Museum of Natural History (F:AM); British Museum of Natural History (BMNH); The Field Museum of Natural History (FMNH); Florida Museum of Natural History (UF); Museum of Comparative Zoology, Harvard University (MCZ), Texas Memorial Museum (TMM). Specimens designated P, PM, UM, UT and UC are currently housed at the Field Museum.

Felidae: *Acinonyx*: BMNH 16573; *Dinobastus serus*: UF 22908, UF 22909; *Felis brachygnatha*: BMNH16537; *Felis amnicola*: UF 1933, UF 19351, UF 19352; *Felis aurata*: AMNH 51998, AMNH 51994; *Felis badia*: FMNH 8378; *Felis bengalensis*: FMNH 62894; *Felis chaus*: FMNH 105559; *Felis colo colo*: AMNH 189394, AMNH 16695, FMNH 43291; *Felis rexroadensis*: UF 25067, UF 58308; *Felis serval*: AMNH 34767, AMNH 205151; *Felis viverrina*: FMNH 105562; *Homotherium serum*: UF 24992; *Hoplophoneus occidentalis*: AMNH 102394; *Machaerodus aphanistus*: BMNH 8975; *Machaerodus cultridens*: BMNH 49967A; *Machairodus palanderi*: F:AM 50476, F:AM 50478; *Megantereon cultridens*: AMNH 105446; *Megantereon falconeri*: BMNH 16350, BMNH 16557; *Megantereon hesperus*: UF22890; *Metailurus*: F:AM China L-604, F:AM 95294; *Nimravides*: AMNH 25206, UF 24471, UF 24479,F:AM 61855, F:AM 104044; *Nimravides catacopis*: Kan 45–99, Kan 93–60; *Nimravides galiani*: UF24462; *Panthera leo*: UF 10643, UF 10645; *Panthera onca*: UF 14765, UF 14766, UF 23685; *Panthera paleonca*: TMM 31181–192, TMM 31181–193;

Panthera pardus: AMNH 35522; *Paramachairodus ogygia*: BMNH 1574; *Paramachaerodus orientalis*: BMNH 8959; *Proaelurus lemanansis*: BMNH 1646, AMNH 105065, AMNH 101931, AMNH 107658; *Proailurus medius*: BMNH 9636, BMNH 9640; *Pseudaelurus*: AMNH 18007, AMNH 27318, AMNH 27446, AMNH 27447, AMNH 61938, AMNH 62129, AMNH 62190, AMNH 62192, F:AM 61925, AMNH 27451-A, BMNH 9633; *Pseudaelurus marshi*: F:AM 27453, F:AM 27457; *Pseudailurus intermedius*: BMNH 2375; *Smilodon californicus*: UF 167140, UF 167141; *Smilodon floridanus*: UF 22704, UF 22705 *Smilodon gracilis*: UF 81700; *Xenosmilus hodsonae*: UF 60000

Nimravidae: *Barbourofelis*: P15811; *Barbourofelis fricki*: AMNH 103202, AMNH 108193, F:AM 61982; *Barbourofelis lovei*: UF 24447, UF 24429, UF 36858, UF 37052; *Barbourofelis morrisi*: AMNH 25201, F:AM 79999; *Barbourofelis whitfordi*: AMNH C38A-210, F:AM 69454, F:AM 69455; *Dinictis cyclops*: AMNH 6937; *Dinictis felina*: AMNH 38805 P12004 PM 21039; *Dinictis*: UF 155216, UF 207947; *Eusmilus cerebralis*: AMNH 6941; *Eusmilus*: F: AM 99259, F:AM 98189; *Hoplophoneus*: F:AM 69344, UC 1754; *Hoplophoneus primaevus latidens*: UM 420, UM 701; *Hoplophoneus oharrai* AMNH 27798, AMNH 82911; *Hoplophoneus oreodontis*: AMNH 9764; *Nanosmilus kurteni*: UF 207943; *Nimravus*: F:AM 62151; *Nimravus sectator*: AMNH 12882; *Nimravus brachyops*: AMNH 6930; *Nimravus gomphodus*: AMNH 6935; *Pogonodon*: AMNH 1403, F:AM 69369; AMNH 1398; *Pogonodon platycopis*: AMNH 6938; *Sansanosmilus*: AMNH 26608; *Vampyrictis vipera*: T 3335

Hyaenidae: *Adrocuta eximia*: AMNH 26372, BMNH 8971, M8968, M901; *Chasmaporthetes* AMNH 99788; *Chasmaporthetes exilelus*: AMNH 26369; *Chasmaporthetes lunensisi*: F:AM China 94 B-1046, F:AM China 96B 1054, AMNH 10261, AMNH 26955; *Chasmaporthetes ossifragus*: AMNH 108691, AMNH 95208; *Crocuta crocuta*: AMNH 187771, FMNH 98952 UF 5665; *Hyaena brunnea*: FMNH 34584; *Hyaena hyaena dubbah*: FMNH 140216; *Ictitherium pannonicum*: BMNH 8983; *Ictitherium Robustum*: BMNH 8987, BMNH 8988; *Ictitherium viverrinum*: F:AM China (G) – L100; *Leecyaena bosei*: BMNH 1554, BMNH 37133, BMNH 578; *Lycyaena chaeretis*: F:AM China 26-B47, F:AM China 52-L495, F:AM China 56-L560; *Lycyaena choeretis*: BMNH 8978, BMNH 8979a; *Lycyaena crusafonti*: AMNH 108175, AMNH 116120; *Pachycrocuta perrieri*: AMNH 27756, F:AM 107766, F: AM 107767, AMNH 27757; *Palhyaena reperta*: F:AM China 42-L338, F:AM China 51-L443; *Pliocrocuta*: BMNH 16565; *Plioviverrops*: AMNH 99607; *Proteles cristatus*: FMNH 127833; *Thalassictis hyaenoides*: F:AM China 14-L344, F:AM China 14-L35; *Thalassictis wangii*: AMNH 20555, AMNH 20586; *Tungurictis spocki*: AMNH 26600, AMNH 26610;

Viverridae: *Arctogalidia trivirgata stigmatica*: FMNH 68709; *Chrotogale owstoni*: FMNH 41597; *Cryptoprocta ferox*: AMNH 30035, FMNH 161707, FMNH 161793, FMNH 33950, FMNH 5655; *Cryptoprocta ferox spelaea*: BMNH 9949; *Eupleres goudotii*: FMNH 30492, AMNH 188211; *Fossa fossa*: AMNH 188209, AMNH 188210, FMNH 85196; *Genetta genetta senegalesis*: FMNH 140213; *Genetta maculata*: FMNH 153697; *Nandinia binotata*: FMNH 25306; *Prionodon linsang*: FMNH 8371; *Viverra zibetina picta*: FMNH 75883; *Viverricula indica babistae*: FMNH 75815, FMNH 75816;

Canidae: Hesperocyoninae: *Cynodesmus thooides*: AMNH 129531; *Ectopocynus antiquus*: AMNH 63376; *Ectopocynus simplicidens*: F:AM 25426, F:AM 25431; *Enhydrocyon basilatus*: AMNH 129549, F:AM 54072; *Enhydrocyon crassidens*: AMNH 12886, AMNH 27579, AMNH 59574; *Enhydrocyon pahinsintewakpa*: AMNH 129535; *Hesperocyon gregarious*: AMNH 9313; *Mesocyon coryphaeus*: AMNH 6859; *Mesocyon temnodon*: F:AM 63367; *Osbornodon fricki*: AMNH 27363; *Osbornodon*: AMNH 54325; 92 *Parenhydrocyon*: AMNH 81086; *Parenhydrocyon josephi*: F:AM 54115; *Philotrox condoni*: AMNH 32796, F:AM 63383; *Prohesperocyon wilsoni*: AMNH 12712; *Sunkahetanka geringensis*: AMNH 96714

Canidae: Borophaginae: *Aelurodon taxoides*: F:AM 61781; *Borophagus diversidens*: AMNH 67364; *Borophagus secundus*: AMNH 61640; *Epicyon haydeni*: F:AM 61461; *Epicyon saevus*: F:AM 61432; *Euoplocyon praedator*: AMNH 18261; *Euoplocyon*: AMNH 25443, AMNH 27315; *Paratomarctos euthos*: F:AM 61101; *Desmocyon thomasii*: AMNH 12874

Mustelidae: *Arctonyx collaris*: AMNH 57373; *Eira barbara*: AMNH 128127, AMNH 29597, UF 3194; *Enhydra lutra*: UF 24196; *Galictis cuja*: AMNH 33281; *Galictis vittata* UF 29310; *Gulo gulo*: AMNH 35054, FMNH 14026, AMNH 169501; *Ictonyx*: AMNH 165812; *Lutra canadensis*: UF 24007; *Martes americana*: UF 13212; *Martes cauvina*: UF 5642; *Martes foina*: UF 29046, AMNH 70182; *Martes pennanti*: UF 23316; *Mustela altaica* UF 26514; *Mustela erminea*: UF 3982, UF 1417; *Mustela felipei*: AMNH 63839, FMNH 86745; *Mustela frenata*: UF 26144, UF 4779; *Mustela kathiah*: FMNH 32502; AMNH 150090; *Mustela nigripes*: AMNH 22894; *Mustela nivalis*: UF 1418; *Mustela putorius*: UF 1425, UF 14422; *Mustela sibirica*: AMNH 114878, F:AM 104392; *Mustela vison*: UF 4569, UF 4724, UF 8108; *Mydaus javanensis*: AMNH 102701; *Poecilogale albinucha*: AMNH 86491; *Taxidea taxus*: UF 384; *Vormala peregusna negans*: AMNH 60103

'Paleo' mustelids: *Aelurocyon*: F:AM 25430; *Aelurocyon brevifacies*: P12283, P12154 P26051; *Brachysypsalis*: AMNH 25295, AMNH 25299, AMNH 27307, F:AM 25284, F:AM 25363, AMNH 27424; *Oligobunis crassivultus*: AMNH 6903, PM 537; *Oligobunis darbyi*: P 25609; *Oligobunis floridanus*: MCZ 4064; *Paroligobunis frazieri*: UF 23928; *Plesictis cf. Pygmaeus*: M27815; *Plesiogulo marshalli*: UF 19253; *Promartes lepidus*: P12155; *Zodialestes*: F:AM 27599, F:AM 27600; *Zodialestes daimonelixensis*: P12032

8

The biogeography of carnivore ecomorphology

LARS WERDELIN AND GINA D. WESLEY-HUNT

Introduction

Traditional studies of biodiversity are mainly concerned with patterns of taxonomic richness. In neontology, particularly conservation biology, the focus is generally at the species level (Reid, 1998; Mittermeier *et al.*, 2005), while in paleontology, the genus and family levels are often used as proxies (Sepkoski, 1988; Bambach *et al.*, 2004). However, there are of course other aspects to diversity, including genetic diversity (e.g. Petit *et al.*, 2003) and phylogenetic diversity (Faith, 1992). A further type of diversity that has generated some interest over the past decade or so is morphological diversity, often referred to as disparity (Gould, 1991; Foote, 1997). This kind of diversity, which, importantly, does not necessarily covary with richness measures, takes as its study the variation in morphology or morphological types in a study group at a particular time or place. The focal level is generally a higher taxonomic category, such as a Family or Order, but can also be a non-monophyletic adaptive category such as carnivore or herbivore, as the object is not in the first instance to trace the evolution of a specific clade, but to investigate the range of adaptations realised by a group of organisms in a particular setting, or, in other words, the totality of their context-specific ecomorphology.

Such studies of ecomorphology can be used to investigate differences in ecological structure in time and space and help differentiate between processes such as selective or random extinctions. It leads to a much fuller depiction of biological diversity than richness alone. Ecomorphology and analysis of disparity has been used at various scales to study the diversification of vertebrates (Van Valkenburgh, 1989, 1994; Jernvall *et al.*, 1996; Werdelin, 1996; Wesley-Hunt, 2005), invertebrates (Foote, 1994, 1997; Wills *et al.*, 1994; Wills, 1998; Roy *et al.*, 2001), and plants (Lupia, 1999) over their evolutionary history.

In this study we investigate the disparity and distribution of ecomorphology among extant members of the Order Carnivora. This order is particularly

Carnivoran Evolution: New Views on Phylogeny, Form, and Function, ed. A. Goswami and A. Friscia. Published by Cambridge University Press. © Cambridge University Press 2010.

interesting for such studies, due to the broad range of body size, and dietary and locomotor adaptations within the group. Carnivora has the greatest range of body mass of any mammalian Order, ranging from the least weasel (average body mass <0.05 kg) to the polar bear (average body mass 400–500 kg). Diet varies within the family from nearly exclusively herbivorous in the giant panda (*Ailuropoda melanoleuca*) to nearly exclusively carnivorous in the cheetah (*Acinonyx jubatus*) to insectivorous in the aardwolf (*Proteles cristatus*). Locomotor adaptations vary from the extremely cursorial African hunting dog *Canis* (*Lycaon*) *pictus* to the ambulatory panda to the arboreal, slow-moving binturong (*Arctictis binturong*) and fast-moving fossa (*Cryptoprocta ferox*). The Carnivora is also one of the most speciose of mammal orders.

Our main goal with this research is to characterise the disparity patterns of carnivorans across families and continents. This will provide a firm foundation for future studies of the evolution of carnivoran disparity through time.

Material and methods

The analysis on which this study is based was carried out on a set of 216 extant species, encompassing 85% of modern carnivoran species around the world. For each species, one specimen was selected to measure and code for 16 dental characters and body size (see Appendix 8.1), following the protocol detailed in Wesley-Hunt (2005). The specimen was selected after studying a larger sample to ensure that the data collected represent an average individual, with no morphologic abnormalities. Due to the nature of the characters, which were designed to allow the extremes of Carnivora to be compared (a polar bear to a weasel), the variation among individuals of a species would only rarely result in a slight difference in the code, and a concomitant minute difference in the position in morphospace. The specimens are housed in the following museums: Field Museum, Chicago; National Museum of Natural History, Smithsonian, Washington DC; American Museum of Natural History, New York; Museum für Naturkunde, Berlin; Museo Nacional De Ciencias Naturales, Madrid; Swedish Museum of Natural History, Stockholm.

To analyse the data we have here chosen to use correspondence analysis (e.g. Benzécri and Benzécri, 1980). This method of data reduction is especially suited to discrete characters and data where the assumption of normality inherent in, for example, principal components analysis is not met. Since we are not here dealing with abundance data and no inherent underlying trends are expected, we have not applied detrended or canonical correspondence analysis (e.g. Hill and Gauch, 1980; Legendre and Legendre, 1998), although detrended correspondence analysis gives an almost identical result to the one presented here.

Correspondence analysis finds the eigenvalues and eigenvectors for a matrix of the Chi-squared distances between the datapoints. An advantage of correspondence analysis is the ability to plot R-mode and Q-mode in the same coordinate system.

A considerable number of measures of disparity have been suggested (see e.g. Foote, 1992; Wills et al., 1994). Here we have settled on the measure suggested by Foote (1992), the mean pairwise dissimilarity between species, scaled for multistate characters and normalised for the number of characters compared. This is identical to a normalised Hamming distance (Hamming, 1950) and calculated as such.

Statistical analyses were carried out using PAST, version 1.82 (Hammer and Harper, 2001, available for download free from: http://folk.uio.no/ohammer/past/). Additional analyses were carried out in Systat Ver. 11. All graphing was done in Aabel Ver. 2.4.2.

Results

The mean disparity per taxonomic group is given in Table 8.1, and plotted together with 99.9% confidence intervals in Figure 8.1a. Tests of the significance of the means and variances of Table 8.1 are shown in Table 8.2. When combined with the morphospace occupation shown in Figure 8.2, these disparities give a good picture of the variability within each family of

Table 8.1 Disparity measured as mean normalised Hamming distance between species for each family of carnivoran. N_s is the number of species and N is the number of pairwise comparisons. The comparison of a species with itself has been eliminated. The variances are approximations based on 1000 bootstrap replications. The table excludes Nandiniidae, Prionodontidae, and Ailuridae, for which insufficient species are available ($N_s \leq 2$).

Family	N_s	N	Mean	Variance
Canidae	31	465	0.194	0.0238
Eupleridae	7	21	0.518	0.0213
Felidae	35	595	0.134	0.00561
Herpestidae	27	351	0.422	0.0257
Hyaenidae	4	6	0.549	0.0121
Mephitidae	11	55	0.295	0.0249
Mustelidae	51	1275	0.391	0.0240
Procyonidae	16	105	0.349	0.0334
Ursidae	8	28	0.401	0.0163
Viverridae	23	253	0.490	0.0316

Table 8.2 Tests of differences between the means (above diagonal) and variances (below diagonal) of the mean disparities per family given in Table 8.1. The values above the diagonal are values of t and the significance levels are $^*p_{eq} \leq 0.05$, $^{**}p_{eq} \leq 0.01$, $^{***}p_{eq} \leq 0.001$, NS: $p_{eq} > 0.05$. Values in italics are cases where the variances are unequal. In this case the t value is the Welch t, which is an approximate solution. The values below the diagonal are values of F and the significance levels are the same as before.

	Canidae	Eupleridae	Felidae	Herpestidae	Hyaenidae	Mephitidae	Mustelidae	Procyonidae	Ursidae	Viverridae
Canidae	—	−9.4345***	7.7068***	−20.552***	−5.6153***	−4.5803***	−23.499***	−8.9733***	−6.956***	−23.255***
Eupleridae	1.1174 NS	—	12.002***	2.678**	−0.4801 NS	5.6201***	3.7295***	3.9866***	2.9857**	0.7021 NS
Felidae	4.2424***	3.7968***	—	−31.679***	−13.439***	−7.4891***	−48.351***	−11.88***	−10.978***	−30.716***
Herpestidae	1.0798 NS	1.2066 NS	4.5811***	—	−1.9313 NS	5.4741***	3.2947*	3.9601***	0.67597 NS	−5.796***
Hyaenidae	1.9669 NS	1.7603 NS	2.1569 NS	2.124 NS	—	3.8282***	2.4948**	2.646**	2.6203*	0.8084 NS
Mephitidae	1.0462 NS	1.169 NS	4.4385***	1.0321 NS	2.0579 NS	—	−4.4962***	−1.8578 NS	−3.076**	−7.5152***
Mustelidae	1.0084 NS	1.1268 NS	4.2781***	1.0708 NS	1.9835 NS	1.0375 NS	—	2.6316**	−0.3390 NS	−9.0514***
Procyonidae	1.4034**	1.5681 NS	5.9537***	1.2966 NS	2.7603 NS	1.3414 NS	1.3917 NS	—	−1.7331 NS	−6.7765***
Ursidae	1.4601 NS	1.3067 NS	2.9055***	1.5767 NS	1.3471 NS	1.5276 NS	1.4724 NS	2.0491*	—	−3.3471**
Viverridae	1.3277**	1.4836 NS	5.6328***	1.5767***	2.6116 NS	1.2691 NS	1.3167 NS	1.057 NS	1.9387*	—

Table 8.3 Disparity measured as mean normalised Hamming distance between species on each continent. N is the number of pairwise comparisons, not the number of species. The comparison of a species with itself has been eliminated. The variances are approximations based on 1000 bootstrap replications.

Continent	N	Mean	Variance
Africa	1891	0.588	0.0385
Eurasia	4095	0.580	0.0410
North America	1128	0.581	0.0379
South America	946	0.595	0.0463

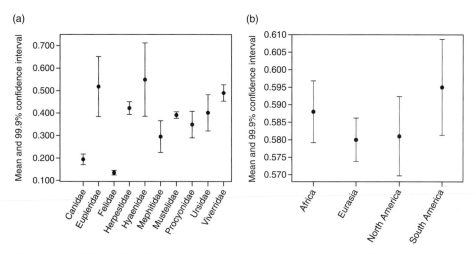

Figure 8.1 a, Mean disparities and 99.9% confidence intervals for all carnivoran families with N (number of species) ≥ 4. b, The same for all continents.

carnivoran. There are clear groupings in the disparity measures. Felidae and Canidae have the lowest disparities, while Herpestidae, Mustelidae, Procyonidae, Ursidae, and Viverridae have considerably higher disparities. Mephitidae is somewhat intermediate, while Eupleridae and Hyaenidae have very high disparities. The significance of these differences will be discussed below. It is also worth noting that the Felidae have a very low variance, while the variances of the other families only occasionally show significant differences.

Table 8.3 and Figure 8.1B provide the same information parsed by continent. No disparity differences are significant (Table 8.4). In fact, the disparities are very similar indeed on the four continents.

Table 8.4 Tests of differences between the means (above diagonal) and variances (below diagonal) of the mean disparities per continent given in Table 8.3. The values above the diagonal are values of t and the significance levels are $*p_{eq} \leq 0.05$, $**p_{eq} \leq 0.01$, $***p_{eq} \leq 0.001$, NS: $p_{eq} > 0.05$. Values in italics are cases where the variances are unequal. In this case the t value is the Welch t, which is an approximate solution. The values below the diagonal are values of F and the significance levels are the same as before. These tests should be considered slightly suspect, as the samples are not entirely independent because the same species can appear in more than one continent. Therefore, similarities may be somewhat exaggerated, though they must still be considered very close.

	Africa	Eurasia	North America	South America
Africa	–	1.4349 NS	0.9511 NS	−0.8670 NS
Eurasia	1.0649 NS	–	−0.1481 NS	−1.9536 NS
North America	1.0158 NS	1.0818 NS	–	−1.541 NS
South America	1.2026 NS	1.1293*	1.2216*	–

The results of the correspondence analysis of the 216 carnivoran taxa are shown in Figure 8.2. The two correspondence factor axes shown together account for approximately 61.5% of the total difference in the data (Table 8.5). In the figure we have differentiated the carnivoran families by colour as indicated. The figure also includes the locations of the variables with the strongest associations with the axes. Factor 1 describes the transition from hypercarnivory on the left (negative) side to hypocarnivory on the right (positive) side. The negative end of this factor is associated with variables K (angle α, lower carnassial), M (shape of upper first molar) and P (grinding area of lower molars), while the positive end is associated with variables J (angle β, upper carnassial) and N (number of upper molars). All of these variables are in some way associated with the hypercarnivory/hypocarnivory spectrum. Factor 2 illustrates a more subtle continuum. Its positive end is associated with variables C (number of premolars anterior to carnassial), D (shape of largest upper premolar anterior to carnassial), and N (number of upper molars), while its negative end is associated with variables J (angle β) and Q (body mass). (Variables J and N appear twice since they plot at the upper and lower right of the diagram.) Thus, Factor 2 represents a continuum of taxa from relatively small forms with a slender broadest premolar (chiefly canids) to relatively large forms with a broad broadest premolar (chiefly ursids and some mustelids).

The majority of taxa within each family are fairly well clumped together in space (indicating relatively low within-family disparity). It is clear even from a first visual inspection that species of Felidae and Canidae are

Table 8.5 Eigenvalues obtained from the correspondence analysis and the percentages of the variation in the data explained.

Eigenvalue	Percent explained	Cumulative percent
0.054901	42.92	42.92
0.0237315	18.55	61.47
0.0113907	8.90	70.37
0.00841681	6.58	76.95
0.00545706	4.27	81.22
0.00454123	3.55	84.77
0.0039409	3.08	87.85
0.0032308	2.53	90.38
0.00280487	2.19	92.57
0.00237986	1.86	94.43
0.00169709	1.33	95.76
0.00152374	1.19	96.95
0.00143757	1.12	98.07
0.00128777	1.01	99.08
0.000712554	0.56	99.64
0.000466158	0.36	100.00

closer together than species in other families, and this is confirmed by the analysis of disparity (Figure 8.1), showing significantly lower disparity in these families. Despite this clumping in morphospace, many families have one or more outliers, some of which are highlighted in Figure 8.2. In Canidae, the most distant outlier is *Otocyon megalotis*, which is more hypocarnivorous than other canids. Diametrically opposite it, but not as distant from the main group, is *Canis pictus*, the African hunting dog, which is the most hypercarnivorous canid species. In Eupleridae, both *Eupleres* and *Cryptoprocta* are outliers, the latter not unexpectedly plotting among the Felidae. Among Mephitidae, the two species of stink badgers, *Mydaus*, are outliers, while among Procyonidae the two species of *Bassariscus* are outliers, plotting in a much less hypocarnivorous part of the morphospace than the other procyonids. Finally, among Ursidae *Melursus* is an outlier due to its smaller body mass and more slender largest premolar, while among Hyaenidae, *Proteles* is aberrant, but not a very distant outlier. Other families, e.g. Mustelidae and Viverridae, have no obvious outliers, although *Enhydra* is shown as the most aberrant mustelid. Also designated in Figure 8.2 are the species of Nandiniidae, Prionodontidae, and Ailuridae that would otherwise be difficult to detect.

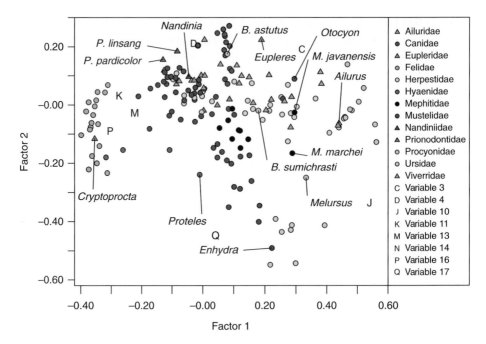

Figure 8.2 For colour version, see Plate 9. Plot of the first two factors of the correspondence analysis. This is the primary morphospace discussed in the text. Species are denoted by coloured symbols, variables by letters. The association between high negative values along Factor 1 and variables *K*, *M*, and *P*, between high positive values on Factor 1 and variables *C* and *J*, between high negative values on Factor 2 and variable *Q*, and between high positive values on Factor 2 and variables *C* and *D* are all clearly in evidence. Species that are outliers in their families or otherwise of interest are noted in the diagram.

Discussion

In the following we shall first consider disparity patterns across families and continents. After this we will discuss morphospace occupation and how this compares between families and continents. To our knowledge, there have not been any previous studies that have considered disparity and morphospace occupation at such a broad scale for a single time period. Previous studies of a similar kind have for the most part focused on changes in either disparity or morphospace occupation (or both) over time (Foote, 1992; Wills, 1998; Wesley-Hunt, 2005) while largely ignoring biogeographic patterns. Partial exceptions are the seminal studies of Van Valkenburgh (1985, 1988, 1991) that used morphospace occupation to compare some fossil guilds of carnivorans with a variety of modern guilds. However, these studies were concerned with guilds of co-occurring species, not with large-scale patterns of whole families or continents. As a result, there is little to compare our results with, although we shall attempt to make connections with other work whenever possible.

Disparity

Although most differences in mean disparity between families are significant (Figure 8.1a, Tables 8.1 and 8.2; this is to a large extent due to the large sample sizes leading to small standard errors), the families can be separated into a few groups. Families with clearly lower disparities than the rest are Felidae and Canidae, while Eupleridae and Hyaenidae have high disparities. The remainder of the families have intermediate mean disparities, with Mephitidae having the lowest of them and Viverridae the highest.

That Felidae have low mean disparity is an expected result matching the findings of Holliday and Steppan (2004) and stems from the demonstration by these authors of constraints on the morphology and evolution of hypercarnivores. The latter are also reflected in the very low variance seen in the Felidae, indicating that not only is the average distance between members of the family small, but there are no outliers affecting this pattern. We shall return to this issue in the discussion of morphospace occupation.

The low mean disparity seen in Canidae is rather more surprising as this family, unlike the Felidae, is not generally noted to have a very uniform morphology. However, a second look at the extant members of this family does indicate a strong uniformity of dental morphology and function, with the smallest, such as the fennec fox, *Vulpes zerda*, being very similar in overall dental morphology to the largest, such as the wolf, *Canis lupus*. This uniformity seems to be a characteristic of the subfamily Caninae. A look into the fossil record of the Canidae indicates that neither the Hesperocyoninae nor the Borophaginae show the same type of uniformity (Werdelin, 1989; Wang, 1994; Wang *et al.*, 1999). A full analysis of this issue may provide clues as to why Caninae appear to have lower disparity than their extinct confamilials.

The two families with the highest mean disparities are also those with the smallest number of species. It might be argued that, despite evidence to the contrary (Foote, 1992), the Hamming distance used as a disparity measure here is sensitive to sample size. However, a closer look at these two families suggests that the high disparities may reflect real evolutionary patterns and not be artefacts of small sample size. The Eupleridae are the only carnivorans on the island of Madagascar, and as such may be expected to have evolved to occupy niches that are elsewhere divided up among several families. Thus, the Eupleridae include a very felid-like species, *Cryptoprocta ferox*, as well as a specialised invertebrate feeder, *Eupleres goudotii*. This opportunity for members of the family to evolve into a variety of niches would necessarily lead to an increase in mean disparity as specialisation leads to movement away from the mean morphology of the clade. In the case of Hyaenidae, the presence of the specialised termite-feeder *Proteles cristatus* naturally increases disparity within the family. However, the disparity increase may be more due to the relatively recent (late Miocene, *ca* 7–5 Mya) extinction of the 'canid-like' hyaenids (Werdelin and Solounias, 1991, 1996). Addition of these taxa would reduce the mean distance between species, and the disparity of late Miocene Hyaenidae can on this basis be predicted to be more similar to that of Canidae today.

Mean disparities are remarkably uniform between continents (Figure 8.1b, Tables 8.3 and 8.4). This is true despite the vastly different areal extent of the continents, the different numbers of species included, and the varying numbers of habitat types present in each continent. A simple hypothesis to account for this phenomenon is that all these factors cancel each other out and what remains is simply an average of everything. Such a hypothesis is difficult to entirely refute, but in the present case it is gainsaid by the fact that carnivoran mean disparities are also uniform between habitat types (Werdelin and Wesley-Hunt, unpublished data). This demonstrates that the uniform mean disparities for continents reported here are not due to an averaging of everything, but are rather a pattern inherent to carnivorans. It suggests that carnivoran guilds primarily evolve by interordinal species interactions, rather than by interactions with the environment, a topic that will be further developed in future work.

Morphospace occupation

The overall pattern of morphospace occupation seen in Figure 8.2 confirms and complements the results of the disparity analysis. At the left of the diagram, the Felidae can be seen to be tightly packed into the two-dimensional morphospace, with no outliers. The disparity analysis shows that this result can be generalised to the 17-dimensional space of all variables. The dense packing of felids is in line with the results of the study of hypercarnivore evolution by Holliday and Steppan (2004).

The Canidae can also be seen to be quite tightly packed into the morphospace, although as noted before, *Otocyon megalotis* is an outlier that is closer to some Viverridae and Herpestidae than to other Canidae. Among the Eupleridae the outliers *Cryptoprocta* and *Eupleres* can be clearly seen, while other Eupleridae lie close together in morphospace. This tripartite division explains the high mean disparity of this family. The Hyaenidae can be seen to be spread out in morphospace despite only including four species. This suggests that extinct species

Table 8.6 Pairwise Mantel tests of the matrices underlying the distributions of taxa shown in Figure 8.4. Values of R are given above the diagonal and values of p below it. Values of p below the standard significance values (*0.05, **0.001, ***0.001) indicate that the matrices are significantly associated.

	Africa	Eurasia	North America	South America
Africa	–	0.9627	0.3691	0.5861
Eurasia	0.0262**	–	0.2078	0.4288
North America	0.2147 NS	0.211 NS	–	0.9330
South America	0.0591 NS	0.2033 NS	0.0174**	–

[such as *Pachycrocuta* and *Pliocrocuta* (Werdelin and Solounias, 1991; Werdelin, 1999)] may have filled the gaps that are currently present.

The Mustelidae, which has the largest number of species, occupies a large swathe of morphospace in the centre of Factor 1, with the constituent species being neither extreme hypercarnivores (although some, such as *Mustela* spp., are relatively hypercarnivorous) nor extreme hypocarnivores (although e.g. *Meles* spp. and *Taxidea* spp. are relatively hypocarnivorous). Disparity seems to increase as one goes from positive (including genera such as *Mustela* and *Martes*) to negative values (mainly members of the Lutrinae) along Factor 2. A discussion of such detailed, within-family patterns would lead too far in the present context and will be considered further elsewhere (an example is provided by Wesley-Hunt *et al.*, this volume).

The most interesting patterns emerge when the morphospace occupations of the different continents are compared, all the while keeping in mind that the mean disparities are nearly identical. Figure 8.3 shows the morphospace occupation pattern in Figure 8.2 broken down by continent. Although some elements in the morphospace are ubiquitous, such as the Felidae and Canidae, it is clear that the overall patterns are strikingly different between the different continents. In Figure 8.4 we show the different morphospace distributions on the four continents in gridded form, while Table 8.6 shows tests of significant differences between the continents. These indicate that there are two groups in the data, Africa/Eurasia and North America/South America. All comparisons between groups are significant while comparisons within groups are not.

Looking in greater detail at differences, Africa and Eurasia have no or almost no representation in the part of the morphospace that Procyonidae occupy in North and South America. Two possible reasons for this may be suggested. First, it is possible that the families present in Africa and Eurasia are constrained in such a way that they are unable to evolve morphologies that would

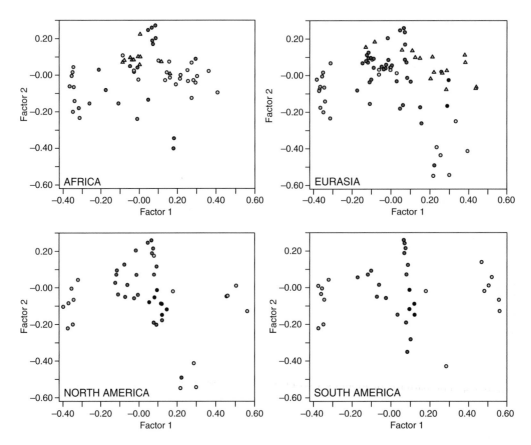

Figure 8.3 For colour version, see Plate 10. The morphospace occupation of Figure 8.2 parsed by continent. Note the general similarities between the two upper patterns and between the two lower patterns and the differences between the upper and lower rows. Symbols are as in Figure 8.2.

fill this space. The alternative explanation is that the niches for which this morphospace is an adaptation are occupied in Africa and Eurasia by non-carnivoran taxa. Which taxa these might be is not apparent and requires a thorough study of the adaptations of Procyonidae and a subsequent analysis of which Eurasian and African taxa may have similar ecologies.

A second, similar pattern is the lack of representation in Africa and South America in the most negative part of the distribution along Factor 2, where bears as well as *Enhydra* are present in Eurasia and North America. The only exceptions are *Tremarctos ornatus* in South America (a North American immigrant) and two species of *Aonyx* in Africa that all occupy positions at the margin of this area. The possibility that other families may not be able to evolve into this part of the morphospace is probably more valid here, but on the other hand

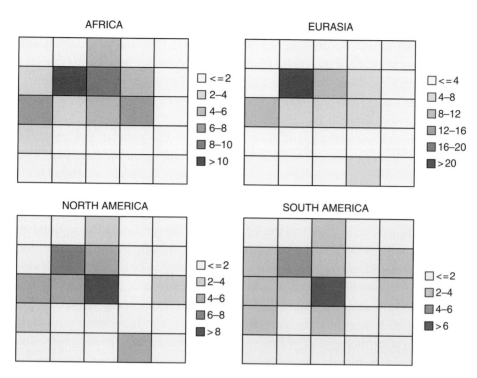

Figure 8.4 The patterns of Figure 8.3 gridded and quantified. The results of Mantel tests of pairwise comparisons of these patterns are given in Table 8.6. Note that the shading scales differ between continents since the continents have varying numbers of species on them.

does not explain the absence of bears from Africa (completely) and South America (partially). From paleontological data (Werdelin and Lewis, 2005; Werdelin and Peigné, 2010) we know that bears were present in Africa from the late Miocene to the late Pliocene. This presence extended at different times from Morocco to the Cape Province, South Africa. The family went extinct in Africa in the late Pliocene for unknown reasons. Later, the family migrated to the continent again, but then only to North Africa, where *Ursus arctos* became extinct in the Holocene. Thus, it appears that niche space for bears has been sporadically present in the past in Africa but has intermittently disappeared, presumably due to environmental change. The original extinction of the family in the late Pliocene coincides with a major environmental change that led to a rise to dominance of C4 grasslands. One can speculate that this was not suitable habitat for bears and that they were unable to adapt to the rapid environmental change. A slightly different scenario surely applies to South America, where Central America would act as a filter limiting the southward spread of Holarctic species after the closing of the land bridge between North and South America between 2 and 3 million years ago. Thus, the absence of bears of the genus *Ursus* in South America may not be due to a lack of suitable habitat, but rather to a lack of opportunities to reach them.

The third, and most interesting difference between the morphospace distributions on the different continents lies in the distribution gap in North and South America just to the left of the main cluster of Procyonidae on Factor 1. In Africa and Eurasia this gap is filled by either Herpestidae or Viverridae, two families with their main distributions in the Afrotropical and Oriental regions that never reached the New World. What makes this gap particularly interesting is that it has, in fact, been bridged by the Procyonidae, since the two species of *Bassariscus* are to be found to the left of it and the remaining Procyonidae to the right, with no species in between. It seems unlikely in the extreme that procyonids could not evolve an intermediate morphology (and a methodological problem would seem to be ruled out by the presence of Herpestidae and Viverridae in that morphospace). The alternative explanation is that species occupying this morphospace existed in the past but have become extinct. A number of extinct genera of Procyonidae are known (e.g. Baskin, 1982) and their position in morphospace may explain the present disjunct distribution, although not why species in this particular part of the morphospace would have selectively gone extinct.

Conclusions

The different families of Carnivora differ in mean disparity (Tables 8.1 and 8.2). Felidae and Canidae have the lowest disparities, which in Felidae can be attributed to their status as extreme hypercarnivores (Holliday and Steppan, 2004), while the reason for the low disparity among Canidae is less well understood and seems specific to the extant subfamily Caninae. Eupleridae and Hyaenidae have high disparity, with Ursidae not far behind. In Eupleridae, this can be attributed to their status as the sole carnivoran family in Madagascar and that they therefore occupy a broader range of niches than mainland carnivorans. In Hyaenidae, the high disparity may at least in part be due to the mass extinction of Hyaenidae in the late Miocene, along with a subsequent, more gradual extinction of post-Miocene hyenas. Ursidae are another family with a broad ranges of niches, which may explain their relatively high mean disparity.

The families differ considerably in morphospace occupation (Figure 8.2). Some families, such as Felidae, Canidae, Ursidae, and Procyonidae, occupy relatively distinct parts of the morphospace, while others, such as Herpestidae,

Viverridae, and Mustelidae, are distributed more widely across morphospace. Again, the reason for this is fairly well understood for Felidae (Holliday and Steppan, 2004; Holliday, this volume), but much research remains to be done to gain a full understanding of why some families have broader morphospace occupations than others. In addition, there are detectable interactions between families with regard to morphospace occupation that need to be explored (e.g. Wesley-Hunt *et al.*, this volume).

There are no detectable differences in mean disparity between continents (Tables 8.3 and 8.4). Despite this, the morphospace occupations of carnivorans on the different continents differ considerably (Figures 8.3 and 8.4). The reason for this is surely a combination of many factors, but the main contributors seem to be factors intrinsic to the taxa, i.e. constraints on the evolution of morphology (which come in many versions that we cannot go into here; Antonovics and van Tienderen, 1991, but see Maynard Smith *et al.*, 1985), and historical factors such as tectonic events leading to dispersal and phylogenetic history. On the other hand, there is little indication of a close connection between disparity and morphospace occupation on the one hand and the environment on the other. Thus, at the scale of investigation, morphospace occupation differs widely in areas where the environments are largely similar, and these differences are mainly due to which taxa happen for historical reasons to be present in the different areas. At a different scale of analysis, Van Valkenburgh (1985, 1988, 1989) found that species richness in carnivoran guilds was greater in environments where the biomass and richness of prey species was greater. This may indicate lower disparity in such environments (although it need not), but is not incompatible with our results, since the scale of analysis is vastly different. Future work will consider this issue (Werdelin and Wesley-Hunt, unpublished data). At the continental scale, however, our results indicate that the main driving force behind the disparity and morphospace occupation patterns we see in carnivorans is species interactions within the order rather than interactions between carnivoran species and their non-carnivoran environment.

Acknowledgements

We would like to thank staff at the Museum für Naturkunde, Berlin; Museo Nacional De Ciencias Naturales, Madrid, National Museum of Natural History, Smithsonian, Washington DC, American Museum of Natural History, New York, Field Museum of Natural History, Chicago, and Swedish Museum of Natural History, Stockholm, for their time and assistance. Funding for this project was provided by grants to LW from the biodiversity programme of the Swedish Research Council.

REFERENCES

Antonovics, J. and van Tienderen, P. H. (1991). Ontoecogenophyloconstrains? The chaos of constraint terminology. *Trends in Ecology and Evolution*, **6**, 166–68.

Bambach, R. K., Knoll, A. H. and Wang, S. C. (2004). Origination, extinction, and mass depletions of marine diversity. *Paleobiology*, **30**, 522–42.

Baskin, J. A. (1982). Tertiary Procyoninae (Mammalia: Carnivora) of North America. *Journal of Vertebrate Paleontology*, **2**, 71–93.

Benzécri, J.-P. and Benzécri, F. (1980). *Pratique de L'Analyse des Données. 1. Analyse des Correspondances. Exposé Élémentaire*. Paris: Dunod.

Crusafont Pairó, M. and Truyols Santonja, J. (1956). A biometric study of the evolution of fissiped carnivores. *Evolution*, **10**, 314–32.

Crusafont Pairó, M. and Truyols Santonja, J. (1957). Estudios masterométricos en la evolución de los Fissipedos. *Boletin del Instituto Geológico i Minero de España*, **68**, 1–140.

Faith, D. P. (1992). Conservation evaluation and phylogenetic diversity. *Biological Conservation*, **61**, 1–10.

Foote, M. (1992). Paleozoic record of morphological diversity in blastozoan echinoderms. *Proceedings of the National Academy of Sciences*, **89**, 7325–29.

Foote, M. (1994). Morphological disparity in Ordovician–Devonian crinoids and the early saturation of morphological space. *Paleobiology*, **20**, 320–44.

Foote, M. (1995). Morphological diversification of Paleozoic crinoids. *Paleobiology*, **21**, 273–99.

Foote, M. (1997). The evolution of morphological diversity. *Annual Review of Ecology and Systematics*, **28**, 129–52.

Gittleman, J. L. (1986). Carnivore life history patterns: allometric, phylogenetic, and ecological associations. *American Naturalist*, **127**, 744–71.

Gould, S. J. (1991). The disparity of the Burgess Shale arthropod fauna and the limits of cladistic analysis: why we must strive to quantify morphospace. *Paleobiology*, **17**, 411–23.

Hammer, Ø. and Harper, D. A. T. (2001). PAST: Palaeontological Statistics software package for education and data analysis. *Palaeontologia Electronica*, **4**(1), 9.

Hamming, R. W. (1950). Error detecting and error detecting codes. *Bell System Technical Journal*, **26**, 147–60.

Hill, M. O. and Gauch, H. G. J. (1980). Detrended correspondence analysis: an improved ordination technique. *Vegetatio*, **42**, 47–58.

Holliday, J. A. and Steppan, S. J. (2004). Evolution of hypercarnivory: the effect of specialization on morphological and taxonomic diversity. *Paleobiology*, **30**, 108–28.

Jernvall, J. (1995). Mammalian molar cusp patterns: developmental mechanisms of diversity. *Acta Zoologica Fennica*, **198**, 1–61.

Jernvall, J., Hunter, J. P. and Fortelius, M. (1996). Molar tooth diversity, disparity, and ecology in Cenozoic ungulate radiations. *Science*, **274**, 1489–92.

Legendre, P. and Legendre, L. (1998). *Numerical Ecology*. New York, NY: Elsevier.

Lucas, P. W. (1979). The dental–dietary adaptations of mammals. *Neues Jahrbuch für Geologie, Paläontologie und Mineralogie, Monatshefte*, **1979**(8), 486–512.

Lucas, P. W. and Peters, C. R. (2000). Function of postcanine tooth crown shape in mammals. In *Development, Function and Evolution of Teeth*, ed. M. F. Teaford, M. M. Smith and M. W. J. Ferguson. Cambridge: Cambridge University Press, pp. 282–89.

Lupia, R. (1999). Discordant morphological disparity and taxonomic diversity during the Cretaceous angiosperm radiation: North American pollen record. *Paleobiology*, **25**, 1–28.

Maynard Smith, J., Burian, R., Kauffman, S., *et al.* (1985). Developmental constraints and evolution. *The Quarterly Review of Biology*, **6**, 265–87.

McNab, B. K. (1971). On the ecological significance of Bergmann's rule. *Ecology*, **52**, 845–54.

McNab, B. K. (1989). Basal rate of metabolism, body size, and food habits in the order Carnivora. In *Carnivore Behavior, Ecology, and Evolution*, ed. J. L. Gittleman. Ithaca, NY: Cornell University Press, pp. 335–54.

Mittermeier, R. A., Robles Gil, P., Hoffman, M., *et al.* (2005). *Hotspots Revisited: Earth's Biologically Richest and Most Endangered Terrestrial Ecoregions*. Arlington, VA: Conservation International.

Petit, R. J., Aguinagalde, I., de Beaulieu, J.-L., *et al.* (2003). Glacial refugia: hotspots but not melting pots of genetic diversity. *Science*, **300**, 1563–65.

Reid, W. V. (1998). Biodiversity hotspots. *Trends in Ecology and Evolution*, **13**, 275–80.

Roy, K., Balch, D. P. and Hellberg, M. E. (2001). Spatial patterns of morphological diversity across the Indo-Pacific: analyses using strombid gastropods. *Proceedings of the Royal Society B*, **268**, 2503–08.

Sepkoski, J. J. Jr. (1988). Alpha, beta, or gamma: where does all the diversity go? *Paleobiology*, **14**, 221–34.

Van Valkenburgh, B. (1985). Locomotor diversity within past and present guilds of large predatory mammals. *Paleobiology*, **11**, 406–28.

Van Valkenburgh, B. (1988). Trophic diversity in past and present guilds of large predatory mammals. *Paleobiology*, **14**, 155–73.

Van Valkenburgh, B. (1989). Carnivore dental adaptations and diet: a study of trophic diversity within guilds. In *Carnivore Behavior, Ecology, and Evolution*, ed. J. L. Gittleman. Ithaca, NY: Cornell University Press, pp. 410–36.

Van Valkenburgh, B. (1990). Skeletal and dental predictors of body mass in carnivores. In: *Body Size in Mammalian Paleobiology: Estimation and Biological Implications*, ed. J. Damuth and B. J. MacFadden. Cambridge: Cambridge University Press, pp. 181–205.

Van Valkenburgh, B. (1991). Iterative evolution of hypercarnivory in canids (Mammalia: Carnivora): evolutionary interactions among sympatric predators. *Paleobiology*, **17**, 340–62.

Van Valkenburgh, B. (1994). Ecomorphological analysis of fossil vertebrates and their communities. In *Ecomorphology*, ed. P. C. Wainwright and S. M. Reilly. Chicago, IL: University of Chicago Press, pp. 140–66.

Van Valkenburgh, B. (1996). Feeding behavior in free-ranging large African carnivores. *Journal of Mammalogy*, **77**, 240–54.

Wang, X. (1994). Phylogenetic systematics of the Hesperocyoninae (Carnivora: Canidae). *Bulletin of the American Museum of Natural History*, **221**, 1–207.

Wang, X., Tedford, R. H. and Taylor, B. E. (1999). Phylogenetic systematics of the Bor-ophaginae (Carnivora: Canidae). *Bulletin of the American Museum of Natural History*, **243**, 1–391.

Werdelin, L. (1989). Constraint and adaptation in the bone-cracking canid *Osteoborus* (Mammalia: Canidae). *Paleobiology*, **15**, 387–401.

Werdelin, L. (1996). Carnivoran ecomorphology: a phylogenetic perspective. In *Carnivore Behavior, Ecology, and Evolution. Volume 2*, ed. J. L. Gittleman. Ithaca, NY, Cornell University Press, pp. 582–624.

Werdelin, L. (1999). *Pachycrocuta* (hyaenids) from the Pliocene of east Africa. *Paläontologisches Zeitschrift*, **73**, 157–65.

Werdelin, L. and Lewis, M. E. (2005). Plio-Pleistocene Carnivora of eastern Africa: species richness and turnover patterns. *Zoological Journal of the Linnean Society*, **144**, 121–44.

Werdelin, L. and Peigné, S. (2010). Carnivora. In *Cenozoic Mammals of Africa*, ed. L. Werdelin and W. J. Sanders. Berkeley, CA: University of California Press, pp. 609–30.

Werdelin, L. and Solounias, N. (1991). The Hyaenidae: taxonomy, systematics and evolution. *Fossils and Strata*, **30**, 1–104.

Werdelin, L. and Solounias, N. (1996). The evolutionary history of hyaenas in Europe and western Asia during the Miocene. In *The Evolution of Western Eurasian Neogene Mammal Faunas*, ed. R. L. Bernor, V. Fahlbusch and H.-W. Mittmann. New York, NY: Columbia University Press, pp. 290–306.

Wesley-Hunt, G. D. (2005). The morphological diversification of carnivores in North America. *Paleobiology*, **31**, 35–55.

Wills, M. A., Briggs, D. E. G. and Fortey, R. A. (1994). Disparity as an evolutionary index: a comparison of Cambrian and Recent arthropods. *Paleobiology*, **20**, 93–130.

Wills, M. A. (1998). Crustacean disparity through the Phanerozoic: comparing morphological and stratigraphic data. *Biological Journal of the Linnean Society*, **65**, 455–500.

Appendix 8.1

Character descriptions (modified from Wesley-Hunt 2005). See Wesley-Hunt (2005) for discussion and illustrations of characters.

1. *Incisor row: parabolic or straight.* (1) parabolic organisation, (2) straight incisor row (Van Valkenburgh, 1996).

2. *Canine: length over width*, measured at the enamel–dentine junction. This is an ordered, continuous character. The continuum was divided into five categories based on the distribution and the extreme morphology of sabre-toothed forms: (1) $X \leq 1.2$, (2) $1.2 < X \leq 1.35$, (3) $1.35 < X \leq 1.5$, (4) $1.5 < X \leq 1.7$, (5) $X > 1.7$. The first category includes the main mode characterising relatively round canines. Taxa in the last category have elongate canines.

3. *Number of upper premolars anterior to the carnassial.* Vestigial premolars (those that appear to be non-functional – a small 'nub', globular in morphology and without distinct cusps) are not included. This character is discrete and ordered.

4. *Largest upper premolar anterior to the carnassial: length over width.* This is an ordered, continuous character. The continuum was divided into three categories based on its distribution. The exact cut-off between categories was arbitrary. The three bins are defined as: (1) $X \leq 1.7$, (2) $1.7 < X \leq 2.3$, (3) $X > 2.3$. The first bin describes relatively round premolars. The second bin contains relatively long premolars. The third bin contains few taxa and describes the extremely elongate premolar condition.

5. *Upper premolar spacing: close or spaced.* Premolars were characterised as spaced if gaps were present between the first, second, and third upper premolar. Premolars were characterised as close under the following conditions: (1) no space between the second and third premolar; however, there may be space present between the canine and the first premolar, or between the first and second premolar, but not both; or (2) the first and second premolar are vestigial.

6. *Last lower premolar: length over width.* This character is ordered, and continuous. The continuum was divided into three categories based on its distribution, the exact cut-offs between categories were arbitrary: (1) $X < 1.7$, (2) $1.7 \leq X \geq 2.2$, (3) $X > 2.2$. The first bin loosely defines a shape that is considered 'rounded' in the context of the sample. The second bin contains the largest portion of the sample. The third bin characterises elongate lower premolars.

7. *Shape of the upper carnassial:* (1) square, (2) equilateral triangle, (3) an elongate triangle (approximating a right, scalene triangle), (4) linear. The shape is determined by the outline of the occlusal surface. To distinguish between a triangular and linear outline of the upper carnassial, the carnassial is classified as linear if the protocone participates in the shear and there is no shelf lingual to the protocone (Van Valkenburgh, 1991).

8. *Blade length of upper carnassial:* length of shearing blade compared to total length of the upper carnassial. (1) no blade present, (2) the blade 1/3 of total length, (3) the blade 1/2 of total length, and (4) the blade 2/3 or greater of total tooth length. This character is ordered.

9. *Relative blade length of lower carnassial:* ratio of the anteroposterior length of the trigonid, measured on the buccal side, over the total maximum length of the tooth. This character is ordered and continuous in distribution. The distribution is divided into five categories: (1) $X = 0$ (2) $0 < X < 0.55$ (3) $0.55 \leq X \leq 0.75$, (4) $0.75 \leq X \leq 0.9$, (5) $X > 0.9$ (Van Valkenburgh 1988, 1989).

10. *Angle β, upper carnassial.* β is the angle between a line drawn from the metacone to the most anterior projection of the parastyle, and a line drawn from the metacone to the apex of the protocone, with the tooth positioned in full occlusal view

(Crusafont Pairó and Truyols Santonja, 1956, 1957). This character is ordered. The continuum was divided into four categories based on the distribution, and distinguishing the extremes: (1) $X < 24°$, (2) $25° \leq X < 30°$, (3) $30° \leq X < 40°$, (4) $X \geq 40°$. Category 4 contains the most hypocarnivorous taxa, while category 1 generally contains the specialised meat-eaters.

11. *Angle α, lower carnassial.* α is the angle between a line drawn along the base of the tooth crown, above the roots, and a line drawn tangential to the protoconid and to the highest point on the talonid with the tooth positioned in full occlusal view (Crusafont Pairó and Truyols Santonja, 1956, 1957). This character is ordered. The continuum was divided into five categories based on the distribution: (1) $0° < X < 15°$, (2) $15° \leq X < 30°$ (3) $30° \leq X < 50°$ (4) $50° \leq X < 70°$ (5) $X \leq 70°$. The categories begin with omnivores or vegetarians and end with the extreme hypercarnivores.

12. *Angle γ, lower carnassial.* γ is the angle between the paralophid and protolophid of the trigonid of the lower carnassial positioned in occlusal view. This character is ordered. The continuum was divided into 5 categories: (1) $X = 0°$, (2) $0° < X \leq 40°$, (3) $40° < X \leq 80°$, (4) $80° < X \leq 130°$, (5) $X = 180°$. If the metaconid is absent there is no protolophid, and the angle is coded as $180°$ (category 5) which is characteristic of hypercarnivorous forms. Hypocarnivores tend to have closed trigonids (low γ values).

13. *Shape of upper first molar:* (1) square or longitudinal rectangle, (2) transverse rectangle, (3) triangle, (4) absent. Shape was determined by the outline of the occlusal surface. If the anteroposterior length was equal to or greater than the width, it was considered square or a longitudinal rectangle. A molar that is mediolaterally wider than long, and has a hypocone or posterior shelf, was coded as a transverse rectangle. The triangle category was only applied to molars that were distinctly triangular – one lingual cusp, and no hypocone, hypocone shelf, or enlarged cingulum around the protocone. If the molar is reduced, and not counted in the number of upper molars, it was coded as absent. This character addresses the amount of grinding area on the upper tooth row, and in conjunction with other characters, describes how the upper molars are organised.

14. *Number of upper molars.* For a molar to be considered present, its occlusal surface area must be equal to at least one half of the surface area of the first upper molar. This character is ordered. The reduction of molars is always associated with increasing carnivory. In non-carnivoramorphan taxa, in which the first upper molar may be the carnassial, only molars with a grinding surface are counted. For example, a taxon in which the first molar is the carnassial, and there are no molars with a grinding surface, the taxon is coded as having no molars.

15. *Cusp shape.* Cusps on the upper first molar were classified as round or sharp (from Jernvall, 1995). 'Sharp' is defined as a cusp that comes to a point, and possesses

sides with straight slopes. A 'round' cusp has a rounded tip and the sides have curved slopes. This character is somewhat subjective; therefore, only the cusps of the upper first molar are used for consistency within the sample, and the cusps are coded as round only when the condition is unambiguous. Any cusp that was intermediate was coded as sharp due to the artefact of wear. Round cusps are more able to withstand the forces associated with crushing, while sharp cusps are more suited to process soft foods by piercing (Lucas 1979; Lucas and Peters 2000).

16.   *Grinding area of the lower molars.* Calculated as the total occlusal surface area of the lower molars divided by the total grinding surface area: (surface area of m1+m2+m3 …)/(m1 talonid+m2+m3+grinding area of p4). This character is modified from Van Valkenburgh's (1988, 1989) Relative Grinding Area. This character is ordered, and continuous. The continuum was divided into five categories beginning with the entire molar region dedicated to grinding and ending with no grinding area at all: (1) $X=1$, (2) $1 < X < 2.25$, (3) $2.25 \leq X \leq 4$, (4) $4 < X \leq 6$ (5) $X > 6$. The last bin represents all taxa with a grinding area of 0 or a grinding ratio greater than 6. Area measurements were taken digitally from digital images. When a portion of the lower fourth premolar forms a component of the grinding area, it is included in the denominator. In some taxa, the trigonid of the first lower molar is very low, and has become a grinding surface. Therefore the total area and the grinding area are equal.

17.   Body size was estimated from the log of the length of the first lower molar (Van Valkenburgh, 1990). This character is ordered. The distribution of log body size was divided into five categories: (1) $X \leq 3.845$, (2) $3.845 < X < 4.332$, (3) $4.332 \leq X \leq 4.699$, (4) $4.699 < X \leq 5$, (5) $X > 5$. Log body mass is ordered because, due to its important ecological ramifications (McNab 1971, 1989; Gittleman 1986), it is necessary to preserve the linear relationship of the categories.

9

Comparative ecomorphology and biogeography of Herpestidae and Viverridae (Carnivora) in Africa and Asia

GINA D. WESLEY-HUNT, REIHANEH
DEHGHANI, AND LARS WERDELIN

Introduction

Ecological morphology (ecomorphology) is a powerful tool for exploring diversity, ecology, and evolution in concert (Wainwright, 1994, and references therein). Alpha taxonomy and diversity measures based on taxon counting are the most commonly used tools for understanding long-term evolutionary patterns and provide the foundation for all other biological studies above the organismal level. However, this provides insight into only a single dimension of a multidimensional system. As a complement, ecomorphology allows us to describe the diversification and evolution of organisms in terms of their morphology and ecological role. This is accomplished by using quantitative and semi-quantitative characterisation of features of organisms that are important, for example, in niche partitioning or resource utilisation. In this context, diversity is commonly referred to as disparity (Foote, 1993). The process of speciation, for example, can be better understood and hypotheses more rigorously tested if it can be quantitatively demonstrated whether a new species looks very similar to the original taxon or whether its morphology has changed in a specific direction. For example, if a new species of herbivore evolves with increased grinding area in the cheek dentition, it can either occupy the same area of morphospace as previously existing species, suggesting increased resource competition, or it can occupy an area of morphospace that had previously been empty, suggesting evolution into a new niche. This example illustrates a situation where speciation did not just increase the number of taxa, but also morphologic and ecologic diversity. In turn, this quantitative information can be used to test speciation hypotheses in the extant fauna as well as the fossil record suggested by previous studies using molecular data and habitat reconstruction (Gaubert and Begg,

Carnivoran Evolution: New Views on Phylogeny, Form, and Function, ed. A. Goswami and A. Friscia. Published by Cambridge University Press. © Cambridge University Press 2010.

2007). Ecomorphology has been used at various scales to study the diversification of vertebrates (Van Valkenburgh, 1988, 1989; Werdelin, 1996; Wesley-Hunt, 2005), invertebrates (Foote, 1994, 1997), and plants, (Lupia, 1999) over their evolutionary history. It is akin to taxon-free analysis (Damuth, 1992), but whereas that mode of analysis seeks to use morphology to characterise paleoecological features of communities and habitats, the focus in ecomorphology is on diversity of function within habitats or larger taxonomic groups.

In this study, we apply ecomorphological analysis to the small Feliformia of the families Herpestidae and Viverridae. These taxa are all small tropical and subtropical carnivorans, distributionally limited to Africa and South and Southeast Asia (disregarding some introduced populations). In this paper we isolate these taxa, pulling them out of a much larger analysis including all modern terrestrial carnivorans (see Werdelin and Wesley-Hunt, this volume), to focus on their biogeographic and morphospace occupation patterns. The reason we analyse Herpestidae and Viverridae in this study is because of their shared ecology, ancestry, and biogeography. We have not included small Mustelidae, which share some similarities in ecology and morphology with Herpestidae and Viverridae, because although they may have some impact on the distribution patterns of herpestids and viverrids in Asia, in Africa they have not been a significant presence since the Early Miocene (Werdelin and Peigné, 2010). Although competition with herpestids and viverrids (especially the former) may have influenced the representation of small Mustelidae in Africa, such interactions as took place 15–20 million years ago will have had little impact on the morphospace distributions of herpestids and viverrids in Africa today.

Little is known about the evolutionary history and diversification of Herpestidae and Viverridae due to their generally poor fossil record. In addition, compared to the majority of Carnivora, little is known about the biology of their extant representatives due to the difficulties of studying them in the wild. Our aim here is to quantify their morphology and therefore some aspects of their ecology, in order to better understand the processes underlying their current distribution. The ability to distinguish dietary (ecological) categories has been demonstrated using dental and skull morphology in modern carnivorans (Radinsky, 1981a,b, 1982; VanValkenburgh, 1988, and more recently using small carnivoran dentition, Friscia et al., 2007; Friscia and Van Valkenburgh, this volume). The detection of this pattern makes it possible to establish a set of morphological characters to infer the ecology of fossil taxa.

Like many mammalian groups, Herpestidae and Viverridae have undergone extensive phylogenetic revision over the last decade, following the routine use of molecular methods in phylogenetic analysis. Most early authors placed these taxa in the single family Viverridae, with the majority of what are now viewed as Herpestidae placed in a subfamily Herpestinae (Gregory and Hellman, 1939). Later authors (e.g. Hunt, 1974) disagreed on morphological grounds with this placement, and with the introduction of molecular data, it was unequivocally shown that 'Herpestinae' neither belonged within Viverridae, nor was it the sister taxon to that family, and the validity of a separate family Herpestidae was recognised (e.g. Flynn and Nedbal, 1998). Our understanding of the relationships between species groups in the families Viverridae and Herpestidae has also changed drastically (Gaubert *et al.*, 2002, 2004a,b, 2005; Veron *et al.*, 2004; Gaubert and Cordeiro-Estrela, 2006; Perez *et al.*, 2006; Patou *et al.*, 2008; Veron, this volume), and many subfamilies and genera are no longer considered monophyletic. Indeed, the Asiatic linsangs, hitherto considered to be Viverridae, have recently been shown to be the sister-group to the family Felidae (Gaubert and Veron, 2003; Veron, this volume). Further, it has been shown that the debated Malagasy carnivorans, earlier split among Viverridae and Herpestidae, and even Felidae, is a monophyletic sister-group to the Herpestidae (Veron and Catzeflis, 1993; Yoder *et al.*, 2003; Flynn *et al.*, 2005, this volume; Veron, this volume). These taxa are not included in the present analysis, which is concerned only with the monophyletic families Herpestidae and Viverridae, as currently conceived.

The fossil record of both families is extremely sparse and difficult to interpret, compounding the confusion surrounding the evolutionary history of these groups (Werdelin, 2003; Werdelin and Peigné, 2010). Research continues to be done concerning these groups, but in the absence of a well-corroborated taxonomy and phylogeny, it has been very difficult to understand the determinants of their extensive radiation in the Old World.

Ecomorphology uses morphological characters to assess the diversity of morphologies within a higher taxonomic unit. The target of analysis can be a clade, a guild, or a community, and the object is to link the morphology to the diversity of ecologies seen in the target group. The characters used will determine the type of ecological information included (feeding modes, locomotor categories, etc.). No study can address all aspects of an animal's ecomorphology, and here we use characters that describe the dentition, coupled with body mass, the latter being a fundamental ecological parameter (Damuth and MacFadden, 1990; Owen-Smith, 1988). The characters were chosen to maximise the ecological and functional information available from macroscopic study of carnivoran dentition. This approach allows us to explore the radiation of Herpestidae and Viverridae in the absence of detailed knowledge of taxonomy and interrelationships. While our conclusions would be enhanced by better understanding of the evolutionary history of the taxa included, especially to examine if their morphospace occupation in deep time is constrained by their phylogeny, the present study is an important step toward teasing apart the evolutionary history and interconnectedness of these groups.

Methods

The analysis on which this study is based was carried out on a set of 216 species, encompassing 85% of modern carnivoran species around the world. For each species, one specimen was selected to measure and code. This specimen was selected after studying a larger sample to ensure that the data collected represent an average individual, with no morphologic abnormalities. Due to the nature of the characters, which were designed to allow the extremes of Carnivora to be compared (a polar bear to a weasel), the variation among individuals of a species would only rarely result in a slight difference in the code, and a concomitant minute difference in the position in morphospace. From this study, the Herpestidae and Viverridae were extracted and their occupation of morphospace analysed. The species studied include 27 Herpestidae and 23 Viverridae, representing approximately 82% and 80%, respectively, of existing species (Wozencraft, 1993) (Table 9.1). The specimens are housed in the following museums: Field Museum, Chicago; National Museum of Natural History, Smithsonian, Washington DC; American Museum of Natural History, New York; Museum für Naturkunde, Berlin; Museo Nacional De Ciencias Naturales, Madrid; Swedish Museum of Natural History, Stockholm.

The morphologic data include 16 dental characters and one character describing body size (see Appendix 8.1 in Chapter 8). These characters and their rationale are fully discussed in Wesley-Hunt (2005). The dental characters describe the entire tooth row including incisors, canines, premolars, and molars to capture the dental functional complexity present in carnivorans. Both qualitative and quantitative characters are incorporated to describe the diversity of morphology within Carnivora. A dissimilarity matrix was calculated from the coded characters of all 216 taxa. Ordered characters were rescaled so the maximum difference was one and unordered characters were treated as a difference of zero or one. The dissimilarity matrix was plotted in morphospace using the first and second axis of a Principal Coordinate Analysis (PCO) (see Wesley-Hunt, 2005 for discussion). Figures 9.1–9.7 show only the herpestid and viverrid taxa from this larger analysis. Principal Coordinates scores and continent distributions are shown in Table 9.2.

The morphospace occupations of Herpestids and Viverrids in this paper are interpreted with reference to the morphospace occupation of Carnivora as a whole, one form of which (not identical to the one used herein) can be seen in

Table 9.1 Character codings used in analyses. See Appendix 8.1 and Wesley-Hunt (2005) for descriptions of characters.

Family	Genus	Species																		
			Character number																	
			1	2	3	4	5	6	7	8	9	10	11	12	13	14	15	16	17
Herpestidae	*Atilax*	*paludinosus*	2	3	2	1	1	2	3	3	3	3	3	3	2	2	2	2	1
Herpestidae	*Bdeogale*	*crassicauda*	1	3	3	1	1	1	2	2	3	3	2	2	2	2	2	2	1
Herpestidae	*Bdeogale*	*jacksoni*	1	4	2	1	2	2	1	2	3	3	2	2	2	2	2	2	1
Herpestidae	*Bdeogale*	*nigripes*	1	4	2	1	1	2	1	2	2	4	1	2	2	2	2	2	1
Herpestidae	*Crossarchus*	*alexandri*	1	4	2	1	2	1	3	2	2	3	3	2	2	2	2	2	1
Herpestidae	*Crossarchus*	*obscurus*	1	4	2	1	1	1	3	3	3	3	2	3	2	2	2	2	1
Herpestidae	*Cynictis*	*penicillata*	1	4	2	1	1	2	3	3	2	2	3	3	2	2	2	2	1
Herpestidae	*Dologale*	*dybowskii*	1	3	2	1	1	2	3	3	2	3	4	2	2	2	2	2	1
Herpestidae	*Galerella*	*pulverulenta*	2	3	2	1	1	2	4	3	3	2	3	3	3	1	2	2	1
Herpestidae	*Galerella*	*sanguinea*	2	3	2	1	2	2	4	3	3	1	4	3	3	1	2	2	1
Herpestidae	*Helogale*	*hirtula*	1	4	2	1	1	1	3	3	3	3	3	2	3	2	2	2	1
Herpestidae	*Helogale*	*parvula*	1	5	2	1	1	1	2	3	2	3	3	2	3	2	2	2	1
Herpestidae	*Herpestes*	*brachyurus*	2	4	2	1	2	2	4	3	3	2	3	3	3	1	2	2	1
Herpestidae	*Herpestes*	*edwardsii*	2	3	2	1	2	2	4	4	3	1	3	3	3	1	2	3	1
Herpestidae	*Herpestes*	*ichneumon*	2	3	2	1	2	2	4	4	3	2	3	3	3	2	2	2	2
Herpestidae	*Herpestes*	*javanicus*	2	2	2	1	2	2	4	4	3	2	4	3	3	2	2	2	1
Herpestidae	*Herpestes*	*semitorquatus*	2	3	2	1	1	2	4	3	3	2	3	3	2	1	2	2	1
Herpestidae	*Herpestes*	*smithii*	2	2	2	1	1	2	4	4	3	2	3	3	3	1	2	2	1
Herpestidae	*Herpestes*	*urva*	2	3	2	1	2	2	4	3	3	2	3	3	3	1	2	2	1
Herpestidae	*Herpestes*	*vitticollis*	2	3	2	1	1	2	3	3	2	2	2	3	3	1	2	2	1
Herpestidae	*Ichneumia*	*albicauda*	1	4	3	1	2	2	3	3	3	3	2	2	2	2	2	2	1
Herpestidae	*Liberiictis*	*kuhni*	1	4	3	1	2	2	3	2	2	4	2	2	2	2	2	2	1
Herpestidae	*Mungos*	*gambianus*	1	3	2	1	1	1	2	2	2	4	2	3	3	2	2	2	1
Herpestidae	*Mungos*	*mungo*	1	4	2	1	1	1	2	2	2	3	2	2	3	2	2	2	1
Herpestidae	*Paracynictis*	*selousi*	2	3	3	1	2	2	3	2	3	2	2	3	3	2	2	2	1
Herpestidae	*Rhynchogale*	*melleri*	1	3	2	1	2	1	2	1	3	3	2	2	2	2	2	2	1

Family	Genus	species																
Herpestidae	Suricata	suricatta	1	2	2	2	3	2	3	3	2	2	2	1	1	1	2	3
Viverridae	Arctictis	binturong	2	3	1	1	1	3	2	3	1	1	2	2	1	1	3	4
Viverridae	Arctogalidia	trivirgata	2	2	2	2	1	3	1	4	3	3	2	2	2	1	3	4
Viverridae	Civettictis	civetta	1	2	1	2	2	3	2	2	2	3	3	2	2	1	3	2
Viverridae	Cynogale	bennettii	2	2	2	2	1	3	1	4	2	2	3	3	2	3	3	4
Viverridae	Diplogale	hosei	1	2	2	2	3	3	2	3	3	3	3	3	2	2	3	4
Viverridae	Genetta	abyssinica	1	2	2	1	3	3	3	1	3	3	4	3	2	2	3	4
Viverridae	Genetta	angolensis	1	2	2	1	3	3	3	2	4	4	4	3	2	2	2	2
Viverridae	Genetta	genetta	1	2	2	2	3	3	3	3	3	3	4	3	2	2	3	2
Viverridae	Genetta	maculata	1	2	2	2	3	3	3	3	3	3	4	3	2	2	2	2
Viverridae	Genetta	servalina	1	2	2	2	3	3	3	2	3	3	4	3	2	2	2	3
Viverridae	Genetta	thierryi	1	2	2	?	3	3	3	2	4	4	4	2	2	1	3	2
Viverridae	Genetta	tigrina	1	3	2	1	3	3	4	2	3	3	4	3	2	2	2	2
Viverridae	Genetta	victoriae	1	2	2	1	3	3	3	2	3	3	4	2	2	2	2	3
Viverridae	Hemigalus	derbyanus	1	2	1	2	1	2	1	1	2	2	3	3	2	1	3	4
Viverridae	Paguma	larvata	1	2	2	1	2	3	1	3	3	3	3	2	2	1	2	4
Viverridae	Paradoxurus	hermaphroditus	1	1	1	1	2	3	2	2	3	3	3	1	2	2	3	3
Viverridae	Paradoxurus	jerdoni	1	1	1	1	2	3	1	3	3	3	3	2	2	1	3	3
Viverridae	Paradoxurus	zeylonensis	1	1	1	1	2	3	1	3	2	3	3	1	2	1	3	3
Viverridae	Poiana	richardsonii	1	3	1	1	3	4	4	4	3	4	4	3	2	2	2	3
Viverridae	Viverra	megaspila	2	2	2	2	2	3	3	3	3	3	4	3	2	2	3	3
Viverridae	Viverra	tangalunga	2	2	2	2	3	3	3	2	1	3	4	3	2	2	3	3
Viverridae	Viverra	zibetha	1	2	2	2	2	3	3	2	3	4	4	2	2	2	3	4
Viverridae	Viverricula	indica	1	2	2	2	2	3	3	2	3	3	4	2	2	2	3	2
Nandiniidae	Nandinia	binotata	2	3	2	1	3	3	3	1	1	4	4	1	2	1	3	4
Prionodontidae	Prionodon	linsang	2	3	2	1	3	4	4	4	4	4	4	3	2	3	3	4
Prionodontidae	Prionodon	pardicolor	2	4	2	1	4	4	4	4	4	4	4	3	2	3	3	3

Table 9.2 Principal Coordinates scores and continental provenance for taxa included in this study.

#	Family	Genus	Species	PCO1	PCO2	Africa	Asia
1	Herpestidae	*Atilax*	*paludinosus*	−0.152	−0.132	1	0
2	Herpestidae	*Bdeogale*	*crassicauda*	−0.319	−0.056	1	0
3	Herpestidae	*Bdeogale*	*jacksoni*	−0.323	0.039	1	0
4	Herpestidae	*Bdeogale*	*nigripes*	−0.246	−0.079	1	0
5	Herpestidae	*Crossarchus*	*alexandri*	−0.389	0.054	1	0
6	Herpestidae	*Crossarchus*	*obscurus*	−0.341	−0.123	1	0
7	Herpestidae	*Cynictis*	*penicillata*	−0.241	−0.052	1	0
8	Herpestidae	*Dologale*	*dybowskii*	−0.227	−0.131	1	0
9	Herpestidae	*Galerella*	*pulverulenta*	−0.018	−0.133	1	0
10	Herpestidae	*Galerella*	*sanguinea*	−0.023	−0.03	1	0
11	Herpestidae	*Helogale*	*hirtula*	−0.32	−0.053	1	0
12	Herpestidae	*Helogale*	*parvula*	−0.308	−0.104	1	0
13	Herpestidae	*Herpestes*	*brachyurus*	−0.037	−0.084	0	1
14	Herpestidae	*Herpestes*	*edwardsii*	0.063	0.013	0	1
15	Herpestidae	*Herpestes*	*ichneumon*	0.01	0.002	1	0
16	Herpestidae	*Herpestes*	*javanicus*	−0.003	−0.058	0	1
17	Herpestidae	*Herpestes*	*semitorquatus*	−0.025	−0.102	0	1
18	Herpestidae	*Herpestes*	*smithii*	0.048	−0.114	0	1
19	Herpestidae	*Herpestes*	*urva*	−0.08	−0.022	0	1
20	Herpestidae	*Herpestes*	*vitticollis*	−0.09	−0.336	0	1
21	Herpestidae	*Ichneumia*	*albicauda*	−0.351	0.115	1	0
22	Herpestidae	*Liberiictis*	*kuhni*	−0.333	0.113	1	0
23	Herpestidae	*Mungos*	*gambianus*	−0.297	−0.223	1	0
24	Herpestidae	*Mungos*	*mungo*	−0.319	−0.159	1	0
25	Herpestidae	*Paracynictis*	*selousi*	−0.216	−0.019	1	0
26	Herpestidae	*Rhynchogale*	*melleri*	−0.377	−0.034	1	0
27	Herpestidae	*Suricata*	*suricatta*	−0.287	−0.126	1	0
28	Viverridae	*Arctictis*	*binturong*	−0.377	−0.074	0	1
29	Viverridae	*Arctogalidia*	*trivirgata*	−0.34	−0.007	0	1
30	Viverridae	*Civettictis*	*civetta*	−0.326	0.076	1	0
31	Viverridae	*Cynogale*	*bennettii*	−0.265	0.135	0	1
32	Viverridae	*Diplogale*	*hosei*	−0.214	0.068	0	1
33	Viverridae	*Genetta*	*abyssinica*	−0.101	0.277	1	0
34	Viverridae	*Genetta*	*angolensis*	0.047	0.071	1	0
35	Viverridae	*Genetta*	*genetta*	−0.092	0.172	1	0
36	Viverridae	*Genetta*	*maculata*	−0.009	0.058	1	0
37	Viverridae	*Genetta*	*servalina*	−0.024	0.06	1	0
38	Viverridae	*Genetta*	*thierryi*	−0.024	0.099	1	0
39	Viverridae	*Genetta*	*tigrina*	0.042	−0.009	1	0

Table 9.2 (*cont.*)

#	Family	Genus	Species	PCO1	PCO2	Africa	Asia
40	Viverridae	*Genetta*	*victoriae*	−0.095	0.084	1	0
41	Viverridae	*Hemigalus*	*derbyanus*	−0.407	0.116	0	1
42	Viverridae	*Paguma*	*larvata*	−0.438	−0.074	0	1
43	Viverridae	*Paradoxurus*	*hermaphroditus*	−0.4	−0.077	0	1
44	Viverridae	*Paradoxurus*	*jerdoni*	−0.409	−0.099	0	1
45	Viverridae	*Paradoxurus*	*zeylonensis*	−0.408	−0.064	0	1
46	Viverridae	*Poiana*	*richardsonii*	0.11	0.091	1	0
47	Viverridae	*Viverra*	*megaspila*	−0.148	0.28	0	1
48	Viverridae	*Viverra*	*tangalunga*	−0.007	0.26	0	1
49	Viverridae	*Viverra*	*zibetha*	−0.143	0.284	0	1
50	Viverridae	*Viverricula*	*indica*	−0.136	0.224	0	1

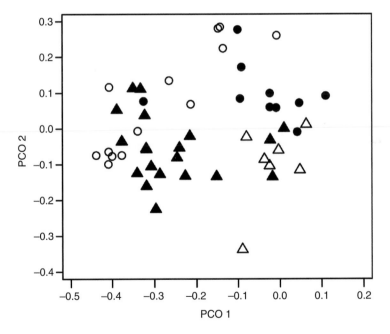

Figure 9.1 Pattern of morphospace occupation based on the first two principal coordinates for Herpestidae and Viverridae. All taxa included in this paper are shown. See Figures 9.6 and 9.7 to identify individual taxa. Symbols: ● African Viverridae; ○ Asian Viverridae; ▲ African Herpestidae; △ Asian Herpestidae.

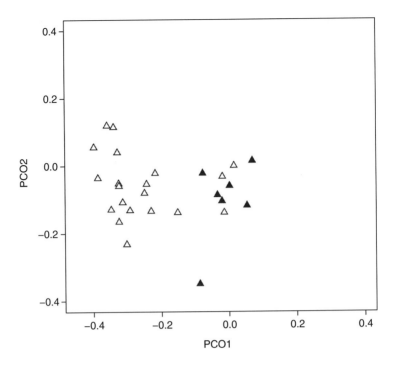

Figure 9.2 Herpestidae. Pattern of morphospace occupation based on the first two principal coordinates, showing the distribution of all herpestid taxa. Symbols: ▲ Asian Herpestidae; △ African Herpestidae.

Werdelin and Wesley-Hunt (this volume: Figure 8.2). Thus, the axes are interpreted with reference to the position of such obvious hypercarnivores as Felidae and hypocarnivores as Procyonidae. None of the taxa herein is as extreme as this, and therefore, hypercarnivore and hypocarnivore herein should be considered relative terms and not based on a comparison of any specific dietary analysis of the taxa involved.

Results

In the morphologic space represented by the first and second Principal Coordinate axes (Figure 9.1), taxa positioned at the top, i.e. with high scores on the second (y-)axis, have extensive slicing blades on their carnassials in combination with large functional grinding surfaces on the molars. Ecologically, these taxa can be characterised as relatively hypercarnivorous. They include members of the viverrid genera *Genetta* and *Viverra*. To the right of centre, i.e. with high scores on the first (x-)axis, can be found other hypercarnivores, although these taxa approach another form of hypercarnivory through a reduced molar grinding

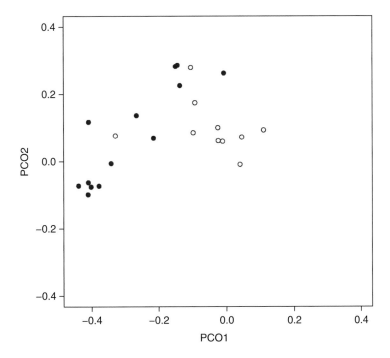

Figure 9.3 Viverridae. Pattern of morphospace occupation based on the first two principal coordinates, showing the distribution of all viverrid taxa. Symbols: ○ Asian Viverridae; ● African Viverridae.

area, and moderately elongated carnassials. Species that are more dietary generalists occupy the left, centre, and lower area of the morphospace. They have a short or no slicing blade on the carnassials and have significant molar grinding surface that in some taxa also extends onto the premolars.

Areas of overlap between Herpestidae and Viverridae are found in the relatively hypocarnivorous region to the left in Figure 9.1, i.e. among taxa with low scores on the first Principal Coordinate, as well as near the middle of the diagram, where *Genetta tigrina* (large-spotted genet) overlaps with some Herpestidae (Figure 9.1). The hypercarnivore space at the top of the distribution is occupied exclusively by viverrids. At the opposite end of the distribution, the hypocarnivorous area at the bottom of the morphospace is occupied only by the Asian herpestid, '*Herpestes*' *vitticollis* (stripe-necked mongoose).

All Herpestidae

As Figure 9.2 shows, African Herpestidae shows a broad range in morphospace occupation, from hypocarnivores to the left in the diagram to

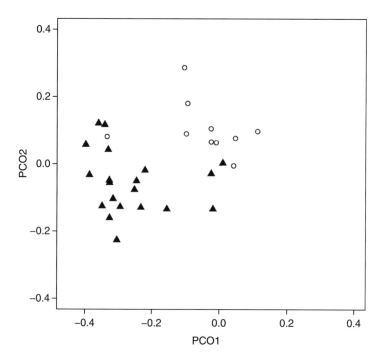

Figure 9.4 Africa. Pattern of morphospace occupation based on the first two principal coordinates, showing the distribution of all Herpestidae (▲) and Viverridae (○) found on Africa.

hypercarnivores to the right. Asian Herpestidae, on the other hand, occupies only the centre and right of the diagram, indicating a narrower ecomorphological range. The only area occupied in Asia that is not represented in Africa is the one taxon at the bottom of the distribution, '*Herpestes*' *vitticollis*, an extreme hypocarnivore that nevertheless differs from the hypocarnivorous African Herpestidae in its large molar region. Taxonomically, Asian herpestids are limited to the genus '*Herpestes*'. [Note: Modern molecular phylogenies (e.g. Veron *et al.*, 2004; Perez *et al.*, 2006) indicate that the genus *Herpestes* is not monophyletic, as the African *Herpestes ichneumon* does not fall in a clade together with Asian species placed in that genus, but instead forms a clade together with some African species often placed in the genus *Galerella*. *H. ichneumon* is the type species of *Herpestes*, and should these results be verified, the nomen *Herpestes* will belong to the later clade, with the Asian species requiring a new genus name. Since this is not a taxonomic contribution, we shall continue to use *Herpestes* in the traditional sense, but mark the taxonomic situation by using quotation marks around the genus name when referring to the Asian clade.] Thus, Herpestidae found in Asia inhabit a small subset of the

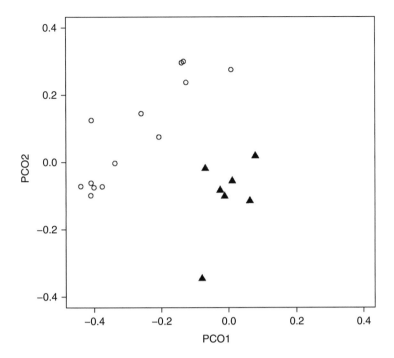

Figure 9.5 Asia. Pattern of morphospace occupation based on the first two principal coordinates, showing the distribution of all Herpestidae (▲) and Viverridae (○) found on Asia.

morphospace occupied by taxa in Africa, just as they form a subclade within the larger clade Herpestidae, which is dominated by African taxa.

All Viverridae

African and Asian viverrids show less overlap in morphospace occupation than their herpestid counterparts (Figure 9.3). African taxa are found primarily in the centre of the morphospace, with the exception of a single taxon at the top, the hypercarnivorous *Genetta abyssinica* (Abyssinian genet), and one in the left part of the morphospace, the omnivorous *Civettictis civetta* (African civet). Asian viverrids occupy an arc along the top and the left, while no Asian viverrid is present near the centre of the overall morphospace.

Africa – Herpestidae and Viverridae

It is not until the pattern of morphospace occupation is analysed by continent that the striking underlying pattern emerges. In Africa, the left side

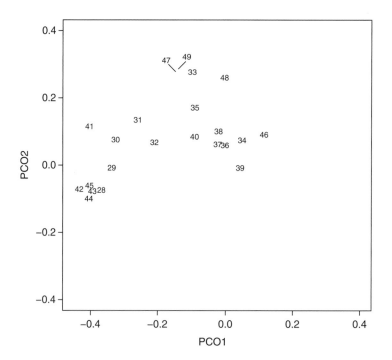

Figure 9.6 Herpestidae. Pattern of morphospace occupation. Numbers identify individual taxa from Table 9.2.

of morphospace (omnivores) is predominantly occupied by herpestids, while the centre and top (carnivores) is predominantly filled by viverrids (Figure 9.4). African viverrids mostly belong to the genus *Genetta* (several species), but also its sister-taxon *Poiana* (Gaubert and Veron, 2003). In the left generalist region, the lone viverrid is the African civet, *Civettictis civetta* (a generalist; Ray, 1995). In the centre, carnivore space, the herpestid minority is represented by *Herpestes ichneumon* (Egyptian mongoose), and the two species of *Galerella*, *G. sanguinea* (slender mongoose) and *G. pulverulenta* (cape grey mongoose). However, these species only slightly overlap the region of morphospace occupied by African Viverridae. Thus, in Africa the overall pattern is one of relatively hypercarnivorous Viverridae and relatively hypocarnivorous Herpestidae.

Asia – Herpestidae and Viverridae

In Asia there is no overlap between the morphospace occupied by viverrids and that occupied by herpestids (Figure 9.5). The left area of morphospace (omnivore/generalists), as well as the top (hypercarnivores) is occupied only by viverrids, and the centre is occupied by herpestids (generalised carnivores). The hypercarnivorous part of the morphospace is occupied by several

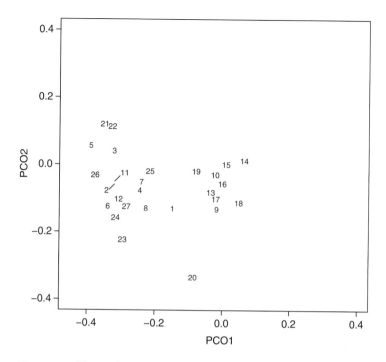

Figure 9.7 Viverridae. Pattern of morphospace occupation. Numbers identify individual taxa from Table 9.2.

viverrids. Thus, the Asian pattern is one in which Viverridae are both relatively hypercarnivorous and relatively hypocarnivorous, while Herpestidae are more generalised mesocarnivores.

The identification of individual taxa in morphospace is indicated in figures 9.6 and 9.7.

Discussion

The distribution of taxa in the morphological space obtained from the analysis herein indicates that there is significant overlap between the Herpestidae and Viverridae when all species are investigated. However, when the distribution of species is analysed continent by continent, almost all overlap disappears. Therefore, while viverrids and herpestids clearly have the potential to occupy the same morphospace, they do not do so in a geographic context, i.e. during conditions of actual or potential sympatry. In Africa, the more generalist species belong to the Herpestidae, while the more carnivorous species belong to the Viverridae (Figure 9.4). In Asia, on the other hand, the Viverridae includes more generalist species and the Herpestidae more carnivorous ones (Figure 9.5). Judging by this analysis, it appears that with a few exceptions,

Herpestidae and Viverridae do not overlap in ecomorphospace when in actual or potential sympatry. It must be recalled that even though sympatry may not always occur in the modern fauna, the pattern may equally be reflective of previous sympatry broken up by distributional changes brought about by extinctions or human-induced habitat change. As the fossil record of these families is poor and hence much of their paleobiology unknown, it is at present not possible to say from the fossil record which family diversified first and on what continent. However, as more pieces of the puzzle are provided – more fossils, resolved modern phylogenies – the ecological and morphological distribution of taxa may help answer this question.

The pattern of morphospace occupation in Africa, as opposed to Asia, brings to light areas of morphospace not (or at least not currently) explored in Africa. Neither herpestids nor viverrids include any African hypercarnivores (top of morphospace – with the single exception of *G. abyssinica*) or hypocarnivores (bottom of morphospace). Since we know from Asia that occupation by herpestids and/or viverrids of this part of the morphospace is morphologically possible, we can begin to look for other reasons why those areas of morphospace are relatively unoccupied by herpestids and viverrids in Africa. In the case of felid-like hypercarnivory, it is tempting to invoke some form of constraint, as the species positioned in Asia are Prionodontidae, which presumably evolved from a different starting point, although this explanation at present is purely ad hoc. The absence of more hypocarnivorous African species does not lend itself to a ready explanation, however. It cannot be due to competition with smaller terrestrial Mustelidae, as the African species fitting that description (*Ictonyx*, two species; *Poecilogale*, one species) are relatively hypercarnivorous. An explanation may be sought from an improved fossil record.

In summary, the geographic analysis of morphospace occupation shows that overlap between Viverridae and Herpestidae, although extensive when the families are viewed in their entirety, is limited or absent when in actual or potential sympatry. In addition, the African and Asian patterns of morphospace occupation differ distinctly from each other: in Africa, Herpestidae occupy the extreme of hypocarnivore space, while in Asia this space is occupied by Viverridae. The possible causes of this pattern remain to be discussed.

Phylogenetic aspects

The fact that the African and Asian patterns differ markedly from each other suggests that the underlying cause is more complex than a simple case of competitive exclusion, where viverrids are always dominant in one part of the morphospace and herpestids in another. Instead, the implication is that historical patterns of migration and phylogeny have shaped the ecological histories of the two families on the respective continents.

In order to begin to understand how these patterns have been formed, it is necessary to have a firm understanding of either the fossil history or the phylogeny of the two families, and preferably both. Unfortunately, as already noted above, the fossil record of both Viverridae and Herpestidae is poor in Africa (e.g. Werdelin and Lewis, 2005; Werdelin and Peigné, 2010) and almost non-existent in Asia, especially Southeast Asia, which today is their main distribution centre. The alternative is to turn to our present understanding of the phylogeny of the two families for information that might inform us of the causes behind the patterns.

Considerable work has recently been carried out documenting the phylogenies of these two families (Herpestidae: Yoder *et al.*, 2003; Veron *et al.*, 2004; Perez *et al.*, 2006; Viverridae: Veron and Heard, 2000; Gaubert and Veron, 2003; Gaubert *et al.*, 2004a,b, 2005; Gaubert and Cordeiro, 2006; Patou *et al.*, 2008; Veron, this volume), although there are still important relationships to be worked out. Composite phylogenies derived from these publications are shown here in Figures 9.8 (Viverridae) and 9.9 (Herpestidae).

Some of the morphospace patterns we see can be explained by phylogenetic relationships within the Viverridae and Herpestidae, in the sense that the phylogenetic background of a species may constrain its morphology, thereby limiting its potential morphospace occupation. In the Viverridae, the African genera *Genetta* and *Poiana* are close together in morphospace (Figures 9.3 and 9.6), as well as in the phylogeny (Figure 9.8), where *Poiana* is the sister-group to the speciose *Genetta*. The Asian viverrid species are somewhat more dispersed in morphospace than their African confamilials, forming an arc in the distribution (Figure 9.3). However, the genera *Paradoxurus*, *Paguma*, *Arctictis* and *Arctogalidia* are grouped together at the lower left of this arc (Figures 9.3 and 9.6). These taxa are closely related (Figure 9.8), forming a viverrid subclade of their own. This pattern of morphospace occupation coupled to phylogeny can be extended to include the genera *Hemigalus*, *Viverra* and *Viverricula*, by comparison with *Paradoxurus*, *Paguma* and *Arctogalidia*, where the former genera are more distantly situated in both morphospace and phylogeny, compared to the latter three.

The morphospace pattern for Herpestidae shows a less close relationship to its phylogeny, but there are some interesting points to be made. The herpestids are divided into two subclades (Figure 9.9 – not counting *Paracynictis*, the phylogenetic position of which has not yet been resolved). All herpestid species that occupy the generalist niche at the centre of morphospace either belong to *Herpestes* or *Galerella* (Figures 9.2 and 9.7). Here we see an overlap in morphospace between African and Asian species. The morphospace distribution of the

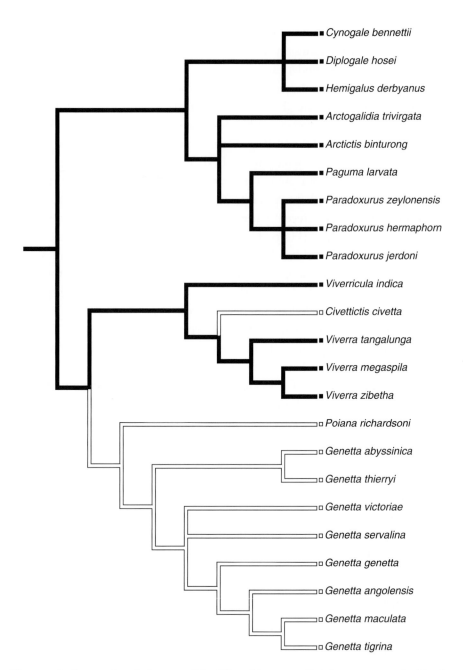

Figure 9.8 Composite phylogeny of the Viverridae, compiled from sources cited in the text. All taxa incuded in this study are represented. Asian species are indicated by black branches, African taxa by white branches.

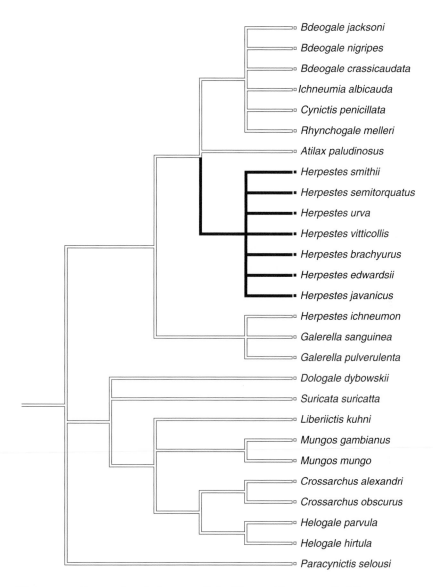

Bdeogale jacksoni
Bdeogale nigripes
Bdeogale crassicaudata
Ichneumia albicauda
Cynictis penicillata
Rhynchogale melleri
Atilax paludinosus
Herpestes smithii
Herpestes semitorquatus
Herpestes urva
Herpestes vitticollis
Herpestes brachyurus
Herpestes edwardsii
Herpestes javanicus
Herpestes ichneumon
Galerella sanguinea
Galerella pulverulenta
Dologale dybowskii
Suricata suricatta
Liberiictis kuhni
Mungos gambianus
Mungos mungo
Crossarchus alexandri
Crossarchus obscurus
Helogale parvula
Helogale hirtula
Paracynictis selousi

Figure 9.9 Composite phylogeny of the Herpestidae, compiled from sources cited in the text. All taxa included in this study are represented. Asian species are indicated by black branches, African taxa by white branches.

remaining herpestid species does not appear to reflect their phylogenetic relationships (Figure 9.9). Instead they are clustered together left of the generalist area of morphospace (Figure 9.2).

One key issue to be answered with regard to the morphospace occupation pattern is whether Herpestidae and Viverridae dispersed to Asia and/or Africa simultaneously or, if not, which family was the first to reach either continent. If this can be established it will be possible to decide whether the present-day pattern is due to displacement of the incumbent by an immigrant or whether 'possession is nine-tenths of the law', as it were, and the immigrant was limited to, in a sense, suboptimal morphospace. Ideally, a good fossil record would assist in determining the timing of historical events. However, as noted, the fossil record of Herpestidae and Viverridae is poor at best. Therefore we must turn to the phylogeny of extant members of these families for clues to the historical pattern.

The general pattern seen in the phylogenies in Figures 9.8 and 9.9 is one of clades either being African or Asian. Thus, e.g. with the exception of *Civettictis*, as further discussed below, African Viverridae form a monophyletic clade within the family and Asian Viverridae, exclusive of Viverrinae, form another (Figure 9.8). The only example of a geographically aberrant member is the African *Civetticitis* within the Asian clade Viverrinae (containing *Viverra* and *Viverricula*). The fossil record, which is better here than for other Viverridae and Herpestidae, indicates that the extant Viverrinae may be relicts of a once much more extensive radiation, with several species suggested to belong to this clade present in the Neogene of Africa [e.g. *Viverra howelli*, *V. leakeyi*, and several undescribed taxa (Werdelin and Peigné, 2010)], as well as in western Eurasia and India [e.g. *Megaviverra*, Kretzoi and Fejfar (1982); *V. bakeri*, *V. durandi* Hunt (1996)]. The relationships of these forms to extant Viverrinae need to be determined before the biogeographic pattern within this subfamily can be established. Nevertheless, the split between the African genets (Genettinae) and their closest Asian relatives the Viverrinae, appears on molecular grounds to have occurred in the mid-Oligocene, while the split of *Civettictis* from within this ostensibly Asian clade appears to have occurred in the Middle Miocene, which is entirely consistent with the fossil record (Gaubert and Cordeiro, 2006).

The phylogeny of the Herpestidae (Figure 9.9) shows Asian herpestids to form a monophyletic clade nested within an African family. The age of the clade encompassing Asian Herpestidae (the genus '*Herpestes*', cf. above) is not well established, but the analysis of Gaubert and Cordeiro (2006) suggests that it is less than 10 Mya, and thus younger than the African clades of Viverridae. The family Herpestidae as a whole, on the other hand, goes back to the Early Oligocene.

These data, coupled with the analysis presented herein, allow us to posit the following scenario. Viverridae and Herpestidae both evolved in the Early Oligocene of Eurasia. Although their earliest centre of radiation is not known, judging from the phylogenies, Viverridae had reached eastern Asia at least by the Late Oligocene, whereas the Herpestidae did not, or at least we have no evidence for their presence there at that time. Thus, a viverrid radiation in Asia was established early on. The two families probably arrived in Africa along with the first wave of Eurasian immigrants to that continent in the latest Oligocene/earliest Miocene. Both families are likely to be present in the oldest well-studied faunas of the African continent (Werdelin and Peigné, 2010). This suggests that the pattern of terrestrial, relatively omnivorous and hypocarnivorous Herpestidae and arboreal or scansorial, relatively carnivorous Viverridae is an old one and established on the basis of a combination of competition and initial founder-effect. The immigration of Herpestidae to Asia is a later event, one that occurred in the face of a pre-established viverrid presence in the region. Whether the pattern of morphospace occupation in Asia is due to replacement of Viverridae by these late-coming Herpestidae in the mesocarnivorous niches, or whether Viverridae blocked the evolution of Herpestidae into hyper- and hypocarnivorous niches, cannot be answered without a better fossil record than is presently available. Nevertheless, the absence of Herpestidae from hypocarnivorous morphospace in Asia, when they are prominently situated in that space in Africa, provides some indication that they may have been excluded from this region of morphospace by the incumbent Asian Viverridae.

Conclusions

When taxa in Viverridae and Herpestidae have the potential to interact, there is little to no overlap in morphospace occupation. This pattern is more striking with the knowledge that the potential for overlap exists. Indeed, when all members of the families are included, regardless of geography, extensive overlap in morphospace occurs. Based on the pattern of morphospace occupation, the ecological role of taxa in the two families differs between Africa and Asia. Understanding the processes underlying the pattern is problematic.

Several confounding factors bear on this question and may affect the scenario suggested above. (1) We have not considered habitat selectivity in this analysis, although we know that Asian Herpestidae are mainly terrestrial and Asian Viverridae mainly arboreal; (2) we have no fossil record to refer to, to determine whether Viverridae in Asia originally occupied mesocarnivorous niches; and (3) we do not know whether extinctions occurred within these clades that would cause us to re-evaluate the patterns based on extant taxa.

With these problems in mind, ecomorphology provides us with tools that allow us to explore these biogeographic and ecological questions. By quantifying this aspect of their ecology and joining it with phylogeny, a new picture of their diversity and pattern of diversification emerges that phylogeny and taxonomic diversity alone could not provide.

Acknowledgements

We would like to thank the editors, A. Goswami and A. Friscia, for the invitation to take part in the symposium that led to this volume. We also thank two reviewers whose comments greatly improved the quality of this contribution. We greatly appreciate the assistance provided by the collection managers and researchers at the following museums: Museum für Naturkunde, Berlin; Museo Nacional De Ciencias Naturales, Madrid; National Museum of Natural History, Smithsonian, Washington DC; American Museum of Natural History, New York; Field Museum of Natural History, Chicago; Swedish Museum of Natural History, Stockholm. Funding for this project was provided by the Swedish Research Council.

REFERENCES

Damuth, J. D. (1992). Taxon-free characterization of animal communities. In *Terrestrial Ecosystems through Time*, ed. A. K. Behrensmeyer, J. D. Damuth, W. A. DiMichele, R. Potts, H.-D. Sues and S. L. Wing. Chicago, IL: University of Chicago Press, pp. 183–203.

Damuth, J. and MacFadden, B. J., eds. (1990). *Body Size in Mammalian Paleobiology: Estimation and Biological Implications*. Cambridge: Cambridge University Press, 397.

Flynn, J. J. and Nedbal, M. A. (1998). Phylogeny of the Carnivora (Mammalia): congruence vs incompatibility among multiple data sets. *Molecular Phylogenetics and Evolution*, **9**, 414–26.

Flynn, J. J., Finarelli, J. A., Zehr, S., Hsu, J. and Nedbal, M. A. (2005). Molecular phylogeny in the Carnivora (Mammalia): assessing the impact of increased sampling on resolving enigmatic relationships. *Systematic Biology*, **54**, 317–37.

Foote, M. (1993). Contributions of individual taxa to overall morphological disparity. *Paleobiology*, **19**, 403–19.

Foote, M. (1994). Morphological disparity in Ordovician–Devonian crinoids and the early saturation of morphological space. *Paleobiology*, **20**, 320–44.

Foote, M. (1997). The evolution of morphological diversity. *Annual Review of Ecology and Systematics*, **28**, 129–52.

Friscia, A. R., Van Valkenburgh, B. and Biknevicius, A. R. (2007). An ecological analysis of extant small carnivorans. *Journal of Zoology*, **272**, 82–100.

Gaubert, P. and Begg, C. M. (2007). Re-assessed molecular phylogeny and evolutionary scenario within genets (Carnivora, Viverridae, Genettinae). *Molecular Phylogenetics and Evolution*, **44**, 920–27.

Gaubert, P. and Cordeiro-Estrela, P. (2006). Phylogenetic systematics and tempo of evolution of the Viverrinae (Mammalia, Carnivora, Viverridae) within feliformians: implications for faunal exchanges between Asia and Africa. *Molecular Phylogenetics and Evolution*, **41**, 266–78.

Gaubert, P. and Veron, G. (2003). Exhaustive sample set among Viverridae reveals the sister-group of felids: the linsangs as a case of extreme morphological convergence within Feliformia. *Proceedings of the Royal Society, London Series B*, **270**, 2523–30.

Gaubert, P., Veron, G. and Tranier, M. (2002). Genets and 'genet-like' taxa (Carnivora, Viverrinae): phylogenetic analysis, systematics and biogeographic implications. *Zoological Journal of the Linnean Society*, **134**, 317–34.

Gaubert, P., Fernandes, C. A., Bruford, M. W. and Veron, G. (2004a). Genets (Carnivora, Viverridae) in Africa: an evolutionary synthesis based on cytochrome *b* sequences and morphological characters. *Biological Journal of the Linnean Society*, **81**, 589–610.

Gaubert, P., Tranier, M., Delmas, A.-S., Colyn, M. and Veron, G. (2004b). First molecular evidence for reassessing phylogenetic affinities between genets (*Genetta*) and the enigmatic genet-like taxa *Osbornictis*, *Poiana*, and *Prionodon* (Carnivora, Viverridae). *Zoologica Scripta*, **32**, 117–29.

Gaubert, P., Wozencraft, W. C., Cordeiro-Estrela, P. and Veron, G. (2005). Mosaics of convergences and noise in morphological phylogenies: what's in a viverrid-like carnivoran? *Systematic Biology*, **54**, 865–94.

Gregory, W. K. and Hellman, M. (1939). On the evolution and major classification of the civets (Viverridae) and allied fossil and recent Carnivora: a phylogenetic study of the skull and dentition. *Proceedings of the American Philosophical Society*, **81**, 309–92.

Hunt, R. M. Jr. (1974). The auditory bulla in Carnivora: an anatomical basis for reappraisal of carnivore evolution. *Journal of Morphology*, **143**, 21–76.

Hunt, R. M. Jr. (1996). Basicranial anatomy of the giant viverrid from 'E' Quarry, Langebaanweg, South Africa. In *Palaeoecology and Palaeoenvironments of Late Cenozoic Mammals: Tributes to the Career of C. S. (Rufus) Churcher*, ed. K. M. Stewart and K. L. Seymour. Toronto: University of Toronto Press, pp. 588–97.

Kretzoi, M. and Fejfar, O. (1982). Viverriden (Carnivora, Mammalia) im europäischen Altpleistozän. *Zeitschrift für geologische Wissenschaften*, **10**, 979–95.

Lupia, R. (1999). Discordant morphological disparity and taxonomic diversity during the Cretaceous angiosperm radiation: North American pollen record. *Paleobiology*, **25**, 1–28.

Owen-Smith, R. N. (1988). *Megaherbivores: The Influence of Very Large Body Size on Ecology*. Cambridge: Cambridge University Press, 369 pp.

Patou, M.-L., Debruyne, R., Jennings, A., Zubaid, A., Rovie-Ryan, J. J. and Veron, G. (2008). Phylogenetic relationships of the Asian palm civets (Hemigalinae & Paradoxurinae, Viverridae, Carnivora). *Molecular Phylogenetics and Evolution*, **47**, 883–92.

Perez, M., Li, B., Tillier, A., Cruaud, A., and Veron, G. (2006). Systematic relationships of the bushy-tailed and black-footed mongooses (genus *Bdeogale*, Herpestidae, Carnivora) based on molecular, chromosomal and morphological evidence. *Journal of Zoological Systematics*, **44**, 251–59.

Radinsky, L. B. (1981a). Evolution of skull shape in carnivores 1. Representative modern carnivores. *Biological Journal of the Linnean Society*, **15**, 369–88.

Radinsky, L. B. (1981b). Evolution of skull shape in carnivores 2. Additional modern carnivores. *Biological Journal of the Linnean Society*, **16**, 337–55.

Radinsky, L. B. (1982). Evolution of skull shape in carnivores. 3. The origin and early radiation of the modern carnivore families. *Paleobiology*, **8**(3), 177–95.

Ray, J. C. (1995). *Civettictis civetta. Mammalian Species*, **488**, 1–7.

Van Valkenburgh, B. (1988). Trophic diversity in past and present guilds of large predatory mammals. *Paleobiology*, **14**, 155–73.

Van Valkenburgh, B. (1989). Carnivore dental adaptations and diet: a study of trophic diversity within guilds. In *Carnivore Behavior, Ecology and Evolution*, ed. J. L. Gittleman. Ithaca, NY: Cornell University Press, pp. 410–36.

Veron, G. and Catzeflis, F. M. (1993). Phylogenetic relationships of the endemic Malagasy carnivore *Cryptoprocta ferox* (Aeluruidea): DNA/DNA hybridization experiments. *Journal of Mammalian Evolution*, **1**, 169–85.

Veron, G. and Heard, S. (2000). Molecular systematics of the Asiatic Viverridae (Carnivora) inferred from mitochondrial cytochrome *b* sequence analysis. *Journal of Zoological Systematics and Evolutionary Research*, **38**, 209–17.

Veron, G., Colyn, M., Dunham, A. E., Taylor, P. and Gaubert, P. (2004). Molecular systematics and origin of sociality in mongooses (Herpestidae, Carnivora). *Molecular Phylogenetics and Evolution*, **30**, 582–98.

Wainwright, P. C. (1994). Functional morphology as a tool in ecological research. In *Ecological Morphology: Integrative Organismal Biology*, ed. P. C. Wainwright and S. M. Reilly. Chicago, IL: The University of Chicago Press, pp. 42–59.

Werdelin, L. (1996). Carnivoran ecomorphology: a phylogenetic perspective. In *Carnivore Behavior, Ecology, and Evolution. Volume 2*, ed. J. L. Gittleman. Ithaca, NY: Cornell University Press, pp. 582–624.

Werdelin, L. (2003). Mio-Pliocene Carnivora from Lothagam, Kenya. In *Lothagam: The Dawn of Humanity in Eastern Africa*, ed. M. G. Leakey and J. M. Harris. New York, NY: Columbia University Press, pp. 261–314.

Werdelin, L. and Lewis, M. E. (2005). Plio-Pleistocene Carnivora of eastern Africa: species richness and turnover patterns. *Zoological Journal of the Linnean Society*, **144**, 121–44.

Werdelin, L. and Peigné, S. (2010). Carnivora. In *Cenozoic Mammals of Africa*, ed. L. Werdelin and W. J. Sanders. Berkeley, CA: University of California Press, pp. 609–63.

Wesley-Hunt, G. D. (2005). The morphological diversification of carnivores in North America. *Paleobiology*, **31**, 35–55.

Wozencraft, W. C. (1993). Order Carnivora. In *Mammal Species of the World — A Taxonomic and Geographic Reference*, ed. D. E. Wilson and D. M. Reeder. Washington, DC: Smithsonian Institution Press, pp. 279–348.

Yoder, A. D., Burns, M. M., Zehr, S., *et al.* (2003). Single origin of Malagasy Carnivora from an African ancestor. *Nature*, **421**, 734–37.

10

Ecomorphological analysis of carnivore guilds in the Eocene through Miocene of Laurasia

MICHAEL MORLO, GREGG F. GUNNELL, AND DORIS NAGEL

Introduction

Quantitative analyses of guild structures of living and fossil mammals have a relatively long history (e.g. Valverde, 1964; Van Valkenburgh, 1988; Legendre, 1989; Gunnell *et al.*, 1995), although carnivores have often been excluded from older studies. However, some studies have been published dealing with general carnivore ecomorphology (e.g. Van Valkenburgh, 1992, 1999; Werdelin, 1996; Van Valkenburgh *et al.*, 2004; Wesley-Hunt, 2005), or structures of single guilds (e.g. Dayan *et al.*, 1989; Viranta and Andrews, 1995; Dayan and Simberloff, 1996; Jones, 2003; Hertler and Volmer, 2008). Few of these studies, however, have combined more than two parameters (e.g. body mass and diet or body mass and locomotion). In addition to body mass, diet and locomotor patterns can satisfactorily be estimated for fossil taxa (see Morlo, 1999, for an example using these three parameters in an analysis of creodont guilds). A similar methodological approach has been applied to compare several carnivore guilds (Morlo, 1999; Nagel and Morlo, 2000, 2003; Morlo and Gunnell, 2003, 2005a,b, 2006; Nagel *et al.*, 2005; Stefen *et al.*, 2005; Morlo and Nagel, 2007). In this chapter, we augment these studies with the addition of a set of guild analyses from the Paleocene to the Recent. Having guild structure established on the three parameters, two guilds can be tested against each other by principal component analysis (PCA) to clarify which parameters are mainly responsible for the differences.

The aims of this chapter are twofold. First we examine the usefulness of carnivore guild structure for estimations of palaeoclimate and palaeoenvironment, and second, we explore the evolution of ecomorphological patterns in carnivore taxa by use of taxon-free methodologies. To achieve these aims, we compare carnivore guilds from four different perspectives: (1) effects of large-scale faunal interchange

Carnivoran Evolution: New Views on Phylogeny, Form, and Function, ed. A. Goswami and A. Friscia. Published by Cambridge University Press. © Cambridge University Press 2010.

on carnivore guild structure is examined through a comparison of guilds before and after the Paleocene–Eocene boundary in North America; (2) the potential effects of climate dependence of carnivore guilds are examined by comparing the continental middle Eocene of North America (middle Bridgerian carnivore guild) to the insular middle Eocene of Europe (Geiseltal carnivore guild); (3) the potential effects of environmental dependence of carnivore guilds are exemplified by comparing the middle Miocene of Southeast Europe (Çandir) to the middle Miocene of Central Europe (Sandelzhausen, Sansan, and Steinheim); (4) detailed differences in guilds from similar environments but different times are exemplified by comparison of open landscape guilds from the middle Eocene of North America, early Oligocene of Mongolia, middle Miocene of South Eastern Europe, and Recent of Africa (Serengeti). As Paleogene carnivore guilds not only contain Carnivora, but oxyaenid and hyaenodontid creodonts, mesonychians, pantolestids, arctocyonids, and didymoconids as well, we include those taxa in our analysis.

Methods

Parameter class identification

Following Morlo (1999), we used three parameters to estimate the ecomorphology of a single species: body mass, diet type, and locomotor pattern. Combinations of ecomorphological types within a specific fauna define the structure of the guild and comparisons of these structures provide our results. Additional information may come from the number of guild members sharing the same ecomorphospace. We applied our reconstruction of ecomorphology on species, but the method itself is taxon-free and may be applied to other taxonomical ranks or even to single specimens. Distribution of a certain taxon among the ecomorphospaces of a guild may thus serve as a fifth parameter.

We use number of species sharing the same ecomorphospace and taxonomic distribution within a guild to discuss the decreasing number of carnivore orders through the Cenozoic and to understand differences between guilds which are poorly differentiated by principal components analysis (see below).

Body mass

In several studies, the body mass of carnivorans was calculated by indices based on measurements of the carnassials (Thackeray and Kieser, 1992; Viranta and Andrews, 1995; Legendre and Roth, 1988; Van Valkenburgh, 1990). The same is true for creodonts (Morlo, 1999). Alternatively, limb bone measurements have been used for body mass estimations of carnivores (Gingerich, 1990; Anyonge, 1993; Heinrich and Biknevicius, 1998; Christiansen, 1999) and creodonts (Egi, 2001). If possible, we used limb bone measurements, but for the majority of taxa this was impossible because of a lack of associated postcranial specimens relevant for this purpose. In these cases we used carnassial size to estimate body mass. Body masses of taxa which do not belong to either Carnivora or Creodonta (as mesonychids, pantolestids, arctocyonids, and didymoconids) were determined using the regression Morlo (1999) developed for creodonts.

In order to minimise the possible variation of absolute body mass data due to methodological approach and to cover body mass variation known from living carnivorans, we use body mass classes instead of absolute body masses, thereby following previous studies starting with Morlo (1999). As in these former studies, we used the following body mass classes: 0–1 kg, 1–3 kg, 3–10 kg, 10–30 kg, 30–100 kg, >100 kg.

Diet
Reconstruction of diet types was based on methods developed by Van Valkenburgh (1988) for carnivorans. Based on measurements of teeth, she defined four diet types: meat, meat/bone, meat/non-vertebrate, non-vertebrate/meat. These four diet classes are referred to here as hypercarnivorous, bone/meat, carnivorous, and hypocarnivorous. Hypocarnivores as used here include omnivores, herbivores, and durophagous taxa, while piscivores are regarded as carnivorous. As insectivorous taxa cannot be separated from hypocarnivores by this method, we add a class 'insectivorous' for taxa falling into 'non-vertebrate/meat', but having pointed cusps instead of blunt ones (another method to identify insectivorous carnivores was recently published by Friscia et al., 2007). For carnivores other than Carnivora, we used the adjustments of Van Valkenburgh's method provided by Morlo (1999). Pantolestids are generally regarded as piscivorous and thus carnivores based on direct evidence coming from the Messel pantolestid Buxolestes, which shows fish ribs in its stomach contents (Koenigswald, 1980), and on morphological evidence recently provided by Boyer and Georgi (2007).

Locomotor pattern
We used qualitative characters of postcranial morphology as provided by Barnet and Napier (1953), Ginsburg (1961a), Taylor (1974, 1976, 1989), Jenkins and Camazine (1977), Laborde (1987), Bertram and Biewener (1990), Rose (1990), Gebo and Rose (1993), MacLeod and Rose (1993), Wang (1993), Polly (1996), Heinrich and Rose (1997), Heizmann and Morlo (1998), Morlo and Habersetzer (1999), and Andersson (2004) for the reconstruction of locomotor patterns. These include: scapular outline, length and shape of scapular spine, glenoid shape, shape and size of humeral head, strength of deltopectoral and supracondylar crests and size of distal humeral epicondyle, length and orientation of olecranon, shape and size of humeral and radial notches and size of distal radioulnar articular process

at the ulna, height of capitular eminence on radial head and shaft shape of the radius, hand posture, shape and depth of acetabulum and breadth of dorsal iliac spine, shape and orientation of femoral head, size of contact area between tibia and fibula, shape of astragalus and calcaneum, and foot posture. We separate arboreal, scansorial, cursorial, generalised terrestrial, semifossorial, and semi-aquatic taxa. It should be noted, however, that locomotor patterns always include potential comparative bias (Carrano, 1996). A specific taxon might be cursorial if compared to other taxa of its guild, but not if compared to taxa of other guilds (e.g. *Sinopa rapax*, which is the most cursorial taxon of the Br-2 guild, would not be judged to be cursorial if present in the modern Serengeti). Assigning a specific taxon to a locomotor pattern thus includes first the analysis based on the post-cranial characters given above, and second, a relative judgement within its specific guild. In any case, however, a specific taxon is assigned to the same pattern in all guilds in which it occurs (e.g. middle Miocene of Europe).

Statistics

In addition to 3D visualisations of guild structure, we applied principal components analyses (PCA) to a subset of our data in order to evaluate the structure of the relationships among the three variables we chose for analysis (body size, locomotion, diet). We made comparisons between two recent faunas from differing habitat settings (an equatorial rainforest fauna in Guyana vs. an equatorial savannah in the Serengeti), two faunas of different ages but presumably of similar habitat (mixed woodland fauna from the middle Eocene of Wyoming vs. the middle Miocene fauna from Steinheim in Germany), and two fossil faunas from differing habitats but similar ages (a woodland fauna from Sandelzhausen (MN 5) and an open country fauna from Çandir (MN 6), with both localities representing the middle Miocene).

Guilds analysed

The localities and faunal samples used for the analyses done in this chapter were chosen because they are well known and at least one of the co-authors has first-hand knowledge of the carnivores.

Results

Guild change across an epoch boundary

Paleocene Cf-3 vs. Eocene Wa-0 in Wyoming

The Paleocene–Eocene boundary is marked globally by a dramatic increase in mean annual temperature as indicated by carbon isotope geochemistry (Koch *et al.*, 1992; Bowen *et al.*, 2001). In conjunction with this isotope event, there is a major reorganisation of mammalian faunas across the northern continents, including a striking change in carnivore guild structure (Morlo and Gunnell, 2005b, 2006). At the beginning of the Eocene, the most important new carnivorous group to appear is hyaenodontid creodonts. This group is added to carnivorans, oxyaenids, mesonychians, pantolestids, and omnivorous arctocyonids. This event enables us to examine the possible reaction of a carnivore guild to a major faunal turnover. We compared the fauna from the latest Paleocene (Cf-3) with that of the earliest Eocene (Wa-0) from the Sand Coulee area, Wyoming, but included other contemporary taxa from Wyoming as well (e.g. from Heinrich *et al.*, 2008).

Results

Our results suggest that hyaenodontids did not replace previously existing taxa, but instead occupied previously vacant ecomorphospaces. All hyaenodontids from Wa-0 are small (<3 kg), scansorial to terrestrial, and insectivorous to hypercarnivorous. It appears that all six known species of Wa-0 hyaenodontids were added to the guild present in Cf-3 instead of replacing existing taxa, resulting in an increased diversification of the overall carnivore guild, especially in terms of locomotor pattern. Interestingly, this additional partitioning of food resources, instead of an increased competition for food resources, also has been documented in the Recent immigrations of cats to islands. While generally having detrimental effects on the island fauna, 'the competitive effects from feral cats have not emerged' (Phillips *et al.*, 2007: 173). Instead, feral cats mostly coexist with native island carnivores (Phillips *et al.*, 2007). This suggests that competition within a carnivore guild might be less than has been generally regarded, at least for small species. This also was found by Van Valkenburgh (1999) in the analysis of major turnovers among carnivore faunas. Consequently, studies based on strong competitive models (e.g. Hertler and Volmer, 2008) may be in need of re-assessment.

Another result is an overall decrease in mean body mass of the guild (see Table 10.1 for summary of parameters). This is congruent with general mammalian faunal turnover patterns which followed the rapid warming at the Paleocene–Eocene boundary (Clyde and Gingerich, 1998; Gingerich, 2001). Today, tropical carnivore guilds are more diverse than are those from temperate regions because of the presence of more small (arboreal) species (and thus a smaller mean body mass). The increase in carnivore diversity at the Paleocene–Eocene boundary is therefore not unexpected given the expansion of more tropical climatic zones into northern latitudes. This result also suggests that, wherever hyaenodontids ultimately originated, it was likely from a tropical region.

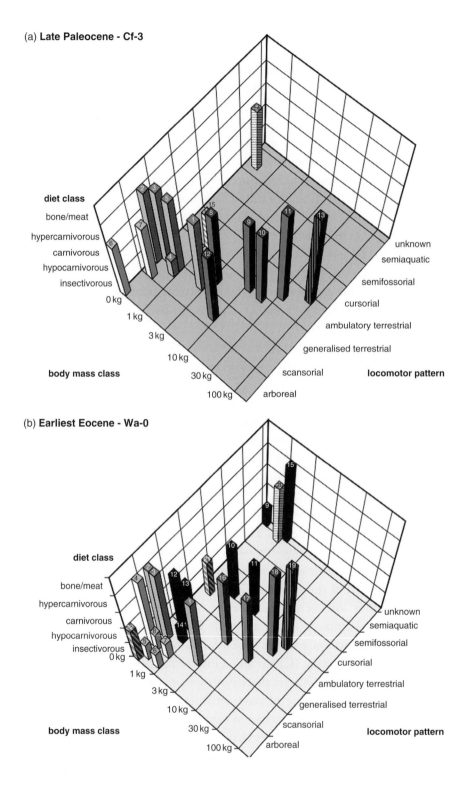

Figure 10.1 Carnivore guild structures at the Paleocene–Eocene boundary of North America. a, Guild structure of Paleocene Cf-3. Viverravidae, 1: *Didymictis leptomylus*, 2:

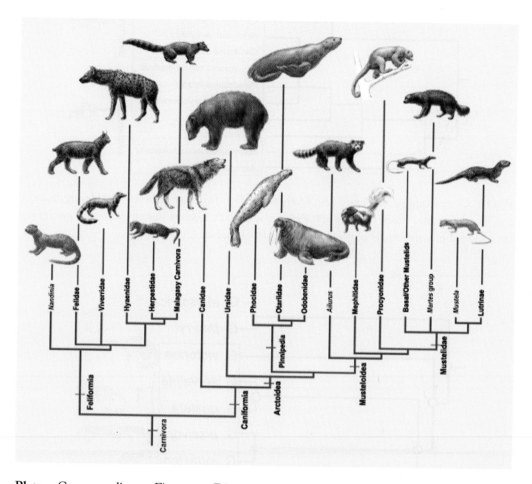

Plate 1 Corresponding to Figure 2.2. Diagrammatic summary of the molecular phylogeny of major clades of living Carnivora (from Flynn *et al.*, 2005). Illustrations of representative taxa for major lineages include (from top): *Nandinia binotata*; Felidae (*Lynx rufus*); Viverridae (*Viverra zibetha*); Hyaenidae (*Crocuta crocuta*); Herpestidae (*Mungos mungo*); Malagasy carnivorans (*Eupleres goudotii*); Canidae (*Canis lupus*); Ursidae (*Ursus americanus*); Phocidae (*Phoca vitulina*); Otariidae (*Zalophus californianus*); Odobenidae (*Odobenus rosmarus*); *Ailurus fulgens*; Mephitidae (*Mephitis mephitis*); Procyonidae (*Potos flavus*); Mustelidae, basal/other mustelids (generalised schematic representing diverse taxa [African polecat and striped marten, badger, etc.]); Mustelidae, *Martes*-group (*Gulo gulo*); Mustelidae, *Mustela* (*Mustela frenata*); Mustelidae, Lutrinae (*Lontra canadensis*).

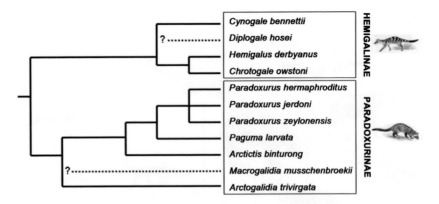

Plate 2 Corresponding to Figure 3.6. Synthetic tree of the Asian palm civets (Hemigalinae and Paradoxurinae) (from Patou, 2008; Patou *et al.*, 2008). Dashed lines indicate the species not yet included in a molecular phylogeny. Images from Francis (2008).

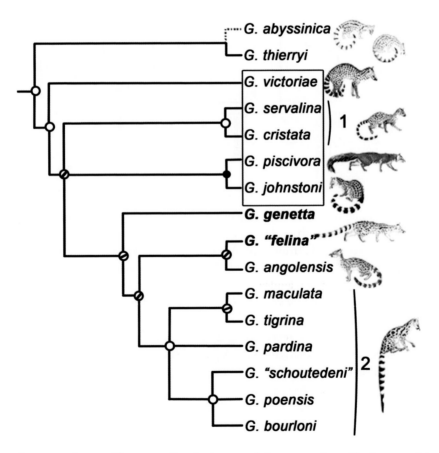

Plate 3 Corresponding to Figure 3.7. Synthetic tree of the genets from Gaubert *et al.* (2004b) (1: true servaline genets, 2: large spotted genet complex; in bold: small spotted genet). Dashed lines indicate the species not yet included in a molecular phylogeny.

Plate 4 Corresponding to Figure 4.1. The only extant species of Ailuridae, the red panda, *Ailurus fulgens*.

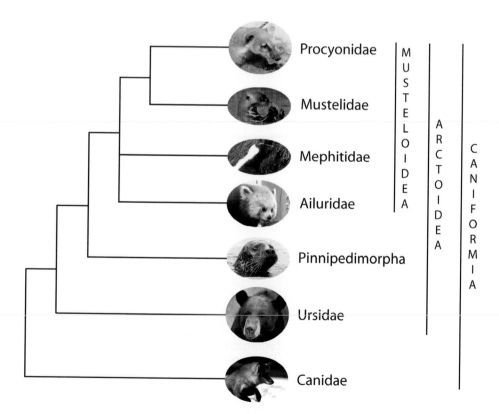

Plate 5 Corresponding to Figure 4.2. Consensus phylogenetic tree of the extant arctoid families, based on molecular data.

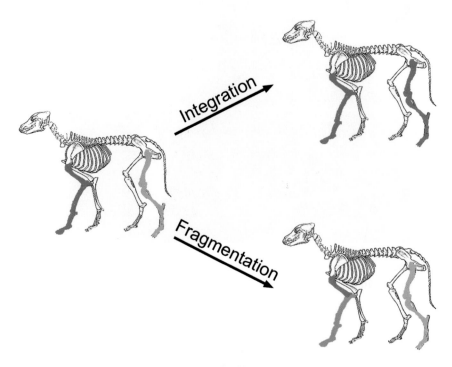

Plate 6 Corresponding to Figure 5.1. The two paths by which modules may evolve: integration, where ancestrally independent modules evolve strong correlations; or fragmentation, where ancestrally correlated traits become independent of each other, shown in the postcranial skeleton of a dog. Elements shaded with the same colour are integrated.

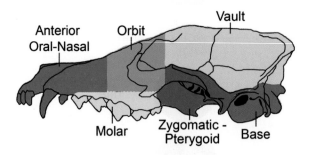

Plate 7 Corresponding to Figure 5.2. The six morphometrically derived cranial modules (Goswami, 2006a) upon which analyses of discrete character evolution are based.

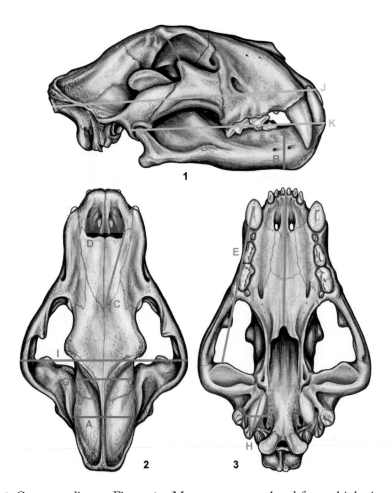

Plate 8 Corresponding to Figure 6.1. Measurements analysed for multiphasic analyses. A, Braincase width (BCW); B, mandibular flange (i.e. symphyseal) depth (MFD); C, nasal bone length (NBL); D, external narial area (NRA); E, facial length measured by length of the palate (PLL) and the length from the glenoid to the tip of the canine (GCL); F, orbit size measured by the distance between the left or right postorbital processes of the frontal and zygoma (LRPP); G, postorbital constriction width (POC); H, auditory bulla size measured as anteroposterior length (TBL) and mesiolateral width (TBW); I, zygomatic arch width measured as the greatest distance across the skull at the zygomatic arches (ZAW); J, skull length (SKL); K, mandible length (GML). The measurements are colour-coded based on whether that measurement should be larger (red), smaller (blue), or ambiguous (purple) in the American lion, according to descriptions in the literature as discussed above. The baseline measurements are shown in green. GML is the allometric baseline for MFD analyses. Skull of *P. l. atrox* illustrated by Emma Schachner, redrawn from Merriam and Stock (1932). Skulls are drawn in 1. lateral, 2. dorsal, and 3. ventral views.

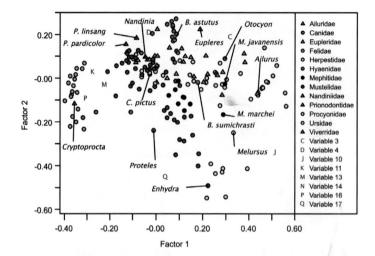

Plate 9 Corresponding to Figure 8.2. Plot of the first two factors of the correspondence analysis. This is the primary morphospace discussed in the text. Species are denoted by coloured symbols, variables by letters. The association between high negative values along Factor 1 and variables *K*, *M*, and *P*, between high positive values on Factor 1 and variables *C* and *J*, between high negative values on Factor 2 and variable *Q*, and between high positive values on Factor 2 and variables *C* and *D* are all clearly in evidence. Species that are outliers in their families or otherwise of interest are noted in the diagram.

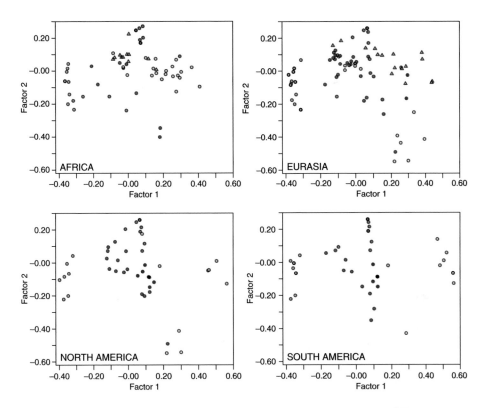

Plate 10 Corresponding to Figure 8.3. The morphospace occupation of Figure 8.2 parsed by continent. Note the general similarities between the two upper patterns and between the two lower patterns and the differences between the upper and lower rows. Symbols are as in Plate 9 and Figure 8.2.

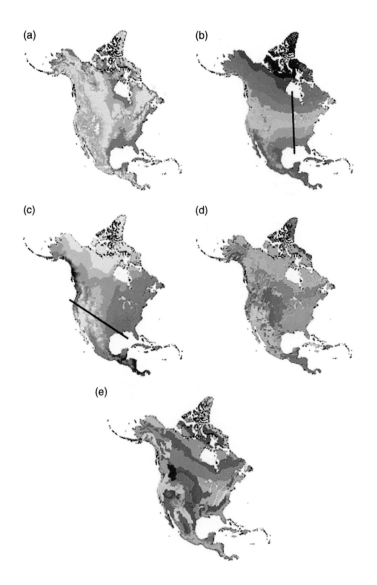

Plate 11 Corresponding to Figure 13.2. Fifty km grid point data. a, Elevation, divided into
10 colour categories using Jenk's natural breaks algorithm. Dark green is the lowest elevation
(1–173 m), white is the highest (2596–3660 m). b, Mean annual temperature, divided into
10 colour categories using Jenk's natural breaks algorithm. Dark blue is the lowest temperature
(−19.9 to −12.5°C) and dark red is the highest (21.4 to 28.6°C). The black line emphasises the
north–south gradient in temperature. c, Mean annual precipitation, divided into 10 colour
categories using Jenk's natural breaks algorithm. Light yellow is the lowest precipitation
(49.1–257.4 mm) and dark blue is the highest (2988.8–5239 mm). The black line emphasises the
east–west gradient in precipitation. d, Matthews' vegetation cover, with a categorical colour
scheme that is roughly ordered from densest vegetation in dark green (tropical evergreen forest
and subtropical evergreen seasonal broad-leaved forest) to sparsest in bright red (ice). e, Bailey's
ecoregion provinces, with an arbitrary colouring scheme.

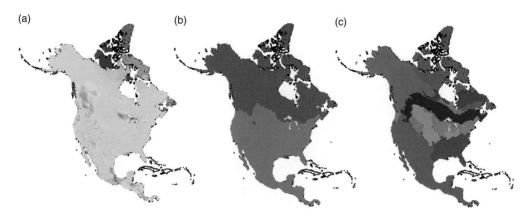

Plate 12 Corresponding to Figure 13.5. Carnivoran species diversity and faunal regions.
a, Number of carnivoran species, ranging from dark blue (1 species) to bright red (20 species).
b, Faunal regions based on carnivoran species using *k*-means clustering of 50 km grid points
and the silhouette test statistic, which identified two clusters. c, Faunal regions based on
carnivoran species using *k*-means clustering of 50 km grid points with the number of clusters
forcibly set at 10.

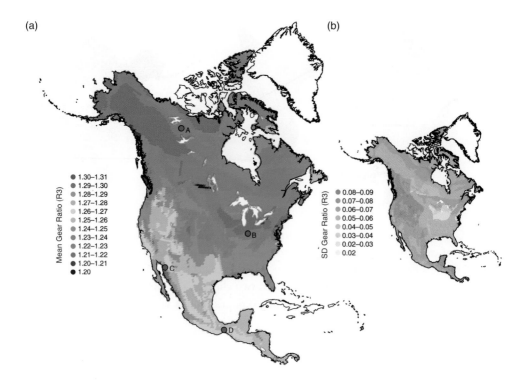

Plate 13 Corresponding to Figure 13.7. a, Mean calcaneal gear ratio (calcaneum length/sustentacular position, R3) and b, its standard deviation for carnivoran species present at 50km grid points across North America. The four lettered points (A–D) show the location of the data presented in Figure 13.8.

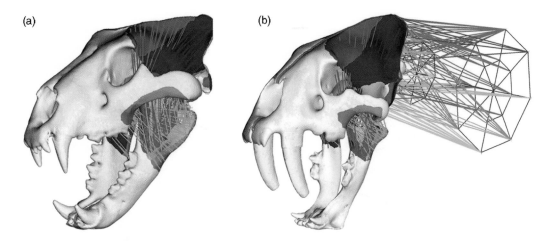

Plate 14 Corresponding to Figure 15.1. Construction of models. a, *Panthera leo* model showing the Temporalis (blue) and Masseter (red) systems, muscle 'beams' and attachment areas. b, *Smilodon fatalis* model showing the neck assembly. Different coloured beams on the neck correspond to neck muscle groups in felids. From McHenry *et al.* (2007).

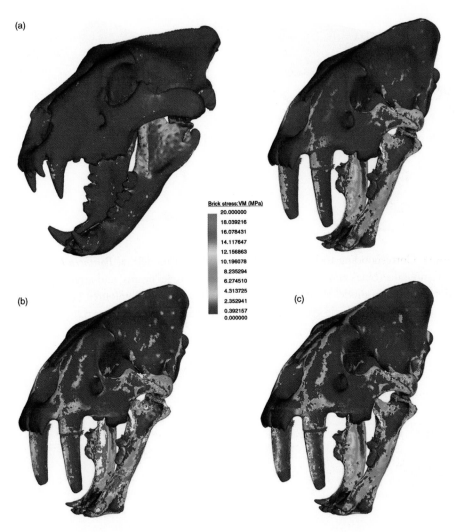

(a)

(b)

(c)

Brick stress:VM (MPa)
20.000000
18.039216
16.078431
14.117647
12.156863
10.196078
8.235294
6.274510
4.313725
2.352941
0.392157
0.000000

Plate 15 Corresponding to Figure 15.2. Von Mises stress under intrinsic loads (bilateral canine bites). a, Bite force predicted by 3D dry skull method, adjusted to account for pennation; shown are lion biting at 3388 N (*Left*) and *S. fatalis* biting at 1104 N (*Right*). b and c, *S. fatalis* biting at the forces calculated using regression of bite force on body mass for a 229-kg felid (2110 N), powered by jaw adductors only (b) and by neck + jaw muscles (c). From McHenry *et al.* (2007).

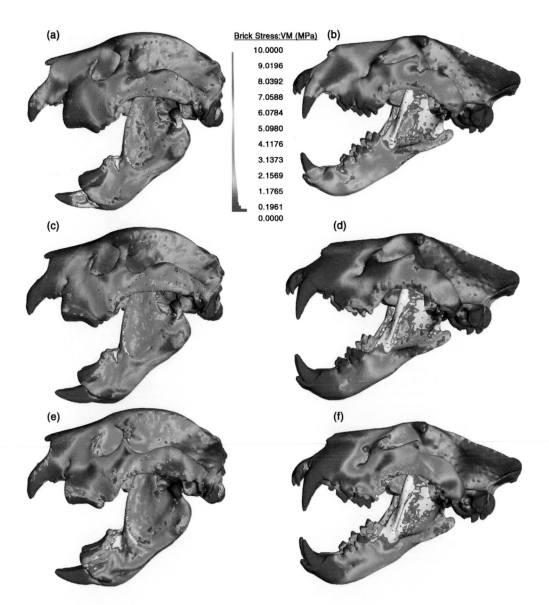

Plate 16 Corresponding to Figure 15.3. Stress (Von Mises) distributions in lateral views for intrinsic loading cases in *Thylacoleo carnifex* (a, c, e) and *Panthera leo* (b, d, f): (a, b) bilateral bite at canines; (c, d) bilateral bite at carnassials; and (e, f) unilateral bite at left carnassial. MPa = mega pascals. From Wroe (2008).

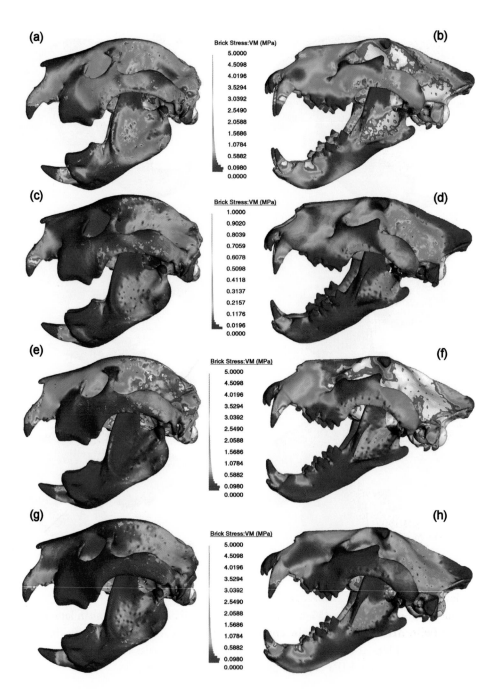

Plate 17 Corresponding to Figure 15.4. Stress (Von Mises) distributions in lateral views for extrinsic loading cases in *Thylacoleo carnifex* (a, c, e, g) and *Panthera leo* (b, d, f, h): (a, b) lateral shake; (c, d) axial twist; (e, f) dorsoventral head depression; and (g, h) pull-back. MPa = mega pascals. From Wroe (2008).

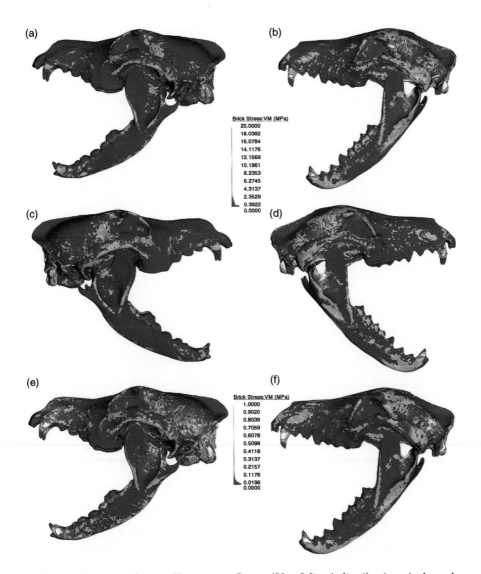

Plate 18 Corresponding to Figure 15.5. Stress (Von Mises) distributions in lateral views of FE models of *Canis lupus dingo* (a–c) and *Thylacinus cynocephalus* (e–g) under two load cases: (a, d) lateral shake (left); (b, e) lateral shake (right); and (c–f) axial twist. MPa = mega pascals. From Wroe *et al.* (2007).

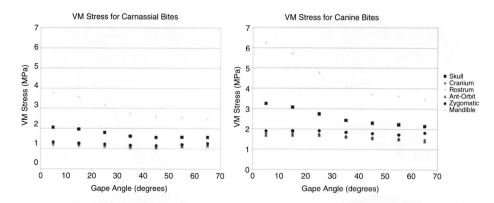

Plate 19 Corresponding to Figure 15.6. Comparison of mean 'brick' stress (Von Mises) distributions in selected cranial regions of *C. l. dingo* in bites at seven different gape angles. From Bourke *et al.* (2008).

Guild structure environmental dependency: continental vs. insular fauna

Middle Eocene North American Br-2 vs. European MP 11–13 of Geiseltal

These two contemporary guilds occurred under very different conditions. While the mammalian fauna of the North American Br-2 lived in an interior continental ecological setting of mixed landscapes, the European Geiseltal fauna comes from the Eocene island of Europe. Analyses of these guilds were independently provided by Morlo (1999) for Br1–3 and Geiseltal and Morlo and Gunnell (2005a) for Br-2. Morlo (1999), however, only included creodonts in his analysis, so we have added the Geiseltal Carnivora, Pantolestida, and Condylarthra.

Results

Obviously, the main difference between the two guilds is the much higher number of taxa (26 vs. 13) in the North American (Br-2) sample. Additionally, even though several spatially and temporally divergent faunas are included in both guilds, the Geiseltal guild is clearly limited with respect to body mass variation and in the number of taxa of a specific ecomorphology. The European guild shows a highly diverse ecomorphospace pattern within the subfamily Proviverrinae, while the top predators are hyaenodontines. Ecologically much less diverse are the North American Proviverrinae, which show just a slight increase in size from the lower to middle Eocene and no hypercarnivore specialisation. The niches of small to middle-sized hypercarnivores are instead occupied by Limnocyoninae, Machairoidinae, and viverravid carnivorans.

Figure 10.1 (*cont.*)

Viverravus bowni, 3: *Viverravus politus*, 4: *Viverravus rosei*, 5: *Viverravus acutus*; Miacidae, 6: *Miacis* sp., 7: *Uintacyon rudis*; Oxyaenidae, 8: *Dipsalidictis platypus*, 9: *Dipsalidictis krausei*, 10: *Dipsalidictis aequidens*, 11: *Palaeonictis peloria*, 12: *Tytthaena* sp.; Mesonychia, 13: *Dissacus* sp.; Pantolesta, 14: *Palaeosinopa* sp.; Arctocyonidae, 15: *Thryptacodon antiquus*. b, Guild structure of Eocene Wa-0. Viverravidae, 1: *Didymictis leptomylus*, 2: *Viverravus bowni*, 3: *Viverravus politus*; Miacidae, 4: *Uintacyon gingerichi*, 5: *Miacis winkleri*, 6: *Miacis deutschi*, 7: *Miacis rosei*, 8: *Vassacyon bowni*; Hyaenodontidae, 9: *Acarictis ryani*, 10: *Arfia junnei*, 11: *Arfia shoshoniensis*, 12: *Prototomus deimos*, 13: *Prototomus martis*, 14: *Prolimnocyon eerius*, 15: *Galecyon mordax*; Oxyaenidae, 16: *Dipsalidictis platypus*, 17: *Dipsalidictis transiens*, 18: *Palaeonictis occidentalis*; Mesonychia, 19: *Dissacus praenuntius*; Pantolesta, 20: *Thelysia* sp.; Arctocyonidae, 21: *Thryptacodon barae*, 22: *Chriacus badgleyi*.

Table 10.1 List of taxa included in analyses with age, locality, and ecomorphological attributes.

Order	Family	Genus	Species	Locality	Body mass (kg)	Locomotor class	Diet class
Carnivora	Viverravidae	Didymictis	leptomylus	Paleocene Cf3	1–3	Scansorial	Hypercarnivorous
Carnivora	Viverravidae	Viverravus	bowni	Paleocene Cf3	<1	Scansorial	Hypercarnivorous
Carnivora	Viverravidae	Viverravus	politus	Paleocene Cf3	<1	Scansorial	Hypercarnivorous
Carnivora	Viverravidae	Viverravus	rosei	Paleocene Cf3	<1	Scansorial	Insectivorous
Carnivora	Viverravidae	Viverravus	acutus	Paleocene Cf3	<1	Scansorial	Hypercarnivorous
Carnivora	Miacidae	Miacis	sp.	Paleocene Cf3	<1	Arboreal	Carnivorous
Carnivora	Miacidae	Uintacyon	rudis	Paleocene Cf3	<1	Scansorial	Carnivorous
Creodonta	Oxyaenidae	Dipsalidictis	platypus	Paleocene Cf3	3–10	Generalised terrestrial	Hypercarnivorous
Creodonta	Oxyaenidae	Dipsalidictis	krausei	Paleocene Cf3	10–30	Generalised terrestrial	Hypercarnivorous
Creodonta	Oxyaenidae	Dipsalidictis	aequidens	Paleocene Cf3	10–30	Generalised terrestrial	Hypercarnivorous
Creodonta	Oxyaenidae	Palaeonictis	peloria	Paleocene Cf3	30–100	Ambulatory terrestrial	Meat/bone
Creodonta	Oxyaenidae	Tytthaena	sp.	Paleocene Cf3	3–10	Scansorial	Hypercarnivorous
Mesonychia	Mesonychidae	Dissacus	sp.	Paleocene Cf3	30–100	Ambulatory terrestrial	Meat/bone
Cimolesta	Pantolesta	Palaeosinopa	sp.	Paleocene Cf3	<1	Semi-aquatic	Carnivorous
Condylartha	Arctocyonidae	Thryptacodon	antiquus	Paleocene Cf3	1–3	Ambulatory terrestrial	Hypocarnivorous
Carnivora	Viverravidae	Didymictis	leptomylus	Eocene Wa0	1–3	Scansorial	Hypercarnivorous

Order	Family	Genus	species			Locomotion	Diet
Carnivora	Viverravidae	*Viverravus*	*bowni*	Eocene Wao	<1	Scansorial	Hypercarnivorous
Carnivora	Viverravidae	*Viverravus*	*politus*	Eocene Wao	<1	Scansorial	Hypercarnivorous
Carnivora	Miacidae	*Uintacyon*	*gingerichi*	Eocene Wao	<1	Arboreal	Hypercarnivorous
Carnivora	Miacidae	*Miacis*	*winkleri*	Eocene Wao	<1	Arboreal	Insectivorous
Carnivora	Miacidae	*Miacis*	*deutschi*	Eocene Wao	<1	Arboreal	Insectivorous
Carnivora	Miacidae	*Miacis*	*rosei*	Eocene Wao	<1	Arboreal	Insectivorous
Carnivora	Miacidae	*Vassacyon*	*bowni*	Eocene Wao	<1	Arboreal	Insectivorous
Creodonta	Hyaenodontidae	*Acarictis*	*ryani*	Eocene Wao	<1	Unknown	Insectivorous
Creodonta	Hyaenodontidae	*Arfia*	*junnei*	Eocene Wao	1–3	Cursorial	Carnivorous
Creodonta	Hyaenodontidae	*Arfia*	*shoshoniensis*	Eocene Wao	3–10	Cursorial	Carnivorous
Creodonta	Hyaenodontidae	*Prototomus*	*deimos*	Eocene Wao	<1	Generalised terrestrial	Carnivorous
Creodonta	Hyaenodontidae	*Prototomus*	*martis*	Eocene Wao	<1	Generalised terrestrial	Carnivorous
Creodonta	Hyaenodontidae	*Prolimnocyon*	*eerius*	Eocene Wao	<1	Scansorial	Insectivorous
Creodonta	Hyaenodontidae	*Galecyon*	*mordax*	Eocene Wao	1–3	Unknown	Hypercarnivorous
Creodonta	Oxyaenidae	*Dipsalidictis*	*platypus*	Eocene Wao	3–10	Generalised terrestrial	Hypercarnivorous
Creodonta	Oxyaenidae	*Dipsalidictis*	*transiens*	Eocene Wao	10–30	Generalised terrestrial	Hypercarnivorous
Creodonta	Oxyaenidae	*Palaeonictis*	*occidentalis*	Eocene Wao	10–30	Ambulatory terrestrial	Bone/meat
Mesonychia	Mesonychidae	*Dissacus*	*praenuntius*	Eocene Wao	10–30	Ambulatory terrestrial	Bone/meat
Cimolesta	Pantolesta	*Thelysia*	sp.	Eocene Wao	<1	Semi-aquatic	Carnivorous
Condylartha	Arctocyonidae	*Thryptacodon*	*barae*	Eocene Wao	<1	Ambulatory terrestrial	Hypocarnivorous

Table 10.1 (*cont.*)

Order	Family	Genus	Species	Locality	Body mass (kg)	Locomotor class	Diet class
Condylartha	Arctocyonidae	*Chriacus*	*badgleyi*	Eocene Wa0	<1	Arboreal	Hypocarnivorous
Carnivora	Viverravidae	*Viverravus*	*gracilis*	Eocene Br2	<1	Scansorial	Insectivorous
Carnivora	Viverravidae	*Viverravus*	*minutus*	Eocene Br2	<1	Scansorial	Insectivorous
Carnivora	Miacidae	*Miacis*	*parvivorus*	Eocene Br2	<1	Arboreal	Carnivorous
Carnivora	Miacidae	*Vulpavus*	*palustris*	Eocene Br2	3–10	Arboreal	Carnivorous
Carnivora	Miacidae	*Vulpavus*	*profectus*	Eocene Br2	1–3	Arboreal	Carnivorous
Carnivora	Miacidae	*Vulpavus*	*ovatus*	Eocene Br2	1–3	Arboreal	Carnivorous
Carnivora	Miacidae	*Palaearctonyx*	*meadi*	Eocene Br2	<1	Arboreal	Hypocarnivorous
Carnivora	Miacidae	*Oodectes*	*herpestoides*	Eocene Br2	<1	Arboreal	Carnivorous
Carnivora	Miacidae	*Oodectes*	*proximus*	Eocene Br2	<1	Arboreal	Carnivorous
Carnivora	Miacidae	*Uintacyon*	*edax*	Eocene Br2	<1	Generalised terrestrial	Carnivorous
Carnivora	Miacidae	*Uintacyon*	*vorax*	Eocene Br2	1–3	Generalised terrestrial	Carnivorous
Carnivora	Miacidae	*Uintacyon*	*major*	Eocene Br2	10–30	Generalised terrestrial	Carnivorous
Mesonychia	Mesonychidae	*Mesonyx*	*obtusidens*	Eocene Br2	10–30	Generalised terrestrial	Bone/meat
Mesonychia	Mesonychidae	*Harpagolestes*	*macrocephalus*	Eocene Br2	30–100	Generalised terrestrial	Bone/meat
Creodonta	Hyaenodontidae	*Sinopa*	*grangeri*	Eocene Br2	3–10	Cursorial	Carnivorous
Creodonta	Hyaenodontidae	*Sinopa*	*major*	Eocene Br2	3–10	Cursorial	Carnivorous
Creodonta	Hyaenodontidae	*Sinopa*	*minor*	Eocene Br2	1–3	Cursorial	Carnivorous

Order	Family	Genus	species	Age	Size	Locomotion	Diet
Creodonta	Hyaenodontidae	*Sinopa*	*pungens*	Eocene Br2	1–3	Cursorial	Carnivorous
Creodonta	Hyaenodontidae	*Sinopa*	*rapax*	Eocene Br2	3–10	Cursorial	Carnivorous
Creodonta	Hyaenodontidae	*Tritemnodon*	*agilis*	Eocene Br2	3–10	Generalised terrestrial	Hypercarnivorous
Creodonta	Hyaenodontidae	*Limnocyon*	*verus*	Eocene Br2	3–10	Generalised terrestrial	Carnivorous
Creodonta	Hyaenodontidae	*Limnocyon*	*cuspidens*	Eocene Br2	3–10	Generalised terrestrial	Carnivorous
Creodonta	Hyaenodontidae	*Thinocyon*	*medius*	Eocene Br2	1–3	Generalised terrestrial	Hypercarnivorous
Creodonta	Hyaenodontidae	*Thinocyon*	*velox*	Eocene Br2	1–3	Generalised terrestrial	Hypercarnivorous
Creodonta	Hyaenodontidae	*Machaeroides*	*eothen*	Eocene Br2	3–10	Generalised terrestrial	Hypercarnivorous
Creodonta	Oxyaenidae	*Patriofelis*	*ulta*	Eocene Br2	30–100	Ambulatory terrestrial	Hypercarnivorous
Creodonta	Oxyaenidae	*Patriofelis*	*ferox*	Eocene Br2	30–100	Ambulatory terrestrial	Hypercarnivorous
Cimolesta	Paroxyclaenidae	*Vulpavoides*	*germanica*	Geiseltal MP 11–13	1–3	Arboreal	Hypocarnivorous
Cimolesta	Paroxyclaenidae	*Pugiodens*	*mirus*	Geiseltal MP 11–13	1–3	Arboreal	Hypocarnivorous
Cimolesta	Pantolesta	*Buxolestes*	sp.	Geiseltal MP 11–13	3–10	Semi-aquatic	Carnivorous
Carnivora	Miacidae	*Querygale*	*helvetica*	Geiseltal MP 11–13	3–10	Scansorial	Carnivorous

Table 10.1 (*cont.*)

Order	Family	Genus	Species	Locality	Body mass (kg)	Locomotor class	Diet class
Creodonta	Hyaenodontidae	*Oxyaenoides*	*biscuspidens*	Geiseltal MP 11–13	10–30	Cursorial	Hypercarnivorous
Creodonta	Hyaenodontidae	*Cynohyaenodon*	*ruetimeyeri*	Geiseltal MP 11–13	3–10	Generalised terrestrial	Hypercarnivorous
Creodonta	Hyaenodontidae	*Cynohyaenodon*	*trux*	Geiseltal MP 11–13	1–3	Generalised terrestrial	Carnivorous
Creodonta	Hyaenodontidae	*Eurotherium*	*matthesi*	Geiseltal MP 11–13	3–10	Unknown	Hypercarnivorous
Creodonta	Hyaenodontidae	*Eurotherium*	*theriodis*	Geiseltal MP 11–13	3–10	Unknown	Hypercarnivorous
Creodonta	Hyaenodontidae	*Leonhardtina*	*gracilis*	Geiseltal MP 11–13	1–3	Unknown	Carnivorous
Creodonta	Hyaenodontidae	*Matthodon*	*tritens*	Geiseltal MP 11–13	10–30	Ambulatory terrestrial	Hypercarnivorous
Creodonta	Hyaenodontidae	*Prodissopsalis*	*eocaenicus*	Geiseltal MP 11–13	10–30	Unknown	Hypercarnivorous
Creodonta	Hyaenodontidae	*Proviverra*	*typica*	Geiseltal MP 11–13	<1	Unknown	Carnivorous
Carnivora	Amphicyonidae	*Amphicyon*	cf. *major*	Sandelzhausen MN-5	>100	Ambulatory terrestrial	Carnivorous
Carnivora	Ursidae	*Hemicyon*	*stehlini*	Sandelzhausen MN-5	30–100	Cursorial	Carnivorous
Carnivora	Ursidae	*Pseudarctos*	*bavaricus*	Sandelzhausen MN-5	10–30	Ambulatory terrestrial	Hypocarnivorous

Order	Family	Genus	species	Locality/Age	Body mass	Locomotion	Diet
Carnivora	Mustelidae	*cf. Proputorius*	*pusillus*	Sandelzhausen MN-5	<1	Unknown	Carnivorous
Carnivora	Mustelidae	*Ischyrictis*	*zibethoides*	Sandelzhausen MN-5	10–30	Generalised terrestrial	Hypercarnivorous
Carnivora	Mustelidae	*Martes*	*cf. munki*	Sandelzhausen MN-5	<1	Unknown	Carnivorous
Carnivora	Viverridae	*Leptoplesictis*	*sp.*	Sandelzhausen MN-5	<1	Scansorial	Insectivorous
Carnivora	Barbourofelidae	*Prosansanosmilus*	*eggeri*	Sandelzhausen MN-5	10–30	Unknown	Hypercarnivorous
Carnivora	Felidae	*Pseudaelurus*	*lorteti*	Sandelzhausen MN-5	10–30	Scansorial	Hypercarnivorous
Carnivora	Canidae	*Canis*	*aureus*	Serengeti–Recent	3–10	Cursorial	Carnivorous
Carnivora	Canidae	*Canis*	*adustus*	Serengeti–Recent	10–30	Cursorial	Carnivorous
Carnivora	Canidae	*Canis*	*mesomelas*	Serengeti–Recent	10–30	Cursorial	Carnivorous
Carnivora	Canidae	*Lycaon*	*pictus*	Serengeti–Recent	10–30	Cursorial	Carnivorous
Carnivora	Canidae	*Otocyon*	*megalotis*	Serengeti–Recent	3–10	Cursorial	Carnivorous
Carnivora	Felidae	*Felis*	*serval*	Serengeti–Recent	10–30	Scansorial	Hypercarnivorous
Carnivora	Felidae	*Caracal*	*caracal*	Serengeti–Recent	10–30	Scansorial	Hypercarnivorous
Carnivora	Felidae	*Acinonyx*	*jubatus*	Serengeti–Recent	30–100	Cursorial	Hypercarnivorous

Table 10.1 (*cont.*)

Order	Family	Genus	Species	Locality	Body mass (kg)	Locomotor class	Diet class
Carnivora	Felidae	*Panthera*	*pardus*	Serengeti-Recent	30–100	Scansorial	Hypercarnivorous
Carnivora	Felidae	*Panthera*	*leo*	Serengeti-Recent	>100	Scansorial	Hypercarnivorous
Carnivora	Herpestidae	*Helogale*	*parvula*	Serengeti-Recent	<1	Semi-fossorial	Insectivorous
Carnivora	Herpestidae	*Mungos*	*mungo*	Serengeti-Recent	1–3	Semi-fossorial	Insectivorous
Carnivora	Herpestidae	*Herpestes*	*ichneumon*	Serengeti-Recent	1–3	Semi-fossorial	Insectivorous
Carnivora	Herpestidae	*Ichneumon*	*albicauda*	Serengeti-Recent	1–3	Semi-fossorial	Insectivorous
Carnivora	Herpestidae	*Atilax*	*paludinosus*	Serengeti-Recent	3–10	Generalised terrestrial	Carnivorous
Carnivora	Hyaenidae	*Hyaena*	*hyaena*	Serengeti-Recent	30–100	Cursorial	Bone/meat
Carnivora	Hyaenidae	*Crocuta*	*crocuta*	Serengeti-Recent	30–100	Cursorial	Bone/meat
Carnivora	Mustelidae	*Mellivora*	*capensis*	Serengeti-Recent	3–10	Ambulatorial terrestrial	Carnivorous
Carnivora	Viverridae	*Genetta*	*genetta*	Serengeti-Recent	1–3	Generalised terrestrial	Carnivorous
Carnivora	Viverridae	*Civettictis*	*civetta*	Serengeti-Recent	10–30	Generalised terrestrial	Carnivorous

Carnivora	Canidae	*Speothos*	*venaticus*	Guyana–Recent	3–10	Cursorial	Carnivorous
Carnivora	Felidae	*Leopardus*	*tigrinus*	Guyana–Recent	3–10	Arboreal	Hypercarnivorous
Carnivora	Felidae	*Leopardus*	*pardalis*	Guyana–Recent	3–10	Scansorial	Hypercarnivorous
Carnivora	Felidae	*Leopardus*	*wiedi*	Guyana–Recent	3–10	Scansorial	Hypercarnivorous
Carnivora	Felidae	*Herpailurus*	*yaguarondi*	Guyana–Recent	3–10	Scansorial	Hypercarnivorous
Carnivora	Felidae	*Puma*	*concolor*	Guyana–Recent	30–100	Scansorial	Hypercarnivorous
Carnivora	Felidae	*Panthera*	*onca*	Guyana–Recent	30–100	Scansorial	Hypercarnivorous
Carnivora	Mustelidae	*Galictis*	*cuja*	Guyana–Recent	1–3	Generalised terrestrial	Carnivorous
Carnivora	Mustelidae	*Pteronura*	*brasiliensis*	Guyana–Recent	30–100	Semi-aquatic	Carnivorous
Carnivora	Mustelidae	*Lontra*	*longicaudis*	Guyana–Recent	3–10	Semi-aquatic	Carnivorous
Carnivora	Mustelidae	*Eira*	*barbara*	Guyana–Recent	3–10	Scansorial	Carnivorous
Carnivora	Procyonidae	*Potus*	*flavus*	Guyana–Recent	3–10	Arboreal	Hypocarnivorous
Carnivora	Procyonidae	*Nasua*	*nasua*	Guyana–Recent	3–10	Arboreal	Hypocarnivorous

Table 10.1 (*cont.*)

Order	Family	Genus	Species	Locality	Body mass (kg)	Locomotor class	Diet class
Carnivora	Procyonidae	*Bassaricyon*	*gabbii*	Guyana–Recent	1–3	Arboreal	Hypocarnivorous
Carnivora	Amphicyonidae	*Pseudocyon*	*sansaniensis*	Sansan MN-6	>100	Generalised terrestrial	Bone/meat
Carnivora	Amphicyonidae	*Amphicyon*	*major*	Sansan MN-6	>100	Ambulatory terrestrial	Carnivorous
Carnivora	Amphicyonidae	*Pseudarctos*	*bavaricus*	Sansan MN-6	10–30	Ambulatory terrestrial	Hypocarnivorous
Carnivora	Ursidae	*Plithocyon*	*armagnacensis*	Sansan MN-6	>100	Ambulatory terrestrial	Carnivorous
Carnivora	Ursidae	*Hemicyon*	*sansaniensis*	Sansan MN-6	30–100	Cursorial	Carnivorous
Carnivora	Viverridae	*Viverrictis*	*modica*	Sansan MN-6	<1	Scansorial	Insectivorous
Carnivora	Viverridae	*Leptoplesictis*	*aurelianensis*	Sansan MN-6	<1	Scansorial	Insectivorous
Carnivora	Viverridae	*Semigenetta*	*sansaniensis*	Sansan MN-6	3–10	Scansorial	Carnivorous
Carnivora	Lophocyonidae	*Schlossericyon*	*viverroides*	Sansan MN-6	3–10	Arboreal	Hypocarnivorous
Carnivora	Lophocyonidae	*Sivanasua*	indet.	Sansan MN-6	3–10	Arboreal	Hypocarnivorous
Carnivora	Ailuridae	*Alopecocyon*	*leptorhynchus*	Sansan MN-6	3–10	Scansorial	Carnivorous
Carnivora	Barbourofelidae	*Sansanosmilus*	*palmidens*	Sansan MN-6	30–100	Generalised terrestrial	Hypercarnivorous
Carnivora	Felidae	*Pseudaelurus*	*lorteti*	Sansan MN-6	10–30	Scansorial	Hypercarnivorous
Carnivora	Felidae	*Pseudaelurus*	*quadridentatus*	Sansan MN-6	30–100	Scansorial	Hypercarnivorous
Carnivora	Mustelidae	*Lartetictis*	*dubia*	Sansan MN-6	10–30	Semi-aquatic	Carnivorous
Carnivora	Mustelidae	*Ischyrictis*	*zibethoides*	Sansan MN-6	10–30	Generalised terrestrial	Hypercarnivorous

Carnivora	Mustelidae	*Mellidelavus*	*leptorynchus*	Sansan MN-6	1–3	Generalised terrestrial	Hypocarnivorous
Carnivora	Mustelidae	*Taxodon*	*sansaniensis*	Sansan MN-6	3–10	Generalised terrestrial	Carnivorous
Carnivora	Mustelidae	*Proputorius*	*sansaniensis*	Sansan MN-6	3–10	Scansorial	Carnivorous
Carnivora	Mustelidae	*Martes*	*sansaniensis*	Sansan MN-6	3–10	Scansorial	Carnivorous
Carnivora	Amphicyonidae	*Pseudocyon*	*steinheimensis*	Steinheim MN-7	>100	Ambulatory terrestrial	Bone/meat
Carnivora	Amphicyonidae	*Pseudocyon*	*sansaniensis*	Steinheim MN-7	>100	Ambulatory terrestrial	Bone/meat
Carnivora	Amphicyonidae	*Amphicynopsis*	*serus*	Steinheim MN-7	>100g	Ambulatory terrestrial	Hypercarnivorous
Carnivora	Felidae	*Pseudaelurus*	*lorteti*	Steinheim MN-7	10–30	Scansorial	Hypercarnivorous
Carnivora	Felidae	*Pseudaelurus*	*quadridentatus*	Steinheim MN-7	30–100	Scansorial	Hypercarnivorous
Carnivora	Mustelidae	*Martes*	cf. *filholi*	Steinheim MN-7	1–3	Scansorial	Carnivorous
Carnivora	Mustelidae	*Trocharion*	*albanense*	Steinheim MN-7	1–3	Semi-fossorial	Carnivorous
Carnivora	Mustelidae	*Paralutra*	*jaegeri*	Steinheim MN-7	3–10	Semi-aquatic	Carnivorous
Carnivora	Mustelidae	*Ischyrictis*	*mustelinus*	Steinheim MN-7	3–10	Generalised terrestrial	Hypercarnivorous
Carnivora	Mustelidae	*Proputorius*	sp.	Steinheim MN-7	3–10	Scansorial	Carnivorous
Carnivora	Mustelidae	*Trochotherium*	*cyamoides*	Steinheim MN-7	3–10	Semi-aquatic	Hypocarnivorous

Table 10.1 (*cont.*)

Order	Family	Genus	Species	Locality	Body mass (kg)	Locomotor class	Diet class
Carnivora	Barbourofelidae	*Sansanosmilus*	*jourdani*	Steinheim MN-7	>100	Ambulatory terrestrial	Hypercarnivorous
Carnivora	Ursidae	*Ursavus*	cf. *intermedius*	Steinheim MN-7	30–100	Ambulatory terrestrial	Hypocarnivorous
Carnivora	Ursidae	*Hemicyon*	*goeriachensis*	Steinheim MN-7	>100	Cursorial	Carnivorous
Carnivora	Viverridae	*Semigenetta*	*sansaniensis*	Steinheim MN-7	3–10	Scansorial	Carnivorous
Carnivora	Amphicyonidae	*Amphicyon*	*major*	Candir MN-6	>100	Ambulatory terrestrial	Carnivorous
Carnivora	Ursidae	*Hemicyon*	*sansaniensis*	Candir MN-6	30–100	Cursorial	Carnivorous
Carnivora	Ailuridae	*Amphictis*	*cuspida*	Candir MN-6	3–10	Unknown	Hypocarnivorous
Carnivora	Mustelidae	cf. *Trochictis*	*depereti*	Candir MN-6	3–10	Scansorial	Carnivorous
Carnivora	Mustelidae	*Ischyrictis*	*anatolicus*	Candir MN-6	10–30	Generalised terrestrial	Hypercarnivorous
Carnivora	Mustelidae	*Paralutra*	cf. *jaegeri*	Candir MN-6	3–10	Semi-aquatic	Carnivorous
Carnivora	Mustelidae	*Proputorius*	indet.	Candir MN-6	3–10	Scansorial	Carnivorous
Carnivora	Hyaenidae	*Protictitherium*	*intermedium*	Candir MN-6	3–10	Generalised terrestrial	Insectivorous
Carnivora	Hyaenidae	*Protictitherium*	cf. *gaillardi*	Candir MN-6	3–10	Generalised terrestrial	Insectivorous
Carnivora	Percrocutidae	*Percrocuta*	*miocenica*	Candir MN-6	30–100	Cursorial	Bone/meat
Carnivora	Percrocutidae	*Percrocuta*	indet.	Candir MN-6	30–100	Cursorial	Bone/meat

Carnivora	Felidae	*Pseudaelurus*	cf. *quadri-dentatus*	Candir MN-6	30–100	Scansorial	Hypercarnivorous
Creodonta	Hyaenodontidae	*Hyaenodon*	cf. *Gigas*	Mongolia, Oligocene	>100	Cursorial	Bone/meat
Creodonta	Hyaenodontidae	*Hyaenodon*	*mongoliensis*	Mongolia, Oligocene	>100	Unknown	Bone/meat
Creodonta	Hyaenodontidae	*Hyaenodon*	*incertus*	Mongolia, Oligocene	>100	Cursorial	Bone/meat
Creodonta	Hyaenodontidae	*Hyaenodon*	*pervagus*	Mongolia, Oligocene	30–100	Cursorial	Bone/meat
Creodonta	Hyaenodontidae	*Hyaenodon*	*eminus*	Mongolia, Oligocene	10–30	Unknown	Hypercarnivorous
Carnivora	Hemicyonidae	*Cephalogale*	sp.	Mongolia, Oligocene	3–10	Generalised terrestrial	Hypocarnivorous
Carnivora	Amphicyno-dontidae	*Amphicynodon*	*teilhardi*	Mongolia, Oligocene	1–3	Scansorial	Carnivorous
Carnivora	Semantoridae	*Amphicticeps*	*makhchinus*	Mongolia, Oligocene	3–10	Unknown	Carnivorous
Carnivora	Semantoridae	*Amphicticeps*	*dorog*	Mongolia, Oligocene	3–10	Unknown	Carnivorous
Carnivora	Semantoridae	*Amphicticeps*	*shakelfordi*	Mongolia, Oligocene	3–10	Generalised terrestrial	Carnivorous
Carnivora	Semantoridae	*Amphicticeps*	small species	Mongolia, Oligocene	1–3	Unknown	Carnivorous
Carnivora	Mustelidae	*Pyctis*	*inamatus*	Mongolia, Oligocene	1–3	Unknown	Hypercarnivorous

Table 10.1 (*cont.*)

Order	Family	Genus	Species	Locality	Body mass (kg)	Locomotor class	Diet class
Carnivora	Nimravidae	*Nimravus*	*mongoliensis*	Mongolia, Oligocene	30–100	Ambulatory terrestrial	Hypercarnivorous
Carnivora	Nimravidae	Nimravidae	indet.	Mongolia, Oligocene	10–30	Unknown	Hypercarnivorous
Carnivora	Stenoplesictidae	*Shandgolictis*	*elegans*	Mongolia, Oligocene	1–3	Scansorial	Carnivorous
Carnivora	Stenoplesictidae	*Asiavorator*	*altidens*	Mongolia, Oligocene	1–3	Scansorial	Carnivorous
Carnivora	Stenoplesictidae	*Asiavorator*	sp. nov.	Mongolia, Oligocene	<1	Unkown	Carnivorous
Carnivora	Viverravidae?	*Palaeogale*	*sectoria*	Mongolia, Oligocene	<1	Scansorial	Hypercarnivorous
Leptictida	Didymoconida	*Didymoconus*	*colgatei*	Mongolia, Oligocene	<1	Semi-fossorial	Insectivorous
Leptictida	Didymoconida	*Didymoconus*	*berkeyi*	Mongolia, Oligocene	<1	Semi-fossorial	Insectivorous
Leptictida	Didymoconida	cf. *Ergilictis*	sp.	Mongolia, Oligocene	<1	Unknown	Insectivorous

The top predators of the North American middle Eocene are the Oxyaenidae and Mesonychia. The different ecological importance of proviverrines may be explained by the competition they were confronted with in North America by groups such as limnocyonines, machairoidines, and oxyaenids, animals that were not present in the European middle Eocene. Additionally, the European record of Eocene carnivorans (Miacidae and Viverravidae) is much smaller than in North America with only one viverravid and no miacids presently known from Geiseltal, while in Br-2 carnivorans play a major role in the guild.

Guild structure environmental dependency: open landscape vs. woodland

Çandir vs. Sandelzhausen/Sansan/Steinheim (Middle Miocene of Europe)

The middle Miocene fossil mammals of Europe are among the best known in the world, including carnivores. Sansan (Ginsburg, 1961b) and Steinheim (Heizmann, 1973) are classical localities that include a high number of carnivores, nearly all of which are Carnivora. Çandir (Nagel, 2003) and Sandelzhausen (Nagel et al., 2009) have been more recently studied. These latter two are also the only middle Miocene carnivore guilds of Europe which have been analysed at least briefly (Çandir [open landscape], Nagel and Morlo, 2000; Sandelzhausen [woodland], Nagel et al., 2005).

Results

Clear evidence of environmental differences can be documented by the presence of insectivorous hyaenids and bone-cracking percrocutids in the open landscape guild of Çandir in Southeastern Europe and the higher diversity of small mustelids in the woodland Central European guilds. The PCA, however, tells another story, which is discussed below.

Guild structures in open landscapes from Eocene to Recent

Here we focus on guilds of an open landscape environment to examine the influence of environment type on guild structure through time. Besides the middle Miocene MN 6 guild from Çandir documented above, we additionally compare the guilds of the early middle Eocene of North America (Br-2), the lower Oligocene of Mongolia (Morlo and Nagel, 2007), and the Recent Serengeti guild as a modern example.

Results

All open landscape faunas show a high diversity of large (>30 kg), terrestrial, hypercarnivorous to bone/meat taxa. This becomes especially clear when

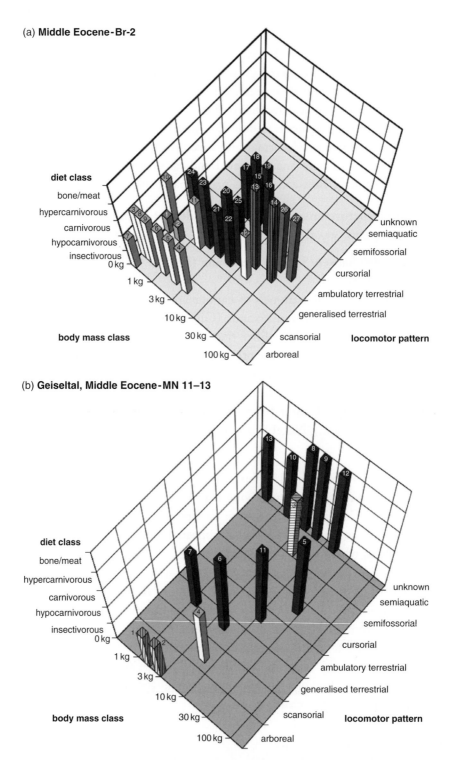

Figure 10.2 Carnivore guild structures in the Middle Eocene of North America and Europe. a, Guild structure of Br-2 from Wyoming (continental climate). Viverravidae,

comparing the open landscape guilds with woodland faunas (Geiseltal, San-delzhausen, Sansan, Guyana). While the Tertiary guilds are rather uniform in their structure, Serengeti differs in having the largest predators being scan-sorial instead of terrestrial. This result is clearly affected by taxonomic history – while the largest Tertiary predators were oxyaenids or arctoids, the largest forms are represented at Serengeti by pantherine felids instead. It is well known that the postcranial morphology of felids is phylogenetically very conservative (Day and Jayne, 2007) and does not differ substantially among Recent pantherines (except maybe in cheetahs; Flynn *et al.*, 1988; Bertram and Biewener, 1990).

Another result is the absence of arboreal taxa in open landscape faunas. The presence of arboreal taxa and a large diversity of large predators indicates a mixture of faunas and habitats. This community structure is exactly what is represented in the middle Eocene Br-2 community from Wyoming (Figure 10.4a) where our samples come from several localities representing differing habitats (near-shore forested to more open, patchy forest settings). Interestingly, the middle Miocene guild from Steinheim shows a structure similar to the one from Br-2, but unlike Br-2 the sample clearly comes from a single locality. However, the presence of taxa obviously coming from different environments (arboreal vs. large terrestrial predators) fits with the reconstruction of the Steinheim environment as a dense, but small, woodland area surrounding a crater lake and an open landscape with patchy woodlands being located farther away. The lake itself may have played an important role as a water resource even for the animals living away from its shores.

Figure 10.2 (*cont.*)

1: *Viverravus gracilis*, 2: *Viverravus minutus*; Miacidae, 3: *Miacis parvivorus*, 4: *Vulpavus palustris*, 5: *Vulpavus profectus*, 6: *Vulpavus ovatus*, 7: *Palaearctonyx meadi*, 8: *Oodectes herpestoides*, 9: *Oodectes proximus*, 10: *Uintacyon edax*, 11: *Uintacyon vorax*, 12: *Uintacyon major*; Mesonychia, 13: *Mesonyx obtusidens*, 14: *Harpagolestes macrocephalus*; Hyaenodontidae, 15: *Sinopa grangeri*, 16: *Sinopa major*, 17: *Sinopa minor*, 18: *Sinopa pungens*, 19: *Sinopa rapax*, 20: *Tritemnodon agilis*, 21: *Limnocyon verus*, 22: *Limnocyon cuspidens*, 23: *Thinocyon medius*, 24: *Thinocyon velox*, 25: *Machaeroides eothen*; Oxyaenidae, 26: *Patriofelis ulta*, 27: *Patriofelis ferox*. b, Guild structure of MP 11–13 from the Geiseltal (insular climate). Paroxyclaenida, 1: *Vulpavoides germanica*, 2: *Pugiodens mirus*; Pantolesta, 3: *Buxolestes* sp.; Miacidae, 4: *Quercygale helvetica*; Hyaenodontidae, 5: *Oxyaenoides bicuspidens*, 6: *Cynohyaenodon ruetimeyeri*, 7: *Cynohyaenodon trux*, 8: *Eurotherium matthesi*, 9: *Eurotherium theriodis*, 10: *Leonhardtina gracilis*, 11: *Matthodon tritens*, 12: *Prodissopsalis eocaenicus*, 13: *Proviverra typica*.

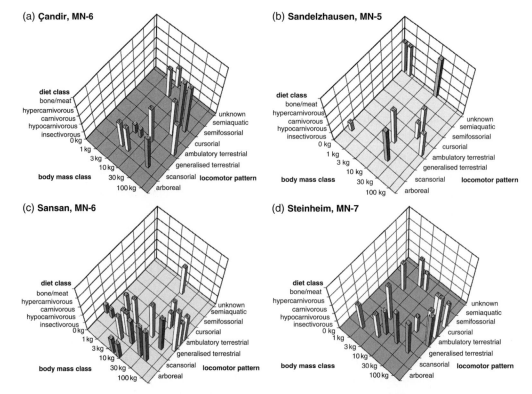

Figure 10.3 Carnivore guild structures in the middle Miocene of Europe. a, Çandir (open landscape, MN 6). Amphicyonidae, 1: *Amphicyon major*; Ursidae, 2: *Hemicyon sansaniensis*; Ailuridae, 3: *Amphictis cuspida*; Mustelidae, 4: cf. *Trochictis depereti*, 5: *Ischyrictis anatolicus*, 6: *Paralutra* cf. *jaegeri*, 7: *Proputorius* indet.; Hyaenidae, 8: *Proticititherium intermedium*, 9: *Proticititherium* cf. *gaillardi*; Percrocutidae, 10: *Percrocuta miocaenica*, 11: *Percrocuta* indet.; Felidae, 12: *Pseudaelurus* cf. *quadridentatus*. b, Sandelzhausen (woodland, MN 5). Amphicyonidae: 1: *Amphicyon* cf. *major*; Ursidae: 2: *Hemicyon stehlini*, 3: *Pseudarctos bavaricus*; Mustelidae, 4: *Proputorius pusillus* 5: *Ischyrictis zibethoides*, 6: *Martes munki*; Viverridae, 7: *Leptoplesictis*; Barbourofelidae 8: *Prosansanosmilus eggeri*; Felidae, 9: *Pseudaelurus lorteti*. c, Sansan (woodland, MN 6). Ampicyonidae, 1: *Pseudocyon sansaniensis*, 2: *Amphiyon major*; Ursidae, 3: *Pseudarctos bavaricus*, 4: *Plithocyon armagnacensis*, 5: *Hemicyon sansaniensis*; Viverridae, 6: *Viverrictis modica*, 7: *Leptoplesictis aurelianensis*, 8: *Semigenetta sansaniensis*; Lophocyonidae: 9: *Schlossericyon viverroides*, 10: *Sivanasua* indet.; Ailuridae, 11: *Alopecocyon leptorhynchus*; Barbourofelidae, 12: *Sansanosmilus palmidens*; Felidae, 13: *Pseudaelurus lorteti*, 14: *Pseudaelurus quadridentatus*; Mustelidae, 15: *Lartetictis dubia*, 16: *Ischyrictis zibethoides*, 17: *Mellidelavus leptorhynchus*, 18: *Taxodon sansaniensis*, 19: *Proputorius sansaniensis*, 20: *Martes sansaniensis*. d, Steinheim (mixed fauna, MN 7). Amphicyoniae, 1: *Pseudocyon steinheimensis*, 2: *Pseudocyon sansaniensis*, 3: *Amphicyonopsis serus*; Felidae, 4: *Pseudaelurus lorteti*, 5: *Pseudaelurus quadridentatus*; Mustelidae, 6: *Martes* cf. *filholi*, 7: *Trocharion albanense*, 8: *Paralutra jaegeri*, 9: *Ischyrictis mustelinus*, 10: *Proputorius* sp., 11: *Trochotheirum cyamoides*; Barbourofelidae, 12: *Sansanosmilus jourdani*; Ursidae, 13: *Ursavus* cf. *intermedius*, 14: *Hemicyon goeriachensis*; Viverridae, 15: *Semigenetta sansaniensis*.

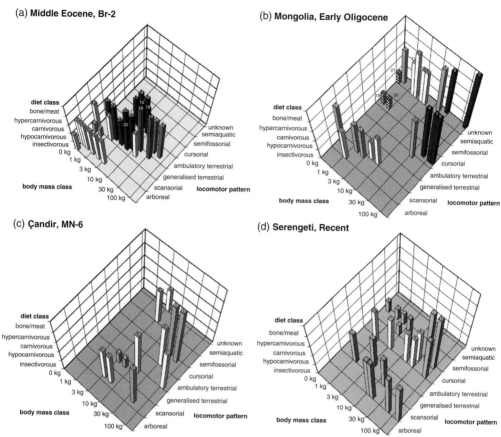

Figure 10.4 Carnivore guild structures of open landscapes. a, Wyoming, middle Eocene, Br-2. Viverravidae, 1: *Viverravus gracilis*, 2: *Viverravus minutus*; Miacidae, 3: *Miacis parvivorus*, 4: *Vulpavus palustris*, 5: *Vulpavus profectus*, 6: *Vulpavus ovatus*, 7: *Palaearctonyx meadi*, 8: *Oodectes herpestoides*, 9: *Oodectes proximus*, 10: *Uintacyon edax*, 11: *Uintacyon vorax*, 12: *Uintacyon major*; Mesonychia, 13: *Mesonyx obtusidens*, 14: *Harpagolestes macrocephalus*; Hyaenodontidae, 15: *Sinopa grangeri*, 16: *Sinopa major*, 17: *Sinopa minor*, 18: *Sinopa pungens*, 19: *Sinopa rapax*, 20: *Tritemnodon agilis*, 21: *Limnocyon verus*, 22: *Limnocyon cuspidens*, 23: *Thinocyon medius*, 24: *Thinocyon velox*, 25: *Machaeroides eothen*; Oxyaenidae, 26: *Patriofelis ulta*, 27: *Patriofelis ferox*.
b, Mongolia, lower Oligocene. Hyaenodontidae, 1: *Hyaenodon* cf. *gigas*, 2: *Hyaenodon mongoliensis*, 3: *Hyaenodon incertus*, 4: *Hyaenodon pervagus*, 5: *Hyaenodon eminus*; Amphicyonidae, 6: *Cephalogale* sp.; Amphicynodontidae, 7: *Amphicynodon teilhardi*; Semantoridae, 8: *Amphicticeps makhchinus*, 9: *Amphicticeps dorog*, 10: *Amphicticeps shakelfordi*, 11: *Amphicticeps* new small species; Mustelidae, 12: *Pyctis inamatus*; Nimravidae, 13: *Nimravus mongoliensis*, 14: Nimravidae indet.; Stenoplesictidae, 15: *Shandgolicits elegans*, 16: *Asiavorator altidens*, 17: *Asiavorator* sp. nov.; Viverravidae?, 18: *Palaeogale sectoria*; Didymoconidae, 19: *Didymoconus colgatei*, 20: *Didymoconus berkeyi*, 21: cf. *Ergilictis*. c, Çandir, middle Miocene (MN 6). Amphicyonidae,

Principal components analysis (PCA) – Recent

We conducted a PCA using recent species within carnivore guilds from an equatorial tropical rainforest in South America (Guyana) and an equatorial savannah in East Africa (Serengeti) in order to better visualise the ecomorphospace utilised by each guild. Each species was categorised based on the three parameters (diet, body size, locomotor capability) we used in the multivariate analyses with the same divisions within parameters (see Table 10.1 for a listing of all data for each species and locality). The sample from South America consisted of 14 species in four families (Mustelidae, Canidae, Procyonidae, Felidae) while the sample from East Africa included 20 species in 6 families (Felidae, Viverridae, Herpestidae, Canidae, Mustelidae, Hyaenidae).

The first principal component (PCA I) reflects the range of body size in the sample and explains 61% of the observed variability. PCA II reflects differences in locomotor capabilities and explains 31% of the variability. The remaining 8% of the variability is accounted for on PCA III, which is a measure of dietary habit.

In general, the carnivore guild from Guyana tends to consist mostly of intermediate-sized forms, while that of Serengeti is more broadly distributed across the size spectrum with very large (*Panthera leo*) and very small (*Helogale*, *Herpestes*) taxa being common. In terms of locomotor adaptations, the opposite pattern holds with the Serengeti taxa tending to be mostly terrestrial or large scansorial forms. In Guyana there are very few strictly terrestrial taxa, with all but four being either arboreal or scansorial – not a surprising finding in a tropical rainforest. Of the terrestrial forms, two are semi-aquatic, one is a specialised cursor, while the last is a small, generalised form (*Galictis*). There are no consistent dietary patterns between the two guilds except that there are no insectivores in the Guyana guild and no hypocarnivores in the Serengeti guild.

Caption for figure 10.4 (*cont.*)

1: *Amphicyon major*; Ursidae, 2: *Hemicyon sansaniensis*; Ailuridae, 3: *Amphictis cuspida*; Mustelidae, 4: cf. *Trochictis depereti*, 5: *Ischyrictis anatolicus*, 6: *Paralutra* cf. *jaegeri*, 7: *Proputorius* indet.; Hyaenidae, 8: *Protictitherium intermedium*, 9: *Protictitherium* cf. *gaillardi*; Percrocutidae, 10: *Percrocuta miocaenica*, 11: *Percrocuta* indet.; Felidae, 12: *Pseudaelurus* cf. *quadridentatus*. d, Serengeti, Recent. Canidae, 1: *Canis aureus*, 2: *Canis adustus*, 3: *Canis mesomelas*, 4: *Lycaon pictus*; 5: *Otocyon megalotis*, Felidae, 6: *Felis serval*, 7: *Caracal caracal*, 8: *Acinonyx jubatus*, 9: *Panthera pardus*, 10: *Panthera leo*; Herpestidae, 11: *Helogale parvula*, 12: *Mungos mungo*, 13: *Herpestes ichneumon*, 14: *Ichneumon albicauda*, 15: *Atilax paludinosus*; Hyaenidae, 16: *Hyaena hyaena*, 17: *Crocuta crocuta*; Mustelidae, 18: *Mellivora capensis*; Viverridae, 19: *Genetta genetta*, 20: *Civettictis civetta*.

In the PCA comparisons between Serengeti and Guyana, several different taxa were found to share similar ecomorphospaces. For example, the mustelid *Galicitis cuja* from Guyana occupies the same ecomorphospace as the viverrid *Genetta genetta* in Serengeti. In a similar manner, the canid *Canis aureus* (Serengeti) and the felids *Panthera onca* (Guyana) and *Panthera pardus* (Serengeti) also share a similar ecomorphospace.

It must be kept in mind that different workers may characterise a particular species in a different manner than we have. For example, *Speothos* might be characterised as being hypercarnivorous and semi-aquatic rather than carnivorous and cursorial as we have done (see Table 10.1). Obviously, if the parameters of the species are altered, its position within the ecomorphospace will also change.

Principal components analysis (PCA) – Eocene (Green River Basin) vs. Miocene (Steinheim)

The comparison between the middle Bridgerian fauna (Br-2) of the Green River Basin (GRB) and Steinheim was conducted to look at carnivore guilds from similar habitats (woodland and patchy closed forest) but different time periods and continents. As in the comparisons with the recent guilds, body size (PCA I) was the most influential ecomorphological component, accounting for 57% of the observed variability. Locomotion (PCA II) was responsible for 32% of the variability, while the diet component accounted for 11%.

In terms of body size, the Eocene guild in general was dominated by smaller taxa than the Miocene guild, with over 80% of taxa being under 10 kg in the former and less than 50% being under 10 kg in the latter. Both guilds have arboreal/scansorial components, but as befits the mixed woodland environmental reconstruction for both (see Discussion), there are a lot of terrestrial taxa as well. In the Eocene, the terrestrial component is dominated by cursorial and generalised terrestrial forms mostly due to the high diversity of hyaenodontid creodonts present in the North American middle Eocene. The Miocene terrestrial component is dominated by large forms such as ursids, amphicyonids, and barbourofelids. The only obvious outliers among Miocene taxa in terms of locomotion are the semi-aquatic (*Paralutra* and *Trochotherium*) and semi-fossorial (*Trocharion*) forms. Except for two very small, insectivorous viverravids in the GRB guild, the dietary components of both guilds are distributed among predominantly carnivorous and hypercarnivorous forms.

Other than the differences in overall body size, the two guilds are fairly similar to one another indicating an overprinting of environmental factors on guild structure. Interesting cross-guild comparisons can be made in terms of ecomorphology, especially given the phylogenetic distance separating

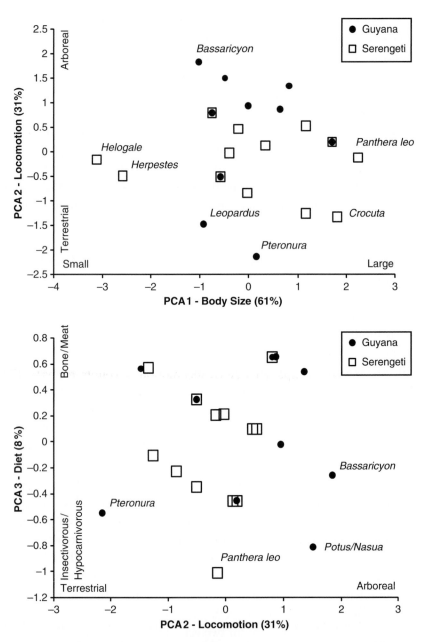

Figure 10.5 Principal components analysis of two recent carnivore guilds, one from an equatorial rainforest setting in Guyana, the other from an equatorial savannah at Serengeti. Plots present bivariate comparisons of principal component scores for axes 1 and 2 (top) and 2 and 3 (bottom). Note that differences in body size and locomotion account for 92% of observed variability. The rainforest guild tends to have more animals of intermediate sizes and more

most of the Eocene and Miocene taxa. Two relatively small, generalised terrestrial, hypercarnivorous Eocene hyaenodontid creodonts (*Tritemnodon* and *Machaeroides*) share the same ecomorphospace as the small Miocene mustelid *Ischyrictis*, suggesting similar roles within the guild structure of each sample. Some species of the Eocene miacid *Vulpavus* appear to share similar ecomorphospace with the Miocene felid *Pseudaelurus*, and the same is true for the Eocene miacid *Uintacyon major* and Miocene *Amphicynopsis* and *Sansanosmilus*, and the Eocene miacid *Palaearctonyx* and Miocene *Proputorius* and *Semigenetta*. However, such apparent similar ecomorphospaces can be misleading if not carefully evaluated because the plots shown in Figures 10.5–10.7 are bivariate comparisons of principal components that are actually parts of a multidimensional morphospace.

Principal components analysis (PCA) – Miocene (Sandelzhausen vs. Çandir)

A comparison was made between two Miocene localities in order to compare carnivore guilds of differing habitats but from a similar time. Sandelzhausen (MN 5, southern Germany) preserves a woodland fauna, while Çandir (MN 6, northwestern Turkey) represents a more open country faunal assemblage. PCA I, the body size component, is responsible for 60% of the variability between the two samples, PCA II, locomotion, accounts for 27%; PCA III, diet, accounts for 13% of the observed variability.

In terms of body size, there is a general sense that the carnivore guild from Çandir is made up of smaller-bodied forms. While this is true overall (50% of Çandir carnivores are under 10 kg, 67% of Sandelzhausen taxa are over 10 kg), the distribution is a bit misleading, because two of the smallest carnivores (less than 1 kg) from Sandelzhausen (*Martes* and cf. *Proputorius*) were not included in the PCA results because their locomotor adaptations are unknown (any taxon that does not have a complete ecomorphological profile is eliminated from the analysis). So, in essence, there is little differentiation across the body mass spectrum, although the woodland Sandelzhausen guild in general preserves larger taxa which could be viewed as unusual.

Figure 10.5 (*cont.*)
arboreal taxa, while the savannah guild has more extreme sizes (both large and small) and most taxa are terrestrial (although some large felids are classified as scansorial). Carnivore community of Iwokrama, Recent of Guyana (see Figure 10.4 for Serengeti carnivore community). Caniformia, Canidae, *Speothos venaticus*; Procyonidae, *Potos flavus, Nasua nasua, Bassaricyon gabbii*; Mustelidae, *Galictis cuja, Pteronura brasiliensis, Lontra longicaudis, Eira barbara*; Feliformia, Felidae, *Leopardus tigrinus, Leopardus pardalis, Leopardus wiedi, Herpailurus yaguarondi, Puma concolor, Panthera onca*.

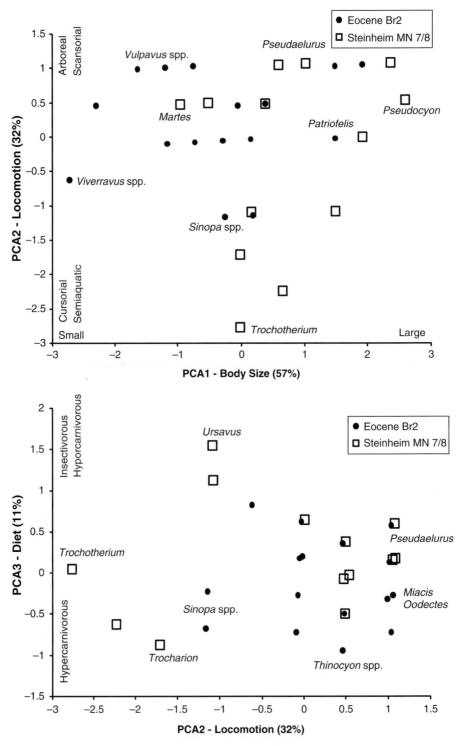

Figure 10.6 Principal components analysis comparing a middle Eocene North American carnivore guild (Br2) from a mixed forest setting with a European middle

Locomotor patterns also do not differentiate the two Miocene faunas too much as both are distributed across the ecomorphological spectrum. Both guilds are dominated by terrestrial forms (64% at Çandir, 67% at Sandelzhausen, for forms of known locomotor adaptation) with the main outlier being the semiaquatic *Paralutra* from Çandir. Both faunal samples have two scansorial forms (*Trochictis* and *Proputorius* from Çandir and *Leptoplesictis* and *Pseudaelurus* from Sandelzhausen).

The distribution of taxa across PCA III (diet) is also fairly uniform for both Miocene guilds. Çandir has a slightly broader dietary spectrum, mostly due to the present of two species of *Protictitherium*, which are insectivorous, and two species of *Percrocuta*, which are bone/meat eaters; *Leptoplesictis* represents the lone insectivorous form at Sandelzhausen.

Cross-guild comparisons reveal that some taxa are held in common between the two localities at the species (*Amphicyon major*) and generic (*Ischyrictis, Pseudaelurus, Hemicyon, Proputorius*) levels. In addition to the obvious shared ecomorphospace of the two *Amphicyon major* occurrences, the differing species of *Hemicyon* and *Ischyrictis* also share the same ecomorphospaces between Sandelzhausen and Çandir. Overall, despite the differing habitats proposed for these two Miocene localities, they are surprisingly similar in ecomorphological pattern, with most differences being very subtly expressed.

Discussion

The methods used to analyse guild structure presented in this chapter are based on the assignment of each species to a specific body mass, diet, and locomotor class. These three parameters together can be viewed as representing the ecomorphology of a species. Two additional parameters implicitly involved are taxonomic diversity within a guild and number of taxa sharing the same ecomorphospace. We do not discuss these two aspects in detail here, but it

Figure 10.6 (*cont.*)
Miocene guild (Steinheim) from a similar inferred habitat. Not included in the analyses due to unknown locomotor pattern are *Prosansanosmilus*, cf. *Proputorius*, and *Martes*, all from Steinheim. Plots present bivariate comparisons of principal component scores for axes 1 and 2 (top) and 2 and 3 (bottom). Note that diet explains only slightly more of the observed variability (11%) between the samples compared to the modern comparisons presented in Figure 10.5. In this case, body size and locomotion account for 89% of the observed variability. Note that the Eocene taxa tend to be smaller, while the Miocene taxa generally are larger. Additionally, the Miocene guild has a broad distribution of locomotor types.

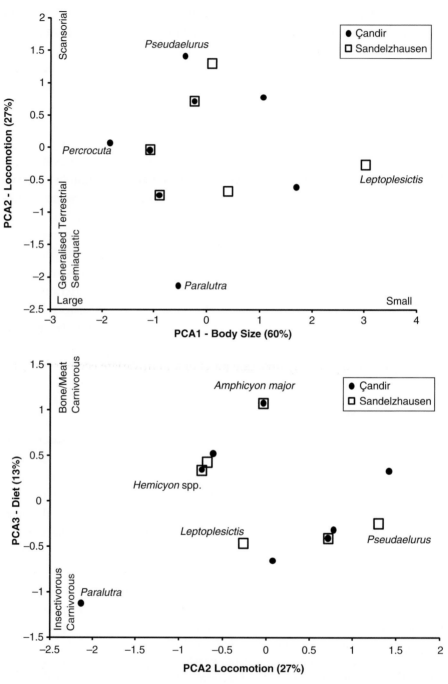

Figure 10.7 Principal components analysis comparing two middle Miocene European guilds, one from a woodland setting (Sandelzhausen, MN5) and the other from an open country habitat (Çandir, MN6). Plots present bivariate comparisons of principal

should be noted that the number of carnivore higher taxa decreased sharply (mostly because of the decline and extinction of Creodonta) throughout the Cenozoic while the number of ecomorphospaces present in a guild or number of taxa in the same ecomorphospace remained relatively constant. In short, this means that guilds remain complex throughout, but that complexity changed from being represented by relatively high taxonomic diversity of more or less morphologically uniform groups, to a single, morphologically complex order Carnivora.

It is not surprising that evidence of this taxonomic change can be seen in the Oligocene of Mongolia (Figure 10.4b) and continuing in the middle Miocene of Steinheim (Figure 10.3d), as the replacement of non-carnivorans by Carnivora is a well-documented pattern (Van Valkenburgh, 1992; Wesley-Hunt, 2005; Friscia, this volume). However, the rather broad classes we used to define an ecomorphospace make species-to-species comparison a task which has to be performed with caution. That the middle Eocene (Br-2) hyaenodontids *Tritemnodon* and *Machaeroides* are placed into the same ecomorphospace as the Miocene mustelid *Ischyrictis* or that the Guyanan mustelid *Galictis* parallels the Serengeti viverrid *Genetta* is not surprising, as these taxa share some morphological characters. However, it is important to recognise that ecomorphospace as depicted by bivariate plots of a multivariate ecomorphospace must be interpreted with caution. For example, the middle Eocene miacid *Uintacyon* and middle Miocene barbourofelid *Sansanosmilus* appear to share similar morphospace but only on the plot depicting PCA 2 (Locomotion) vs. PCA 3 (Diet). In this case, the PCA poorly differentiates between generalised terrestrial and carnivorous (*Uintacyon*) and ambulatory terrestrial and hypercarnivorous (*Sansanosmilus*). However, when the third factor (Body Mass) is added into a multidimensional ecomorphospace, the large (>100 kg) *Sansanosmilus* is easily differentiated from the relatively small (10–30 kg) *Uintacyon*.

After reconstructing guild structures and performing PCAs, the results show that body mass is always the main factor differentiating members within

Figure 10.7 (*cont.*)
component scores for axes 1 and 2 (top) and 2 and 3 (bottom). Like the comparisons between the middle Eocene and middle Miocene (Figure 10.6), dietary differences account for a somewhat higher percentage (13%) of observed variability than is the case in the recent comparisons (Figure 10.5). As might be expected, the Sandelzhausen guild consists of mostly intermediate body size forms, many of which are scansorial. The distribution of body sizes is more uniform in the Çandir guild as is that of locomotion.

carnivore guilds in our analyses. This is not unexpected, given the importance body mass has in influencing the structure of mammalian communities today (Valverde, 1964; Eisenberg, 1981; Legendre, 1989; Legendre and Roth, 1988; Gingerich, 2003). It should be noted that body mass is the only parameter which may vary intraspecifically as the class borders are artificial and the assignment is obtained by quantitative measures. A specific taxon might be larger than 30 kg in one guild, but smaller in another.

More surprising than the prominent role of body mass in guild structure differences is that locomotor pattern is often very important. Of course, this parameter is highly connected to habitat: arboreal taxa can only be present if trees are present and very big terrestrial carnivores are normally restricted to open landscapes. This result also suggests that a guild analysis of carnivores is of restricted usefulness without having data on the locomotor pattern of each species. We obtained some results from the comparison of Br-2 with that of the Geiseltal from which data on the locomotor pattern of a couple of species is lacking. Even if a PCA is not useful in this case (all species lacking any data are omitted from the analysis, see above), the comparison demonstrates differences between a continental and an insular fauna, with the latter being reduced in absolute number of taxa, in lacking very large taxa, and in having less taxonomic diversity within a single ecomorphospace. Of course, in some instances (as in the case of the comparisons with Çandir cited above), diet is as useful as locomotor pattern in differentiating guilds.

However, in most cases studied here, diet class proved to be relatively less important than other parameters, even if the differences between some middle Miocene European faunal samples can be explained partly (13%) by dietary differences. That diet, which has been given a lot of attention in the past, is the least important of the three components used in these comparisons of carnivore guilds, is a surprising result but may be relatively straightforward to interpret: carnivores feed on meat and this resource is available in all environments (Wesley-Hunt, 2005; Van Valkenburgh, 2008). As might be expected, studies on herbivores suggest that diet is of greater importance, and because of this diet has been used as a factor in reconstructing habitats based on dental morphology.

The assignment of any taxon to a locomotor class is obviously dependent on the presence of postcranial specimens of that species in a faunal sample, which in turn is directly affected by taphonomic influences. Taphonomy may play an important role in some other results as well. As we demonstrated, guilds composed of faunas from different environments can be identified easily by the presence (or absence) of arboreal taxa and a large variety of big, terrestrial, and hypercarnivorous to bone/meat eating carnivores. However, the PCA of

Çandir vs. Sandelzhausen shows limited differentiation (see above). It remains unclear whether: (1) taphonomic factors such as size sorting or differing (or similar) accumulation vectors are responsible for this lack of expected difference, (2) the habitat reconstructions (open landscape vs. woodland) are less well supported than previously thought, or (3) middle Miocene carnivore guilds of Europe are more uniform and less affected by environment than expected, e.g. because the guild members are not specialised to a certain habitat. That the third possibility has a major impact is corroborated by the number of species and genera shared by Çandir and all of the other middle Miocene localities, including Sandelzhausen (Nagel, 2003). Some differences are present between these guilds, though, as becomes clear when examining taxonomic subgroups, e.g. hyaenids and percrocutids, which are mostly interpreted as typical elements of open landscape faunas. Both families occur only in Çandir and they occur with two taxa in diet classes (insectivorous and bone/meat, respectively). Only a single insectivorous taxon has been documented from Sandelzhausen. The comparison of the Çandir and Sandelzhausen guilds is an example in which the three ecomorphological parameters alone are not sufficient to separate the two guilds (as PCA shows). Instead, this can be achieved by looking at parameters of taxonomy and numbers of taxa within a specific ecomorphospace.

Among the guilds we analysed, the Br-2 and the Steinheim guilds are unusual because they contain arboreal taxa as well as a large variety of large, terrestrial hypercarnivorous to bone/meat eating carnivores. Based on the discussion above, a conclusion could be drawn that these guilds include faunal elements from different habitats. Moreover, the PCA shows that these two guilds are structured very similarly, even if Br-2 on average consists of smaller taxa. An even closer look reveals that this is due to very different reasons. While the Br-2 guild is composed of taxa from different localities, which obviously were located in different environments, Steinheim is a single locality that accumulated bones from a variety of differing nearby habitats.

A completely separate set of results is based on the observation that similar ecomorphological profiles are developed by taxonomically very different taxa. Small to medium-sized, extremely arboreal fruit-eaters are found today in two carnivoran groups: the African viverrid *Arctictis* and the South American procyonid *Potos*, both equipped with a prehensile tail. In the Eocene, this ecomorphology was present as well, but was developed by either condylarths (*Chriacus*, Wa-0) or paroxyclaenids (Geiseltal). In none of our other guilds did this ecomorphology occur, as it is obviously restricted to woodlands. However, in the early Miocene, the controversial lophocyonines (viverrid Carnivora after Fejfar and Schmidt-Kittler, 1984 or hyaenodontid Creodonta after Ginsburg and Morales, 1999) may have played this role. Starting with the middle Miocene, ailurine ailurid carnivorans developed this specialised ecomorphology (Morlo and Peigné, this book). Small to medium-sized arboreal, hypocarnivorous taxa therefore were present in all of the Tertiary woodland samples we analysed.

Today, the largest known carnivore is the omnivorous (in our sense, hypocarnivorous) brown bear. In the past, the largest carnivores of a specific guild are mostly bone/meat-eaters: *Mesonyx*, *Harpagolestes* and *Patriofelis* (Br-2), *Hyaenodon gigas* (Mongolia), *Hyainailouros* (Sansan). The interesting fact is that all of these taxa represented the last or nearly the last of their kind on their respective continents: *Harpagolestes* and *Mesonyx* are the last mesonychids known in North America. In Asia, the supposed middle Eocene mesonychian *Andrewsarchus*, the largest known mammalian predator of all time, is the end member of its clade. It co-occurred with *Sarkastodon*, the last and largest oxyaenid, the latest and largest taxon of which in North America is *Patriofelis* (Br-2). In Mongolia, the largest *Hyaenodon*, *H. gigas*, lasted until the earliest Miocene, longer than this genus did on any other continent. The other creodont clade producing gigantic taxa, Hyainailourinae, is represented in the middle Miocene of Europe and Asia by *Hyainailouros* and Africa by *Megistotherium*.

The reasons for this pattern are unclear, but one can hypothesise that getting larger is a logical reaction to competition by slightly smaller taxa (Holliday and Steppan, 2004; Van Valkenburgh *et al.*, 2004). Staying larger than the next competing taxon and having bone-cracking capabilities opens the possibility to behave as a scavenger on the kills of smaller predators. Today, the cheetah loses 10–12% of its kills to scavengers such as lions and hyenas (Hunter *et al.*, 2007) and the presence of such scavengers affects its disposition to hunt (Cooper *et al.*, 2007). A similar behaviour may also explain the lack of a carnivore bone/meat-eater in the Geiseltal guild: in the middle Eocene of Europe the largest predator was the bird *Diatryma* which definitely was capable of crushing bones (Wittmer and Rose, 1991).

A final example of taxa belonging to different orders but evolving the same ecomorphology is the development of a semi-aquatic lifestyle. While this was done by pantolestids in the Eocene (Wa-0, Br-2, Geiseltal), hyaenodontids did the same in the Oligocene (*Quercitherium* in Europe, Morlo, 1999) and Miocene (*Teratodon* in Africa, Lewis and Morlo, 2010). The first semi-aquatic carnivorans are known in the Oligocene with semantorids (Wolsan, 1993; Wang *et al.*, 2005), and there are four different groups of musteloids that become semi-aquatic in the middle Miocene of Europe (Heizmann and Morlo, 1998). Steinheim is the first guild where both diet-types of semi-aquatic carnivores occur: a fish-eater (carnivorous) and a shell-eater (hypocarnivorous).

Conclusions

In summary, it is clear that a comparison of carnivore guild structures based on the parameters of body mass, diet class, and locomotor class reveals results on very different levels: it aids in clarifying the influence that time depth (across a faunal turn over), climate (continental vs. insular), and environment (woodland vs. open landscapes) have on guild structure. Comparing guilds of similar environments through the whole Cenozoic shows that the guilds differ enormously in terms of taxonomic composition and diversity (number of higher taxa), but were structurally fairly similar. In some cases, specific differences can be explained by phylogenetic patterns (scansoriality of large cats in the Serengeti).

It is also clear that, after body mass, locomotor pattern plays the most important role in differentiating carnivore guilds. As a consequence, it is paramount to include these data in guild comparisons and in the reconstruction of ecomorphospace. However, quantitative methods at present are lacking to confidently recognise locomotor classes and more rigorous methods of defining and recognising locomotor classes need to be developed.

Acknowledgements

We thank Anjali Goswami and Tony Friscia for inviting us to participate in this book and the symposium that preceeded the book. We thank Blaire Van Valkenburgh and K. E. Beth Townsend for their comments on an earlier draft of this paper. This research was supported by the United States National Science Foundation (NSF # OISE 0436295 to GFG).

REFERENCES

Andersson, K. (2004). Elbow-joint morphology as a guide to forearm function and foraging behaviour in mammalian carnivores. *Zoological Journal of the Linnean Society*, **142**, 91–104.

Anyonge, W. (1993). Body mass in large extant and extinct carnivores. *Journal of the Linnean Society*, **11**, 177–205.

Barnet, C. H. and Napier, J. R. (1953). The rotary mobility of the fibula in eutherian mammals. *Journal of Anatomy*, **87**, 11–21.

Bertram, J. E. and Biewener, A. A. (1990). Differential scaling of the long bones in the terrestrial Carnivora and other mammals. *Journal of Morphology*, **204**, 157–69.

Bowen, G. J., Koch, P. L., Gingerich, P. D., Norris, R. D., Bains, S. and Corfield, R. M. (2001). Refined isotope stratigraphy across the continental Paleocene–Eocene boundary on Polecat Bench in the northern Bighorn Basin. In *Paleocene–Eocene*

Stratigraphy and Biotic Change in the Bighorn and Clarks Fork Basins, Wyoming, ed. P. D. Gingerich. *University of Michigan, Papers on Paleontology*, **33**, 73–88.

Boyer D. M. and Georgi, J. A. (2007). Cranial morphology of a pantolestid eutherian mammal from the Eocene Bridger Formation, Wyoming, USA: implications for relationship and habitat. *Journal of Mammalian Evolution*, **14**, 239–80.

Carrano M. T. (1996). What, if anything, is a cursor? Categories versus continua for describing locomotor performance in terrestrial amniotes. *Journal of Vertebrate Paleontology*, **16** (Suppl. 3), **26**A.

Christiansen, P. (1999). Scaling of the limb long bones to body mass in terrestrial mammals. *Journal of Morphology*, **239**, 167–90.

Clyde, W. C. and Gingerich, P. D. (1998). Mammalian community response to the latest Paleocene therman maximum: an isotaphonomic study in the northern Bighorn Basin, Wyoming. *Geology*, **26**, 1011–14.

Cooper A. B., Pettorelli N. and Durant S. M. (2007). Large carnivore menus: factors affecting hunting decision by cheetahs in the Serengeti. *Animal Behaviour*, **73**, 651–59.

Day L. M. and Jayne B. C. (2007). Interspecific scaling of the morphology and posture of the limbs during the locomotion of cats (Felidae). *The Journal of Experimental Biology*, **210**, 642–54.

Dayan, T. and Simberloff, D. (1996). Patterns of size separation in carnivore communities. In *Carnivore Behavior, Ecology, and Evolution*, Vol. 2, ed. J. L. Gittleman. Ithaca, NY: Cornell University Press, pp. 243–66.

Dayan, T., Tchernov, E., Yom-Tov, Y. and Simberloff, D. (1989). Ecological character displacement in Saharo-Arabian *Vulpes*: outfoxing Bergmann's rule. *Oikos*, **55**, 263–72.

Egi, N. (2001). Body mass estimates in extinct mammals from limb bone dimensions: the case of North American hyaenodontids. *Palaeontology*, **44**, 497–528.

Eisenberg, J. F. (1981). *The Mammalian Radiations – An Analysis of Trends in Evolution, Adaptation, and Behavior*. Chicago, IL: University of Chicago Press.

Fejfar, O. and Schmidt-Kittler, N. (1984). *Sivanasua* und *Euboictis* n. gen. – zwei pflanzenfressende Schleichkatzenvorläufer (Viverridae, Carnivora, Mammalia) im Europäischen Untermiozän. *Mainzer geowissenschaftliche Mitteilungen*, **13**, 49–72.

Flynn, J. J., Neff, N. A. and Tedford, R. H. (1988). Phylogeny of the Carnivora. In *Phylogeny and Classification of Tetrapods*, vol. **2**, ed. M. Benton. Oxford: Oxford University Press, pp. 73–115.

Friscia, A. R., Van Valkenburg, B. and Biknevicius, A. R. (2007). An ecomorphological analysis of extant small carnivorans. *Journal of Zoology*, **272**, 82–100.

Gebo, D. L. and Rose, K. D. (1993). Skeletal morphology and locomotor adaptation in *Prolimnocyon atavus*, an early Eocene hyaenodontid creodont. *Journal of Vertebrate Paleontology*, **13**, 125–44.

Gingerich, P. D. (1990). Prediction of body mass in mammalian species from long bone lengths and diameters. *Contributions from the Museum of Paleontology, University of Michigan*, **28**, 79–92.

Gingerich, P. D. (2001). Biostratigraphy of the contienental Paleocene–Eocene boundary interval on Polecat Bench in the northern Bighorn Basin. In *Paleocene–Eocene*

Stratigraphy and Biotic Change in the Bighorn and Clarks Fork Basins, Wyoming, ed. P. D. Gingerich. *University of Michigan Papers on Paleontology*, **33**, 37–71.

Gingerich, P. D. (2003). Land-to-sea transition in early whales: evolution of Eocene Archaeoceti (Cetacea) in relation to skeletal proportions and locomotion of living semiaquatic mammals. *Paleobiology*, **29**, 429–54.

Ginsburg, L. (1961a). Plantigradie et digitigradie chez les carnivores fissipèdes. *Mammalia*, **25**, 1–21.

Ginsburg, L. (1961b). La faune des carnivores miocènes de Sansan (Gers). *Mémoires du Muséum National d'Histoire naturelle, n.s. C*, **9**, 1–190.

Ginsburg, L. and Morales J. (1999). Le genre *Sivanasua* (Lophocyoninae, Hyaenodontidae, Creodonta, Mammalia). *Estudios Geológicos*, **55**, 173–80.

Gunnell, G. F., Morgan, M. E., Maas, M. C. and Gingerich, P. D. (1995). Comparative paleoecology of Paleogene and Neogene mammalian faunas: trophic structure and composition. *Palaeogeography, Palaeoclimatology, Palaeoecology*, **115**, 265–86.

Heinrich, R. E. and Biknevicius, A. R. (1998). Skeletal allometry and interlimb scaling patterns in mustelid carnivorans. *Journal of Morphology*, **235**, 121–34.

Heinrich, R. E. and Rose, K. D. (1997). Postcranial morphology and locomotor behaviour of two Early Eocene miacoid carnivorans, *Vulpavus* and *Didymictis*. *Palaeontology*, **40**, 279–305.

Heinrich, R. E., Strait, S. G. and Houde, P. (2008). Earliest Eocene Miacidae (Mammalia, Carnivora) from northwestern Wyoming. *Journal of Paleontology*, **82**, 154–62.

Heizmann, E. P. J. (1973). Die tertiären Wirbeltiere des Steinheimer Beckens. B. Ursidae, Felidae, Viverridae sowie Ergänzungen und Nachträge zu den Mustelidae. *Palaeontographica*, **8**, 1–95.

Heizmann, E. P. J. and Morlo, M. (1998). Die semiaquatische *Lartetictis dubia* (Mustelinae, Carnivora, Mammalia) vom Goldberg/Ries (Baden-Württemberg). In *Festschrift zum 70. Geburtstag von Prof. Dr. Karlheinz Rothausen*, ed. K. I. Grimm, M. C. Grimm and M. Morlo. *Mainzer naturwissenschaftliches Archiv*, **21**, 141–53.

Hertler, C. and Volmer, R. (2008). Assessing prey competition in fossil carnivore communities – a scenario for prey competition and its evolutionary consequences for tigers in Pleistocene Java. *Palaeogeography, Palaeoclimatology, Palaeoecology*, **257**, 67–80.

Holliday J. A. and Steppan, S. J. (2004). Evolution of hypercarnivory: the effect of specialization on morphology and taxonomic diversity. *Paleobiology*, **30**, 108–28.

Hunter J. S., Durant S. M. and Caro T. M. (2007). To flee or not to flee: predator avoidance by cheetahs at kills. *Behavior, Ecology, and Sociobiology*, **61**, 1033–42.

Jenkins, F. A. Jr., and Camazine, S. M. (1977). Hip structure and locomotion in ambulatory and cursorial carnivores. *Journal of Zoology*, **181**, 351–70.

Jones, M. (2003). Convergence in ecomorphology and guild structure among marsupial and placental carnivores. In *Predators with Pouches*, ed. M. Jones, C. Dickman and M. Archer. Collingwood, Australia: CSIRO Publishing, pp. 285–96.

Koch, P. L., Zachos, J. C. and Gingerich, P. D. (1992). Correlation between isotope records in marine and continental carbon reservoirs near the Palaeocene–Eocene boundary. *Nature*, **358**, 319–22.

Koenigswald, W. von. (1980). Das Skelett eines Pantolestiden (Proteutheria, Mamm.) aus dem mittleren Eozän von Messel bei Darmstadt. *Paläontologische Zeitschrift*, **54**, 267–87.

Laborde, C. (1987). Caractères d'aptation des membres au mode de vie arboricole chez *Cryptoprocta ferox* par comparaison d'autres Carnivores Viverridés. *Annales des Sciences Naturelles, Zoologie, Série* **13**(8), 25–39.

Legendré, S. (1989). Les communautés de mammières du Paléogéne (Eocène supérieur et Oligocéne) d'Europe occidentale: structures, milieux et Évolution. *Münchner Geowissenschaftliche Abhandlungen, Reihe A, Geologie und Paläontologie*, **16**, 1–110.

Legendré, S. and Roth, C. (1988). Correlation of carnassial tooth size and body weight in Recent carnivores (Mammalia). *Historical Biology*, **1**, 85–98.

Lewis, M. and Morlo, M. (2010). Creodonta. In *Cenozoic Mammals of Africa*, ed. L. Werdelin and W. J. Sanders. Berkeley, CA: University of California Press.

MacLeod, N. and Rose, K. D. (1993). Inferring locomotor behavior in Paleogene mammals via Eigenshape analysis. *American Journal of Sciences*, **293**A, 300–55.

Morlo, M. (1999). Niche structure and evolution in creodont (Mammalia) faunas of the European and North American Eocene. *Géobios*, **32**, 297–305.

Morlo, M. and Gunnell, G. F. (2003). Small Limnocyoninae (Hyaenodontidae, Mammalia) from the Bridgerian, middle Eocene of Wyoming: *Thinocyon, Iridodon* n. gen., and *Prolimnocyon. Contributions from the Museum of Paleontology, The University of Michigan*, **31**, 43–78.

Morlo, M. and Gunnell, G. F. (2005a). New species of *Limnocyon* (Mammalia, Creodonta) from the Bridgerian (middle Eocene). *Journal of Vertebrate Paleontology*, **25**, 247–51.

Morlo, M. and Gunnell, G. F. (2005b). Comparison of carnivore guild structure across the Paleocene–Eocene boundary in North America. *Journal of Vertebrate Paleontology*, **25** (Suppl. 3), 93A.

Morlo, M. and Gunnell, G. F. (2006). Carnivore guild structure and abundance across the Paleocene–Eocene boundary in North America. *Climate and Biota of the Early Paleogene Conference, Bilbao 12th to 18th June 2006, Abstracts*, 87. Bilbao: University of the Basque Country.

Morlo, M. and Habersetzer, J. (1999). The Hyaenodontidae (Creodonta, Mammalia) from the lower Middle Eocene (MP 11) of Messel (Germany) with special remarks on new X-ray methods. *Courier Forschungsinstitut Senckenberg*, **216**, 31–73.

Morlo, M. and Nagel, D. (2007). The carnivore guild of the Taatsiin Gol area: Hyaenodontidae (Creodonta), Carnivora, and Didymoconida from the Oligocene of Central Mongolia. *Annalen des Naturhistorischen Museums Wien*, **108**A, 217–31.

Nagel, D. (2003). Carnivora from the middle Miocene hominoid locality of Çandir. *Courier Forschungsinstitut Senckenberg*, **240**, 113–31.

Nagel, D. and Morlo, M. (2000). Middle Miocene carnivore guilds of Europe: 1. The south-eastern fauna of Çandir/Anatolia. Environments and Ecosystem Dynamics of the Eurasian Neogene (EEDEN), 'State of the Art' Workshop, Lyon, France, 16–18 November 2000, 46. Lyon: University of Lyon.

Nagel, D. and Morlo, M. (2003). Guild structure of the carnivorous mammals (Creodonta, Carnivora) from the Taatsiin Gol Area, Lower Oligocene of Central Mongolia. *DEINSEA*, **10**, 419–29.

Nagel, D., Morlo, M. and Stefen, C. (2005). Sandelzhausen, a unique carnivore guild in the Middle Miocene (MN 5) of Europe. *Berichte des Instituts für Erdwissenschaften, Karl-Franzens-Universität Graz*, **10**, 78–79.

Nagel, D., Stefen, C. and Morlo, M. (2009). The carnivoran community of Sandelzhausen, (Germany). *Paläontologische Zeitschrift*, **80**/2, 107–11.

Phillips, R. B., Winchell, C. S. and Schmidt, R. H. (2007). Dietary overlap of an alien and native carnivore on San Clemente Island, California. *Journal of Mammalogy*, **88**, 173–80.

Polly, P. D. (1996). The skeleton of *Gazinocyon vulpeculus* gen. et comb. nov. and the cladistic relationships of Hyaenodontidae (Eutheria, Mammalia). *Journal of Vertebrate Paleontology*, **16**, 303–19.

Rose, K. D. (1990). Postcranial skeletal remains and adaptations in Early Eocene mammals from the Willwood Formation, Bighorn Basin, Wyoming. *Geological Society of America, Special Paper*, **243**, 107–33.

Stefen, C., Morlo, M. and Nagel, D. (2005). Carnivore guild structure in the Middle Miocene of the Mediterranean area. 12th Congress Regional Committee on Mediterranean Neogene Stratigraphy, 6–11 September 2005, Vienna. Program, Abstracts, Participants, 212. Vienna: Naturhistorisches Museum.

Taylor, M. E. (1974). The functional anatomy of the forelimb of some African Viverridae (Carnivora). *Journal of Morphology*, **143**, 307–36.

Taylor, M. E. (1976). The functional anatomy of the hindlimb of some African Viverridae (Carnivora). *Journal of Morphology*, **148**, 227–53.

Taylor, M. E. (1989). Locomotor adaptations by carnivores. In *Carnivore Behavior, Ecology, and Evolution*, ed. J. L. Gittleman. Ithaca, NY: Comstock Publication Association, pp. 382–409.

Thackeray, J. F. and Kieser, J. A. (1992). Body mass and carnassial length in modern and fossil carnivores. *Annals of the Transvaal Museum*, **35**, 337–41.

Valverde, J. A. (1964). Remarque sur la structure et l'évolution des communautés terrestres. I. Structure d'une communauté. II. Rapports entre prédateurs et proies. *Terre et Vie*, **1964**, 121–54.

Van Valkenburgh, B. (1988). Trophic diversity in past and present guilds of large predatory mammals. *Paleobiology*, **14**, 155–73.

Van Valkenburgh, B. (1990). Skeletal and dental predictors of body mass in carnivores. In *Body Size in Mammalian Paleobiology: Estimation and Biological Implications*, ed. J. Damuth and B. J. MacFadden. New York, NY: Cambridge University Press, pp. 181–206.

Van Valkenburgh, B. (1992). Tracking ecology over geological time: evolution within guilds of vertebrates. *Trends in Ecology and Evolution*, **10**, 71–76.

Van Valkenburgh, B. (1999). Major patterns in the history of carnivorous mammals. *Annual Review of Earth and Planetary Sciences*, **27**, 463–93.

Van Valkenburgh, B. (2008). *Déjà vu*: the evolution of feeding morphologies in the Carnivora. *Integrative and Comparative Biology*, **47**, 147–63.

Van Valkenburgh, B., Wang, X. and Damuth, J. (2004). Cope's rule, hypercarnivory, and extinction in North American canids. *Science*, **306**, 101–04.

Viranta, S. and Andrews, P. (1995). Carnivore guild structure in the Paşalar Miocene fauna. *Journal of Human Evolution*, **28**, 359–72.

Wang, X. (1993). Transformation from plantigrady to digitigrady: functional morphology of locomotion in *Hesperocyon* (Canidae: Carnivora). *American Museum Novitates*, **3069**, 1–23.

Wang, X., McKenna, M. C. and Dashzeveg, D. (2005). *Amphicticeps* and *Amphicynodon* (Arctoidea, Carnivora) from Hsanda Gol Formation, Central Mongolia and phylogeny of basal arctoids with comments on zoogeography. *American Museum Novitates*, **3483**, 1–57.

Werdelin, L. (1996). Carnivoran ecomorphology: a phylogenetic perspective. In *Carnivore Behavior, Ecology, and Evolution*, ed. J. L Gittleman. Ithaca, NY: Cornell University Press, pp. 582–624.

Wesley-Hunt, G. D. (2005). The morphological diversification of carnivores in North America. *Paleobiology*, **31**, 35–55.

Wittmer, L. M. and Rose, K. D. (1991). Biomechanics of the jaw apparatus of the gigantic Eocene bird *Diatryma*: implications for diet and mode of life. *Paleobiology*, **17**, 95–120.

Wolsan, M. (1993). Phylogeny and classification of early European Mustelida (Mammalia: Carnivora). *Acta Theriologica*, **38**, 345–84.

11

Ecomorphology of North American Eocene carnivores: evidence for competition between Carnivorans and Creodonts

ANTHONY R. FRISCIA AND BLAIRE VAN VALKENBURGH

Introduction

Evolutionary history is characterised by numerous occurrences of 'double-wedge' patterns of diversification and decline, wherein one taxon rises in diversity, but then declines alongside an increase in diversity of a second group (Figure 11.1). In some instances, temporal overlap of these two diversity curves has been taken to imply competitive replacement. In a review of the subject, Benton (1987) discussed the problem of distinguishing true competitive replacement from turnover events that result from some extrinsic factor such as environmental change. This distinction between intrinsic and extrinsic factors is at the heart of the debate over detection of competition in the fossil record.

Intrinsic factors imply a competitive advantage that one group has over another, but defining that advantage is difficult. Replacement can be caused by direct competition, such as uneven resource gathering capabilities (Sepkoski, 1996; Schluter, 2000) or interference competition (e.g. carcass theft and interspecific killing) (Palomares and Caro, 1999; Van Valkenburgh, 2001). The successful group in this direct competition may possess an adaptation (Rosenzweig and McCord, 1991) that allows it to outcompete the declining group. Benton (1987) termed turnover events derived from direct competition 'active replacement' or 'ecological replacement', and found very few convincing cases in the literature for this type of replacement (Benton, 1996).

Alternatively, extrinsic factors can trigger turnover events and often involve an abiotic agent (Janis, 1989; Kemp, 1999; Barnosky, 2001). For example, a climatic change or cataclysmic event can cause one taxon to decline in numbers,

Carnivoran Evolution: New Views on Phylogeny, Form, and Function, ed. A. Goswami and A. Friscia. Published by Cambridge University Press. © Cambridge University Press 2010.

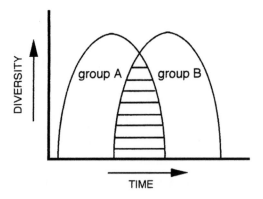

Figure 11.1 Hypothetical example of 'double-wedge' diversity curve, time vs. taxonomic diversity. Modified from Van Valkenburgh (1999).

while another, possibly with an adaptation that allows it to flourish in the new conditions, is better able to resist or recover from this change and consequently outcompetes the declining taxon. This type of turnover has been dubbed a 'passive replacement' (see McShea, 2000, for a discussion of 'active vs. passive' evolutionary trends). A final sort of replacement involves a stochastic event, which by chance affects one group less than another, and does not involve adaptation to new conditions or direct competition.

Distinguishing between these three types of replacement – active, passive, and stochastic – in the fossil record is difficult, although it has been attempted. Classic studies on mammalian ecological replacement have been done on the turnover between plesiadapiform primates and rodents in the Paleogene (Maas *et al.*, 1988), and on ungulates (Meng and McKenna, 1998), with the best documented being the decline of perissodactyls and the rise of artiodactyls (Janis *et al.*, 1994). In these instances, the ecological similarity of the two groups, as well as identification of specific characters that led to the increased taxonomic diversity of the new group, strengthened the case for active eco-logical replacement. Maas *et al.* (1988), in their study of primates and rodents, established four criteria for determining whether ecological replacement took place: change in taxonomic diversity, change in relative abundance, ecological similarity, and paleobiogeographic overlap.

One of the earliest mammalian replacement events involves carnivorous mammals of the Paleogene (Figure 11.2). In the Paleocene (65–55.5 million years ago, Mya) the order Creodonta dominated the carnivore niche. Creodonts were clearly carnivorous, possessing carnassials (specialised meat-shearing teeth) like true carnivorans (members of the extant order Carnivora). In contrast, though, creodonts possessed a variable pair of carnassials (unlike the distinguishing

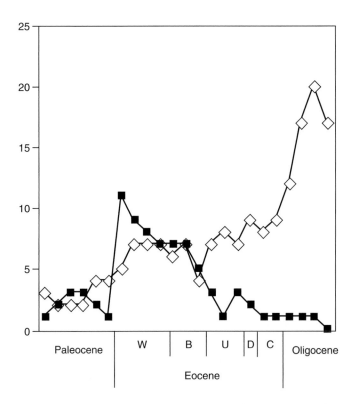

Figure 11.2 Diversity curve for Carnivora (open diamonds) and Creodonta (closed squares) for the mid-Paleocene to Oligocene. The Eocene is divided into North American Land Mammal Ages (NALMA): 'W', Wasatchian; 'B', Bridgerian; 'U', Uintan; 'D', Duchesnean; 'C', Chadronian. The relative lengths of the ages along the *x*-axis are not to scale. The *y*-axis represents number of genera. Modified from Van Valkenburgh (1999).

upper last premolar/lower first molar pair that carnivorans possess), and many creodonts even had multiple sets of carnassials in their tooth row. Creodonts continued to dominate the early part of the Eocene epoch (55.5–34 Mya), and some creodonts at this time attained large body sizes (Gunnell, 1998). The taxonomy of creodonts is not particularly well-defined, although by the middle Eocene they are represented by two or three families (Hyaenodontidae, Limnocyonidae, and possibly the Oxyaenidae). Soon after, they declined in diversity, and by Oligocene times (34–24 Mya) the creodonts were relegated to a few hyaenodontid species.

Members of the extant order Carnivora are first found in the Paleocene, and they steadily increase in diversity until the mid-Eocene, when they are represented by two families: the Viverravidae (a possible sister-group to all other carnivorans) and the 'Miacidae' (a possible basal paraphyletic carnivoran grouping) (Flynn, 1998; Wesley-Hunt and Flynn, 2005, Flynn et al., this volume). Although most of the carnivoran forms remained small and generalised, they did increase in taxonomic diversity until the end of the Eocene, when the first identifiable members of a crown group of Carnivora are recognised (Van Valkenburgh, 1999).

Competition has been proposed as a cause for the turnover of creodonts to carnivorans in the Eocene (Flynn, 1998). Direct competition has already been demonstrated among modern carnivoran species (Palomares and Caro, 1999; Van Valkenburgh, 2001), and modern carnivorans have provided some of the best examples of character displacement as a result of competition and resource partitioning (Dayan and Simberloff, 1996; Werdelin, 1996b). Although climatic change can play a role in replacement events, it is likely to play a smaller part in carnivore turnovers than in herbivore turnovers because, unlike plants, the physical qualities of meat as a resource change little in relation to the environment (Van Valkenburgh, 1999; Schluter, 2000).

Turning to the criteria for ecological replacement established by Maas et al. (1998), taxonomic turnover has been established for the creodont to carnivoran transition (Van Valkenburgh, 1999). In addition, we know that there was geographic overlap of the two groups. North America has an excellent carnivore fossil record, for both groups, and numerous stratigraphic and phylogenetic analyses have established their distributions (Flynn, 1998; Gunnell, 1998). All three creodont families are represented: the Oxyaenidae, which may have originated in North America; the Hyaenodontidae, probable immigrants from Africa via Eurasia; and the Limnocyonidae, a possibly endemic North American group (Gunnell, 1998). Carnivorans may have originated in North America, and at least one modern carnivoran family originated in North America in the Eocene (the Canidae) (Wang and Tedford, 1996; Flynn, 1998; Munthe, 1998). Relative abundance of these orders is often hard to determine, because the small proportion of carnivorans in most communities (Van Valkenburgh and Janis, 1993) translates to poor fossil representation. Because we could not assess relative abundance with confidence, it will not be included in this study.

Data on the details of ecological similarity between the orders (the final criterion of Maas et al., 1998), including body size and diet, are lacking. Wesley-Hunt (2005) addressed the issue of morphological diversification of mammalian carnivores in North America throughout the Cenozoic, concluding that carnivoran morphological diversity was not suppressed during their overlap with creodonts, although the implications for competition between the two orders are unclear. Morlo and others (Morlo, 1999; Morlo and Gunnell, 2003; Nagel and Morlo, 2003; Morlo et al., this volume) used qualitative variables, as described in Van Valkenburgh (1988), to classify creodonts into broad 'classes' of dietary and locomotory trends, but none of these studies looked at the potential for ecological overlap between creodonts and carnivorans in a quantitative way.

To better understand the possible ecological interactions of creodonts and carnivorans over evolutionary time, we need estimates of their diets and body sizes. Diet must be inferred from dental morphology, and thus it is critical to establish the association between diet and morphology in modern analogues of these Paleogene taxa. In the early Tertiary, most mammalian carnivores, both creodonts and carnivorans, were restricted to small body sizes (Werdelin, 1996a; Janis et al., 1998), and thus their closest modern analogues are found among the viverrids, herpestids, mustelids, procyonids, and smaller canids (Friscia et al., 2007). Recently, Friscia et al. (2007) showed that dietary groups of small-bodied (<10 kg) modern carnivorans can be distinguished using morphological characters. Three dietary groups (carnivores, insectivores, and omnivores/ hard-object feeders) were largely distinguished on the basis of the relative proportions of cutting-blade length to grinding molar area of the dentition and premolar shape. Carnivores have reduced grinding areas, larger cutting blades, and narrower premolars than omnivores/hard-object feeders. Insectivores are not consistently distinguished from other dietary types, although they do tend to have weaker dentaries and shorter temporalis muscle moment arms. Usually, insectivores group with the omnivores/hard-object feeders in relation to carni- vores, or occupy a middle ground between what seem to be two extremes of dietary specialisation within extant Carnivora. Since the differences between groups were often small, the patterns cannot be used as a precise predictive tool for species with unknown diets, such as fossils, but they can help interpret morphologies that include taxa of uncertain ecologies.

Temporal studies of large carnivorans have suggested that lineages are driven to large body size and hypercarnivory, by a process termed a 'macroevolutionary ratchet' (Carbone et al., 1999, 2007; Van Valkenburgh, 1999; Hunt, 2004; Hone and Benton, 2005; Van Valkenburgh et al., 2004). Evidence for this has not been explored in small carnivore lineages, although a study across modern caniforms has shown this trend to be true (Finarelli and Flynn, 2006). An investigation of Eocene taxa will help answer this question for the earliest mammalian carnivores.

Putative ecologies of fossil groups can be determined through comparison with the modern groups, and then compared amongst themselves and across time. Morphological variables that correspond with diet can be used in a multivariate analysis to produce a morphospace that represents dietary variation among taxa. The position of creodonts and carnivorans can then be compared in this space at time intervals encompassing the turnover event. If there was a case of ecological replacement, the space occupied by the creodonts will be occupied by the carnivorans over time. General patterns of disparity over time can also be investigated within this morphospace, i.e. how many different forms each group takes (as opposed to diversity, which is taxonomically based). In turnover events, the incumbent group may show a decline in disparity as taxa disappear, while the incoming group is likely to become increasingly disparate over time as it evolves into unoccupied or recently vacated niches (Van Valkenburgh, 1985). This increase in disparity could be driven by an evolution to the edge of morphospace, where groups tend to be less diverse than those at the middle of the defined space (Vermeij, 1977; Van Valkenburgh, 1991; Foote, 1996; Erwin, 1998). This study will investigate diet and body size diversity in Eocene carnivores over time to explore the potential for ecological overlap, and hence competition between creodonts and carnivorans.

Materials and methods

Data were collected on 15 carnivoran and 17 creodont genera across the Eocene (approximately 55 and 35 species, respectively). Analyses were performed at the generic level due to the uncertain species designations for some fossil taxa (Wagner, 1995; Roy et al., 1996; Sepkoski, 1998). In a previous study of modern taxa (Friscia et al., 2007), results at both the genus and species levels were similar, because all species within a genus tend to have similar ecomorphology, and differences that distinguish species are usually limited to those of size and characters under sexual selection (e.g. pelage) (Nowak, 1991; Kingdon, 1997). Fossil genera were placed into temporal bins corresponding to the North American Land Mammal Ages (NALMA) of the Eocene (Prothero and Emry, 2004; Robinson et al., 2004): Wasatchian (55.5–50 Mya); Bridgerian (50–45 Mya); Uintan (45–40 Mya); and Chadronian (37–34 Mya). The Duchesnean NALMA (40–37 Mya) was not used due to its depauperate fossil record (Prothero, 1998; Robinson et al., 2004). While the NALMA are not of equal size, any systematic error this could engender would be equally applied to both carnivorans and creodonts, so comparisons between them should be unaffected. If a genus was found in multiple NALMA, it was included in all ages that it occurred, but measurements for only those specimens found in a given NALMA were used for that age (Table 11.1). Specimens were examined from the following collections: American Museum of Natural History (New York, NY); Carnegie Museum (Pittsburgh, PA); Field Museum (Chicago, IL); University of Michigan, Museum of Paleontology (Ann Arbor, MI); San Diego Museum of Natural History (San Diego, CA); Washington University in St. Louis (St. Louis, MO); and Yale-Peabody Museum (New Haven, CT).

Table 11.1 List of taxa and ratios used in this analysis. A 'o' for a taxa in a measurement column indicates the taxa does not display this feature (e.g. upper grinding areas in some taxa); a '–' indicates the feature is present, but was missing from the specimens examined; a blank indicates that the ratio was not computed for that taxon (i.e. M2S and UM2I for creodonts). 'mL' is the largest molar in that creodont taxon, used for body size estimation; all carnivorans used m1. 'mLs' is the area of the molar used as an estimator of body size. Ratio abbreviations are detailed in Table 11.2.

Genus	mL	mLs	RBL	RLGA	RUGA	M1BS	M2S	1xP4	1xM2	MAT	MAM	C1	P4P	UM2I	P4S	PmZ	P4Z
WASATCHIAN																	
Carnivora																	
Didymictis		68.79	0.54	1.22	1.01	0.07	0.62	0.70	0.80	3.25	2.47	0.58	0.36	0.58	0.38	3.39	1.32
Miacis		15.03	0.58	1.42	0.99	–	0.75	0.95	1.04	–	–	–	0.36	0.69	0.44	–	1.20
Oodectes		8.64	0.63	1.14	0.99	–	0.74	0.85	0.90	–	–	–	0.43	0.67	0.52	3.33	1.17
Uintacyon		21.92	0.62	1.26	1.21	–	0.78	0.83	0.85	–	–	0.61	0.45	0.74	0.46	2.82	1.10
Vassacyon		41.71	0.62	1.25	–	–	0.76	0.97	1.02	–	–	–	–	–	0.47	–	1.04
Viverravus		14.87	0.60	1.13	1.05	–	0.72	0.65	0.74	–	–	–	0.33	0.75	0.40	–	1.20
Vulpavus		24.15	0.50	1.90	–	–	0.95	–	0.90	–	–	0.75	–	–	0.47	2.60	1.01
Creodonta																	
Arfia	3	50.02	0.59	0.58	0.85	0.06		0.82	0.90	5.03	3.92	0.71	1.11		0.48	3.10	1.18
Dipsalidictis	2	88.67	0.58	0.71	0.83	0.06		0.79	0.82	2.94	–	0.84	1.01		0.53	2.61	1.19
Oxyaena	2	164.18	0.67	0.54	0.67	0.08		0.87	0.94	–	2.19	0.79	0.82		0.54	2.75	1.13
Palaeonictis	1	110.98	0.63	0.56	–	0.05		1.38	1.39	–	4.34	1.44	–		0.63	3.17	1.34
Prolimnocyon	2	17.63	0.63	–	–	–		0.54	0.69	–	–	–	–		–	–	–
Prototomus	2	28.86	0.58	0.51	0.69	–		0.74	0.81	–	–	0.53	1.08		0.42	3.33	1.17
Pyrocyon	3	30.29	0.57	0.44	–	–		–	–	–	–	–	–		–	4.32	1.15
Tritemnodon	3	44.02	0.64	0.40	–	0.05		0.95	1.05	4.56	3.50	0.77	–		0.49	3.18	1.12

Table 11.1 (*cont.*)

Genus	mL	mLs	RBL	RLGA	RUGA	M1BS	M2S	IxP4	IxM2	MAT	MAM	C1	P4P	UM2I	P4S	PmZ	P4Z
BRIDGERIAN																	
Carnivora																	
Miacis		24.97	0.59	1.10	1.04	0.06	0.67	0.61	0.68	–	2.84	–	0.45	0.64	0.45	2.90	1.12
Miocyon		62.12	0.71	1.05	–	0.07	0.76	0.79	0.90	–	–	0.74	–	–	–	–	–
Oodectes		9.30	0.51	1.57	–	–	0.84	–	0.53	–	–	0.82	–	–	0.52	3.42	1.14
Uintacyon		25.62	0.61	1.26	–	–	0.74	0.61	0.76	–	–	0.63	–	0.33	0.50	2.88	1.07
Viverravus		15.31	0.59	1.23	0.98	–	0.84	0.72	0.76	–	–	0.87	0.38	0.51	0.42	3.54	1.18
Vulpavus		17.99	0.50	1.71	–	0.05	0.86	0.96	0.99	3.85	2.93	0.87	0.41	0.91	0.48	2.71	1.16
Creodonta																	
Limnocyon	2	36.01	0.63	0.54	0.74	–		–	–	4.10	–	0.92	0.84	–	0.48	3.20	1.11
Machaeroides	2	–	–	–	0.76	–		–	–	–	–	–	0.91	–	–	–	–
Patriofelis	2	162.16	0.96	0.12	0.00	–		1.05	1.00	4.08	2.60	1.11	0.49	–	0.51	3.85	1.54
Sinopa	2	34.70	0.56	0.59	0.81	0.06		0.81	0.87	3.69	2.91	0.61	1.00	–	0.42	3.25	1.35
Thinocyon	2	17.65	0.58	0.71	0.66	0.06		0.71	0.73	3.52	1.64	0.71	0.90	–	0.49	2.83	1.09
Tritemnodon	3	38.17	0.62	0.37	0.71	–		0.91	0.96	–	–	0.76	1.00	–	0.38	3.80	1.33
UINTAN																	
Carnivora																	
Daphoenus		52.50	0.63	1.07	–	–	0.62	0.75	0.80	–	–	–	–	–	0.44	–	1.07
Miacis		21.31	0.62	0.99	0.87	–	0.57	0.55	0.64	–	–	–	0.34	0.53	0.46	–	1.10
Miocyon		91.21	0.67	1.20	1.34	–	0.81	–	–	–	–	–	0.49	0.85	0.60	–	–
Procynodictis		30.58	0.67	0.85	0.90	–	0.59	0.63	0.70	2.71	2.32	0.52	0.36	0.51	0.40	2.67	1.02
Tapocyon		78.39	0.69	0.74	0.83	0.07	0.51	0.73	0.83	3.61	2.93	0.66	0.37	0.36	0.45	2.50	1.13

	n															
Creodonta																
Apataelurus	2	101.07	0.99	0.10	—	0.13	0.90	0.91	3.19	1.38	—	—	—	0.53	—	0.97
Mimocyon	2	—	—	—	—	—	—	—	—	—	—	—	—	0.43	—	—
Limnocyon	2	66.38	0.64	0.59	0.61	0.07	1.03	0.93	—	2.17	0.85	0.86	—	0.50	2.96	1.13
Oxyaenodon	2	42.88	0.68	0.45	0.65	—	0.84	0.90	—	—	0.72	0.88	—	0.54	—	1.13
CHADRONIAN																
Carnivora																
Brachyrhynchocyon		68.03	0.56	1.22	—	—	0.71	0.79	0.87	—	0.79	—	—	0.46	2.82	1.20
Daphoenus		77.98	0.59	1.10	1.15	—	0.69	0.60	0.65	—	0.62	0.37	0.69	0.37	2.93	1.154
Dinictis		80.94	0.82	—	0.48	—	—	0.67	0.71	—	0.65	0.35	0.00	0.42	—	1.32
Hesperocyon		29.33	0.65	0.84	0.99	0.10	0.61	0.56	0.65	2.02	0.39	0.38	0.55	0.42	2.69	1.02
Parictis		32.30	0.61	0.98	1.03	—	0.64	0.66	0.70	—	0.61	0.43	0.62	0.51	2.80	1.05
Creodonta																
Hyaenodon	3	—	—	—	0.00	—	—	—	—	—	0.66	—	—	—	—	—

Taxonomy was taken from museum tags and modified following McKenna and Bell (1997), Flynn (1998), Gunnell (1998), and the National Center for Ecological Analysis and Synthesis (NCEAS) Paleobiology Database (http://paleodb.org/). Most of the genera found throughout the Eocene were represented: 8 of 13 known creodont and 7 of 7 known carnivoran Wasatchian genera; 6 of 8 known creodont and 6 of 9 known carnivoran Bridgerian genera; 4 of 4 known creodont and 5 of 8 known carnivoran Uintan genera; and 1 of 2 known creodont and 5 of 9 known carnivoran Chadronian genera (Janis et al., 1998).

Measurements of the dentitions of each specimen were taken either on actual specimens with digital calipers or from digital pictures of the specimens (as shown in Friscia et al., 2007: figure 2). Digital pictures were taken in standard positions with a scale bar included. Measurements taken from the pictures were scaled using the ImageJ program (available as freeware from NIH, http://rsb. info.nih.gov/ij/). These raw measurements were then combined to form ratios that have previously been found to be indicative of ecomorphology of carnivorous taxa (Table 11.2) (Van Valkenburgh and Koepfli, 1993; Sacco and Van Valkenburgh, 2004; Friscia et al., 2007). The ratios for all specimens from a given genus in a given NALMA were then averaged to give a value for the entire taxon for that age, describing the morphology of a 'chronotaxon' (Table 11.1) (Walker, 1980; Radinsky, 1981).

Working with fossils generally, and creodonts specifically, posed problems that are not confronted with studies of extant taxa. The fragmentary nature of fossils limited sample sizes for some measurements, such as length of the dentary, which is rarely complete in fossils. Functionally equivalent measures had to be used, such as size of the lower first molar as a proxy for body size in lieu of dentary length. Creodonts were problematic because the meat-cutting function of their teeth was not limited to one carnassial locus with non-carnassial grinding teeth located posterior to the carnassial, as in carnivorans. Instead, all creodont molars were often similarly carnassialised, and often their first lower molar was premolariform. To account for this, many of the ratios had to be modified to take this more diffuse meat-cutting function into account (Table 11.2). For example, the largest molar was used for measures of blade length and as a landmark for moment arms, and grinding areas were combined across multiple molars. These modifications are similar to those made by Morlo (1999) in dealing with creodonts, as well as changes that were made by Jones (2003) in studies of marsupial carnivores that also possess multiple carnassials. In addition, the size of the largest lower molar was used as a proxy for body size for creodonts (Table 11.1). Morlo (1999) used an average length of all molars as his proxy for body size in creodonts, after using correlation analysis to post-cranial measures of size. Because very few specimens possessed all their molars,

Table 11.2 Ratios used in this study and how each was measured in carnivorans and creodonts. Ratios modified from Van Valkenburgh and Koepfli (1993) and Friscia *et al.* (2007).

Ratio	Carnivora	Creodonta
RBL – Relative Blade Length	Measured as the ratio of trigonid length to total anteroposterior length of m1	Measured as the ratio of trigonid length to total anteroposterior length of the largest molar as shown in Table 11.1
RLGA – Relative Lower Grinding Area	Measured as the square root of the summed areas of the m1 talonid and m2 (if present) divided by the length of the m1 trigonid. Area was estimated as the product of maximum width and length of the talonid of m1 and m2, respectively	Measured as the square root of the summed areas of the m2 and m3 (if present) talonids divided by the length of the m2 and m3 trigonid. Area was estimated as the sum of the products of maximum width and length of the talonids of m2 and m3
RUGA – Relative Upper Grinding Area	Measured as the square root of the summed areas of M1 and M2 (if present) divided by the anteroposterior length of P4 (carnassial). Area was estimated by the product of width and length of M1 and M2, respectively	Measured as the square root of the area of M3 (if present) divided by the anteroposterior length of M2. Area was estimated by the product of width and length of M3
M1BS – carnassial blade size	Blade size relative to dentary length, measured as the length of the trigonid of m1 (carnassial) divided by dentary length. Dentary length was measured as the distance between the posterior margin of the mandibular condyle and the anterior margin of the canine	Blade size relative to dentary length, measured as the length of the trigonid of the largest molar divided by dentary length. Dentary length was measured as the distance between the posterior margin of the mandibular condyle and the anterior margin of the canine
M2S – relative size of second lower molar	m2 size relative to dentary length, measured as the square root of m2 area	Not computed

Table 11.2 (*cont.*)

Ratio	Carnivora	Creodonta
	(if present) divided by the area of m1. Molar area measured as in RLGA. If no m2 was present in the taxon M2S was recorded as zero	
IXP4 – Bending strength at p3/p4 junction	Second moment of area of the dentary at the interdental gap between third and fourth lower premolars relative to m1 area. Moment area is used as an estimate of resistance of the dentary to bending. Second moment of area was calculated using the formula $I_x = (\pi D_x D_y^3)/64$, where D_x is maximum dentary width and D_y is maximum dentary height at the p3–p4 interdental gap. I_x relative to m1 size was then estimated as the fourth root of I_x divided by the square root of m1 area, measured as the product of maximum length and maximum breadth	Second moment of area of the dentary at the interdental gap between third and fourth lower premolars relative to area of the largest molar. Moment area is used as an estimate of resistance of the dentary to bending. Second moment of area was calculated using the formula $I_x = (\pi D_x D_y^3)/64$, where D_x is maximum dentary width and D_y is maximum dentary height at the p3–p4 interdental gap. I_x relative to m1 size was then estimated as the fourth root of I_x divided by the square root of the area of the largest molar, measured as the product of maximum length and maximum breadth
IXM2 – Bending strength at p4/m1 junction	Estimate of resistance of dentary to bending, as measured by the second moment of area at the interdental gap between the first and second molars (or posterior to the first molar if no second molar was present). Measured as IXP4, except maximum dentary width and height were taken at the m1–m2 interdental gap (or posterior to m1 if m2 was not present in the taxon)	
MAT – Relative Size of the moment arm	Measured as the distance from the mandibular condyle to the apex of the coronoid	Measured as the distance from the mandibular condyle to the apex of the coronoid

Table 11.2 (*cont.*)

Ratio	Carnivora	Creodonta
of the Temporalis muscle	process divided by the size of m_1 (measured as in IXP_4)	process divided by the size of the largest molar (measured as in IXP_4)
MAM – Relative Size of the moment arm of the Masseter muscle	Measured as the distance from the mandibular condyle to the ventral border of the mandibular angle divided by the size of m_1 (measured as in IXP_4)	Measured as the distance from the mandibular condyle to the ventral border of the mandibular angle divided by the size of the largest molar (measured as in IXP_4)
C_1 – Relative Size of the lower canine	Measured by the square root of the basal area of c_1 divided by square root of the size of the lower first molar (carnassial). c_1 area was calculated as the product of length and width of the tooth. m_1 size was measured as IXP_4	Measured by the square root of the basal area of c_1 divided by square root of the size of the largest lower molar. c_1 area was calculated as the product of length and width of the tooth. Largest molar size was measured as IXP_4
P_4P – upper carnassial shape	Measured as the ratio of width of P_4 divided by length of P_4	Measured as the average of the ratios of width and length of M_1 and M_2
UM_{21} – Relative Size of upper molars	Square root of upper second molar area (if present) divided by square root of upper first molar area. Areas estimated as $RUGA$. If no M_2 was present in a taxon UM_{21} was recorded as zero	Not computed
P_4S – lower premolar shape	Measured as width of p_4 divided by its length	
PMZ – Relative total length of premolars	Measured as the sum of the lengths of p_2, p_3, p_4 divided by size of m_1 (measured as in IXP_4)	Measured as the sum of the lengths of p_2, p_3, p_4 divided by size of the largest molar (measured as in IXP_4)
P_4Z – Relative length of fourth lower premolar	Relative length of fourth lower premolar, measured as the length of p_4 divided by size of m_1 (measured as in IXP_4)	Relative length of fourth lower premolar, measured as the length of p_4 divided by size of the largest molar (measured as in IXP_4)

and because postcranial material was not included in this analysis, a single tooth seemed the better solution, and the results are probably not greatly affected either way in a study of this scale (Van Valkenburgh, 1990). The modifications to the ratios for fossils generally, and creodonts specifically, made it impossible to make direct comparisons with the results of the previous work of extant small carnivorans (e.g. Friscia et al., 2007), and fossil carnivorans and creodonts had to be treated in separate analyses. Nevertheless, trends within the groups can be compared.

The ratio averages on each chronotaxa were standardised to z-scores in each order across each Land Mammal Age (LMA) and were then analysed using non-metric multidimensional scaling (NMDS) in the open-source statistical program R (using the 'isoMDS' function in the MASS package of R 2.9.0 for Mac OSX) (Venables and Ripley, 2002). An analysis was performed on all the taxa for each clade (i.e. all carnivorans from all LMAs were analysed together, and the same was done for the creodonts), and then taxa of a given age were compared graphically. NMDS first creates a dissimilarity matrix of all taxa using Euclidean distances. Because NMDS only relies on these distances, it is ideal for working with fossil taxa because missing values are ignored in the distance computation, as opposed to other multivariate techniques that remove all cases with missing data (e.g. principal components analysis (PCA) or discriminant function analysis) (Carrano et al., 1999). The 'dist' function in R was used to generate the dissimilarity matrices, as this function corrects for missing data by scaling the distances according to the number of overlapping measures. If there are taxon pairs for which there are no measures in common, then a distance cannot be computed and one of the taxa must be removed (see below). This dissimilariy matrix is then entered into an optimisation algorithm which finds the distribution of points that minimises a 'stress' function to create a multidimensional representation of points that best represents the original calcu-lated dissimilarities (Kruskal, 1964a,b; Kruskal and Wish, 1978; Clark, 1993). NMDS analyses can be solved for any number of dimensions, but in this study the analysis was limited to a solution of only two coordinate axes since only two dietary types (carnivorous and omnivorous) could reasonably be distinguished in previous studies of small mammalian carnivores (Friscia et al., 2007). The coordinates on this two-dimensional plane for each chronotaxa were then used in SPSS (v.11 in Mac OSX) to create maps of taxa in each order in every LMA.

As a test of the usefulness of NMDS to characterise data in a similar way to variable-driven analyses, an analysis was performed on the Friscia et al. (2007) data set for living carnivorans, and the results were similar to results obtained using principal components analysis. In addition, a random subsample of this data set was analysed with simulated missing data similar to the fossil data sets.

Only carnivores and omnivores were used in this subset, as these were the two dietary types that were best distinguished by Friscia *et al.* (2007). Again, NMDS was able to distinguish these two dietary types in a manner similar to PCA results.

Body size was estimated for each chronotaxon, using lower first molar for carnivorans and largest lower molar for creodonts, to investigate whether the two orders overlapped in size. Body size has already proven to be an important factor in niche partitioning among carnivorous taxa (McNab, 1989; Dayan *et al.*, 1989a, 1989b; Dayan and Simberloff, 1994, 1996).

Results

Generic diversity

The taxa used in this study follow the pattern previously observed across mammalian carnivore taxa from the Eocene, i.e. a fall in diversity of creodonts and concomitant rise in the diversity of carnivorans. Because there are differing numbers of total taxa in each LMA, mostly due to preservational and taphonomic differences, it is best to represent this as percentages of the total carnivore genera examined. Overall there is a decrease in mammalian carnivore generic diversity ($20 \rightarrow 17 \rightarrow 12 \rightarrow 11$), and an increase in the percentage of carnivorans ($35\% \rightarrow 53\% \rightarrow 67\% \rightarrow 82\%$) in each LMA included in this study.

Dietary diversity

The NMDS analysis produced a two-dimensional plot of all the taxa in each order. The dissimilarity matrix used to run the NMDS analysis had missing values for distances between three creodont taxa (*Hyaenodon*, *Mimocyon*, and *Machaeroides*) and various other taxa, meaning these taxa did not have any variables in common with some of the other taxa. Since the NMDS algorithm used does not allow for missing values in the dissimilarity matrix, these taxa were removed from further NMDS analysis. The coordinates on each axis in the NMDS plane are dimensionless so the entire plot can be rotated arbitrarily (Kruskal and Wish, 1978). This rotation does not change the ordination of the points, nor their relative positions, it only rotates the points around the origin of the axes. Because a plot with axes that correspond to functional differences is easier to interpret, the coordinates along the *x*-axis in each plot were regressed against Relative Lower Grinding Area (RLGA). The slope of this regression was then used as a rotation factor and the NMDS-generated coordinates were rescored in this rotated plane. RLGA has already been found to be reduced in

Table 11.3 Correlations of the original ratios with the coordinates of the taxa in the rotated NMDS plane for both carnivorans and creodonts. "*" indicates a correlation significant at the 0.05 level; "**" indicates a correlation significant at the 0.01 level.

RATIOS	CARNIVORA		CREODONTA	
	X	Y	X	Y
RBL	0.710**	−0.109	0.658**	0.716**
RLGA	−0.910**	−0.122	−0.644**	−0.419**
RUGA	−0.776**	0.650**	−0.932**	−0.642*
M₁BS	0.734	0.577	0.236	0.829**
M2S	−0.900**	0.042	−	−
IXP4	−0.781**	−0.388	0.863**	−0.262
IXM1	−0.532*	−0.485*	0.797**	−0.415
MAT	−0.657	−0.549	0.181	−0.606
MAM	−0.563	−0.434	0.247	−0.870**
C1	−0.641**	−0.265	0.830**	−0.336
P4P	−0.539*	0.647**	−0.859**	−0.754**
UM21	−0.784**	0.341	−	−
P4S	−0.482*	0.633**	0.483	0.044
PMZ	−0.147	−0.177	0.328	−0.439
P4Z	0.222	−0.743**	0.496	−0.264

carnivorous taxa relative to more omnivorous taxa or hard-object feeders (Friscia et al., 2007; Van Valkenburgh and Koepfli, 1993). Because the x-axis in both creodonts and carnivorans was negatively correlated with RLGA, carnivorous taxa have higher values on the rotated x-axis (Figure 11.3).

A correlation matrix was generated, correlating each of the original ratios with the new, rotated, NMDS axes (Table 11.3). The correlation coefficients generated this way are analogous to axis loadings in PCA. In both creodonts and carnivorans, the x-axis has significant positive correlation with measures of carnivory (RBL, M₁BS) and significant negative ones with measures associated with omnivory or hard-object feeding (RLGA, RUGA, P4P for both orders, and M2S, UM21, P4S for carnivorans). The fact that RLGA is correlated with this axis is not surprising, since it was used as the rotation factor for the axes, but the fact that other measures that distinguish carnivory and omnivory also correlate indicates that this axis is useful for distinguishing diet. Interestingly, the x-axis for carnivorans and creodonts also displays a significant correlation with measures of bending strength of the jaw (IXP4, IXM2) and the lower canine (C1), although the correlations are different in sign, i.e. creodonts with

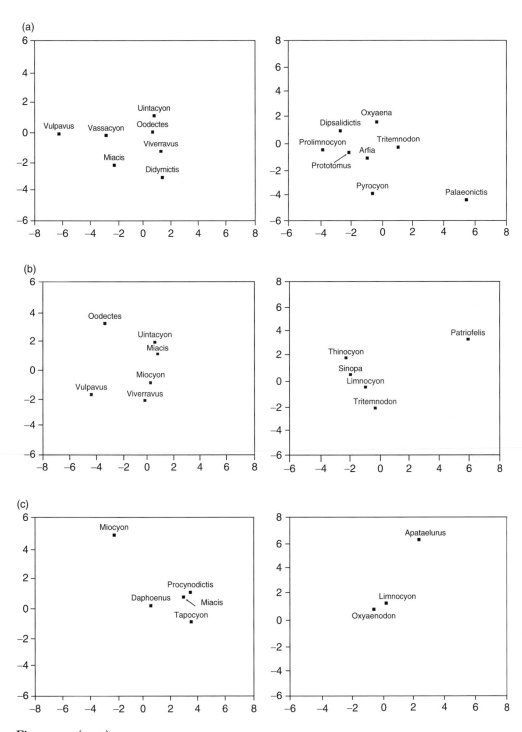

Figure 11.3 (*cont.*)

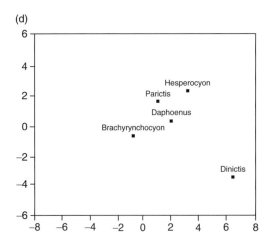

Figure 11.3 (*cont.*) Plots of scores from non-metric multidimensional scaling analysis for creodonta and carnivora (plotted separately) for each North American Land Mammal Age in the Eocene used in this study. Axes are dimensionless and not necessarily equivalent between carnivoran and creodont plots (see text for explanation). In both carnivorans (left) and creodonts (right), more carnivorous taxa have higher scores on the *x*-axis. For carnivorans, carnivorous taxa display lower scores on the *y*-axis, while carnivorous creodonts have larger scores on the *y*-axis. a, Wasatchian taxa; b, Bridgerian taxa; c, Uintan taxa; d, Chadronian taxa.

little grinding area tend to have deeper, stronger jaws, whereas the opposite is true of carnivorans. This may imply different prey selection and killing strategies for these orders (see Discussion).

The *y*-axis for both orders does not have as many significant correlations with the original ratios, nor are the correlations particularly strong. For creodonts, it seems to be another measure of carnivory, similar to the *x*-axis for both taxa, with significant positive correlations with blade size (RBL, M₁BS) and negative correlations with grinding areas (RLGA, RUGA, P4P), as well as a significant negative correlation with moment arm of the masseter (MAM). The masseter is associated with grinding motions of the lower jaw, and a strong masseter is often found in omnivorous and hard-object eating taxa, so this fits the pattern in the other variables. The *y*-axis of carnivorans has significant positive correlations with upper grinding area (RUGA) and premolar shape (P4P and P4S), and significant negative correlations with bending strength of the jaw (IXM₁) and premolar size (P4Z). Positive values along this axis may be thought to indicate omnivory, although the association is not as strong as seen on the *x*-axis.

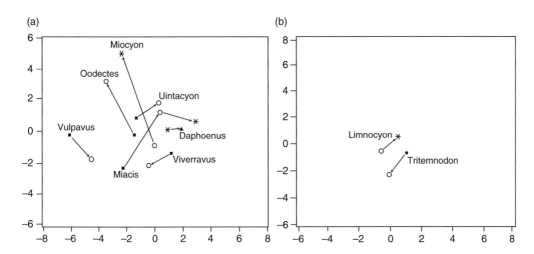

Figure 11.4 Plots of scores from non-metric multidimensional scaling analysis for a, carnivorans and b, creodonts, limited to those taxa that are found in more than one age. Arrows connect chronotaxa of a given genus. Scores of a genera in a certain age are indicated by filled squares (■) for Wasatchian, open circles (o) for Bridgerian, stars (∗) for Uintan, and filled triangles (▲) for Chadronian.

A comparison of the positions of the taxa in each LMA reveals some interesting patterns. In the Wasatchian (Figure 11.3a), carnivorans and creodonts occupy similar positions along their respective x-axes, although creodonts include a particularly carnivorous taxon with a deep jaw, *Palaeonictis*. In the Bridgerian (Figure 11.3b), the situation is similar, although the carnivorous creodont taxon is *Patriofelis*. In the Uintan (Figure 11.3c), a number of more predaceous carnivoran taxa appear, and the taxon *Miocyon* has moved up along the omnivorous y-axis. The few creodont taxa are centrally located, although the sabre-toothed *Apataelurus* (Scott, 1938; Gunnell, 1998) demonstrates a carnivorous tendency with a high score on the y-axis. Unfortunately, because the only Chadronian creodont taxon, *Hyaenodon*, cannot be included in the NMDS analyses (because the examined specimen was only an upper dentition), we cannot compare the positions of creodonts in this age. The positions of the carnivorans (Figure 11.3d) show them to be more dispersed than any previous age, primarily due to the presence of a hypercarnivorous taxon, the nimravid *Dinictis*.

One indication of an increase in morphological diversity over time is a change in morphology within a single lineage. To visualise this, taxa found in more than one NALMA were plotted within the NMDS generated space, and their movement through that space over time is indicated (Figure 11.4).

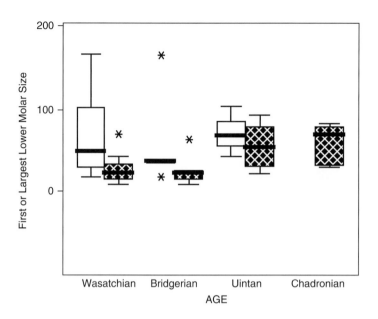

Figure 11.5 Distribution of sizes of first lower molar (for carnivorans – hatched box plots) or largest lower molar size (for creodonts, as indicated in Table 11.1 – open box plots), across the Eocene. Molar size is given in square millimetres (mm²). Stars indicate outliers.

The carnivorans have far more taxa which last through multiple ages. Of these taxa, 4 of 7 of them show an increase along the *x*-axis, indicating an increase in carnivory. This includes the genus *Miacis*, which is found in three ages and shows an increase in carnivory each time. Although *Miacis* is probably paraphyletic (Wesley-Hunt and Flynn, 2005), it does represent a small carnivoran 'morph', and change within this particular morph is still instructive. Two taxa, *Oodectes* and *Miocyon* show large increases along the *y*-axis, possibly indicating a move away from carnivory. The only two creodont taxa to span ages are *Tritemnodon* and *Limnocyon*, and neither shows much change along either axis.

Body size

As a last measure of ecological diversity over time, the distribution of body size was investigated for each order in each LMA (Figure 11.5). Throughout the Eocene the carnivorans never become very large; none possesses a lower first molar area larger than 100 mm², although they do show an average increase in molar size over time ($r^2 = 0.51$, $p = 0.014$). The creodonts do attain some larger body sizes, especially early in the Eocene (*Oxyaena* in the

Wasatchian and *Patriofelis* in the Bridgerian), although most of the taxa overlap the carnivorans in molar size. As in the NMDS studies, there is no creodont taxon for the Chadronian, but measures of Chadronian *Hyaenodon* specimens from the literature (Mellet, 1977) give a range of largest lower molar sizes of 128–150 mm^2 for the two species known (*H. vertus* and *H. montanus*). An average of this range puts the size of Chadronian *Hyaenodon* within the range of carnivorans known from that time. The average largest molar size for creodonts does not change over the time covered in this study ($r^2 = -0.002$, $p > .05$), although the variability in body sizes is larger than in carnivorans (standard deviation [SD] across the entire Eocene for creodonts $= 47.53$ mm^2, for carnivorans, SD $= 26.43$ mm^2).

Discussion

The diversity of creodont and carnivoran genera in the Eocene matches the expected pattern observed across the entirety of both orders and reported in previous studies; that is, the percentage of creodonts declines as carnivorans increase (Van Valkenburgh, 1999; Wesley-Hunt, 2005). In addition, the numbers show an overall decline in carnivorous mammal diversity. This is similar to patterns seen across all mammalian orders during the mid-Eocene (Janis, 1989, 1993; Alroy et al., 2000). A number of other clades show taxonomic turnovers at this time, especially ungulates (Janis, 1989, 1993). An increase in seasonality is often cited as the cause for these declines and replacement events (Alroy et al., 2000). While it is thought that carnivore morphology should not be affected by climatic change (Schluter, 2000; Van Valkenburgh, 1999), a change in habitat could affect them taxonomically, and this seems to be the case in the Eocene. By the end of the Eocene, the carnivorans have started to recover their taxonomic diversity, but creodonts never attain their previous diversity, and this trend continues into the Oligocene (Van Valkenburgh, 1999). The question is whether the climatic change in the mid-Eocene was the trigger that caused the turnover, making it a passive replacement (Benton, 1987), or whether carnivorans were already on their way to replacing creodonts.

The question of ecological competition between early carnivorans and creodonts has always been complicated by the differences in their dentitions. Because both have some sort of carnassial in the tooth row, if not multiple ones in the case of some creodonts, it has always been suspected that they both ate a significant amount of meat, or that they did so primitively (Van Valkenburgh, 1999). The use of morphospace analysis, using quantitative variables as its basis, allows for better visualisation of possible resource usage overlap. In addition, the morphological analysis is relatively taxon-free, which avoids the pitfalls of working with the changing taxonomy of fossil taxa.

In this study, an investigation of the morphospace occupied by carnivorans and creodonts reveals some interesting patterns. At the beginning of the Eocene, carnivorans are limited to the middle of their morphospace (Figure 11.3a) and display a seemingly generalised dentition, similar to some modern viverrids (civets and their relatives) and herpestids (mongooses). By the end of the epoch (Figures 11.3c and d), they have invaded new areas. The trend is toward more carnivorous forms, with some notable exceptions (e.g. *Miocyon* in the Uintan). This trend is true both across the entire order as well as within particular lineages (Figure 11.4), and is not accompanied by the loss of forms in the middle of the morphospace, which persist through the Eocene.

In contrast, the pattern of morphological diversity in creodonts shows major losses of ecomorphological types. Early in the Eocene there is a cluster of taxa near the middle of the morphospace, with one hypercarnivorous taxon outlier. Over time there are taxonomic losses in the middle area, and by the end of the Eocene only a hypercarnivorous form is left, *Hyaenodon*. Although this taxon was not included in the multivariate analysis due to the fragmentary nature of the specimens examined, Mellet (1977) demonstrated that it was an obvious hypercarnivore, having no grinding area on its molars, and carnassials similar to modern cats.

The pattern of size distributions shows a similar pattern in creodonts and carnivorans. The carnivorans increase in size throughout the Eocene, but never reach the large body niche that some creodonts filled since the beginning of the epoch. Carnivorans do eventually reach these sizes in the Oligocene (Van Valkenburgh, 1994). Creodont body size trends show a loss of extremes, i.e. variance or size range, across the epoch. By the Oligocene, only large hyaenodontids remained (Mellet, 1977; Van Valkenburgh, 1994; Hunt, 2004).

Taken together, the investigations of ecological morphology and body size paint a cohesive picture of evolution of mammalian carnivores through the Eocene. Both creodonts and carnivorans tend to evolve toward larger body sizes and more carnivorous forms. This trend has been observed in other studies of more recent carnivorans, and has been termed a 'macroevolutionary ratchet' (Van Valkenburgh, 1999; Van Valkenburgh et al., 2004). In this case, the creodonts attain these larger and more specialised forms while losing generalised morphologies. Carnivorans, on the other hand, retain taxa in the middle of their morphospace as well as small-bodied forms. Wesley-Hunt (2005) previously demonstrated that carnivorans were not limited in their evolution by the presence of creodonts in the Paleogene. The results of the current study are consistent with this, but also suggest that carnivorans may have limited ecological evolution in creodonts. It appears that the large, hypercarnivorous ecomorph was the 'last stand' of creodonts. This is observed on other continents as well, where the final forms, even later than the Eocene, were *Hyaenodon*-like in morphology and often obtained large body sizes, most notably *Megistotherium* of the Miocene of Africa, the largest creodont to exist (Savage, 1978). Evidence from large carnivores has previously suggested this macroevolutionary 'ratchet' that drives lineages to large body size and hypercarnivory is correlated with a subsequent increased risk of extinction (Carbone *et al.*, 1999; Holliday and Steppan, 2004; Hunt, 2004; Hone and Benton, 2005; Van Valkenburgh *et al.*, 2004; Holliday, this volume). The results of this study seem to suggest that this was the case for early carnivorous mammals as well, and may have contributed to the extinction of creodonts.

The question remains as to why the more generalised carnivorans persisted, while the creodont forms did not. The answer may lie in the specialised dentitions of creodonts. No creodont taxon possesses postcarnassial teeth; in all taxa every molar is carnassiform. The lack of grinding area posterior to the carnassials may have limited their effectiveness in more generalised ecologies. A similar argument has been made regarding marsupial carnivores (Werdelin, 1987) whose unique dental eruption sequence, compared to placental carnivores, limits all molars to a carnassial morphology. Although the dental replacement sequence in creodonts is unknown, the complete series of carnassiform molars is similar to marsupial carnivores, and may have limited their evolutionary plasticity. The retention of postcarnassial grinding area has been cited as a key to the success of canid carnivorans because it allows for this plasticity (Ewer, 1973; Van Valkenburgh, 1991), and early carnivorans were similar in their possession of this morphology. This plasticity, which allowed for a broader variety of diets across Carnivora, may have allowed them to survive the environmental change during the Eocene. Whether the turnover would have occurred without this climatic trigger is hard to determine, but the decline in taxonomic diversity in creodonts prior to this climatic change, with a concomitant rise in carnivoran diversity, seems to indicate that it would have.

When examined with respect to the criteria for determining whether ecological replacement is the cause of a turnover event in the fossil record (Maas *et al.*, 1988), the results of this study make a particularly strong case. There is certainly a 'double-wedge' change in taxonomic diversity which has been investigated elsewhere (Van Valkenburgh, 1999; Wesley-Hunt, 2005), as well as an established fossil record showing paleobiogeographic overlap in North America (Flynn, 1998; Gunnell, 1998). The key addition of this study is evidence of ecological similarity. Both groups occupy the middle of their respective morphospaces in the early Eocene, but these ecomorphs disappear throughout the Eocene among the creodonts. Later, both groups display carnivorous morphologies, the creodonts already having some established hypercarnivorous taxa, but the carnviorans retain their generalised forms. Although relative abundance was not examined due to low abundances of carnivores in the fossil record, the above three lines of evidence seem to point to a case of ecological replacement.

Caution must be taken in interpreting these results too broadly. A study of extant small carnivorans (Friscia et al., 2007) has shown that the differences between various diets are subtle, and insectivorous taxa were particularly hard to characterise. An investigation of the correlations of the original ratios used in this study with the axes of the NMDS-defined morphospaces reveals an interesting difference between creodonts and carnivorans. While both show positive correlations with the x-axis and measures thought to be indicative of carnivory (e.g. blade lengths), and corresponding negative correlations with measures of omnivory (e.g. grinding areas), other significant correlations are not similar (Table 11.2). In particular, measures of bending strength of both the jaw and canine correlate positively with this 'carnivory axis' in creodonts, and negatively in carnivorans. This could point to a difference in prey size or type. Previous studies (Friscia et al., 2007; Freeman, 1979, 1988; Strait, 1993a, 1993b) have shown that insectivores have relatively weak jaws, and that canine strength is associated with killing large prey (Van Valkenburgh and Ruff, 1987; Van Valkenburgh, 1989). Possibly Eocene carnivorans were mainly insectivores, as has been suggested before (Van Valen, 1971), similar to some modern-day viverrids and herpestids (e.g. civets and meerkats; Kingdon, 1997), while creodonts often took prey larger than themselves, as do some extant mustelids (e.g. weasels, stoats, and wolverines; King, 1989; Radinsky, 1981; Jones, 2003). This niche division may have alleviated competition between the orders.

Locomotory adaptations may also have allowed carnivorans and creodonts to coexist with minimal or reduced competition. However, studies of locomotory adaptations in carnivorans and creodonts have not shown any great differences in locomotion between the orders. There is good evidence for ambulatory and cursorial creodonts, and some taxa may have been semi-fossorial, scansorial or arboreal (Gebo and Rose, 1993; Gunnell, 1998; Morlo, 1999; Morlo and Gunnell, 2003; Nagel and Morlo, 2003), and early carnivorans show similar morphologies (Heinrich and Rose, 1995, 1997; Flynn, 1998). Studies of later carnivore groups have shown that locomotion, and more specifically how it relates to killing behaviour, is one way that carnivores divide up ecological space (Van Valkenburgh, 1985; Morlo et al., this volume), but this does not seem to have been the case in the Eocene.

Studies of competition in extant carnivorans are typically done at the community level (e.g. Dayan and Simberloff, 1996; Werdelin, 1996b; Nel and Kok, 1999; Ray and Sunquist, 2001). Unfortunately, a broad study such as this one does not allow for that level of analysis. The fragmentary nature of fossils in general makes community level analysis often dubious, as does the time-averaged quality of many fossil localities. The temporal control is lacking that would allow an investigator to confidently say that they are seeing a community 'snapshot'. Nevertheless, a community analysis at the level of a formation within a particular depositional basin may allow for this kind of resolution. Some work has been done on possible communities of carnivorans in fossil deposits (e.g. Viranta and Andrews, 1995; Palmqvist et al., 1999) and this work could be extended to focus on comparing carnivore communities across continents and time (Morlo et al., this volume).

While North America provides an excellent sample of the carnivore diversity across the Eocene, creodonts and carnivorans also coexisted in Europe, where there also is a very good fossil record (Springhorn, 1988; Hunt, 1998; Morlo, 1999; Morlo and Habersetzer, 1999), and in Asia, although the record there is patchier (Tao et al., 1991; Kumar, 1992; Dazhzeveg, 1996; Hunt, 1998). In addition, creodont- and carnivoran-dominated predator communities extend back into the Paleocene, as well as forward to the Miocene, in North America, Europe, Asia, and Africa. Including these times and areas would enlarge the study sample and could reveal the generality of patterns observed in North America.

Conclusions

The case for competition between carnivorans and creodonts was strengthened by this study, and suggests it may have contributed to the eventual extinction of creodonts. Both groups declined in diversity over the Eocene, possibly driven by a climatic change. The decline in creodonts was steeper compared with carnivorans, and creodonts never recovered their taxonomic numbers by the end of the Eocene or into the Oligocene. Both groups tended to evolve larger and more hypercarnivorous forms, although carnivorans never lost the generalised, smaller forms, while the latest creodont taxa were all large hypercarnivores. Although carnivorans never attained the large sizes seen in creodonts in the Eocene, they did so later in the Oligocene. The dental plasticity allowed by the possession of postcarnassial molars may have been the key adaptation that made carnivorans more successful, especially in light of a climatic change that occurred in the middle of the Eocene that affected many mammalian orders.

Acknowledgements

Access to fossil specimens under their care was provided by the following curators, collection managers, and researchers: J. Alexander and R. Tedford (American Museum of Natural History, New York, NY); M. Dawson (Carnegie Museum, Pittsburgh, PA); W. Simpson and J. Flynn (Field Museum, Chicago, IL); G. Gunnell (University of Michigan, Museum of Paleontology, Ann Arbor, MI); T. Demere, K. Randall, H. Wagner, and S. Walsh (San Diego Museum of Natural History. San Diego, CA); D. T. Rasmussen (Washington University in St. Louis, St. Louis, MO); M. A. Turner (Yale-Peabody Museum, New Haven, CT). Many of these people provided insight into the specimens under their care, which was always appreciated. Ray Ingersoll, Dave Jacobs, Bob Wayne, Anjali Goswami, and John Finarelli provided valuable insight in reviewing this chapter. Many other colleagues provided discussion, editing, software education, and thoughts, and their contributions often helped us work through parts of our research. They include P. Adam, G. C. Conroy, K.-P. Koepfli, M. Morris, D. T. Rasmussen, K. Pollard, T. Sacco, R. Shea, G. Slater, J. G. M. Thewissen, K. J. Thomas, K. E. Townsend, and the helpful people at the UCLA Department of Statistics Statistical Consulting Center.

REFERENCES

Alroy, J., Koch, P. L. and Zachos, J. C. (2000). Global climate change and North American mammalian evolution. *Paleobiology*, **26**(supp), 259–88.

Barnosky, A. D. (2001). Distinguishing the effects of the Red Queen and Court Jester on Miocene evolution in the Rocky Mountains. *Journal of Vertebrate Paleontology*, **21**, 172–85.

Benton, M. J. (1987). Progress and competition in macroevolution. *Biological Review of the Cambridge Philosophical Society*, **62**, 305–28.

Benton, M. J. (1996). On the nonprevalence of competitive replacement in the evolution of tetrapods. In *Evolutionary Paleobiology*, ed. D. Jablonski, D. H. Erwin, and J. H. Lipps. Chicago, IL: University of Chicago Press, pp. 185–210.

Carbone, C., Mace, G. M., Roberts, S. C. and Macdonald, D. W. (1999). Energetic constraints on the diet of terrestrial carnivores. *Nature*, **402**, 286–88.

Carbone, C., Teacher, A. and Rowcliff, J. M. (2007). The costs of carnivory. *PLoS Biology*, **5**, 363–68.

Carrano, M. T., Janis, C. M. and Sepkoski, J. J. (1999). Hadrosaurs as ungulate parallels: lost lifestyles and deficient data. *Acta Palaeontologica Polonica*, **44**, 237–61.

Clark, K. R. (1993). Non-parametric analyses of changes in community structure. *Australian Journal of Ecology*, **18**, 117–43.

Dashzeveg, D. (1996). Some carnivorous mammals from the Paleogene of the eastern Gobi Desert, Mongolia, and the application of Oligocene carnivores to stratigraphic correlation. *American Museum Novitates*, **3179**, 1–14.

Dayan, T. and Simberloff, D. (1994). Character displacement, sexual dimorphism, and morphological variation among British and Irish mustelids. *Ecology*, **75**, 1063–73.

Dayan, T. and Simberloff, D. (1996). Patterns of size separation in carnivore communities. In *Carnivore Behavior, Ecology, and Evolution: Volume 2*, ed. J. L. Gittleman. Ithaca, NY: Cornell University Press, pp. 243–66.

Dayan, T., Simberloff, D., Tchernov, E. and Yom-Tov, Y. (1989a). Inter- and intraspecific character displacement in mustelids. *Ecology*, **70**, 1526–39.

Dayan, T., Tchernov, E., Yom-Tov, Y. and Simberloff, D. (1989b). Ecological character displacement in Saharo-Arabian Vulpes: outfoxing Bergmann's rule. *Oikos*, **55**, 263–72.

Erwin, D. H. (1998). The end and the beginning: recoveries from mass extinctions. *Trends in Ecology and Evolution*, **13**, 344–49.

Ewer, R. F. (1973). *The Carnivores*. Ithaca, NY: Cornell University Press.

Finarelli, J. A. and Flynn, J. J. (2006). Ancestral state reconstruction of body size in the Caniformia (Carnivora, Mammalia): the effects of incorporating data from the fossil record. *Systematic Biology*, **55**, 301–13.

Flynn, J. J. (1998). Early Cenozoic Carnivora ('Miacoidea'). In *Evolution of Tertiary Mammals of North America, Volume 1: Terrestrial Carnivores, Ungulates, and Ungulatelike Mammals*, ed. C. M. Janis, K. M. Scott, and L. L. Jacobs. Cambridge: Cambridge University Press, pp. 110–23.

Foote, M. (1996). Models of morphological diversification. In *Evolutionary Paleobiology*, ed. D. Jablonski, D. H. Erwin, and J. H. Lipps. Chicago, IL: University of Chicago Press, pp. 62–86.

Freeman, P. W. (1979). Specialized insectivory: beetle-eating and moth-eating molossid bats. *Journal of Mammalogy*, **60**, 467–79.

Freeman, P. W. (1988). Frugivorous and animalivorous bats (Microchiroptera): dental and cranial adaptations. *Biological Journal of the Linnean Society*, **33**, 249–72.

Friscia, A. R., Van Valkenburgh, B. and Biknevicius, A. R. (2007). An ecomorphological analysis of extant small carnivorans. *Journal of Zoology, London*, **272**, 82–100.

Gebo, D. L. and Rose, K. D. (1993). Skeletal morphology and locomotor adaptation in *Prolimnocyon atavus*, an early Eocene hyaenodontid creodont. *Journal of Vertebrate Paleontology*, **13**, 125–44.

Gunnell, G. F. (1998). Creodonta. In *Evolution of Tertiary Mammals of North America, Volume 1: Terrestrial Carnivores, Ungulates, and Ungulatelike Mammals*, ed. C. M. Janis, K. M. Scott, and L. L. Jacobs. Cambridge: Cambridge University Press, pp. 91–109.

Heinrich, R. E. and Rose, K. D. (1995). Partial skeleton of the primitive carnivoran *Miacis petilus* from the early Eocene of Wyoming. *Journal of Mammalogy*, **76**, 148–62.

Heinrich, R. E. and Rose, K. D. (1997). Postcranial morphology and locomotor behavior of two early Eocene miacoid carnivorans, *Vulpavus* and *Didymictis*. *Palaeontology*, **40**, 279–305.

Holliday, J. A. and Steppan, S. J. (2004). Evolution of hypercarnivory: the effect of specialization on morphological and taxonomic diversity. *Paleobiology*, **30**, 108–28.

Hone, D. W. E. and Benton, H. J. (2005). The evolution of large size: how does Cope's Rule work? *Trends in Ecology and Evolution*, **20**, 4–6.

Hunt, R. M. (1998). Evolution of the aeluroid Carnivora; diversity of the earliest aeluroids from Eurasia (Quercy, Hsanda-Gol) and the origin of felids. *American Museum Novitates*, **3252**, 1–65.

Hunt, R. M. (2004). Global climate and the evolution of large mammalian carnivores during the Later Cenozoic in North America. *Bulletin of the American Museum of Natural History*, **285**, 139–56.

Janis, C. M. (1989). A climatic explanation for patterns of evolutionary diversity in ungulate animals. *Palaeontology*, **32**, 463–81.

Janis, C. M. (1993). Tertiary mammal evolution in the context of changing climates, vegetation, and tectonic events. *Annual Review of Ecology and Systematics*, **24**, 467–500.

Janis, C. M., Gordon, I. J. and Illius, A. W. (1994). Modellling equid/ruminant competition in the fossil record. *Historical Biology*, **8**, 15–29.

Janis, C. M., Scott, K. M. and Jacobs, L. L. (eds.) (1998). *Evolution of Tertiary Mammals of North America, Volume 1: Terrestrial Carnivores, Ungulates, and Ungulatelike Mammals*. Cambridge: Cambridge University Press.

Jones, M. (2003). Convergence in ecomorphology and guild structure among marsupial and placental carnivores. In *Predators with Pouches*, ed. M. Jones, C. Dickman, and M. Archer. Collingwood, Australia: CSIRO Publishing, pp. 285–96.

Kemp, T. S. (1999). *Fossils and Evolution*. New York, NY: Oxford University Press.

King, C. (1989). *The Natural History of Weasels and Stoats*. Ithaca, NY: Cornell University Press.

Kingdon, J. (1997). *The Kingdon Field Guide to African Mammals*. San Diego, CA: Academic Press.

Kruskal, J. B. (1964a). Multidimensional scaling by optimizing goodness of fit to a nonmetric hypothesis. *Psychometrika*, **2**, 1–27.

Kruskal, J. B. (1964b). Nonmetric multidimensional scaling: a numerical method. *Psychometrika*, **29**, 115–29.

Kruskal, J. B. and Wish, M. (1978). *Multidimensional Scaling. Quantitative Applications in the Social Sciences*, vol. 11. Beverly Hills, CA: Sage.

Kumar, K. (1992). *Paratritemnodon indicus* (Creodonta; Mammalia) from the early middle Eocene Subathu Formation, NW Himalaya, India, and the Kalakot mammalian community structure. *Palaeontologische Zeitschrift*, **66**, 387–403.

Maas, M. C., Krause, D. W. and Strait, S. G. (1988). The decline and extinction of Plesiadapiformes (Mammalia: ?Primates) in North America: displacement or replacement? *Paleobiology*, **14**, 410–31.

McKenna, M. C. and Bell, S. K. (1997). *Classification of Mammals Above the Species Level*. New York, NY: Columbia University Press.

McNab, B. K. (1989). Basal rate of metabolism, body size, and food habits in the Order Carnivora. In *Carnivore Behavior, Ecology, and Evolution: volume 1*, ed. J. L. Gittleman. Ithaca, NY: Cornell University Press, pp. 335–54.

McShea, D. W. (2000). Trends, tools, terminology. *Paleobiology*, **26**, 330–33.

Mellett, J. S. (1977). Paleobiology of North American *Hyaenodon* (Mammalia, Creodonta). *Contributions to Vertebrate Evolution*, **1**, 1–134.

Meng, J. and McKenna, M. C. (1998). Faunal turnovers of Palaeogene mammals from the Mongolian Plateau. *Nature*, **394**, 364–67.

Morlo, M. (1999). Niche structure and evolution in creodont (Mammalia) faunas of the European and North American Eocene. *Geobios*, **32**, 297–305.

Morlo, M. and Gunnell, G. F. (2003). Small Limnocyonines (Hyaenodontidae, Mammalia) from the Bridgerian Middle Eocene of Wyoming: *Thinocyon*, *Prolimnocyon*, and *Iridodon*, new genus. *Contributions from the Museum of Paleontology, University of Michigan*, **31**, 43–78.

Morlo, M. and Habersetzer, J. (1999). The Hyaenodontidae (Creodonta, Mammalia) from the lower Middle Eocene (MP 11) of Messel (Germany) with special remarks on new X-ray methods. *Courier Forsch.-Inst.Senckenberg*, **216**, 31–73.

Munthe, K. (1998). Canidae. In *Evolution of Tertiary Mammals of North America, Volume 1: Terrestrial Carnivores, Ungulates, and Ungulatelike Mammals*, ed. C. M. Janis, K. M. Scott, and L. L. Jacobs. Cambridge: Cambridge University Press, pp. 73–90.

Nagel, D. and Morlo, M. (1993). Guild structure of the carnivorous mammals (Creodonta, Carnivora) from the Taatsiin Gol area, Lower Oligocene of Central Mongolia. *DEINSEA*, **10**, 419–29.

Nel, J. A. J. and Kok, O. B. (1999). Diet and foraging group size in the yellow mongoose: a comparison with the suricate and the bat-eared fox. *Ethology, Ecology and Evolution*, **11**, 25–34.

Nowak, R. M. (1991). *Walker's Mammals of the World*, 5th ed. Volume 2. Baltimore, MD: Johns Hopkins University Press.

Palmqvist, P., Arribas, A. and Martínez-Navarro, B. (1999). Ecomorphological study of large canids from the lower Pleistocene of southeastern Spain. *Lethaia*, **32**, 75–88.

Palomares, F. and Caro, T. M. (1999). Interspecific killing among mammalian carnivores. *American Naturalist*, **153**, 492–508.

Prothero, D. R. (1998). The chronological, climatic, and paleogeographic background to North American mammalian evolution. In *Evolution of Tertiary Mammals of North America, Volume 1: Terrestrial Carnivores, Ungulates, and Ungulatelike Mammals*, ed. C. M. Janis, K. M. Scott, and L. L. Jacobs. Cambridge: Cambridge University Press, pp. 9–36.

Prothero, D. R. and Emry, R. J. (2004). The Chadronian, Orellan, and Whitneyan North American Land Mammal Ages. In *Late Cretaceous and Cenozoic Mammals of North America*, ed. M. O. Woodburne. New York, NY: Columbia University Press, pp. 156–68.

Radinsky, L. B. (1981). Evolution of skull shape in carnivores 1. Representative modern carnivores. *Biological Journal of the Linnean Society*, **15**, 369–88.

Ray, J. C. and Sunquist, M. E. (2001). Trophic relations in a community of African rainforest carnivores. *Oecologia*, **127**, 395–408.

Robinson, P., Gunnell, G. F., Walsh, S. L., *et al.* (2004). Wasatchian through Duchesnean biochronology. In *Late Cretaceous and Cenozoic Mammals of North America*, ed. M. O. Woodburne. New York, NY: Columbia University Press, pp. 106–45.

Rosenzweig, M. L. and McCord, R. D. (1991). Incumbent replacement: evidence for long-term evolutionary progress. *Paleobiology*, **13**, 202–13.

Roy, K., Valentine, J. W., Jablonski, D. and Kidwell, S. M. (1996). Scales of climatic variability and time averaging in Pleistocene biotas: implications for ecology and evolution. *Trends in Ecology and Evolution*, **11**, 458–63.

Sacco, T. and Van Valkenburgh, B. (2004). Ecomorphological indicators of feeding behaviour in the bears (Carnivora: Ursidae). *Journal of Zoology, London*, **263**, 41–54.

Savage, R. J. G. (1978). Carnivora. In *Evolution of African Mammals*, ed. V. J. Maglio and H. B. S. Cooke. Cambridge, MA: Harvard University Press, pp. 249–67.

Schluter, D. (2000). Ecological character displacement in adaptive radiation. *American Naturalist*, **156**, S4–16.

Scott, W. B. (1938). Problematical cat-like mandible from the Uinta Eocene, *Apataelurus kayi*, Scott. *Annals of Carnegie Museum*, **27**, 113–20.

Sepkoski, J. J. (1996). Competition in macroevolution: the double wedge revisited. In *Evolutionary Paleobiology*, ed. D. Jablonski, D. H. Erwin, and J. H. Lipps. Chicago, IL: University of Chicago Press, pp. 211–55.

Sepkoski, J. J. (1998). Rates of speciation in the fossil record. *Philosophical Transactions of the Royal Society of London, Series B*, **353**, 315–26.

Springhorn, R. (1988). Carnivorous elements of the Messel fauna. *Courier Forschungsinstitut Senckenberg*, **107**, 291–97.

Strait, S. G. (1993a). Differences in occlusal morphology and molar size in frugivores and faunivores. *Journal of Human Evolution*, **25**, 471–84.

Strait, S. G. (1993b). Molar morphology and food texture among small-bodied insectivorous mammmals. *Journal of Mammalogy*, **74**, 391–402.

Tao, Q., Guanfu, Z. and Yuanqing, W. (1991). Discovery of *Lushilagus* and *Miacis* in Jiangsu and its zoogeographical significance. *Vertebrata PalAsiatica*, **29**, 59–63.

Van Valen, L. (1971). Adaptive zones and the orders of mammals. *Evolution*, **25**, 420–28.

Van Valkenburgh, B. (1985). Locomotor diversity within past and present guilds of large predatory mammals. *Paleobiology*, **11**, 406–28.

Van Valkenburgh, B. (1988). Trophic diversity in past and present guilds of large predatory mammals. *Paleobiology*, **14**, 155–73.

Van Valkenburgh, B. (1989). Carnivore dental adaptations and diet: a study of trophic diversity within guilds. In *Carnivore Behavior, Ecology, and Evolution: volume 1*, ed. J. L. Gittleman. Ithaca, NY: Cornell University Press, pp. 410–36.

Van Valkenburgh, B. (1990). Skeletal and dental predictors of body mass in carnivores. In *Body Size in Mammalian Paleobiology: Estimation and Biological Implications*, ed. J. Damuth and B. J. MacFadden. Cambridge: Cambridge Univeristy Press, pp. 181–205.

Van Valkenburgh, B. (1991). Iterative evolution of hypercarnivory in canids (Mammalia: Carnivora): evolutionary interactions among sympatric predators. *Paleobiology*, **17**, 340–62.

Van Valkenburgh, B. (1994). Extinction and replacement among predatory mammals in the North American Late Eocene and Oligocene: tracking a paleoguild over twelve million years. *Historical Biology*, **8**, 129–50.

Van Valkenburgh, B. (1999). Major patterns in the history of carnivorous. *Annual Review of Earth and Planetary Sciences*, **27**, 463–93.

Van Valkenburgh, B. (2001). The dog-eat-dog world of carnivores: a review of past and present carnivore community dynamics. In *Meat-Eating and Human Evolution*, ed. C. B. Stanford and H. T. Bunn. Oxford: Oxford University Press, pp. 101–21.

Van Valkenburgh, B. and Janis, C. M. (1993). Historical diversity patterns in North American large herbivores and carnivores. In *Species Diversity in Ecological Communities*, ed. R. E. Ricklefs and D. Schluter. Chicago, IL: University of Chicago, pp. 330–40.

Van Valkenburgh, B. and Koepfli, K.-P. (1993). Cranial and dental adaptations to predation in canids. *Symposium of the Zoological Society of London*, **65**, 15–37.

Van Valkenburgh, B. and Ruff, C. B. (1987). Canine tooth strength and killing behavior in large carnivores. *Journal of Zoology, London*, **212**, 379–97.

Van Valkenburgh, B., Wang, X. and Damuth, J. (2004). Cope's Rule, hypercarnivory, and extinction in North American canids. *Science*, **306**, 101–04.

Venables, W. N. and Ripley, B. D. (2002). *Modern Applied Statistics with S*. New York, NY: Springer.

Vermeij, G. J. (1977). The Mesozoic marine revolution: evidence from snails, predators and grazers. *Paleobiology*, **3**, 245–58.

Viranta, S. and Andrews, P. (1995). Carnivore guild structure in the Pasalar Miocene fauna. *Journal of Human Evolution*, **28**, 359–72.

Wagner, P. J. (1995). Diversity patterns among early gastropods: contrasting taxonomic and phylogenetic descriptions. *Paleobiology*, **21**, 410–39.

Walker, A. (1980). Functional anatomy and taphonomy. In *Fossils in the Making: Vertebrate Taphonomy and Paleontology*, ed. A. K. Behrensmeyer and A. P. Hill. Chicago, IL: University of Chicago Press, pp. 182–96.

Wang, X. and Tedford, R. H. (1996). Canidae. In *The Terrestrial Eocene–Oligocene Transition in North America*, ed. D. R. Prothero and R. J. Emry. Cambidge, Cambridge University Press, pp. 433–52.

Werdelin, L. (1987). Jaw geometry and molar morphology in marsupial carnivores: analysis of a constraint and its macroevolutionary consequences. *Paleobiology*, **13**, 342–50.

Werdelin, L. (1996a). Carnivoran ecomorphology: a phylogenetic perspective. In *Carnivore Behavior, Ecology, and Evolution: Volume 2*, ed. J. L. Gittleman. Ithaca, NY: Cornell University Press, pp. 582–624.

Werdelin, L. (1996b). Community-wide character displacement in Miocene hyenas. *Lethaia*, **29**, 97–106.

Wesley-Hunt, G. D. (2005). The morphological diversification of carnivores in North America. *Palaeobiology*, **31**, 35–55.

Wesley-Hunt, G. D. and Flynn, J. J. (2005). Phylogeny of the Carnivora: basal relationships among the carnivoramorphans, and assessment of the position of 'Miacoidea' relative to Carnivora. *Journal of Systematic Paleontology*, **3**, 1–28.

12

Morphometric analysis of cranial morphology in pinnipeds (Mammalia, Carnivora): convergence, ecology, ontogeny, and dimorphism

KATRINA E. JONES AND ANJALI GOSWAMI

Introduction

Pinnipeds are a clade of secondarily aquatic arctoid carnivorans, including 34 extant species dispersed across most of the world's oceans. Extant species are separated into three families (Figure 12.1): Odobenidae (walruses, 1 species), Phocidae (seals, 19 species), and Otariidae (sea lions and fur seals, 14 species) and display a wide range of ecological diversity (Reeves *et al.*, 2002). Predominantly, pinnipeds are generalist feeders. They are opportunistic, and their diets may vary annually, between colonies and between individuals within a colony (King, 1983; Sinclair and Zeppelin, 2002; Williams *et al.*, 2007). However, several species have evolved more specialist feeding techniques: (1) *Odobenus rosmarus* is a suction feeder, using powerful facial musculature to produce forces large enough to extract molluscs from their shells (Adam and Berta, 2002); *Erignathus barbatus* (Phocidae) also uses suction feeding (King, 1983; Marshall *et al.*, 2008); (2) *Lobodon carcinophagus* (Phocidae) is a filter feeder; it uses multicuspidate teeth to sieve out krill as water is expelled from the mouth; (3) *Hyrdrurga leptonyx* (Phocidae) feeds on large, warm-blooded prey such as penguins and seal pups (Adam and Berta, 2002).

Reproductive strategies of the pinnipeds are also diverse. Otariids are universally dimorphic with large harems. Their young are weaned over long periods of up to 2 years whilst learning to forage (Kovacs and Lavigne, 1992; Schulz and Bowen, 2004). On the other hand, phocid young are relatively precocial (4–50 days weaning) and learn foraging skills after leaving their mothers. Phocids also show a diversity of mating strategies and degree of dimorphism (Schulz and Bowen, 2004). It has been hypothesised that this shorter time spent on land has allowed phocids to exploit a broader range of

Carnivoran Evolution: New Views on Phylogeny, Form, and Function, ed. A. Goswami and A. Friscia. Published by Cambridge University Press. © Cambridge University Press 2010.

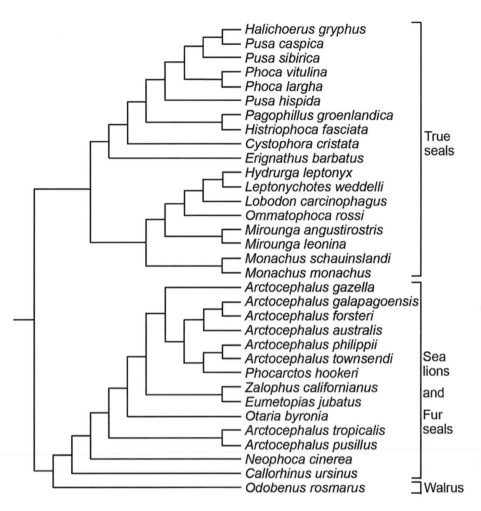

Figure 12.1 Composite phylogeny for extant pinnipeds (Wynen *et al.*, 2001; Arnason *et al.*, 2006).

habitats, including polar regions (Kovacs and Lavigne, 1992; Schulz and Bowen, 2005). Odobenids show extremely long lactation times of three years. During this period, young walruses often accompany mothers on foraging trips.

Despite these many interesting ecological differences, research into pinniped morphology has been fairly limited, and pinnipeds have received much less attention than other marine mammals, such as cetaceans and sirenians. Most research on pinnipeds has focused on taxonomy and phylogenetic relationships, which have been subject to much disagreement. Molecular work (Flynn *et al.*, 2005, this volume; Arnason *et al.*, 2006) suggests a closer relation between otariids and odobenids (forming the Otaroidea clade). Conversely, morphology-based work (Adam and Berta, 2002; Deméré *et al.*, 2003) suggests a closer link between phocids and odobenids (Phocomorpha Clade). The relationship of pinnipeds to other carnivorans has also been contentious, with some morphological studies divided between a closer relationship of Pinnipedia to either Ursidae (Adam and Berta, 2002; Arnason *et al.*, 2006) or Musteloidea (Sato *et al.*, 2006). The latter relationship is also supported by recent molecular analyses across all Carnivora (Flynn *et al.*, 2005).

Perhaps because of early controversies in pinniped relationships, particularly pinniped monophyly (Wyss, 1988), several studies have focused on identifying traits that define pinnipeds. Surprisingly, comparative studies of various traits across Carnivora have indicated that many ecological, life-history and morphological factors fail to discriminate between pinnipeds and a paraphyletic grouping of terrestrial carnivorans (fissipeds; Bininda-Edmonds and Gittleman, 2000; Bininda-Edmonds *et al.*, 2001). Aquatic adaptations found to define pinnipeds were a larger brain size for perception in a 3D environment and longer head and body sizes for a more hydrodynamic form. Smaller litter sizes and shorter interbirth times in pinnipeds were also indicative of a more k-adapted reproductive strategy.

Pinnipeds are first known from late Oligocene (27–25 Mya) fossils of *Enaliarctos mealsi* from the Pacific coast of North America (Berta *et al.*, 1989), although a recently discovered early Miocene pinniped from the Canadian Arctic may represent a more transitional form with webbed feet, rather than flippers like *Enaliarctos* (Rybczynski *et al.*, 2009). The Otarioidea/Phocidae split is placed at around 33 Mya using molecular clock dating, predating the earliest fossils by 5 Mya (Arnason *et al.*, 2006). The Odobenidae/Otariidae divergence was placed at 27 Mya, though the oldest fossils (odobenids) are middle Miocene, ~14 Mya old (Arnason *et al.*, 2006). The earliest otariid fossils are found in the late Miocene, although the first unambiguous crown otariids do not appear until the late Pliocene (Deméré *et al.*, 2003). The basal extant phocid split of monachine and phocine phocids is placed in the early Miocene, ~22 Mya, by molecular estimates, and the oldest fossils that can be clearly assigned to one of these two subclades are late early Miocene (Arnason *et al.*, 2006). The phocid crown group is much older and includes more extinct species than that of crown otariids or crown odobenids, both of which are characterised by more stem taxa.

Many of the studies of early pinniped evolution have focused on paleobiogeography (Muizon, 1982; Deméré *et al.*, 2003), with several events potentially having a vicariant effect on pinniped evolution. For example, early pinniped divergences have been related to the growth of ice during late Oligocene glaciations, which may have caused increased coastal upwelling and ocean stratification. During the Pliocene, the closure of the Isthmus of Panama shut off an east–west dispersal corridor and caused isolation of Pacific and Atlantic pinnipeds, leading to speciation. Further, the adaptation of phocines to cold waters in the Pleistocene caused a high-latitude radiation. This was compounded by glacioeustatic oscillations that acted to isolate colonies and cause more speciation. These examples, and many others, suggest that changing climate and circulation patterns have had a great effect on the evolution of pinnipeds (Deméré *et al.*, 2003).

Fewer studies have focused on the morphological evolution of pinnipeds. Early work (Repenning, 1976), based on observation and qualitative analysis of morphology, noted the importance of adaptive evolution towards a marine lifestyle reflected in fossil and extant forms and the variation in these features between the three extant families. Later, more quantitative methods were used with discrete features (Adam and Berta, 2002), to more accurately link prey capture strategies with anatomy, separating the clade into four groups based on ecology and morphology: pierce feeders, suction feeders, filter feeders, and grip-and-tear feeders.

Studies quantitatively examining morphological diversity of pinnipeds are very limited. A series of 2D traditional morphometric analyses of the cranium in otariid species and subspecies were conducted to examine otariid taxonomy and geographic variation (Brunner, 1998, 2003; Brunner *et al.*, 2002). Another recent investigation used 2D geometric morphometric analyses of the ventral view of the cranium to study the development of dimorphic features in three otariid species: *Arctocephalus australis*, *Callorhinus ursinus* and *Otaria byronia* (Sanfelice and de Freitas, 2008). Other studies focus entirely on individual species (Brunner, 2002; de Oliveira *et al.*, 2005). The authors concluded that dimorphism was achieved through differences in both the rate and the direction of ontogenetic shape change between males and females in each species.

While these studies provide a foundation for quantitative analysis of cranial ontogeny and evolution in pinnipeds, they are relatively limited in phylogenetic breadth. Furthermore, 3D morphometric data are better suited to the complex morphology of the mammalian skull. Here, we use 3D morphometric data to quantitatively examine cranial morphology across the three extant families of pinnipeds. We test hypotheses of phylogenetic and ecological influences on cranial morphology and quantify differences in dimorphism and ontogeny within and among the three families. Specifically, we address the following questions:

1. Do differences in cranial shape correlate with phylogenetic relationships among pinnipeds?
2. Do differences in cranial shape correlate with ecological attributes of pinnipeds?

3. Do differences in cranial ontogeny reflect different reproductive strategies among pinnipeds?
4. Does cranial shape dimorphism reflect established differences in body size dimorphism across pinnipeds?

Methods

Landmarks

An Immersion Microscribe G2X digitiser with 0.2 mm accuracy was used for collecting landmark data from secured skulls. Measurements were taken from the cranium in two different views: dorsal (37 landmarks) and ventral (49 landmarks), which were later merged into a single view with a least-squares algorithm using 10 landmarks common to both views (Table 12.1, Figure 12.2). Landmarks were selected based on clear biological homology across all specimens, with emphasis on sutures, and were chosen so that all

Table 12.1 Cranial landmarks used in analyses. Landmark numbers refer to Figure 12.2. * indicates symmetrical landmarks, gathered from right and left side. + represents overlapping landmarks that were used to unify the dorsal and ventral views.

Number	Landmarks
1	Anterior interpremaxilliary suture[+]
2	Nasal midline
3	Nasal width*
4	Premaxilla–Nasal–Maxilla suture*
5	Nasal–Frontal midline suture
6	Maxilla–Frontal–Nasal suture*
7	Jugal–Maxilla anterior dorsal suture*
8	Antorbital process*
9	Postorbital process/Interorbital width*[+]
10	Jugal–Squamosal anterior suture*
11	Jugal posterodorsal process*
12	Parietal–Occipital midline suture
13	Foramen magnum dorsal extreme[+]
14	Premaxilla–Maxilla venterolateral suture*
15	Canine anterior*
16	Canine posterior*
17	Canine labial*
18	Cheek teeth anterior*

Table 12.1 (*cont.*)

Number	Landmarks
19	Cheek teeth posterior*
20	Maxilla–Premaxilla midline suture
21	Maxilla–Palatine midline suture
22	Palatine–Maxilla lateral suture*
23	Midline between ultimate molars
24	Posterior interpalatine suture
25	Jugal–Maxilla posteroventral suture*[+]
26	Jugal–Squamosal posteroventral suture*[+]
27	External Auditory Meatus lateral extreme*
28	Auditory bulla anteromedial extreme*
29	Auditory bulla posterior extreme*
30	Mastoid process lateral extreme*
31	Mastoid process posterior extreme*
32	Basion
33	Occipital condyle venteromedial*[+]
34	Occipital condyle dorsomedial*

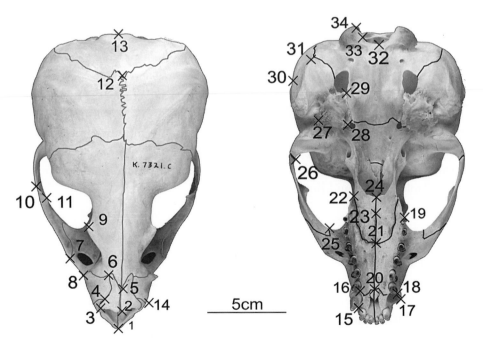

Figure 12.2 Landmarks collected and included in final analysis, shown on *Arctocephalus gazella*. Numbers correspond with landmarks listed in Table 12.1. Symmetrical landmarks are shown on one side only.

regions of the skull were taken into account. Because of the emphasis on points of clear homology, it is possible that structures of ecological or functional importance were not sampled. Analyses based on landmarks, particularly those concentrated on sutures, may well underestimate shape differences between skulls (Macleod, 1999). However, because this study considers both phylogenetic and ecological aspects of shape across a diversity of taxa, the focus on biological homology is justified.

The 10 common landmarks were widely distributed on the *x*, *y*, and *z* axes in order to minimise error when merging the views (Table 12.1, Figure 12.2). Landmarks were repeated 3 times in 7 specimens for error tests, and 18 landmarks with standard deviations greater than 1 mm, on specimens ranging from 20 to 50 cm in skull length, were excluded from further analysis, leaving a total of 58 landmarks.

Specimens

Specimens were measured from the collections at the University Museum of Zoology, Cambridge and the Natural History Museum, London. Of the 34 extant pinniped species, 32 were represented, including all species of phocids and odobenids, covering 20 of 21 genera (Table 12.2). A total of 208 specimens were digitised (Appendix 12.1). Every attempt was made to sample both genders equally, with the final distribution of specimens including 36% male (74 specimens), 29% female (62 specimens), and 35% unsexed (72 specimens). Of the specimens sampled, 26% were infant and juvenile (55 specimens). The young specimens used in this study were primarily identified based on age data during collection. Additional young specimens without original data were identified based on the presence of significantly open sutures. Note that, for many species, particularly phocids, suture closure occurs well after weaning, but before sexual maturity, although more specific information is unavailable (Schulz and Bowen, 2004).

Data analysis

Cranial shape

The dorsal and ventral views were unified into one data set using 10 overlapping landmarks and a least-squares algorithm in Mathematica 6.0.1 (Wolfram Research Inc., Champaign, IL). Next, 12 midline points were used as a mirroring plane to fill in gaps in symmetrical landmarks. Both stages offered an opportunity to measure error and specimens with high error were removed from the analysis. Seventy-two specimens were removed prior to analysis due to high error or missing landmarks, leaving a total of 136 specimens analysed for 58 landmarks

Table 12.2 List of species included in analyses.

Otariidae	Phocidae
Arctocephalus australis	*Hydrurga leptonyx*
Arctocephalus forsteri	*Leptonychotes weddellii*
Arctocephalus galapagoensis	*Lobodon carcinophagus*
Arctocephalus gazella	*Mirounga angustirostris*
Arctocephalus phillippi	*Mirounga leonina*
Arctocephalus pusillus	*Monachus monachus*
Arctocephalus townsendi	*Monachus schauinslandi*
Arctocephalus tropacalis	*Monachus tropacalis*
Callorhinus ursinus	*Ommatophoca rossii*
Eumetopias jubata	*Cystophora cristata*
Neophoca cinerea	*Erignathus barbatus*
Otaria byronia	*Halichoerus grypus*
Phocarctos hookeri	*Histriophoca fasciata*
Zalophus californianus	*Pagophilus groenlandica*
Odobenidae	*Phoca largha*
Odobenus rosmarus	*Phoca vitulina*
	Pusa caspica
	Pusa hispida
	Pusa sibirica

(Appendix 12.1). This unified, mirrored data was then entered into Morphologika 2.5 (O'Higgins and Jones, 2006), in which Generalised Procrustes analysis and principal components analysis were conducted (Zelditch *et al.*, 2004).

Phylogenetic signal

The correlation between phylogenetic relationship and similarity in cranial shape was tested to measure the amount of phylogenetic signal in the pinniped cranium. A patristic distance matrix was constructed using a composite phylogeny. Otariid relationships follow the phylogenetic analysis of Wynen *et al.* (2001; using the position indicated for *Arctocephalus australis* group A), whereas phocid and higher-level pinniped phylogenetic relationships follow Arnason *et al.* (2006) (Otaroidea; Figure 12.1). Euclidean distances between each pair of species were calculated for each significant principal component (Table 12.3, Appendix 12.1) and used to generate four shape distance matrices. Separate distance matrices were generated for male and female specimens, and only adult specimens were included in analyses. Each shape distance matrix was then compared to the patristic distance matrix using Spearman's rank correlation analysis. Analyses were conducted in PAST (Hammer *et al.*, 2001).

Table 12.3 Eigenvalues for each significant PC axis and the five landmarks with the PC loadings that contributed to that axis. Landmark numbers correspond to positions described in Table 12.1 and shown in Figure 12.2.

PC	Eigenvalues (%)	Landmarks with highest PC loadings
1	29.4	13, 5, 33, 4, 19
2	16.8	6, 24, 9, 7, 21
3	10.7	7, 16, 15, 12, 6
4	6.04	32, 7, 19, 17, 16

Table 12.4 Significant ecological correlates of cranial shape for first four principal components using independents contrasts. + indicates significant positive correlation; − indicates significant negative correlation ($p < 0.05$). Sexual size dimorphism was calculated from male body mass divided by female body mass (kg). Marine primary productivity was measured using ^{14}C uptake and simulated fluorescence techniques (g C m^{-2} year^{-1}) (Ferguson and Higdon, 2006). Seasonality was calculated as the annual variation coefficient of the monthly primary productivity, averaged over 20 years, taken from measures of soil evapotranspiration in coastal weather stations (Ferguson, personal communication).

Ecological variable	PC1	PC2	PC3	PC4
Sexual size dimorphism				
Harem size				
Latitude				
Temperature (°C)				
Productivity		−		
Seasonality	+	−		−
Lactation (days)		−		
Female maturity (days)				+
Gestation (days)				
Longevity (months)				+
Interbirth (months)				
Polygamy (yes/no)				
Weaning time (months)				
Neonate (g)		−		

Ecological correlates of cranial shape

To analyse correlations of skull shape with various ecological attributes, data on 14 ecological variables were collected from the literature (Table 12.4; Reeves *et al.*,

2002; Schulz and Bowen, 2004; Ferguson and Higdon, 2006). PC scores were averaged for all specimens of each species, including males and females, and young were excluded from ecology analyses. Because closely related species have the potential to be more similar in morphology or ecology, the independent contrasts method (Felsenstein, 1985) was used. Correlation analyses were conducted in COMPARE 4.6b (Martins, 2004) with the phylogeny shown in Figure 12.1 (Wynen *et al.*, 2001; Arnason *et al.*, 2006) and a significance value of $p < 0.05$.

Ontogenetic shape change

Ontogenetic trajectories were calculated from PC1 and 2 (Figure 12.3). The length and angle were calculated trigonometrically from PC1 and PC2 scores (*X* and *Y* coordinates) of relevant specimens of known age and sex. Only species with both adult and young specimens of the same sex could be included, resulting in a representation of 18 species. Size differences between juvenile

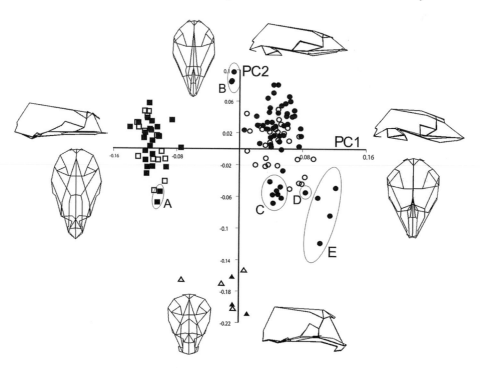

Figure 12.3 Principal components analysis displaying the first two principal components. Wireframes represent the position of landmarks in specimens at the extremes of the axes they are found next to. Symbols represent: ● phocids; ■ otariids; ▲ odobenids; open symbols: young. A, Male *Otaria byronia*; B, *Hydrurga leptonyx*; C, *Erignathus barbatus* and *Halichoerus grypus*; D, Male *Mirounga leonina*; E, *Cystophora cristata*. PC loadings and eigenvectors are provided in Table 12.3.

and adult specimens were measured using centroid size (adult/young). The length of the vector in morphospace was then compared to centroid size ratio of the two specimens with Spearman's rank correlation analysis in PAST (Hammer *et al.*, 2001). This is important to verify that longer ontogenetic trajectories were not simply produced by the uneven sampling of younger (and smaller) specimens. An unbiased data set should not show a significant correlation between ontogenetic trajectory length and centroid size.

Sexual dimorphism

Male–female trajectories on PC1 and PC2 (shape differences due to dimorphism) were calculated using a similar method as in the analysis of ontogenetic trajectories described above. Only adult specimens of known sex were included, eliminating 12 species (Appendix 12.1) from the analysis. Where multiple individuals of each sex of the same species existed all possible trajectories were plotted. An analysis was also conducted to test if the ratio of male to female centroid size correlates with degree of shape dimorphism (vector length) in pinnipeds (i.e. are species that are dimorphic in cranial size also dimorphic in shape?). In addition, shape dimorphism (vector length in PC1 and PC2) between males and females was plotted against published data on body mass dimorphism (Ferguson and Higdon, 2006) to compare cranial and postcranial dimorphism.

Results

Cranial shape

The first four principal components (Table 12.3 and Appendix 12.1) explained significant shape changes in the data set (29%, 17%, 11%, and 6% of the total variance, respectively). The first two principal components (Figure 12.3) primarily reflected phylogeny, as the three families grouped into very distinct areas of the morphospace that did not overlap. PC1 represented otariid-like morphology at the negative end to phocid-like morphology at the positive end. Species with extremely negative scores on PC1, such as *Callorhinus ursinus*, had an enlarged palate, broad interorbit and reduced auditory bullae. At the positive end of the PC1 axis, species, such as *Cystophora cristata*, showed narrow, posteriorly placed nasal and interorbit and inflated auditory bullae. The highest PC loadings for PC1 (Table 12.3) were concentrated in the rostral region and around the occipital region. PC2 (Figure 12.3) (16.8%) represented shape differences between otariids and phocids at the positive end to walruses at the negative end. *Hydrurga leptonyx* represented the positive extreme of PC2, with a pointed snout and more slender nasal and interorbit region. Walrus specimens, which occupied the negative end of PC2, had a wide nasal opening, large canines, and broad nasals. Dominant PC2 loadings were located in the palate and snout (Table 12.3). Suction-feeding species (*Odobenus rosmarus*, *Erignathus barbatus*) had more negative scores on PC2 and another dietary specialist, *Hydrurga leptonyx*, had a more negative PC1 score than the other phocids.

Phocid specimens covered a wider range of morphospace than otariids did, reflecting the greater diversity in cranial morphology in phocids. Although the three pinniped families are clearly distinct in morphospace, there were some species that deviated markedly from their respective clade's space. The phocids *C. cristata*, *Halichoerus grypus*, *E. barbatus* and the otariid *Otaria byronia* had particularly negative PC2 scores. This indicated morphological convergence with the odobenids. One phocid, *H. leptonyx*, had a particularly negative PC1 score, indicating a more otariid-type skull morphology than observed in other phocids.

PC3 and PC4 (Figure 12.4) did not show the strong phylogenetic groupings apparent in PCs 1 and 2. On these axes, odobenids, phocids, and otariids all

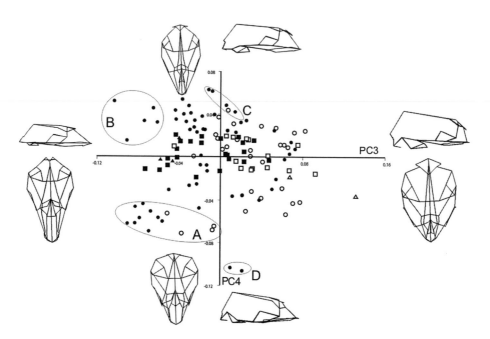

Figure 12.4 Principal components three and four. Symbols represent: ● phocids; ■ otariids; ▲ odobenids; open symbols: young. A, *Hydrurga leptonyx* and *Lobodon carcinophagus*; B, *Halichoerus grypus*; C, *Erignathus barbatus*; D, *Ommatophoca rossi*. PC loadings and eigenvectors are provided in Table 12.3.

occupied similar space. PC3 had a strong ontogenetic component and young from all three clades fell towards the positive end of the axis. At the positive end of PC3 was an odobenid foetus with small canines and a reduced frontal region. *H. grypus*, *H. leptonyx*, and *Lobodon carcinophagus* adults occupied the negative end of PC3. They had a longer skull, enlarged sagittal crest and canines. High PC loadings (Table 12.3) were concentrated on the canines and snout. PC4 (Figure 12.4) was dominated by shape change within the phocids, ensuring otariids and odobenids clustered around zero. *Ommatophoca rossi* was found at the negative end of PC4, and *E. barbatus*, the phocid suction-feeder (Marshall *et al.*, 2008), at the positive end. Dominant PC4 loadings (Table 12.3) involved the basion and dentition, which was reflected in the clear separation of dietary groups on these axes. In addition to *E. barbatus*, filter feeders and large prey feeders formed a cluster away from their sister taxon, *Leptonychotes weddelli*, on the negative end of PC3 and PC4.

Phylogenetic signal

The analyses of the relationship between phylogeny and cranial shape showed several significant correlations. For male cranial shape there were significant correlations between phylogeny and PC1 ($p < 0.001$) and PC4 ($p < 0.001$). Female cranial shape was significantly correlated with phylogeny on PC1 ($p < 0.001$) and PC2 ($p = 0.002$).

Ecological correlates of cranial shape

After removal of phylogenetic effects, seasonality was the only variable to correlate significantly with PC1, showing a positive correlation (Table 12.4). Seasonality and productivity correlated negatively with PC2. The reproductive variables of neonate mass and lactation time were also negatively correlated with PC2 scores.

There were no correlations with PC3 suggesting this axis is not greatly influenced by ecology (Table 12.4). Longevity and age to female maturity were both positively correlated with PC4 scores. Also, seasonality correlated negatively with this axis.

Ontogenetic shape change

The ontogenetic trajectories (arrows drawn from young to adults; Figure 12.5a) varied depending on the distance of the adult members of a species from the mean shape for its respective family. Species with adult morphologies closer to the mean shape for each family (low to moderate

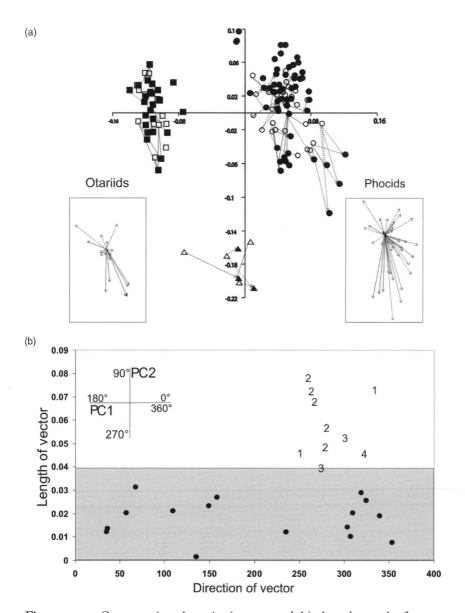

Figure 12.5 a, Ontogenetic trajectories (young to adult) plotted onto the first two principal components axes. Solid arrows are between specimens of the same sex, dotted arrows are between one or two specimens of unknown sex. Only specimens of known sex are included in b. Boxes show all ontogenetic trajectories for the respective family, re-oriented to the same origin. Symbols represent: ● phocids; ■ otariids; ▲ odobenids; open symbols: young. b, Plot of direction of ontogenetic trajectory against degree of ontogenetic shape change. Direction is measured as angle anticlockwise from the positive PC2 axis. A key relating angle to direction on the PC1 and PC2 axes is shown in the top left. Degree of shape change is measured as length of the vector. The grey area marks specimens that show short ontogenetic trajectories in all directions. The white area indicates specimens with long (greater than 0.4) ontogenetic trajectories and are concentrated between 250 and 350 degrees. 1: *Odobenus rosmarus*, 2: *Halichoerus grypus*, 3: *Otaria byronia*, 4: *Lobodon carcinophagus*. *Erignathus barbatus* and *Cystophora cristata* were not included in part B due to lack of sex data.

ontogenetic shape changes) tended to have shorter trajectories. These species also had a wider distribution of directions of the ontogenetic trajectories. However, longer trajectories were consistently oriented in the direction of negative PC2 values (Figure 12.5b). Specifically, for adults with highly negative PC2 scores (convergent on odobenid morphology), the young of those species usually displayed more generalised cranial morphology, near the mean shape for their family, resulting in long trajectories in the direction of negative PC2 (Figure 12.5a).

Plotting length of trajectory against relative size difference between young and adult specimens (ratio of adult to young centroid size) produced no significant correlation (Spearman's $r=0.0054$, $p=$n.s.). This result demonstrated that the ontogenetic patterns observed in Figure 12.5 were not a product of sampling bias.

Sexual dimorphism

Dimorphism vectors (Figure 12.6a) showed patterns similar to those found in ontogenetic trajectories (Figure 12.5a) described above. Low to moderate differences in cranial shape dimorphism were heterogenous in orientation (Figure 12.6b). However, phocids and otariids with negative scores on PC2 because of morphological adaptations relating to mating display showed longer distances between males and females (Figure 12.6a). These species (*C. cristata*, *O. byronia*, *Mirounga leonina*) all showed vectors aligned towards negative PC2 direction, toward odobenid morphospace, in a similar manner to the ontogenetic trajectories described above (Figure 12.6a).

Dimorphism distance was significantly correlated with male/female centroid size ratio ($r=0.45$, $p=0.002$), indicating that species showing large dimorphic differences in cranial size also display large dimorphic differences in cranial shape.

In contrast, results suggested dissociation between cranial shape dimorphism and body mass dimorphism in some species (Figure 12.7). Most phocids and the odobenids displayed low to moderate cranial shape dimorphism (0–0.4) and low body size dimorphism (1–2). On the other hand, most otariids showed higher body size dimorphism (3–4) over a similar range of cranial shape dimorphism (0–0.4). *M. leonina* showed extremely large values on both axes. *C. cristata* and *O. byronia* grouped together as having lower body size dimorphism (1–2) but very high shape dimorphism (0.5–0.8).

Discussion

Pinniped families showed strong phylogenetic signal in their cranial morphology (Figure 12.3). Significant correlations between phylogeny and

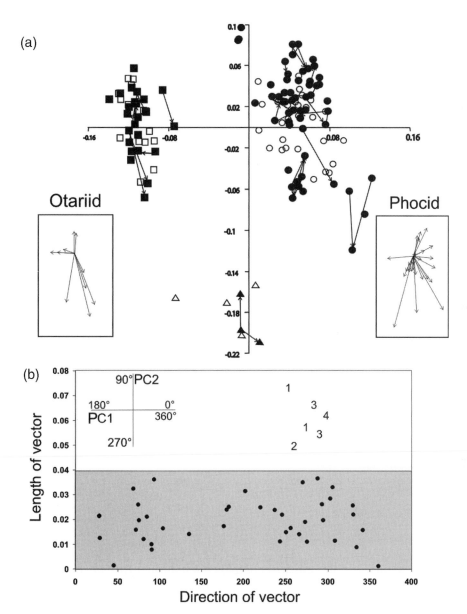

Figure 12.6 a, Distances plotted from females to males onto the first two principal components. Boxes show all vectors for the respective family, re-oriented to the same origin. Symbols represent: ● phocids; ■ otariids; ▲ odobenids; open symbols: young. b, Plot of direction of dimorphic shape difference against degree of dimorphic shape difference. Direction is measured as the angle anticlockwise from the positive PC2 axis. A key relating angle to direction on PC axis is shown in the top left. Degree of difference is measured as the length of the vector. The grey area marks specimens that show short vectors in all directions. The white area indicates specimens with long (greater than 0.4) vectors and are concentrated between 250 and 300 degrees. 1: *Cystophora cristata*, 2: *Arctocephlalus gazella*, 3: *Otaria byronia*, 4: *Mirounga leonina*.

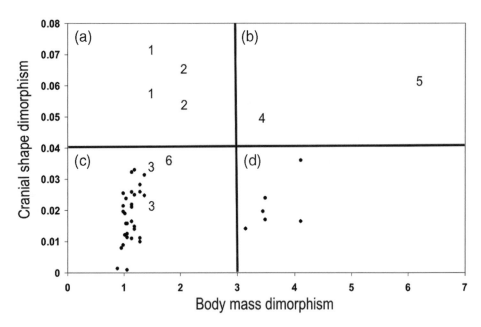

Figure 12.7 Plot showing body mass dimorphism against cranial shape dimorphism. Body mass dimorphism is the ratio of female to male body mass for each species, collected from the literature (Ferguson and Higdon, 2006). Cranial shape dimorphism is measured as the length of the vector between male and female specimens on PC1 and PC2 (Figure 12.6a). Specimens plot into four quadrants: a, high cranial shape dimorphism and low body mass dimorphism; b, high cranial and body mass dimorphism; c, low cranial and body mass dimorphism; and d, low cranial shape dimorphism and high body mass dimorphism. 1: *Cystophora cristata*, 2: *Otaria byronia*, 3: *Odobenus rosmarus*, 4: *Arctocephalus gazella*, 5: *Mirounga leonina*, 6: *Arctocephalus australis*. All phocids except those specified are found in sector C. Sector D contains solely otariids.

cranial shape on multiple principal components indicated that phylogeny was the most dominant influence on cranial morphology. However, cranial morphology did not reflect the considerable ecological overlap between phocids and otariids. This pattern possibly reflects morphological differences that evolved early in the histories of these clades, although fossil taxa need to be included to determine when these distinct areas of morphospace were invaded. The results presented here suggest a number of potential hypotheses that could be tested with fossil data. First, the marked phylogenetic separation among the three clades may be partially due to the loss of intermediate forms, particularly given the relatively large otariid and odobenid stem groups. Second, the smaller range of otariid morphological diversity may also reflect the greater loss of otariid taxonomic diversity through extinction, as the crown group represents a relatively small proportion of the total group for otariids. By contrast, extant phocids represent many basal branches and so the crown group includes many more extinct species than that of the otariids (Demére *et al.*, 2003). Third, the basal split of monachine and phocine phocids (20 Mya) was not reflected in skull morphology on PC1 or 2 (Figure 12.3). For example, *C. cristata* (a phocine) and *M. leonina* (a monachine) plotted very near to each other on PC1 and PC2 (Figure 12.3). This result suggests either that there has been much morphological convergence between these groups since they diverged, or that there was relatively little morphological differentiation associated with their divergence.

Although the families of pinnipeds displayed remarkably different cranial morphology, some species were conspicuously positioned away from position of the standard phylogenetic grouping in morphospace. These species displayed morphological convergence that bridged the morphospace defined by the three pinniped families (Figure 12.3). Convergence was more common in the phocids than in the otariids, reflecting their greater ecological diversity and more extreme specialisations. These examples of convergence can be classified into those due to diet and those due to mating displays.

The most apparent example of morphological convergence reflecting similarity in diet is observed in *E. barbatus*. This species had a particularly low score on the PC2 axis (Figure 12.3) of around −0.06, approaching the region of morphospace occupied by odobenids (the mean for odobenids is −0.18), than other phocids (the mean for phocids is ∼0). This species shares a similar diet with walruses in feeding within the sediment on fish and invertebrates, and a recent study showed that *E. barbatus* uses suction feeding 96.3% of time whilst feeding underwater (Marshall *et al.*, 2008). This result shows there are aspects of cranial morphology adapted for suction feeding that have evolved independently in both phocids and odobenids.

H. grypus also grouped very closely with *E. barbatus* in morphospace (Figure 12.3). However, this species is not solely a sediment feeder, but eats a wider range of fish including bottom-dwellers, crustaceans, and molluscs (King, 1983). The method it uses for feeding on molluscs (crunching or suction-feeding) is unclear, but these results suggest that it has some adaptations for sediment feeding, despite retaining a generalist diet.

In contrast to those species that converge ecologically and morphologically with walruses, *H. leptonyx* represents a phocid that may converge toward otariids. *H. leptonyx* was located in between the phocid and otariid cluster on PC1, with a mean PC score of −0.01, compared to a range of 0.01–0.08 for other phocids and a range of −0.07 to −0.14 for otariids (Figure 12.3).

This position reflects more otariid-like morphology than observed in other phocids. This unusual morphology may relate to the fact that *H. leptonyx* is the only pinniped to specialise on large, warm-blooded prey (Reeves *et al.*, 2002). However, many otariids incorporate large prey as a small part of their diet. The otariid skull is generally more robust than the typical phocid skull, which may reflect the necessity to cope with the large forces associated with large-prey feeding, and may have evolved convergently in *H. leptonyx* for the same purpose.

In addition to convergence relating to diet, several pinnipeds displayed unusual morphologies related to mating displays. The most conspicuous example of cranial adaptations for sexual displays are found in *C. cristata*. This species had the lowest score among phocids on PC2 (−0.012, Figure 12.3). The large dimorphism distance on PC2 between females and males (female average −0.6) (Figure 12.6a) supported the interpretation that this convergence toward odobenid morphospace was related to mating strategy. Male *C. cristata* have a large proboscis that is used in mating displays, including an internal nasal membrane that can be inflated to produce a large, red, facial bladder (Reeves *et al.* 2002). This is facilitated by a wider nasal opening, which is superficially similar to the wide rostrum observed in walruses. However, females also had a much lower score on this axis (Figure 12.3) than the young (−0.03) for reasons that are not apparent.

M. leonina males also fall out relatively low on the PC2 axis (−0.06; Figure 12.3). Male *M. leonina* are the only other phocid species to have evolved a proboscis, convergently with *C. cristata*, although they do not have a facial bladder.

While most unusual morphologies were observed in phocids, *O. byronia* males were significantly more negative on PC2 than the other otariids (−0.07; Figure 12.3), consistent with qualitative reports (King, 1983) describing male *O. byronia* as having a distinctive upturned snout. *O. byronia* feed on seafloor fish and cephalopods (Reeves *et al.*, 2002), and some authors (Adam and Berta, 2002) have suggested *O. byronia* skulls show characteristics associated with suction feeding (lengthening of the hard palate and robust pterygoid hamuli), although experimental confirmation of their feeding mechanism is not available. However, results presented here showing that female and young *O. byronia* cluster near other otariids, around 0.0 on PC2 (Figures 12.3 and 12.6a), support the interpretation of cranial convergence of *O. byronia* with *O. rosmarus* as due to sexual dimorphism and not related to diet.

These results (Figure 12.3) suggest that walrus morphospace was a popular direction for cranial morphological evolution in the pinnipeds. This morphology may have represented a common adaptation for accessing the sediment–water interface (e.g. *E. barbatus*, *H. grypus*), in order to expand their range of feeding opportunities. Alternatively, it may relate to food processing and stages in the independent evolution of suction feeding. Modifications to the cranium for mating displays were also concentrated in the rostral region, often resulting in dimorphic convergence of males in the direction of walrus morphospace.

This repeated pattern of phocids and otariids converging in the direction of walrus morphospace for modifications related to both diet and sexual display suggests that the pinniped morphology may be constrained from exploring other regions of cranial morphospace. Possibly, this repeated modification of primarily the rostral region across all extant pinniped groups, and for both feeding and mating displays, may reflect constraints to maintain a hydrodynamic form. More specifically, transformations of the rostral region may have occurred independently multiple times because vertebrates with postcrania that are highly adapted for swimming cannot drastically modify their skeleton for use in mating displays or prey capture. This constraint would explain why mating displays in pinnipeds are limited to the nasal region. Furthermore, cranial dimorphism may be especially significant for species in which size dimorphism is limited by adaptations for large female body size due to low temperatures, such as *O. byronia* (Ferguson and Higdon, 2006), discussed further below.

Ecological correlates of cranial shape

Seasonality correlated significantly with three of the four significant principal component axes (PC1, 2 and 4; Table 12.4). This may reflect differences between ice-breeding species, living in highly seasonal environments, and those living in more temperate conditions. On PC1 this correlation most likely reflected the presence of phocids at high latitudes. Phocids are primarily found at higher latitudes and in polar regions, while otariids inhabit primarily equatorial to mid-latitude regions (Ferguson and Higdon, 2006). This geographical pattern reflects the fact that otariids are excluded from the most high-latitude environments by their inability to breed on ice, while many phocids are ice-breeders (Schulz and Bowen, 2004).

Productivity as well as seasonality correlated with PC2, suggesting that cranial shape was highly influenced by availability of resources at the base of the food chain. In addition, reproductive factors were correlated with cranial morphology on PC2 (Table 12.4). Large male, female, and neonate body masses were demonstrated in those species converging on odobenid space. Longer lactation times accompanied convergence in some species and this result is likely to be driven by extremely long lactation in walruses. Lactation time is related to female mass as larger fat stores facilitate a longer period of milk production (Schulz and Bowen, 2004).

PC3 scores did not correlate with any ecological variables, perhaps reflecting the strong influence of ontogeny on this axis. Correlation of PC4 scores with age to maturity and longevity suggest life history is an important influence on cranial morphology.

It is interesting to note that ecological specialisations are found exclusively in those pinnipeds native to the high latitudes (above 70° north or south; Ferguson and Higdon, 2006) and high seasonality environments. This may present an environment in which a specialist feeding habit is favourable to the generalist approach seen in all other locations. Resources at higher latitudes are scarce, so specialised ecologies and related morphologies may allow these pinnipeds to exploit the food sources available to them more effectively than their generalist relatives can.

Ontogenetic shape change

The analysis of ontogenetic shape change (Figure 12.5) revealed that there was no consistent shape trajectory for cranial growth across pinnipeds or within phocids and otariids. It did, however, highlight an interesting relationship between morphological convergence on odobenid space and length and direction of ontogenetic trajectory in phocids and otariids. First, it showed that all these species were undergoing similar skull shape changes during growth (Figure 12.5a). Second, the results showed that species that converge towards walrus cranial shape had a greater difference in young and adult morphology than non-convergent species (Figure 12.5b). In these species, young specimens showed morphology more similar to that of the rest of their family (Figure 12.5a). This pattern means that the unusual morphology observed in adults was only achieved after weaning and required extreme modifications in cranial shape during growth. This shift suggests that the morphological traits observed are only required during the later stages of their lives, possibly representing sexual selection or differences in juvenile and adult diets.

In the cases of *C. cristata*, *M. leonina*, and *O. byronia*, the morphological shift may reflect the development of sexual characteristics. In *E. barbatus* and *H. grypus*, however, which converge on odobenid space due to diet, it is possible that these ontogenetic shifts reflect changes in diet after weaning. For sediment feeders, the ability to dive deeply and for prolonged periods may require further development and hence a change in diet. Unfortunately, there are currently little data available regarding the post-weaning diets of pinniped young that would be required to test this hypothesis.

Sexual dimorphism

The analyses of cranial dimorphism (Figure 12.6) demonstrated that there is a strong positive correlation between size and shape dimorphism of pinniped crania. While the data presented here lack the ontogenetic resolution needed for testing specific hypotheses of allometry, the correlation between cranial shape and cranial size dimorphism may simply reflect allometric differences between adult males and females. In at least a few cases, such as *M. leonina*, overlapping ontogenetic (Figure 12.5a) and dimorphism (Figure 12.6a) trajectories provide tentative support for this hypothesis. However, in many cases, such as *C. cristata*, the trajectories are not coordinated, suggesting that shape dimorphism is not simply a consequence of allometric differences between adult males and females.

A recent study of ontogeny and dimorphism in three species of otariids, *O. byronia*, *C. ursinus*, and *A. australis* specifically tested the role of allometry in generating cranial shape dimorphism. Their results demonstrate that shape dimorphism may simply reflect allometry in *C. ursinus*, but that allometry alone cannot explain the shape differences observed in *O. byronia* and *A. australis* (Sanfelice and de Freitas, 2008). Their detailed ontogenetic study showed that dimorphism in *O. byronia* is achieved very early in development, resulting in shape differences even between male and female juveniles. Strikingly, the authors report that the rate of male cranial growth is three times greater than that of females, implicating a strong heterochronic shift in the evolution of cranial dimorphism in *O. byronia*. Improved data from ontogenetic series of a diverse sample of pinnipeds, particularly those species highlighted in this study for converging on odobenid morphology through ontogenetic shape changes, will be essential to rigorously test the role of allometry and heterochrony in cranial shape dimorphism.

It is also notable that *C. cristata*, *M. leonina* and *O. byronia* all showed remarkably similar trajectories of shape dimorphism (Figure 12.6a), despite representing a wide phylogenetic range (otariid and monachine and phocine phocid). These species also displayed the most marked differences in male–female morphology (Figure 12.6b). As noted above, in some of the species that converge on walrus cranial morphology, such as *O. byronia*, dimorphic shape differences (Figure 12.6a) were similar in direction to the ontogenetic trajectories (Figure 12.5a), with adult females that are similar in cranial morphology to juvenile specimens. This result is consistent with previous analyses demonstrating that adult female *O. byronia* share a very similar morphology with subadults of both sexes, whereas adult male morphological traits arise well before adulthood (Sanfelice and de Freitas, 2008). In the results presented here, these differences between males and females primarily reflect the development of sexual characteristics, such as a proboscis or facial bladder, in adult males, driving their convergence on odobenid cranial morphology.

Interestingly, our study demonstrated that large dimorphic shape differences in *C. cristata* and *O. byronia* were not accompanied by increased body mass dimorphism (Figure 12.7). In fact, these species showed amongst the lowest body size dimorphism, suggesting that body mass dimorphism and cranial shape dimorphism may represent alternative strategies. For example, *O. byronia* inhabits environments with very low temperatures (−14°C average, Ferguson and Higdon, 2006) and also has relatively large female body size. One possibility is that the observed low body mass dimorphism (2.08) (Figure 12.7) may relate to a lower limit to female body size due to colder environments. Alternatively, the trade-off between cranial shape dimorphism and body size dimorphism may relate to the relative importance of display to fighting in male competition. Cranial shape dimorphism is expected to be more pronounced in species using elaborate male displays, such as facial bladders, while body mass dimorphism may be more common in species in which male fighting dominates.

Lastly, the analyses of dimorphism presented here demonstrated that intraspecific shape differences among males and females were large compared even to interspecific differences (Figure 12.6a). While most terrestrial carnivorans express dimorphism through size differences, the large cranial shape dimorphism observed in pinnipeds here emphasises the importance of cranial morphology to multiple purposes in pinniped evolution. Pinnipeds may place unusual emphasis on the cranium for mating displays and prey-capture adaptations, such as suction or filter feeding, because the extreme specialisation of the postcranium for swimming reduces its utility in other tasks. Further analyses including fossil taxa would be essential to understanding the shift of multiple functions, such as prey-capture and mating displays, to the cranium during the terrestrial to marine transition in pinniped evolution.

Conclusions

The most striking pattern observed in this quantitative analysis of cranial morphology across extant pinnipeds is the repeated evolution of feeding and mating specialisations that converge towards odobenid morphology. The common evolutionary and developmental trajectories observed here suggest that specialisations for an aquatic lifestyle may constrain the range of functionally viable morphospace available to pinnipeds, reflected in their concentration on adaptations in the rostral region. While the three extant pinniped families occupy distinct areas of morphospace, multiple phocids and otariids independ-ently converge toward odobenid cranial morphology in adaptations related to both diet and mating display. Ontogenetic analyses suggest that these shifts occur primarily during the juvenile growth phase, requiring large alterations in morphology during development, likely due to dietary changes or sexual maturation. Lastly, some species illustrate a trade-off between body size dimorphism and cranial shape dimorphism, perhaps related to differences in mating behaviour or habitat among pinnipeds.

Secondary adaptations to the aquatic realm include some of the most compelling examples in vertebrate evolution (Uhen, 2007). However, pinniped evolutionary morphology remains understudied in comparison to other aquatic mammals. This study demonstrates that the unique reproductive and ecological strategies pursued by pinnipeds are matched by several interesting patterns in the morphological evolution of the pinniped cranium, providing a promising avenue for future studies of major evolutionary transitions.

Acknowledgements

We are indebted to Matt Lowe (University Museum of Zoology Cambridge) and Richard Sabin (Natural History Museum, London) for access to specimens and extensive help during data collection. S. H. Ferguson kindly provided details of ecological measures. A. Piotrowski provided helpful discussion of climatic measures. V. Weisbecker generously read and commented on an early version of this manuscript.

REFERENCES

Adam, P. J. and Berta, A. (2002). Evolution of prey capture strategies and diet in the Pinnipedimorpha (Mammalia, Carnivora). *Oryctos*, **4**, 83–107.

Arnason, U., Gullberg, A., Janke, A., *et al.* (2006). Pinniped phylogeny and a new hypothesis for their origin and dispersal. *Molecular Phylogenetics and Evolution*, **41**, 345–54.

Berta, A. and Adam, P. J. (2001). Evolutionary biology of pinnipeds. In *Secondary Adaptation of Tetrapods to Life in Water*, ed. J. -M. Mazin and V. de Buffrénil. Munich: Verlag Dr. Friedrich Pfeil, pp. 235–60.

Berta, A., Ray, C. E. and Wyss, A. R. (1989). Skeleton of the oldest known pinniped, *Enaliarctos mealsi. Science*, **244**, 60–62.

Bininda-Edmonds, O. R. P. and Gittleman, J. L. (2000). Are pinnipeds functionally different from fissiped carnivores? The importance of phylogenetic comparative analyses. *Evolution*, **54**, 1011–23.

Bininda-Emonds, O. R. P., Gittleman, J. L. and Kelly, C. K. (2001). Flippers versus feet: comparative trends in aquatic and non-aquatic carnivores. *Journal of Animal Ecology*, **70**, 386–400.

Brunner, S. (1998). Cranial morphometrics of the southern fur seals *Arctocephalus forsteri* and *A. pusillus* (Carnivora: Otariidae). *Australian Journal of Zoology*, **46**, 67–108.

Brunner, S. (2002). Geographic variation in skull morphology of adult steller sea lions (*Eumetopias jubatus*). *Marine Mammal Science*, **18**, 206–22.

Brunner, S. (2003). Fur seals and sea lions (Otariidae): identification of species and taxonomic review. *Systematics and Biodiversity*, **1**, 339–439.

Brunner, S., Shaughnessy, P. D. and Bryden, M. M. (2002). Geographic variation in skull characters of fur seals and sea lions (family Otariidae). *Australian Journal of Zoology*, **50**, 415–38.

de Oliveira, L. R., Hingst-Zaher, E. and Stenghel Morgante, J. (2005). Size and shape sexual dimorphism in the skull of the South American fur seal, *Arctocephalus australis* (Zimmerman, 1783) (Carnivora: Otariidae). *Latin American Journal of Aquatic Mammals*, **4**, 27–40.

Deméré, T. A., Berta, A. and Adam, P. J. (2003). Pinnipedimorph evolutionary biogeography. *Bulletin of the American Museum of Natural History*, **279**, 32–76.

Felsenstein, J. (1985). Phylogenies and the comparative method. *American Naturalist*, **125**, 1–15.

Ferguson, S. H. and Higdon, J. W. (2006). How seals divide up the world: environment, life history, and conservation. *Oecologia*, **150**, 318–29.

Flynn, J. J., Finarelli, J. A., Zehr, S., Hsu, J. and Nedbal, M. A. (2005). Molecular phylogeny of the Carnivora (Mammalia): assessing the impact of increased sampling on resolving enigmatic relationships. *Systematic Biology*, **54**, 317–37.

Hammer, O., Harper, D. A. T. and Ryan, P. D. (2001). PAST: palaeontological statistics software package for education and data analysis. *Palaeontologia Electronica*, **4**, 9 pp.

King, J. E. (1983). *Seals of the World*. London: British Museum (Natural History), 154 pp.

Kovacs, K. M. and Lavigne, D. M. (1992). Maternal investment in otariid seals and walruses. *Canadian Journal of Zoology – Revue Canadienne de Zoologie*, **70**, 1953–64.

Macleod, N. (1999). Generalizing and extending the eigenshape method of shape space visualization and analysis. *Paleobiology*, **25**, 107–38.

Marshall, C., Kovacs, K. M. and Lydersen, C. (2008). Feeding kinematics, suction and hydraulic jetting capabilities in bearded seals (*Erignathus barbatus*). *Journal of Experimental Biology*, **211**, 699–708.

Martins, E. P. (2004). COMPARE, version 4.6b: computer programmes for the statistical analysis of comparative data. Department of Biology, Indiana University, Bloomington, IN.

Muizon, C. D. E. (1982). Phocid phylogeny and dispersal. *Annals of the South African Museum*, **89**, 175–213.

O'Higgins, P. and Jones, N. (2006). Morphologika: tools for statistical shape analysis. Hull: York Medical School. http://hyms.fme.googlepages.com/resources.

Reeves, R., Stewart, B. S., Clapham, P. J. and Powell, J. A. (2002). *Sea Mammals of the World*. London: A & C Black Publishers, 528 pp.

Repenning, C. A. (1976). Adaptive evolution of sea lions and walruses. *Systematic Zoology,* **25**, 375–90.

Rybczynski, N., Dawson, M. R. and Tedford, R. H. (2009). A semi-aquatic mammalian carnivore from the Miocene epoch and origin of Pinnipedia. *Nature,* **458**, 1021–24.

Sanfelice, D. and de Freitas, T. R. O. (2008). A comparative description of dimorphism in skull ontogeny of *Arctocephalus australis, Callorhinus ursinus* and *Otaria byronia* (Carnivora: Otariidae). *Journal of Mammalogy,* **89**, 336–46.

Sato, J., Wolsan, M., Suzuki, H., *et al.* (2006). Evidence from nuclear DNA sequences sheds light on the phylogenetic relationships of Pinnipedia: single origin with affinity to Musteloidea. *Zoological Science,* **23**, 125–46.

Schulz, T. M. and Bowen, W. D. (2004). Pinniped lactation strategies: evaluation of data on maternal and offspring life history traits. *Marine Mammal Science,* **20**, 86–114.

Schulz, T. M. and Bowen, W. D. (2005). The evolution of lactation strategies in pinnipeds: a phylogenetic analysis. *Ecological Monographs,* **75**, 159–77.

Sinclair, E. H. and Zeppelin, T. K. (2002). Seasonal and spatial differences in diet in the western stock of Stellar sea lions (*Eumetopias jubatus*). *Journal of Mammalogy,* **83**, 973–90.

Uhen, M. D. (2007). Evolution of marine mammals: back to the sea after 300 million years. *Anatomical Record,* **290**, 514–22.

Williams, T. M., Rutishauser, M., Long, B., *et al.* (2007). Seasonal variability in otariid energetics: implications for the effects of predators on localized prey resources. *Physiological and Biochemical Zoology,* **80**, 433–43.

Wynen, L. P., Goldsworthy, S. D., Insley, S. J., *et al.* (2001). Phylogenetic relationships within the eared seals (Otariidae: Carnivora): implications for the historical biogeography of the family. *Molecular Phylogenetics and Evolution,* **21**, 270–84.

Wyss, A. R. (1988). Evidence from flipper structure for a single origin of pinnipeds. *Nature,* **334**, 427–28.

Zelditch, M., Swiderski, D., Sheets, H. D. and Fink, W. (2004). *Geometric Morphometrics for Biologists: A Primer.* Boston, MA: Elsevier Academic Press, 416 pp.

Appendix 12.1 Table of specimens included in analyses and PC scores for each specimen on the first four principal component axes.

Genus	Species	Sex	Age	Specimen	PC1	PC2	PC3	PC4
Arctocephalus	*australis*	F	Adult	01984919	-0.0859	0.0363	-0.0250	-0.0249
Arctocephalus	*australis*	M	Adult	01950111141	0.0746	0.0009	-0.0192	0.0106
Arctocephalus	*forsteri*	?	Young	K7422	-0.1212	0.0474	0.0439	-0.0029
Arctocephalus	*gazella*	F	Adult	K7321D	-0.1122	0.0343	0.0059	-0.0083
Arctocephalus	*gazella*	F	Subadult	K7321C	-0.1176	0.0425	0.0300	-0.0123
Arctocephalus	*gazella*	F	Subadult	K7321A	-0.1049	0.0165	0.0138	-0.0075
Arctocephalus	*gazella*	M	Adult	K7342IL	-0.1032	0.0159	-0.0021	-0.0172
Arctocephalus	*gazella*	M	Adult	K7321M	-0.1201	-0.0152	-0.0014	-0.0171
Arctocephalus	*philippi*	?	Adult	018831181	-0.1283	0.0330	-0.0299	0.0114
Arctocephalus	*pusillus*	?	?	K7429	-0.1141	0.0244	0.0206	0.0002
Arctocephalus	*pusillus*	F	Adult	K7361	-0.1140	0.0024	0.0039	0.0013
Arctocephalus	*pusillus*	F	Adult	01927728	-0.1126	0.0226	-0.0005	0.0152
Arctocephalus	*pusillus*	M	Adult	K7426	-0.1168	0.0382	-0.0189	0.0169
Arctocephalus	*tropacalis*	F	Adult	01953148	-0.1106	0.0098	0.0303	0.0108
Arctocephalus	*tropacalis*	M	Adult	019574231I	-0.1204	0.0194	0.0048	-0.0035
Artcocephalus	*galapogoensis*	F	Adult	019912	-0.1053	0.0273	0.0197	0.0099
Callorhinus	*ursinus*	?	Young	K72272	-0.0995	-0.0041	0.0918	-0.0161
Callorhinus	*ursinus*	F	Adult	01960522	-0.1397	0.0269	0.0159	0.0150
Callorhinus	*ursinus*	M	Subadult	K7221	-0.1301	0.0328	0.0045	0.0169
Cystophora	*cristata*	?	Young	K7750	0.0823	-0.0362	0.0775	-0.0446
Cystophora	*cristata*	?	?	K7741	0.1141	-0.0850	-0.0302	-0.0337
Cystophora	*cristata*	?	Young	K7742	0.0930	-0.0131	0.0391	-0.0445

Genus	species	sex	age	specimen				
Cystophora	cristata	F	Adult	K7745	0.1210	−0.0496	−0.0126	−0.0182
Cystophora	cristata	F	Adult	1844623I	0.0987	−0.0627	−0.0231	−0.0321
Cystophora	cristata	M	Adult	332h	0.1024	−0.1201	−0.0499	−0.0387
Erignathus	barbaratus	?	Young	K8022	0.0638	−0.0513	−0.0107	0.0612
Erignathus	barbaratus	?	?	K8023	0.0530	−0.0626	−0.0112	0.0640
Erignathus	barbaratus	?	?	K8021	0.0495	−0.0579	−0.0072	0.0639
Erignathus	barbaratus	?	Young	1878619I	0.0327	−0.0226	0.0079	0.0480
Erignathus	barbaratus	F	Adult	193710239	0.0395	−0.0417	0.0056	0.0506
Eumetopias	jubatus	?	?	K7081	−0.1211	0.0133	−0.0513	−0.0053
Eumetopias	jubatus	?	?	0195032912	−0.1180	−0.0321	−0.0517	−0.0113
Eumetopias	jubatus	F	Young	0195032910	−0.1203	0.0217	−0.0453	0.0082
Eumetopias	jubatus	F	Adult	019251o832	−0.0927	−0.0244	−0.0413	−0.0014
Eumetopias	jubatus	M	Adult	019507214	−0.1173	−0.0240	−0.0451	0.0070
Eumetopias	jubatus	M	Adult	0196889I	−0.1108	−0.0229	0.0470	−0.0002
Eumetopias	jubatus	M	Young	0190310I8	−0.1097	−0.0143	0.0308	−0.0081
Halichoerus	grypus	F	Adult	K7943	0.0477	−0.0532	−0.0659	0.0415
Halichoerus	grypus	F	Young	19615I82o	0.0562	0.0185	−0.0262	−0.0016
Halichoerus	grypus	F	Adult	19615I836	0.0528	−0.0491	−0.0612	0.0304
Halichoerus	grypus	F	Adult	19615I832	0.0429	−0.0584	−0.0737	0.0312
Halichoerus	grypus	M	Young	19615I82	0.0337	−0.0130	0.0021	0.0086
Halichoerus	grypus	M	Adult	19615I8II	0.0551	−0.0280	−0.0908	0.0122
Halichoerus	grypus	M	Young	1939114I	0.0367	−0.0210	0.0298	0.0080
Halichoerus	grypus	M	Adult	196236I	0.0438	−0.0697	−0.1032	0.0505

Appendix 12.1 (*cont.*)

Genus	Species	Sex	Age	Specimen	PC1	PC2	PC3	PC4
Histriophoca	*fasciata*	M	Subadult	1661272	0.0487	0.0111	0.0616	0.0107
Histrophoca	*fasciata*	F	Adult	1957197	0.0565	0.0294	0.0681	0.0157
Histrophoca	*fasciata*	F	Adult	1957199	0.0764	0.0159	0.0509	0.0232
Histrophoca	*fasciata*	F	Subadult	1657195	0.0627	0.0095	0.0819	0.0257
Histrophoca	*fasciata*	M	Adult	1637196	0.0753	0.0026	0.0724	0.0072
Histrophoca	*fasciata*	M	Adult	1637197	0.0507	0.0151	0.0665	0.0023
Histrophoca	*fasciata*	M	Subadult	19657191O	0.0759	0.0157	0.0713	0.0232
Hydruga	*leptonyx*	?.	?.	K7864	-0.0067	0.0961	-0.0760	-0.0555
Hydruga	*leptonyx*	F	Adult	19404641	-0.0096	0.0837	-0.0813	-0.0459
Hydruga	*leptonyx*	M	Adult	1901415	-0.0087	0.0850	-0.0731	-0.0423
Leptonychotes	*weddelli*	?.	?.	K7881	0.0387	0.0516	0.0491	-0.0316
Leptonychotes	*weddelli*	?.	?.	K7884	0.0330	0.0249	0.0541	-0.0202
Leptonychotes	*weddelli*	F	Adult	19404640	0.0359	0.0637	0.0378	-0.0408
Leptonychotes	*weddelli*	F	Adult	19404604	0.0352	0.0331	0.0214	-0.0364
Leptonychotes	*weddelli*	M	Adult	K7883	0.0377	0.0454	0.0115	-0.0388
Leptonychotes	*weddelli*	M	Young	1915111	0.0266	0.0281	0.0599	-0.0285
Lobodon	*carcinophaga*	?.	?.	K7903	0.0218	0.0405	-0.0543	-0.0577
Lobodon	*carcinophaga*	?.	Foetus	1586185	0.0104	-0.0041	-0.0030	-0.0589
Lobodon	*carcinophaga*	M	Adult	1935329I	0.0315	0.0251	-0.0800	-0.0589
Lobodon	*carcinophagus*	F	Adult	19404613	0.0262	0.0060	-0.0654	-0.0478
Lobodon	*carcinophagus*	F	Young	18464519	0.0138	0.0214	-0.0045	-0.0614
Lobodon	*carcinophagus*	F	Adult	1959289	0.0233	0.0286	-0.0583	-0.0554
Lobodon	*carcinophagus*	M	Young	18464520	0.0099	0.0438	-0.0346	-0.0668
Lobodon	*carcinophagus*	M	Adult	1959282	0.0452	0.0161	-0.0659	-0.0645
Mirounga	*angustirostris*	F	Young	19661O243	0.0486	-0.0213	0.0304	-0.0346
Mirounga	*leonina*	?.	Young	19381232O	0.0780	-0.0441	0.0638	-0.0575

Mirounga	leonina	F	Adult	1957775	0.0533	−0.0016	−0.0201	−0.0538
Mirounga	leonina	F	Subadult	1943162	0.0736	−0.0137	0.0257	−0.0542
Mirounga	leonina	M	Adult	1939520I	0.0838	−0.0559	−0.0888	−0.0702
Mirounga	leonina	M	Young	19452038	0.0752	−0.0438	0.0718	−0.0433
Monachus	monachus	?	?	K7781	0.0068	0.0233	−0.0132	−0.0080
Monachus	monachus	?	Young	1892117I	0.0206	−0.0204	0.0416	−0.0016
Monachus	monachus	?	Young	1892117I	0.0206	−0.0204	0.0416	−0.0016
Monachus	monachus	F	Adult	18947272	0.0397	0.0135	−0.0181	0.0005
Monachus	monachus	M	Adult	1863411	0.0391	0.0051	−0.0205	−0.0123
Monachus	schauinslandi	M	Young	19581126I	0.0288	0.0288	0.0060	0.0102
Neophoca	cinerea	M	Adult	019391212122	−0.1100	−0.0073	0.0235	−0.0016
Odobenus	rosmarus	?	Young	K7499	−0.0227	−0.1716	0.0147	0.0201
Odobenus	rosmarus	?	Young	K7481	−0.0075	−0.2030	−0.0275	0.0213
Odobenus	rosmarus	F	Adult	K7501	−0.0080	−0.1977	−0.0573	−0.0021
Odobenus	rosmarus	F	Foetus	K7503	−0.0744	−0.1660	0.1332	−0.0306
Odobenus	rosmarus	F	Young	K7490	0.0068	−0.1540	0.0691	−0.0127
Odobenus	rosmarus	M	Adult	K7495	−0.0090	−0.1626	−0.0399	0.0174
Odobenus	rosmarus	M	Adult	K7483	0.0109	−0.2095	−0.0451	−0.0019
Ommatophoca	rossi	M	Adult	1961224 3	0.0666	0.0406	0.0107	−0.1071
Ommatophoca	rossi	M	Adult	190822049	0.0685	0.0417	0.0216	−0.1098
Otaria	byronia	F	Young	01931121I8	−0.0942	−0.0133	0.0451	−0.0095
Otaria	byronia	F	Adult	019391219O	−0.1193	−0.0036	−0.0427	0.0046
Otaria	byronia	M	Young	019507211I	−0.1028	−0.0150	0.0208	−0.0081

Appendix 12.1 (*cont.*)

Genus	Species	Sex	Age	Specimen	PC₁	PC₂	PC₃	PC₄
Otaria	*byronia*	M	Young	019391211O	-0.1082	-0.0535	0.0667	-0.0106
Otaria	*byronia*	M	Young	019082053	-0.1281	-0.0105	0.0279	0.0195
Otaria	*byronia*	M	Adult	0193912168	-0.1007	-0.0547	-0.0594	-0.0166
Otaria	*byronia*	M	Adult	K7030	-0.1038	-0.0680	-0.0734	-0.0168
Phagophilus	*groenlandica*	M	Adult	1963719I	0.0438	0.0264	-0.0306	0.0472
Phagophilus	*groenlandica*	M	Young	1938I262	0.0450	0.0246	-0.0019	0.0374
Phoca	*groenlandica*	?	Young	1843I076	0.0383	0.0082	0.0531	0.0284
Phoca	*hispida*	F	Adult	193710232	0.0411	0.0330	0.0241	0.0337
Phoca	*hispida*	F	Adult	1938I264	0.0701	0.0483	0.0155	0.0251
Phoca	*hispida*	M	Young	193710234	0.0602	0.0263	0.0392	0.0313
Phoca	*hispida*	M	Adult	193710231	0.0564	0.0283	0.0194	0.0325
Phoca	*hispida*	M	Young	193710234	0.0602	0.0263	0.0392	0.0313
Phoca	*largha*	F	Adult	1965719I2	0.0597	0.0412	-0.0068	0.0175
Phoca	*largha*	F	Adult	1965719II	0.0495	0.0453	-0.0250	0.0290
Phoca	*largha*	F	Subadult	1965719I5	0.0494	0.0496	-0.0053	0.0207
Phoca	*largha*	M	Adult	1965719I3	0.0305	0.0290	-0.0338	0.0316
Phoca	*vitulina*	?	Young	K8087	0.0484	0.0214	0.0587	0.0147
Phoca	*vitulina*	?	?	K8092	0.0505	0.0086	-0.0257	0.0317
Phoca	*vitulina*	?	?	K80863	0.0447	0.0284	-0.0040	0.0137
Phoca	*vitulina*	?	Young	184632327	0.0282	0.0241	0.0290	-0.0001
Phoca	*vitulina*	M	Adult	184732238	0.0356	0.0244	0.0063	0.0242
Phoca	*vitulina*	?	Young	18863182	0.0386	0.0207	-0.0030	0.0166
Phoca	*vitulina*	?	?	K8I73	0.0554	0.0166	-0.0223	0.0218
Phoca	*vitulina*	?	Infant	1004f	0.0324	0.0039	0.0658	0.0118
Phoca	*vitulina*	M	Adult	329I	0.0402	0.0023	-0.0161	0.0237
Pusa	*caspica*	?	?	K824I	0.0634	0.0652	-0.0286	0.0261

Pusa	caspica	F	Adult	1963T1910	0.0488	0.0800	−0.0432	0.0342
Pusa	caspica	F	Adult	1965T192	0.0434	0.0798	−0.0356	0.0394
Pusa	caspica	M	Adult	1965T191	0.0434	0.0704	−0.0436	0.0415
Pusa	caspica	M	Adult	1963T1914	0.0584	0.0563	−0.0369	0.0543
Pusa	hispida	?	?	K8205	0.0613	0.0308	0.0152	0.0444
Pusa	hispida	?	?	K8201	0.0770	0.0248	0.0637	0.0016
Pusa	sibirica	F	Adult	1963T198	0.0597	0.0447	−0.0068	0.0371
Pusa	sibirica	F	Adult	1965961	0.0537	0.0536	−0.0179	0.0381
Pusa	sibirica	M	Adult	1963T199	0.0645	0.0597	−0.0175	0.0236
Pusa	sibirica	M	Young	1965T194	0.0524	0.0311	0.0134	0.0326
Pusa	sibirica	M	Young	1965T193	0.0499	0.0367	0.0169	0.0175
Pusa	sibirica	M	Adult	1965962	0.0605	0.0446	−0.0281	0.0401
Zalophus	californianus	F	Adult	01903101I4	−0.1148	0.0572	−0.0443	0.0172
Zalophus	californianus	M	?	K7122m	−0.1021	0.0145	−0.0391	0.0117
Zalophus	californianus	M	Young	01898IIII	−0.1278	0.0267	0.0250	0.0177
Zalophus	californianus	M	Young	01952827I	−0.1151	0.0467	0.0126	0.0199

13

Tiptoeing through the trophics: geographic variation in carnivoran locomotor ecomorphology in relation to environment

P. DAVID POLLY

Introduction

How do communities and species respond to environmental change? For the palaeontologist, the answer to this question is key to addressing its converse: how can we measure palaeoenvironmental change from fossil species and assemblages? This paper examines the association between community-level carnivoran locomotor morphology and climatic parameters to determine whether the average locomotor habits of carnivoran communities are associated closely enough with vegetation cover, topography, and related climatic factors to be used as an independent estimator of palaeoenvironment.

Community-level morphology has the potential to be a powerful indicator of climate. When a particular morphological feature mediates between an organism and its environment – the structure of the foot in relation to the substrate, for example – the average morphology of that feature can be expected to follow whatever environmental gradient is most closely associated with its function (Valverde, 1964; Fortelius *et al.*, 2002). Such a distribution will arise by the effects of climate on individual species, either through local adaptation (evolution by natural selection), by geographic range sorting (migration to more palatable regions), by extinction (Hughes, 2000; Lister, 2004; Davis *et al.*, 2005), or by the interaction of adaptation and range changes (Holt, 2003). All three kinds of species-level change will affect the community's composition and, therefore, the mean morphology of the community. The cumulative effect of climate on the community's mean morphology is likely to be more predictable than the effect on any one species.

Carnivoran Evolution: New Views on Phylogeny, Form, and Function, ed. A. Goswami and A. Friscia. Published by Cambridge University Press. © Cambridge University Press 2010.

In principle, climate change could be measured from the morphology of individual species as it responds adaptively to changing conditions. In practice, however, the adaptive changes are often so small and so tempered by geographic range shifts and extinction that attempts to detect morphological responses, even to large-scale changes like glacial–interglacial cycles, have been frustratingly ambiguous. For example, the morphology of marmot molars (*Marmota*, Sciuridae, Rodentia) is significantly associated in the modern world with diet, local vegetation cover, precipitation, temperature, and elevation, presumably as the result of range sorting and local adaptation (Caumul and Polly, 2005). Yet, no response in marmot tooth shape to Pleistocene climate cycles has been found in lineages that pass stratigraphically through glacial–interglacial episodes (Polly, 2003; Barnosky *et al.*, 2004). The association between phenotype and any one climatic parameter is often so weak and the changes so small, even over hundreds of thousands of years, that the morphology of a single species is only a feeble indicator of climatic change.

At the community level, however, adaptive responses in many species combine with environmentally driven changes in community composition to produce what should be a strong indicator of climate and environment (Valverde, 1964; Legendre, 1986; Brown and Nicoletto, 1991; Montuire, 1999; Millien *et al.*, 2006). In principle, every species in a community responds to climate change through phenotypic adaptation, migration, or extinction (Thompson, 2005). The effects of climate change on a phenotypic trait that can be measured in many members of a community are, thus, likely to be amplified by the combined signals from adaptive change in those species and from the gain and loss of community members that do not or cannot adapt to the new climate. In New World forests, for example, leaf margin characteristics, when averaged across species in reasonably diverse floras, are good indicators of mean annual temperature, even though the same metric in individual species and low-diversity assemblages is not (Wilf, 1997). Likewise, in mammalian herbivore communities, average molar tooth crown height is significantly correlated with mean annual precipitation (Janis and Fortelius, 1988; Damuth and Fortelius, 2001; Damuth *et al.*, 2002; Fortelius *et al.*, 2002). Note that a unified adaptive response to climate change is not to be expected when the morphological trait is linked to competitive displacement (Dayan and Simberloff, 1996, 2005).

The study of community-wide patterns in morphological traits in relation to environment will be referred to here as 'community' or 'faunal' ecomorphology. Faunal ecomorphology is, necessarily, the study of a limited number of species from a larger ecological community: morphological traits are unlikely to be shared by all species (no morphological traits are shared between plants and animals, for example). Faunal ecomorphology is, therefore, the study of traits in a subset of a community, perhaps an ecological guild whose members share a common way of life, perhaps a taxonomic group whose members share a common anatomical plan due to common ancestry, or perhaps those members of a taxonomic group that belong to the same guild. Such a restricted definition of 'community' is explicit or implicit in nearly all studies of ecomorphology (e.g. Valverde, 1964; Van Valkenburgh, 1985; Wilf, 1997; Fortelius et al., 2002; Dayan and Simberloff, 2005.).

This study looks at faunal ecomorphology of the locomotor system of living carnivorans (Carnivora, Mammalia) in the context of modern climate. It explores whether limb mechanics in carnivorans, specifically the 'gear ratio' of the calcaneum, is correlated with vegetation cover, elevation, precipitation, temperature, or ecoregion. Foot posture and locomotor efficiency are expected to be associated with the openness of an environment and, therefore, with those climatic factors that affect its openness. A highly digitigrade posture – in which the carpus and tarsus are positioned well above the substrate and the body weight is supported through the ends of the metapodials – is associated in carnivorans with large body size and cursorial habits, features more likely found in species that inhabit prairies, steppes, and deserts. A more plantigrade posture – in which the carpus and tarsus rest directly on the substrate and transmit some portion of the animal's weight – is often associated with arboreal or generalised ambulatory locomotion characteristic of species inhabiting wooded or forested environments. This study assesses three morphological indices to determine their comparative effectiveness for representing locomotor ecomorphology. The favoured index, the calcaneal gear ratio (calcaneum length/sustentacular position), is then measured in all North American species of Carnivora. The geographic distribution of the gear ratio is estimated by breaking the continent into 50 km² points, determining which carnivoran species are found in each point, and calculating the mean for each 50 km fauna. The resulting ecomorphological data is then tested for correlation with climatic factors to determine whether calcaneal ecomorphology can be used as a proxy for aspects of climate that impact terrestrial ecosystems.

In this paper, the terms climate and environment will be used almost interchangeably to refer to the six factors chosen for study: precipitation, ambient temperature, elevation, latitude, vegetation cover, and ecoregion. Where a distinction is made between the two terms, climate refers to the first four factors and environment refers to all six. The restricted usage of these terms is merely for convenience of expression in this paper and is not meant to imply either that climate and environment are identical or that other climatic or environmental factors are not important, although precipitation and temperature are often viewed as the two most important climatic determinants of terrestrial ecosystems because of their strong effect on the physiognomy of local vegetation cover (Whittaker, 1970; Bailey, 1998).

Data and methods

Taxa

Data were collected from 135 extant species of Carnivora (Appendix 13.1). This sample represents 47% of living carnivoran species (*sensu* Wilson and Reeder, 2005). The 415 skeletons measured in this study are housed in the William R. Adams Zooarchaeology Lab (Indiana University, Bloomington, IN, USA), the Indiana State Museum (Indianapolis, IN, USA), the Field Museum of Natural History (Chicago, IL, USA), the American Museum of Natural History (New York, NY, USA), the University of Michigan Museum of Zoology (Ann Arbor, MI, USA), and the Naturhistoriska Riksmuseet (Stockholm, Sweden). Where possible, measurements were collected on at least two individuals from each species, male and female, to minimise the effects of sexual dimorphism.

The 45 North American species (asterisked taxa in Appendix 13.1) were used to investigate geographic patterns of locomotor ecomorphology. This sample represents 94% of the 48 terrestrial carnivoran species living on the continent. The three missing species are confined to Central America: *Bassaricyon lasius* (Harris's Olingo), *Procyon gloveralleni* (Barbados raccoon), and *Spilogale pygmaea* (Pygmy spotted skunk). These missing species are less than 1% of the North American 50 km grid points (37 out of 9699 points: *B. lasius*: 1 point; *P. gloveralleni*, 1 point; and *S. pygmaea*, 35 points). See below for explanation of the grid point system.

Osteological measurements

Six measurements were taken from the hind limb: (1) maximum proximodistal length of the femur, from ball to condyle; (2) maximum proximodistal length of metatarsal III; (3) length of the distal calcaneum, from centre of the calcaneoastragalar facet to cuboid facet; (4) length of the proximal part of the calcaneum, from medial tubercle to anteroposterior centre of the calcaneoastragalar facet; (5) maximum length of the calcaneum, from medial tubercle to cuboid facet; and (6) position of the sustentacular facet, from medial tubercle to distal margin of the sustentacular process where it intersects the body of the calcaneum (Figure 13.1).

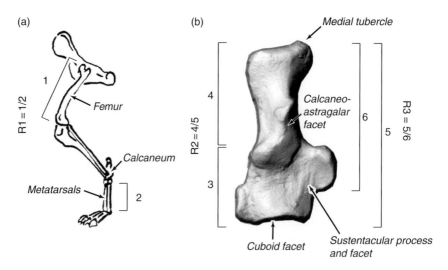

Figure 13.1 Osteological measurements. a, Hind limb. (1) Maximum length of the femur. (2) Maximum length of the third metatarsal. b, Calcaneum, dorsal view. (3) Maximum length of the calcaneum. (4) Length from proximal end of the calcaneum to the centre of the calcaneoastragalar facet. (5) Length from the centre of the calcaneoastragalar facet to distal end of the calcaneum. (6) Length from the proximal end of the calcaneum to distal margin of the sustentacular process.

Ecomorphological indices

Three ratios were calculated from the six osteological measurements. The first ratio (R1) is the metatarsal III to femur ratio (measurement 2 / measurement 1), a classic measure of digitigrady (e.g. Gregory, 1912; Garland and Janis, 1992). The second ratio (R2) is length of the distal to proximal calcaneum (measurement 3 / measurement 4). The third ratio (R3) is the length of the calcaneum to the position of the sustentacular facet (measurement 5 / measurement 6). Ratios R2 and R3 are 'gear ratios' (*sensu* Carrier *et al.*, 1994; Gregersen and Carrier, 2004) related to the lever mechanics of the foot, where the proximal tuber of the calcaneum forms the in-lever for plantarflexion of the foot, the distal part of the calcaneum forms the out-lever along with the metatarsals and phalanges, and the astragalus forms the fulcrum; the two ratios are derived from the two contact points between astragalus and calcaneum at the calcaneoastragalar and sustentacular facets. The last ratio (R3) is derived from a 3D analysis of carnivoran tarsals that found that the position of the sustentacular process is one of the primary components of variance in calcaneum shape and is highly correlated with locomotor stance (Polly, 2008). All three ratios get larger with increasing 'digitigrady'.

Fifty km grids

Sampling points were laid out in a 50 km grid across the whole of North America. A global 50 km grid was first created by laying out points every 0.4491574° latitude along the equator (based on the circumference of the Earth estimated at 40,075.04 km), then moving north and south by the same number of degrees and decreasing the longitudinal spacing proportional to the sin of the latitude. This procedure assumes that the Earth is perfectly spherical, which it is not. Nevertheless, spacing between points was close to 50 km, off at most by 2 or 3 km based on spot checks on Ellesmere Island using ArcGIS measuring tools and Clark 1966, WGS 1984, and North Pole Azimuthal Equidistant projections. North American points were extracted from the global grid by clipping it with the continental outline supplied in the ESRI 2006 World base map data series. There were 9699 grid points in North America.

Species ranges

Geographic ranges of North American carnivorans were taken from the digital species range data set provided by NatureServe (those data were produced by a collaboration between Bruce Patterson, Wes Sechrest, Marcelo Tognelli, Gerardo Ceballos, The Nature Conservancy Migratory Bird Program, Conservation International CABS, World Wildlife Fund US, and Environment Canada WILDSPACE; Patterson *et al.*, 2005). These data were compiled from published scientific sources, notably including Hall (1981) and Wilson and Ruff (1999) (the complete list can be obtained with the data). The geographic ranges in the data set are historical and include areas where species have since been extirpated, although the ranges of some species, such as the wolf, were probably even more extensive in pre-Columbian times. Ranges of species that extend into South America were clipped using the North American continental outline as was done with the grid points. The carnivoran fauna for each North American grid point was tabulated by intersecting it with the species range polygons. Introduced species were excluded. The number of species in each grid point is shown in Figure 13.2a.

Locomotor categories

All species were assigned to a locomotor and posture category (Appendix 13.1). Three posture categories were used: plantigrade (in which the distal heel is in contact with the ground at normal resting stance and during walking locomotion), semi-digitigrade (in which the heel may or may not be in contact with the ground during normal walking locomotion), and digitigrade (in which the heel is well above the ground in normal resting

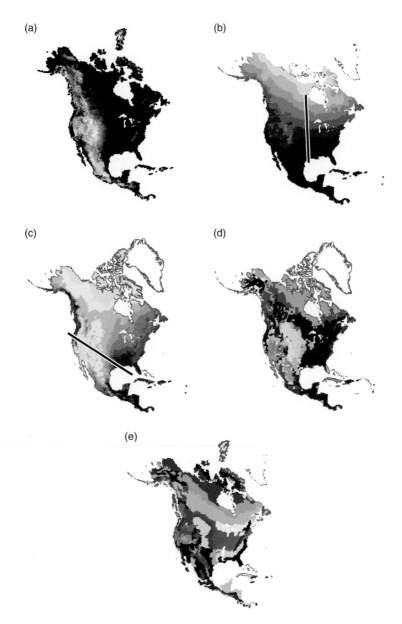

Figure 13.2 For colour version, see Plate 11. Fifty km grid point data. a, Elevation, divided into 10 shading categories using Jenk's natural breaks algorithm. The lowest elevation is darkest (1–173 m), the highest is lightest (2596–3660 m). b, Mean annual temperature, divided into 10 shading categories using Jenk's natural breaks algorithm. The lowest temperature is lightest (−19.9 to −12.5°C) and the highest is darkest (21.4 to 28.6°C). The black line emphasises the north–south gradient in temperature. c, Mean annual precipitation, divided into 10 shading categories using Jenk's natural breaks algorithm. The lowest precipitation is lightest (49.1–257.4 mm) and the highest

stance and does not come into contact with the ground during normal locomotion). These posture types have arbitrary boundaries and are best viewed as a continuum (Carrano, 1997). In this study, locomotion and posture categories were used for the sole purpose of comparing the effectiveness of continuous osteological indices.

Whenever possible, locomotion and posture categories followed Carrano (1997) and Van Valkenburgh (1985). Species that were not studied by those authors were categorised using descriptions and photographs from the following sources: Paradiso and Nowak, 1972; Ewer, 1973; Mech, 1974; Clark, 1975, 1987; Bekoff, 1977; Chorn and Hoffmann, 1978; Powell, 1981; Roberts and Gittleman, 1984; Berta, 1986; Goldman, 1987; Poglayen-Neuwall and Toweill, 1988; Nellis, 1989; Goldman and Taylor, 1990; Baker, 1992; Cavallini, 1992; Van Rompaey and Colyn, 1992; Sillero-Zubiri and Gottelli, 1994; Larivière and Walton, 1997; Murray and Gardner, 1997; Storz and Wozencraft, 1999; Larivière, 2001a, 2001b, 2005; Larivière and Calzada, 2001; Belcher and Lee, 2002; Hwang and Larivière, 2003; Walton and Joly, 2003; Yensen and Tarifa, 2003a, 2003b.

Elevation

Elevation was estimated for each 50 km grid point from the Terrain-Base data set (Row and Hastings, 1994). TerrainBase contains elevation and ocean depth data in metres from mean sea level at 5-min grid resolution. An elevation was assigned to each 50 km grid points from the value of the nearest neighbour point in the TerrainBase data. The resulting elevations are rendered in Figure 13.2a.

Precipitation and temperature

Monthly and annual mean air temperature and precipitation were estimated for each 50 km grid point from the database of Willmott and Legate (1998). Willmott and Legate interpolated their data from weather stations (24,941 for temperature and 26,858 for precipitation) to a 0.5 × 0.5 degree grid

Figure 13.2 (*cont.*)
precipitation is darkest (2988.8–5239 mm). The black line emphasises the east–west gradient in precipitation. d, Matthews' vegetation cover, with a categorical shading scheme that is roughly ordered from densest vegetation in dark (tropical evergreen forest and subtropical evergreen seasonal broad-leaved forest) to sparsest in white (ice). e, Bailey's ecoregion provinces, with an arbitrary shading scheme.

using Shepard's distance-weighting method. Data were assigned to the 50 km grid points using the value of the nearest-neighbour point in Willmott and Legate's data. The resulting mean annual temperature and precipitation are rendered in Figures 13.2b and 13.2c, respectively.

Vegetation

Local vegetation cover was estimated for each 50 km grid point from Matthews' Global Distribution of Vegetation data set (Matthews, 1983, 1984). This vegetation data set reports dominant vegetation cover at 1° resolution categorised using the UNESCO forest classification system which divides vegetation cover into 31 categories such as tropical evergreen rainforest, cold-deciduous forest with evergreens, xeromorphic shrubland, or desert. The Matthews data set classifies vegetation by what existed prior to human modification to the extent that historical data exist. Data were assigned to the 50 km grid points using the value in the nearest-neighbour point in the vegetation data set. The vegetation data are rendered in Figure 13.2d.

Ecological regions

Each 50 km grid point was assigned to one of Bailey's North American ecoregions (1998, 2005). These ecoregions are macroscale climatic areas defined primarily by seasonal interactions between temperature and precipitation and secondarily by dominant vegetation type. The regions are hierarchically arranged into Domains (4 in North America), Divisions (28 in North America), and Provinces (59 in North America). For example, the eastern Kansas prairies belong to the humid temperate domain, the prairie division, and the forest-steppes and prairies division, whereas the east-central Texas prairies just to the south of the ones in Kansas belong to the humid temperate domain and the prairie division, but to the prairies and savannas province. Bailey's ecoregion system, especially its larger hierarchical categories, is derived from earlier work by Köppen (1931), Dice (1943), and Trewartha (1968). Ecoregions were assigned to the 50 km grid points by intersecting them with the ecoregion GIS layer available from the USDA Forest Service. The ecoregion data are rendered in Figure 13.2e.

Statistical analysis

The relative effectiveness of the three ecomorphological index ratios was tested using analysis of variance (ANOVA). The mean of each index was calculated for each of the 134 species in the larger data set and the indices were tested for association with locomotion and posture using ANOVA (index as dependent variable, locomotor category as independent variable) and an F-test for significance. The indices were compared using adjusted R^2, the proportion of ecomorphological variance explained by the locomotor factor. An ANOVA using taxonomic family as the independent categorical variable was also performed to explore the relationship between the indices and phylogeny, although phylogeny was not a primary concern of this paper.

Carnivoran faunal regions were estimated by clustering the 50 km grid points based on the species present in each following the method of Heikinheimo *et al.* (2007). The fauna of each grid point was characterised with a vector of o s and 1 s indicating absence or presence, respectively, of each of the 48 North American species. The points were then clustered based on Euclidean distances using an optimisation method that first builds a k set of objects, clustering around them until an optimal arrangement is found. Clustering was done twice, once using a significance test to find the best number of clusters given the data set, and a second time forcing there to be an arbitrary number of 10 clusters. The silhouette test statistic, which measures how well an item is assigned to its cluster, was used to find the optimal number of clusters in the first run (Kaufman and Rousseeuw, 1989). The silhouette method divides the data into k clusters where k is the number that maximises the silhouette test statistic, which is the average distance from one point to other points in other clusters minus the average distance to points in the same cluster over whichever of the two averages is larger.

Calcaneal gear ratio (calcaneum length/sustentacular position index ratio, R3; the index that performed best in the above tests) was averaged for every fauna at the 50 km grid points. The standard deviation was also calculated as a measure of local faunal diversity in gear ratio (points with only one species present were excluded).

The geographic distribution of mean calcaneal gear ratio (R3) was tested for association with species number, elevation, mean annual temperature, mean annual precipitation, local vegetation cover, and ecoregion province using the product–moment correlation (R) and R^2 for continuous variables and an ANOVA-derived R^2 for categorical variables. Continuous variables were transformed to make their distribution approximately normal as follows: species number was squared, elevation was transformed to its square root, temperature was not transformed, and precipitation was transformed to its natural log.

Randomisation and bootstrap were used to calculate significance and confidence intervals for R and R^2. The calcaneal gear ratio index (R3) was randomised among the 50 km grid points and the two statistics recalculated 1000 times to generate a distribution with no real association. The maximum of this distribution was taken to be the cutoff for significant difference from zero (a more stringent cutoff than taking the 95th percentile value). Confidence intervals on R and R^2 were determined by bootstrapping. Data were resampled with replacement and the two statistics were recalculated 1000 times. The 50th and 950th values of the resulting distribution were taken to be the confidence intervals, giving the 5% and 95% intervals, respectively.

Results

Ecomorphological indices

The three indices had a small and only sometimes significant association with locomotor category (Figures 13.3a–c, Table 13.1). Locomotor category

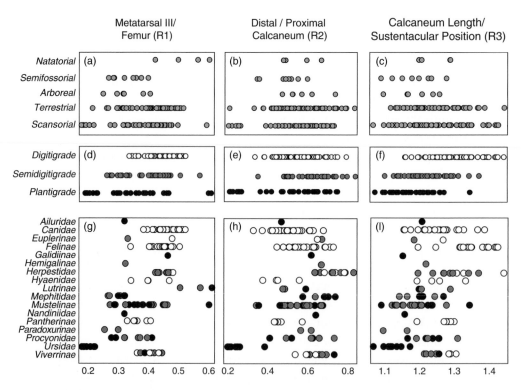

Figure 13.3 Association of the three index ratios with locomotor category (a–c), posture category (d–f), and taxonomic family (g–i). Colour coding in a–c distinguishes the five locomotor categories. Colour coding in d–i distinguishes the three posture categories (red, digitigrade; gold, semi-digitigrade; green, plantigrade).

Table 13.1 ANOVA results for three index ratios against the categorical variables locomotion, posture, and family. Adjusted R^2 is the ratio of the variance explained by the model to the total variance in the index.

	Error d.f.	F-ratio	P	R^2
Metatarsal III/Femur (R1)				
Locomotion	130	6.59	<0.000	0.15
Posture	132	23.7	<0.000	0.25
Family	118	22.86	<0.000	0.73
Distal/Proximal Calcaneum (R2)				
Locomotion	130	0.79	0.529	−0.01
Posture	132	9.41	<0.000	0.11
Family	118	15.23	<0.000	0.63
Calcaneum Length/Sustentacular Position (R3)				
Locomotion	130	2.53	0.043	0.04
Posture	132	37.05	<0.000	0.35
Family	118	12.35	<0.000	0.58

explained 15% of the variance in the metatarsal-to-femur index (R1) and 0% and 4% in the two calcaneal gear ratios (R2 and R3, respectively). R1 and R3 had statistically significant associations with locomotor category. Note that both the terrestrial and scansorial categories have tremendous variance in all three ratios, which explains the poor association between that ratio and locomotor category. The natatorial category drives the significance of the femur-to-metatarsal ratio.

The association between the ratios and posture was higher, with 25%, 11%, and 35% of the variance explained in R1 through R3, respectively (Figures 13.3d–f, Table 13.1). R1 and R3 were linearly related to the gradation from plantigrade through semi-digitigrade to digitigrade, which is a heuristically useful property of these ratios. In contrast, R2 was higher in semi-digitigrade species than in digitigrade or plantigrades. One of the outliers with a high metatarsal-to-femur ratio (R1) in the plantigrade category is *Enhydra lutris*, the sea otter, which has elongated feet specialised for paddling.

All three indices were strongly associated with taxonomic group (family membership), especially the metatarsal-to-femur ratio (R1) where family membership explained 73% of its variance (Figures 13.3g–i, Table 13.1). A high phylogenetic correlation does not in itself diminish the effectiveness of the indices, because locomotion and posture are themselves correlated with phylogeny (e.g. canids as a group are generally more specialised for cursoriality than are procyonids). In this study, it is the taxon-free correlation between mean locomotor ecomorphology and environment that is of interest, regardless of whether faunal sorting is phylogenetically biased because entire clades share certain specialisations.

Because the second calcaneal gear ratio (calcaneum length/sustentacular position, R3) had a high correlation with posture and a low correlation with taxonomic group, this ratio was used as the index of ecomorphology for the remaining analyses. Not only is the calcaneal gear ratio a good predictor of posture, but it can also be measured from a single bone, the calcaneum, that is frequently preserved in the fossil record. Even though R1 performed well as an indicator of posture, it is less practical for use in palaeontology because few fossil specimens have an unbroken femur and metatarsal III.

Interestingly, the three ratios are significantly correlated with one another, but only loosely so (Figure 13.4). The calcaneal gear ratio (R3) has the strongest

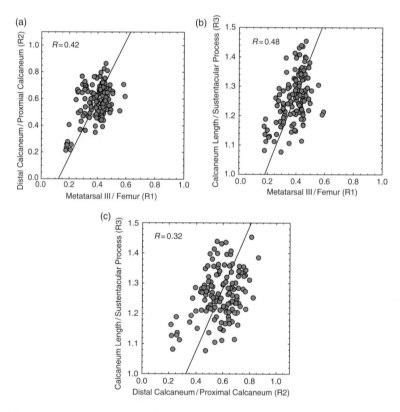

Figure 13.4 Associations among the three index ratios. a, Metatarsal III to femur (R1) and distal to proximal calcaneum length (R2). b, R1 and calcaneum length to sustentacular process position (R3). c, R2 and R3. The correlation coefficient (R) and a major axis regression line are presented for each pair of ratios.

correlation with metatarsal-to-femur ratio (R2). The two calcaneum ratios are more poorly correlated with each other than either is to the metatarsal-to-femur ratio.

Species richness and geographic clustering of carnivoran faunas

Of the 9699 50-km grid points in North America, 88% have carnivorans. The only places without carnivorans are parts of the northernmost Canadian islands and Greenland. The number of species per point ranged from 1 to 20 with a mean of 11.32 species (Figure 13.5a).

The greatest diversity was found in the mountains of Central America, the mountains of the Pacific Northwest, and the mixed coniferous–deciduous forests west of Lake Superior (red and orange). Species richness in these areas is probably enhanced by mixing of species from adjacent ecological domains, between the polar and humid temperate domains and between the dry and humid tropical domains (see Bailey, 1998 and Figure 13.2e for domain boundaries). The lowest species diversity was found in the high arctic and, secondarily, in the northern Midwest broadleaved forests, the northern high plains, and the desert southwest.

Two carnivoran faunal clusters were identified when the number of clusters was limited by the silhouette test statistic (Figure 13.5b). These two regions

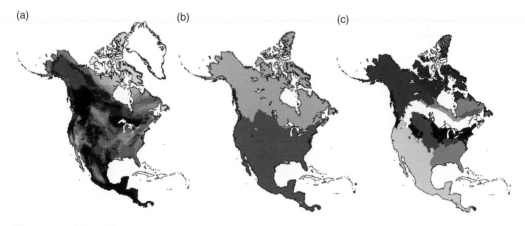

(a)   (b)   (c)

Figure 13.5 For colour version, see Plate 12. Carnivoran species diversity and faunal regions. a, Number of carnivoran species, ranging from light (1 species) to dark (20 species). b, Faunal regions based on carnivoran species using *k*-means clustering of 50 km grid points and the silhouette test statistic, which identified two clusters. c, Faunal regions based on carnivoran species using *k*-means clustering of 50 km grid points with the number of clusters forcibly set at 10.

roughly correspond to the polar domain ecoregion (Bailey, 1998) and the rest of North America. The polar cluster had two outlying areas in the Yellowstone region of Wyoming (due to shared species like *Lynx canadensis* and *Martes americana*) and in the Mississippi River delta (due to lack of species characteristic of the more southern cluster like *Bassariscus astutus*, *Lynx rufus*, *Mephitis mephitis*, and *Urocyon cinereoargenteus*).

When the number of faunal clusters was manually increased to 10, many of their boundaries corresponded with major climatic gradients (Figure 13.5c). The southeastern United States was separated from the more arid western regions (cf. Figure 13.2c), the northern part of the continent is divided latitudinally along the mean annual temperature gradient (cf. Figure 13.2b). Many of the cluster boundaries were, in fact, similar to the boundaries of ecoregion provinces (cf. Figure 13.2e), which are based on the interaction of temperature and precipitation, confirming that carnivore species assemblages are correlated with the same climatic factors as vegetation physiognomy.

Calcaneal gear ratio in North American species

Calcaneal gear ratio (calcaneum length/sustentacular position, $R3$) in the 44 North American species is shown in Figure 13.6. With exceptions, this ordering roughly corresponds to digitigrady as observed in living animals: canids and felids cluster towards the high end of the spectrum and ursids and mustelines towards the low end. The species with the lowest gear ratio are two bears and two weasels; the highest values were cats. The mean and standard deviation for North American species were 1.25 and 0.07.

Geographic variation in average calcaneal gear ratio

The geographic distribution of mean calcalcaneal gear ratio is shown in Figure 13.7a. The colouring on this map was produced by classifying each point into one of 14 bins using Jenks natural breaks algorithm (Jenks, 1977). The minimum mean gear ratio for a 50 km fauna was 1.148, the maximum was 1.1.31, and the mean was 1.25. The general pattern is clear: mean calcaneal gear ratio is highest in Mexico and the desert Southwest and lowest in the arctic and boreal regions.

Variation of mean calcaneal gear ratio at each grid point is shown in Figure 13.7b. This map shows the standard deviation of $R3$ in each local fauna. Darker colours mean greater variation between species with the highest and lowest index in the fauna, lighter colours mean less variation. Variation in calcaneal gear ratio was highest in the west, which has a heterogeneous environment of mountain forests and desert valleys, in Central America, and

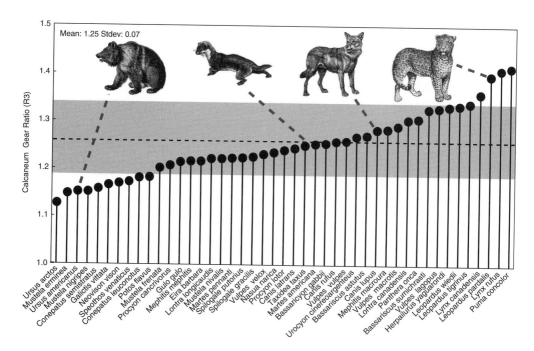

Figure 13.6 Rank order plot of calcaneal gear ratio (calcaneum length/sustentacular position, R3) in North American species. The mean is shown by the horizontal broken line and one standard deviation on either side of the mean is indicated by grey shading. Pictures of four representative species are shown. (Animal pictures are copyright-free illustrations from Large and Weller, 2004).

the high arctic. Variation was lowest in the broad-leaf forest regions of the midwest and northeastern Atlantic seaboard.

Correlation between calcaneal gear ratio and environmental factors

Mean calcaneal gear ratio was significantly correlated with all the factors tested (Table 13.2). The correlation with number of species (Figure 13.5a), elevation (Figure 13.2a), and mean annual precipitation (Figure 13.2c) were all low. None of these factors accounted for more than 10% of the geographic variance in ecomorphology. The 2 faunal clusters in Figure 13.5b explain only 31% of the variance in gear ratio, but the 10 clusters in Figure 13.5c explain 62%. Vegetation cover (Figure 13.2d) and mean annual temperature (Figure 13.2b) are both strongly correlated with calcaneal gear ratio, explaining 49% and 48% of its variance, respectively. The strongest correlation was with ecological province (Figure 13.2e), which explained 70% of the variance in mean calcaneal gear ratio.

Table 13.2 Association between mean calcaneal gear ratio (calcaneum length/ sustentacular position, R3) and environmental and faunal factors. *R* is the product-moment correlation and R^2 is the ratio of the variance explained by the factor. The significance cut-off is the maximum value of *R* when there is zero correlation.

	R	(95% CI)	Significance cut-off	R^2	(95% CI)
Number of species	0.22	(0.19–0.23)	0.037	0.05	(0.03–0.05)
Elevation	0.26	(0.24–0.27)	0.037	0.07	(0.06–0.07)
Mean annual temperature	0.69	(0.68–0.70)	0.049	0.48	(0.46–0.49)
Mean annual precipitation	0.12	(0.10–0.14)	0.029	0.01	(0.01–0.02)
Vegetation cover				0.49	(0.47–0.51)
Ecological province				0.70	(0.69–0.72)
Faunal clusters (*k* = 2)				0.31	(0.30–0.33)
Faunal clusters (*k* = 10)				0.62	(0.60–0.63)

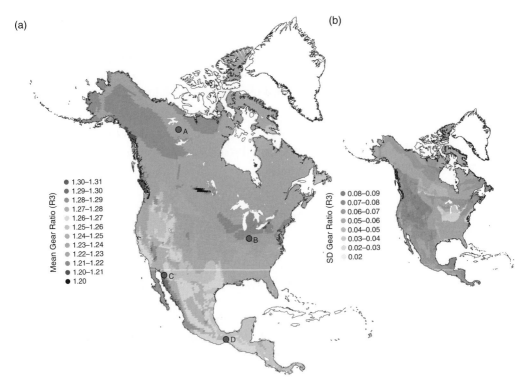

Figure 13.7 For colour version, see Plate 13. a, Mean calcaneal gear ratio (calcaneum length/ sustentacular position, R3) and b, its standard deviation for carnivoran species present at 50 km grid points across North America. The four lettered points (A–D) show the location of the data presented in Figure 13.8.

Discussion

Measuring locomotor ecomorphology

Carrano (1999) argued that categories for locomotor morphology, such as 'cursorial' and 'arboreal' or 'plantigrade' and 'digitigrade', are arbitrarily delimited and might better be treated as quantitative continua. Carrano's suggestion is adopted here by using a continuous quantitative ratio to measure locomotor ecomorphology that is related to but not identical with these standard categories. The results of this study seem to vindicate a continuous approach, because all three index ratios are significantly associated with the categorical posture variables, as expected on bio-mechanical grounds, and the ordering of species by these ratios broadly conforms to the way they would be ordered by a qualitative assessment of their digitigrady.

Carrano (1997) argued that single osteological measurements are unlikely to adequately characterise the full complexity of locomotor habit and foot posture and that multivariate data from several parts of the skeleton are more reliable. While Carrano is certainly correct that multivariate data will capture more of the subtle differentiation in locomotor habit, bivariate ratios were adopted here for practical reasons. Simple data can be collected quickly, which is an advantage for a study on continental scale. Within-species geographic variation was not measured in this study, but it should be, and when it is, the number of specimens for which data will need to be collected will increase drastically. More importantly, simple data can be collected from a larger number of specimens. Fossils especially are often fragmentary, meaning that only a limited number of measurements can be taken from any one individual. Limiting measurements to a single bivariate ratio helps maximise the sample that can be studied. Calcaneal gear ratio (R_3) is based on two measurements from the calcaneum happened to be the best predictor of posture category, but the choice of it over the more traditional metatarsal-to-femur ratio is strategic in that the calcaneum is a single bone that is more likely to be preserved intact in the fossil record than are the femur or metatarsals, thus further maximising the applicability to the fossil record.

Within-species variation in locomotor ecomorphology

Locomotor habit and posture are usually studied as properties of species rather than individuals. Nevertheless, some data exist about geographic differentiation in functionally relevant limb features.

Klein *et al.* (1987) found that hind foot length in the Reindeer, *Rangifer tarandus*, had a latitudinal cline associated with temperature, snowfall, and available forage. They found that leg length was correlated with migration distances and snowfall. Selectively, leg length was a trade-off: speed, efficiency,

Table 13.3 Within-species variation in calcaneal gear ratio (calcaneum length/ sustentacular position, R3). Sample size (*N*), mean, standard deviation (SD), and coefficient of variation (CV) are reported for species with sample size of ten or more.

Species	*N*	Range	Mean	SD	CV
Lynx rufus	10	(1.34–1.43)	1.41	0.030	0.02
Procyon lotor	42	(1.14–1.58)	1.24	0.109	0.09
Urocyon cinereoargenteus	49	(1.18–1.57)	1.27	0.058	0.05
Vulpes vulpes	28	(1.14–1.45)	1.26	0.059	0.05
Mustela frenata	26	(1.06–1.40)	1.20	0.068	0.06
Neovison vison	22	(1.08–1.35)	1.16	0.067	0.06

and ability to cope with deep snow all are likely to selectively favour longer legs, whereas growth costs, thermoregulation, and efficiency of foraging ground-cover plants are likely to favour shorter legs.

A similar study was carried out by Murray and Larivière (2002) on variation in canid foot size in relation to environment. These authors investigated the area of the foot to establish whether it was proportionally greater in areas with deeper, softer snow, which might be expected because of selection for a lower footload. They found that foot area was associated with latitude in coyotes, *Canis latrans*, and red foxes, *Vulpes vulpes*. They also found that foot area was correlated with snowfall in red foxes and arctic foxes, *Alopex lagopus*, but not in coyotes. These authors concluded that environmental conditions have had a selective within-species effect on canid feet, but that the effects have been weak, as one might expect given the short time for in-situ selection of local populations since the last deglaciation, and that selection on feet was not consistent in its effects among the species they investigated. Importantly for this study, canid feet are known to have some geographical variation in foot morphology that is at least weakly associated with aspects of climate.

How much within-species variation is there in calcaneal gear ratio (R3)? Most of the species samples in this study were too small to address the question, but six samples had enough individuals to provide some indication (Table 13.3). Most species had a gear ratio range around 0.3, a substantial range given that the range among species is about 0.45. If this within-species variation is itself sorted geographically, it may also be correlated with environmental variation, which is likely to strengthen the association between mean calcaneal gear ratio and climate that was found here when species were treated as geographically homogeneous. These issues deserve to be explored further in future studies.

Faunal ecomorphology in North American carnivorans

The similarity between the geographic distributions of mean calcaneal gear ratio (Figure 13.7a) and vegetation cover (Figure 13.2d) is striking. The boreal forest region, the maritime northwest, the Great Basin interior, the deciduous and coniferous forest regions of the southeastern US, the highlands of Mexico are all nearly as well delimited in the gear ratio map as in the vegetation map. The similarity between mean gear ratio and Bailey's ecoregion provinces is even closer. High mean gear ratio in the southwest of the continent is driven primarily by felids, the low digitigrady in the north by mustelids and ursids. Some regions seem like unusual outliers. Newfoundland, for example, has a much higher mean digitigrady than adjacent areas of mainland Canada, but this is an island artefact due to the smaller-bodied mustelids not being present in Newfoundland, thus raising the mean gear ratio.

The high variation in gear ratio within faunas in the arctic regions is due to the rather dimorphic faunas there, which are composed of extremely plantigrade mustelids and ursids along with extremely digitigrade canids and felids. The high variation in Central America is similarly due to the rather plantigrade procyonids and mustelids and the digitigrade felids with only a small number of mid-digitigrade species. The comparatively low variation in the mid-continent is due to a high proportion of mid-digitigrade species.

It is instructive to look at the details of the distribution of calcaneal gear ratio in representative areas (Figure 13.8). Point A in the Northwest Territories of Canada, is part of Bailey's Subarctic Division where the mean annual temperature is cold and seasonal and precipitation is constantly low, an open tundra and woodland environment. The carnivoran fauna is relatively diverse with 14 species from four families. This fauna has a comparatively low mean gear ratio, primarily driven by a large number of very plantigrade species, notably bears and mustelids. Point B in Kentucky is part of Bailey's Hot Continental Division, which has a warmer but still seasonal temperature, but much more precipitation spread evenly through the year, a closed deciduous environment. The carnivoran fauna is less diverse, with only nine species and three families. Mean gear ratio is middle of the range, with most species near the faunal mean. Point C is in Sonora, Mexico, part of the Tropical–Subtropical Desert Division. Temperature is very warm with some seasonality and precipitation is low and seasonal. This fauna, like Point A, has 14 species belonging to 4 families. Mean gear ratio is high, partly due to the large number of felids and canids, but also due to the absence of ursids and many plantigrade mustelids. Point D in Oaxaca, Mexico is part of the Savanna Mountains Division The temperature is warm without seasons, but the

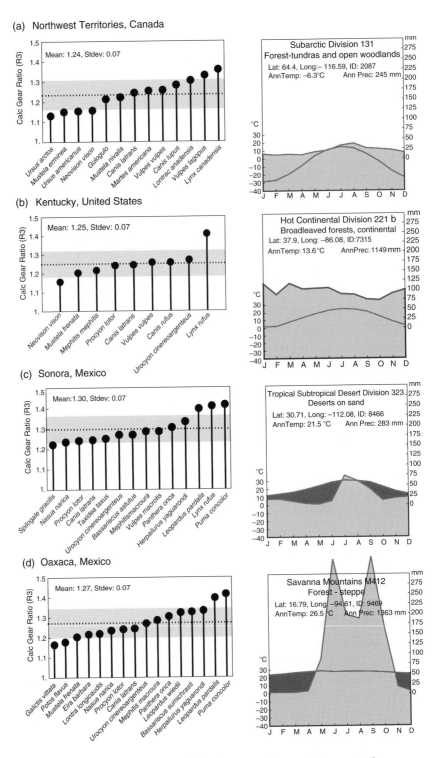

Figure 13.8 Species, gear ratio (R3), climate, and ecoregion at the four representative locations. a, Northwest Territories, Canada, a cold-climate

precipitation is extraordinarily seasonal, with heavy rains in the summer months and nearly none in the winter. The carnivoran fauna has 16 species belonging to 5 families. Three of the species do not have data, but the variance of the others is high, ranging from very digitigrade cats to very plantigrade arboreal procyonids.

The environmental significance of mean gear ratio

The geographic distribution of mean calcaneal gear ratio (R_3) has a clear, strong correlation with the distribution of established ecological regions. Bailey's (1998) ecological provinces, which are themselves derived from older ecoregion classifications, were able to explain 70% of the variance in mean gear ratio. These provinces are defined on the monthly interaction of temperature and precipitation throughout the year, and the interaction is clearly important for the distribution of mean gear ratio because mean annual temperature by itself only explained 48% of the variance in mean gear ratio and mean annual precipitation only 1%. (Note that these percentages were not derived from multiple or partial regression, which would take into account interactions among the climatic factors, meaning that the sum total of variance explained by all the factors is greater than 100%.)

Most likely, the distribution of mean calcaneal gear ratio is not directly influenced by temperature and precipitation. Rather, these factors probably exert an influence indirectly though a combination of local topographic relief, substrate, and openness of vegetation cover. Vegetation cover by itself is capable of explaining 49% of the variance in gear ratio, although elevation on a large scale only explained about 7%.

It is easy to worry that the geographic distribution of calcaneal gear ratio has nothing to do with whether certain locomotor styles are, on average, suited for

Figure 13.8 (*cont.*)
fauna with comparatively low digitigrady. b, Kentucky, USA, a temperate climate fauna with medium digitigrady. c, Sonora, Mexico, a hot-weather climate with high digitigrady. d, Oaxaca, Mexico, a seasonally wet tropical climate with comparatively high digitigrady. Left panels follow the conventions from Figure 13.6. Right panels show mean monthly temperatures (dark fill) and precipitations (light fill). Bailey ecoregion division and province, latitude, longitude, and mean annual temperature and precipitation are reported in each graph (see Figure 13.7 for locations).

locomotion in particular environments, but that it is the chance product of a geographic distribution of a species that is dictated by the environmental consequences on some other phenotypic system, such as features related to diet or body size. While it is true that the distribution of mean gear ratio is a direct product of the geographic distribution of the species, it is not the case that it will have the same geographic distribution as other morphological features that interact with the environment. For example, bears and weasels together contribute to low mean calcaneal gear ratio in the arctic (Figure 13.8a), but together would have quite a different effect on mean body mass because in the arctic they represent both largest and smallest carnivorans on the continent – their net effect on mean digitigrady is to lower it, but their effect on mean body mass would be to keep it close to the continental mean.

The only other faunal ecomorphological systems that have been studied in mammals are body mass and cheek tooth hypsodonty. The association of body mass with temperature is well known, codified as Bergmann's Rule that, on average, body mass will be greater at higher latitudes with colder temperatures (critically reviewed by Millien et al., 2006). While this pattern does not hold for lizards and snakes, whose body mass is higher in hotter climates (Makarieva et al., 2005), it holds for more than 70% of the mammal, bird, and salamander species that have been studied (Millien et al., 2006). The distribution of body mass within mammalian communities is also known to vary with biome type (Valverde, 1964; Legendre, 1986; Croft, 2001), even though the geographic distribution of the average body mass of communities has not been studied in the same way as calcaneal morphology is studied here.

More directly comparable to this study is the finding that average hypsodonty in mammalian herbivore guilds is correlated with mean annual precipitation (Damuth and Fortelius, 2001; Damuth et al., 2002). Fortelius et al. (2002) used this association as a basis for interpreting faunal averages in hypsodonty as a proxy for climate change (specifically for changes in aridity during the late Neogene). Interestingly, this correlation between tooth crown height and precipitation demonstrates that the different kinds of faunal ecomorphology may be correlated with different environmental factors. Calcaneal gear ratio had a poor correlation with precipitation, and so captures a different aspect of climate than does hypsodonty. Whether average dental adaptation in carnivores has an association with environment and whether it is the same as the association between calcaneal gear ratio and environment is an open question. Most likely diet is partitioned differently in a community than is relation to substrate, the latter being potentially common to all members whereas the former is likely partitioned among members.

Conclusions

Climate, environment, topography, and species interact in a geographic mosaic that influences species distributions, results in local morphological adaptation, and determines what species coexist in communities (Thompson, 2005). Common local environments will tend to homogenise environmentally relevant morphological features in all the species that live together in them, whereas community-level competition for resources will tend to differentiate species into mutually exclusive niches. Some environments are more tolerant of phenotypic diversity in any given ecomorphological trait; other environments are more restrictive.

Carnivoran locomotor ecomorphology is an example of a trait that is, on average across all species, distributed according to local environment. The continent-scale geographic distribution of calcaneal gear ratio is closely tied dominant vegetation, mean annual temperature, and ecological province. Gear ratio is on average higher in the dry open areas of Mexico and the southwestern United States and lower in the extensive broadleaf forests of the Midwest and northeastern Atlantic seaboard and the coniferous tundra taiga of Canada and Alaska, suggesting that openness of environment exerts a loose, broad-scale, taxon-free influence on the distribution of locomotor ecomorphology. However, within-community variation among species is highest in heterogenous biomes like the basin and range country of the Great Basin and Mexico and lowest in historically homogeneous biomes like the great broadleaved deciduous forests, suggesting that some environments permit more among-species differentiation in locomotor ecomorphology than others.

Because of its strong correlation with environment, mean calcaneal gear ratio in carnivorans has potential to be a valuable proxy for environment in the fossil record. In the modern world, mean gear ratio is strongly correlated with ecological province, vegetation cover, and temperature. Warmer climes have carnivoran faunas that are more digitigrade, suggesting that calcaneal gear ratio could be used as a coarse palaeothermometer. The translation of mean gear ratio into a proxy for vegetation cover or ecological province is less straightforward, since the latter two are categorical variables so that a particular type of cover or a specific province is difficult to predict from a measure of mean gear ratio; the association is so strong, however, that further investigation is merited.

Acknowledgements

Matthew Rowe, Laura Scheiber, and Susan Spencer at the William R. Adams Zooarchaeology Lab, Indiana University, Ron Richards at the Indiana State Museum, Eileen Westwig at the American Museum of Natural

History, Phil Myers at the University of Michigan, and Bill Stanley at the Field Museum of Natural History provided specimens in their care. Matt Carrano, John Damuth, Jussi Eronen, Mikael Fortelius, Robert Guralnick, Jason Head, Anne Hereford, Christine Janis, Michelle Lawing, Sana Sarwar, and Bruce Shockey commented on or assisted with parts of this work. Two anonymous referees provided helpful comments on the manuscript that improved it. Anjali Goswami and Tony Friscia invited me to participate in their 2007 symposium on carnivoran evolution at the annual meeting of the Society of Vertebrate Paleontology and to contribute a paper to this edited volume. This work was supported by Indiana University and a grant from the US National Science Foundation (EAR-0843935).

REFERENCES

Bailey, R. G. (1998). Ecoregions map of North America. *U.S. Forest Service Miscellaenous Publication*, **1548**, 1–10.

Bailey, R. G. (2005). Identifying ecoregion boundaries. *Environmental Management*, **34**, S14–26.

Baker, C. M. (1992). *Atilax paludinosus. Mammalian Species*, **408**, 1–6.

Barnosky, A. D., Kaplan, M. H. and Carrasco, M. A. (2004). Assessing the effect of Middle Pleistocene climate change on Marmota populations from the Pit Locality. In *Biodiversity Response to Climate Change in the Middle Pleistocene*, ed. A. D. Barnosky. Berkeley, CA: University of California Press, pp. 332–40.

Bekoff, M. (1977). *Canis latrans. Mammalian Species*, **79**, 1–9.

Belcher, R. L. and Lee, T. E. Jr. (2002). *Arctocephalus townsendi. Mammalian Species*, **700**, 1–5.

Berta, A. (1986). *Atelocynus microtis. Mammalian Species*, **256**, 1–3.

Brown, J. H. and Nicoletto, P. F. (1991). Spatial scaling of species compositions: body masses of North American land mammals. *American Naturalist*, **138**, 1478–95.

Carrano, M. T. (1997). Morphological indicators of foot posture in mammals: a statistical and biomechanical analysis. *Zoological Journal of the Linnean Society*, **121**, 77–104.

Carrano, M. T. (1999). What, if anything, is a cursor? Categories versus continua for determining locomotor habit in mammals and dinosaurs. *Journal of Zoology*, **247**, 29–42.

Carrier, D. R., Heglund, N. C. and Earls, K. D. (1994). Variable gearing during locomotion in the human musculoskeletal system. *Science*, **265**, 651–53.

Caumul, R. and Polly, P. D. (2005). Comparative phylogenetic and environmental components of morphological variation: skull, mandible and molar shape in marmots (Marmota, Rodentia). *Evolution*, **59**, 2460–72.

Cavallini, P. (1992). *Herpestes pulverulentus. Mammalian Species*, **409**, 1–4.

Chorn, J. and Hoffmann, R. S. (1978). *Ailuropoda melanoleuca. Mammalian Species*, **110**, 1–6.

Clark, T. W. (1975). *Arctocephalus galapogoensis. Mammalian Species*, **64**, 1–2.

Clark, T. W. (1987). *Martes americana. Mammalian Species*, **64**, 1–8.

Croft, D. A. (2001). Cenozoic environmental change in South America as indicated by mammalian body size distributions (Cenograms). *Diversity and Distributions*, **7**, 271–87.

Damuth, J. and Fortelius, M. (2001). Reconstructing mean annual precipitation based on mammalian dental morphology and local species richness. In *EEDEN Programme Plenary Workshop on Late Miocene to Early Pliocene Environments and Ecosystems*, ed. J. Agustí and O. Oms. Brussels: European Science Foundation, pp. 23–24.

Damuth, J., Fortelius, M., Andrews, P., *et al.* (2002). Reconstructing mean annual precipitation based on mammalian dental morphology and local species richness. *Journal of Vertebrate Paleontology*, **22**, 48A.

Davis, M. B., Shaw, R. G. and Etterson, J. E. (2005). Evolutionary responses to climate change. *Ecology*, **86**, 1704–14.

Dayan, T. and Simberloff, D. (1996). Patterns of size separation in carnivore communities. In *Carnivore Behavior, Ecology, and Evolution*. Vol. **2**, ed. J. L. Gittleman. Ithaca, NY: Comstock-Cornell, pp. 243–66.

Dayan, T. and Simberloff, D. (2005). Ecological and community-wide character displacement: the next generation. *Ecology Letters*, **8**, 875–94.

Dice, L. R. (1943). *The Biotic Provinces of North America*. Ann Arbor, MI: University of Michigan Press.

Ewer, R. G. (1973). *The Carnivores*. Ithaca, NY: Cornell University Press.

Fortelius, M., Eronen, J., Jernvall, J., *et al.* (2002). Fossil mammals resolve regional patterns of Eurasian climate change over 20 million years. *Evolutionary Ecology Research*, **4**, 1005–16.

Garland, T., Jr. and Janis, C. M. (1992). Does metatarsal/femur ratio predict running speed in cursorial mammals? *Journal of Zoology (London)*, **229**, 133–51.

Goldman, C. A. (1987). *Crossarchus obscurus. Mammalian Species*, **348**, 1–3.

Goldman, C. A. and Taylor, M. E. (1990). *Liberiictis kuhni. Mammalian Species*, **348**, 1–3.

Gregersen, C. S. and Carrier, D. R. (2004). Gear ratios at the limb joints of jumping dogs. *Journal of Biomechanics*, **37**, 1011–18.

Gregory, W. K. (1912). Notes on the principles of quadrupedal locomotion and of the mechanism of the limbs in hoofed animals. *Annals of the New York Academy of Sciences*, **22**, 267–94.

Hall, E. R. (1981). *The Mammals of North America*. New York, NY: John Wiley & Sons.

Heikinheimo, H., Fortelius, M., Eronen, J. and Mannila, H. (2007). Biogeography of European land mammals shows environmentally distinct and spatially coherent clusters. *Journal of Biogeography*, **34**, 1053–64.

Holt, R. D. (2003). On the evolutionary ecology of species' ranges. *Evolutionary Ecology Research*, **5**, 159–78.

Hughes, L. (2000). Biological consequences of global warming: is the signal already apparent? *Trends in Ecology and Evolution*, **15**, 56–61.

Hwang, Y. T. and Larivière, S. (2001). *Mephitis macroura. Mammalian Species*, **686**, 1–3.

Janis, C. M. and Fortelius, M. (1988). On the means whereby mammals achieve increased functional durability of their dentitions, with special reference to limiting factors. *Biological Reviews*, **63**, 197–230.

Jenks, G. F. (1977). Optimal data classification for choropleth maps. *University of Kansas Department of Geography Occasional Papers*, **2**, 1–24.

Kaufman, L. and Rousseeuw, P. J. (1989). *Finding Groups in Data: An Introduction to Cluster Analysis*. New York, NY: Wiley-Interscience.

Klein, D. R., Meldgaard, M. and Fancy, S. G. (1987). Factors determining leg length in *Rangifer tarandus*. *Journal of Mammalogy*, **68**, 642–55.

Köppen, W. (1931). *Grudriss der Klimakunde*. Berlin: Walter de Gruyter.

Large, T. and Weller, A. (2004). *Mammals*. Mineola, NY: Dover.

Larivière, S. (2001a). *Ursus americanus*. *Mammalian Species*, **647**, 1–11.

Larivière, S. (2001b). *Poecilogale albinucha*. *Mammalian Species*, **681**, 1–4.

Larivière, S. (2005). *Aonyx capensis*. *Mammalian Species*, **671**, 1–6.

Larivière, S. and Calvada, J. (2001). *Genetta genetta*. *Mammalian Species*, **680**, 1–6.

Larivière, S. and Walton, L. R. (1997). *Lynx rufus*. *Mammalian Species*, **563**, 1–8.

Legendre, S. (1986). Analysis of mammalian communities from the late Eocene and Oligocene of southern France. *Paleovertebrata*, **16**, 191–212.

Lister, A. M. (2004). The impact of Quaternary ice ages on mammalian evolution. *Philosophical Transactions of the Royal Society of London, Series B*, **359**, 221–41.

Makarieva, A. M., Gorshkov, V. G. and Li, B.-L. (2005). Gigantism, temperature and metabolic rate in terrestrial poikilotherms. *Proceedings of the Royal Society B*, **272**, 2325–28.

Matthews, E. (1983). Global vegetation and land use: new high-resolution data bases for climate studies. *Journal of Climatology and Applied Meteorology*, **22**, 474–87.

Matthews, E. (1984). *Prescription of Land-surface Boundary Conditions in GISS GCM II: A Simple Method Based on High-resolution Vegetation Data Sets*. NASA TM-86096. Washington, DC: National Aeronautics and Space Administration.

Mech, L. D. (1974). *Canis lupus*. *Mammalian Species*, **37**, 1–6.

Millien, V., Lyons, S. K., Olson, L., Smith, F. A., Wilson, A. B. and Yom-Tov, Y. (2006). Ecotypic variation in the context of global climate change: revisiting the rules. *Ecology Letters*, **9**, 853–69.

Montuire, S. (1999). Mammalian faunas as indicators of environmental and climatic changes in Spain during the Pliocene–Quaternary transition. *Quaternary Research*, **52**, 129–37.

Murray, D. L. and Larivière, S. (2002). The relationship between foot size of wild canids and regional snow conditions: evidence for selection against a high footload? *Journal of Zoology (London)*, **256**, 289–99.

Murray, J. L. and Gardner, G. L. (1997). *Leopardus pardalis*. *Mammalian Species*, **548**, 1–10.

Nellis, D. W. (1989). *Herpestes auropunctatus*. *Mammalian Species*, **342**, 1–6.

Paradiso, J. L. and Nowak, R. M. (1972). *Canis rufus*. *Mammalian Species*, **22**, 1–4.

Patterson, B. D., Ceballos, G., Sechrest, W., *et al.* (2005). *Digital Distribution Maps of the Mammals of the Western Hemisphere, Version 2.0*. Arlington, VA: NatureServe.

Poglayen-Neuwall, I. and Toweill, D. E. (1988). *Bassariscus astutus. Mammalian Species,* **327**, 1–8.

Polly, P. D. (2003). Paleophylogeography: the tempo of geographic differentiation in marmots (Marmota). *Journal of Mammalogy,* **84**, 369–84.

Polly, P. D. (2008). Adaptive zones and the pinniped ankle: a 3D quantitative analysis of carnivoran tarsal evolution. In *Mammalian Evolutionary Morphology: A Tribute to Frederick S. Szalay,* ed. E. Sargis and M. Dagosto. Dordrecht: Springer, pp. 165–94.

Powell, R. A. (1981). *Martes pennanti. Mammalian Species,* **156**, 1–6.

Roberts, M. S. and Gittleman, J. L. (1984). *Ailurus fulgens. Mammalian Species,* **222**, 1–8.

Row, L. W. and Hastings, D. A. (1994). *TerrainBase worldwide digital terrain data (release 1.0).* Boulder, CO: National Oceanic and Atomospheric Administration, National Geophysical Data Center.

Sillero-Zubiri, C. and Gottelli, D. (1994). *Canis simensis. Mammalian Species,* **485**, 1–6.

Storz, J. F. and Wozencraft, W. C. (1999). *Melogale moschata. Mammalian Species,* **631**, 1–4.

Thompson, J. N. (2005). *The Geographic Mosaic of Coevolution.* Chicago, IL: University of Chicago Press.

Trewartha, G. T. (1968). *An Introduction to Climate.* New York, NY: McGraw-Hill.

Valverde, J. A. (1964). Remarques sur la structure et l'évolution des communautés de vertébrés terrestrés. *Revue d'écologie: La Terre et La Vie,* **III**, 121–54.

Van Rompaey, H. and Colyn, M. (1992). *Crossarchus ansorgei. Mammalian Species,* **402**, 1–3.

Van Valkenburgh, B. (1985). Locomotor diversity within past and present guilds of large predatory mammals. *Paleobiology,* **11**, 406–28.

Walton, L. R. and Joly, D. O. (2003). *Canis mesolmelas. Mammalian Species,* **715**, 1–9.

Whittaker, R. H. (1970). *Communities and Ecosystems.* New York, NY: MacMillan.

Wilf, P. (1997). When are leaves good thermometers? A new case for leaf margin analysis. *Paleobiology,* **23**, 373–90.

Willmott, K. M. and Legates, D. R. (1998). *Global air temperature and precipitation: regridded monthly and annual climatologies (version 2.01).* Newark, DE: Center for Climatic Research, University of Delaware.

Wilson, D. E. and Reeder, D. M. (2005). *Mammal Species of the World. A Taxonomic and Geographic Reference,* 3rd ed. Baltimore, MD: Johns Hopkins University Press.

Wilson, D. E. and Ruff, S. (1999). *The Smithsonian Book of North American Mammals.* Washington, DC: Smithsonian Institution Press.

Yensen, E. and Tarifa, T. (2003a). *Galictis vittata. Mammalian Species,* **727**, 1–8.

Yensen, E. and Tarifa, T. (2003b) *Galictis cuja. Mammalian Species,* **728**, 1–8.

Appendix 13.1 Species included in this study and their associated data. North American species are indicated by an asterisk (*). Locomotor abbreviations: Arb, arboreal; Nata, natatorial; Scan, scansorial; SemFos, semi-fossorial; Terr, terrestrial. Posture categories: P, plantigrade; SD, semi-digitigrade; D, digitigrade.

Taxon	N	Locomotion	Posture	FemLen	CalcLen	AntCalc	PostCalc	SustPos	MetaIII	Met/Fem (R1)	Ant Post (R2)	Sust/Calc (R3)
Nandiniidae												
Nandinia binotata	1	Scan	P	94.0	22.3	9.3	13.8	19.0	29.0	0.31	0.67	1.17
Viverridae												
Paradoxurinae												
Arctictis binturong	1	Arb	SD	143.6	37.8	15.1	25.5	34.0	35.3	0.25	0.59	1.11
Paradoxurus hermaphroditus	4	Arb	SD	93.8	23.2	9.5	14.7	19.7	27.2	0.29	0.65	1.18
Galidiinae												
Salanoia concolor	1	Scan	P	59.2	14.9	5.8	8.9	12.7	26.7	0.45	0.65	1.17
Viverrinae												
Civettictis civetta	1	Scan	D	151.5	36.4	14.1	24.9	28.8	53.2	0.35	0.56	1.26
Genetta genetta	2	Scan	SD	79.9	19.0	8.1	11.4	15.4	34.4	0.43	0.72	1.23
Genetta maculata	2	Scan	SD	85.4	20.9	8.3	13.0	16.6	34.8	0.41	0.64	1.26
Genetta servalina	2	Scan	SD	85.9	21.1	8.8	12.8	17.0	36.3	0.42	0.69	1.24
Genetta victoriae	7	Scan	SD	91.7	22.9	10.2	14.4	18.3	33.2	0.36	0.70	1.25
Poiana richardsonii	2	Arb	P	59.6	14.3	5.9	7.6	11.3	22.3	0.38	0.77	1.27

Viverra tangalunga	2	Scan	D	110.6	25.8	10.3	16.5	19.5	43.7	0.40	0.63	1.32
Viverra zibetha	28	Scan	D	138.2	34.6	13.6	21.9	26.5	54.1	0.39	0.62	1.30
Viverricula indica	2	Scan	D	94.2	24.0	10.1	15.1	18.5	39.6	0.42	0.67	1.30
Euplerinae												
Cryptoprocta ferox	1	Scan	SD	135.1	34.5	15.0	21.5	28.7	43.1	0.32	0.70	1.20
Fossa fossana	2	Terr	D	75.3	19.7	8.3	12.2	15.4	34.8	0.46	0.68	1.28
Herpestidae												
Hemigalinae												
Hemigalus derbyanus	2	Scan	SD	88.3	19.5	8.1	11.6	15.8	27.6	0.31	0.70	1.23
Herpestidae												
Atilax paludinosus	1	Nata	SD	91.4	26.1	11.0	15.9	20.0	37.8	0.41	0.69	1.30
Bdeogale nigripes	1	Terr	D	94.3	27.9	11.8	16.6	21.1	42.7	0.45	0.71	1.32
Crossarchus alexandri	1	Scan	SD	70.7	18.9	8.0	12.0	15.6	28.9	0.41	0.67	1.21
Cynictis penicillata	2	Terr	D	60.7	17.4	7.9	10.3	13.0	25.2	0.42	0.77	1.34
Galerella sanguinea	2	Scan	SD	44.5	12.7	5.3	7.1	9.9	19.8	0.45	0.75	1.29

Appendix 13.1 (cont.)

Taxon	N	Locomotion	Posture	FemLen	CalcLen	AntCalc	PostCalc	SustPos	MetaIII	Met/Fem (R1)	Ant/Post (R2)	Sust/Calc (R3)
Herpestes edwardsi	1	Terr	SD	61.8	17.5	8.4	10.5	13.3	26.1	0.42	0.80	1.31
Herpestes ichneumon	1	Terr	SD	83.3	23.3	9.8	14.8	18.1	34.8	0.42	0.66	1.29
Herpestes javanicus	1	Terr	SD	48.1	13.0	6.2	7.5	10.4	21.4	0.45	0.82	1.25
Herpestes naso	2	Terr	SD	80.3	25.2	12.6	14.5	18.2	36.7	0.46	0.87	1.38
Ichneumia albicauda	2	Terr	D	103.3	30.3	14.1	17.4	20.8	48.3	0.47	0.82	1.46
Suricata suricatta	1	Terr	SD	53.6	14.5	6.4	8.5	10.7	24.2	0.45	0.76	1.36
Hyaenidae												
Crocuta crocuta	1	Terr	D	250.0	60.0	18.6	39.9	44.5	82.6	0.33	0.47	1.35
Hyaena brunnea	2	Terr	D	224.0	50.1	16.0	33.7	39.3	86.6	0.39	0.47	1.27
Hyaena hyaena	2	Terr	D	195.0	44.3	12.9	32.5	36.7	75.0	0.38	0.40	1.21
Proteles cristata	2	Terr	D	145.3	30.9	11.5	19.7	24.7	65.9	0.45	0.58	1.25
Felidae												
Felinae												
Acinonyx jubatus	1	Scan	D	270.0	79.8	29.2	55.2	58.8	116.1	0.43	0.53	1.36
Caracal caracal	2	Scan	D	166.0	39.9	14.1	23.9	27.9	69.7	0.42	0.59	1.43
Catopuma temminckii	2	Scan	D	161.0	42.6	14.3	28.4	30.5	64.6	0.40	0.50	1.40
Felis chaus	1	Scan	D	143.4	35.7	13.2	21.3	24.8	64.1	0.45	0.62	1.44
Felis domesticus	1	Scan	D	104.8	28.3	11.4	18.3	20.2	48.9	0.47	0.63	1.40
Felis manul	1	Scan	D	108.1	26.4	8.2	17.6	20.9	42.5	0.39	0.47	1.26
Felis margarita	1	Scan	D	109.3	24.8	9.1	15.8	20.5	48.5	0.44	0.58	1.21
Felis nigripes	1	Scan	D	77.8	20.3	8.2	12.3	14.8	38.0	0.49	0.67	1.36

	Species												
*	*Herpailurus yaguarondi*	1	Scan	D	110.3	30.8	11.2	20.0	23.1	47.8	0.43	0.56	1.33
	Leopardus geoffroyi	2	Scan	D	131.2	33.6	13.3	20.1	25.1	54.7	0.42	0.66	1.34
*	Leopardus pardalis	2	Scan	D	165.0	40.1	16.3	24.9	28.7	54.3	0.33	0.65	1.40
*	Leopardus tigrinus	1	Scan	D	99.8	25.0	10.7	14.5	18.7	44.0	0.44	0.74	1.34
*	Leopardus wiedii	2	Scan	D	111.6	27.5	12.8	17.0	20.8	44.1	0.39	0.76	1.33
	Leptailurus serval	2	Scan	D	193.5	48.5	17.4	31.4	33.7	84.4	0.44	0.55	1.44
*	Lynx canadensis	2	Scan	D	205.0	47.8	18.9	29.3	35.2	90.5	0.43	0.65	1.36
	Lynx lynx	2	Scan	D	210.5	56.2	21.3	35.0	41.2	96.8	0.46	0.61	1.36
	Lynx pardinus	1	Scan	D	169.0	45.0	16.5	29.5	33.3	73.0	0.43	0.56	1.35
*	Lynx rufus	4	Scan	D	157.0	41.7	16.9	25.7	29.6	66.1	0.42	0.66	1.41
	Pardofelis marmorata	2	Scan	D	106.6	25.9	11.4	15.8	19.3	46.1	0.43	0.72	1.34
	Prionailurus bengalensis	2	Scan	D	117.8	29.3	11.4	15.8	21.6	51.4	0.44	0.72	1.36
	Profelis aurata	1	Scan	D	172.0	45.6	16.0	28.3	33.4	75.6	0.44	0.56	1.36
*	Puma concolor	2	Scan	D	254.5	70.1	25.0	46.9	49.6	99.4	0.39	0.53	1.41
	Pantherinae												

Appendix 13.1 (*cont.*)

Taxon	N	Locomotion	Posture	FemLen	CalcLen	AntCalc	PostCalc	SustPos	MetaIII	Met/Fem (R1)	Ant/Post (R2)	Sust/Calc (R3)
Neofelis nebulosa	1	Scan	D	161.0	44.6	17.4	29.0	34.3	56.6	0.35	0.60	1.30
Panthera leo	2	Terr	D	381.0	90.3	29.2	62.0	83.8	112.9	0.32	0.48	1.32
* *Panthera onca*	1	Scan	D	214.0	60.6	20.7	42.1	46.4	73.6	0.34	0.49	1.31
Panthera pardus	2	Scan	D	230.0	63.0	20.5	43.4	52.1	87.8	0.38	0.47	1.21
Panthera tigris	1	Terr	D	379.0	106.5	34.8	75.6	82.3	129.0	0.34	0.46	1.29
Uncia uncia	3	Scan	D	226.0	60.6	19.0	41.6	47.1	89.0	0.39	0.46	1.29
Canidae												
Atelocynus microtis	1	Terr	D	139.8	33.1	10.5	23.3	26.0	56.7	0.41	0.45	1.27
Canis adustus	3	Terr	D	161.3	37.2	12.3	25.0	28.2	71.3	0.44	0.49	1.32
Canis aureus	2	Terr	D	153.6	36.7	11.9	24.9	28.8	65.9	0.43	0.48	1.27
Canis familiaris dingo	1	Terr	D	192.8	48.4	16.0	33.3	37.6	77.2	0.40	0.48	1.28
* *Canis latrans*	3	Terr	D	174.8	40.9	13.3	28.7	32.9	75.5	0.43	0.46	1.24
* *Canis lupus*	2	Terr	D	245.5	60.8	18.8	43.1	47.4	108.2	0.44	0.44	1.28
Canis mesomelas	2	Terr	D	148.5	34.0	12.0	22.5	27.4	61.9	0.44	0.53	1.28
* *Canis rufus*	2	Terr	D	199.7	49.7	17.1	33.4	39.5	89.0	0.45	0.51	1.26
Canis simensis	2	Terr	D	192.5	45.6	15.8	31.1	37.2	83.3	0.43	0.51	1.23
Cerdocyon thous	1	Terr	D	120.2	27.5	9.9	17.8	19.7	55.4	0.46	0.56	1.40
Chrysocyon brachyurus	2	Terr	D	279.0	59.6	21.1	40.7	43.2	137.6	0.49	0.52	1.38
Cuon alpinus	2	Terr	D	176.0	43.2	13.7	31.3	35.7	74.4	0.42	0.44	1.21
Lycalopex culpaeus	2	Terr	D	137.3	33.4	12.1	21.9	28.1	62.1	0.45	0.55	1.19
Lycalopex griseus	2	Terr	D	114.5	27.1	9.4	18.1	21.0	53.1	0.46	0.52	1.29
	2	Terr	D	136.1	33.3	12.5	20.9	25.7	64.9	0.48	0.60	1.29

	Species												
	Lycalopex gymnocercus												
	Lycalopex sechurae	4	Terr	D	104.4	24.3	9.5	16.3	20.8	46.7	0.45	0.59	1.17
	Lycalopex vetulus	2	Terr	D	92.4	25.0	9.0	16.2	20.1	46.1	0.50	0.56	1.24
	Lycaon pictus	1	Terr	D	209.8	53.9	15.6	39.6	42.0	91.1	0.43	0.39	1.29
	Nyctereutes procyonoides	1	Terr	D	98.9	23.7	7.2	16.3	18.7	42.1	0.42	0.45	1.27
	Otocyon megalotis	1	Terr	D	110.5	25.9	9.6	16.6	19.3	51.2	0.46	0.58	1.34
*	*Speothos venaticus*	2	Terr	D	102.7	25.7	6.6	18.7	21.9	39.4	0.38	0.35	1.17
*	*Urocyon cinereoargenteus*	2	Scan	D	115.7	27.8	11.5	17.7	21.9	54.2	0.47	0.65	1.27
*	*Vulpes lagopus*	1	Terr	D	108.4	27.9	10.2	18.5	21.0	52.2	0.48	0.55	1.33
*	*Vulpes macrotis*	1	Terr	D	94.1	22.5	9.4	14.4	17.5	47.8	0.51	0.66	1.29
*	*Vulpes velox*	2	Terr	D	101.0	25.3	10.4	16.4	20.6	37.5	0.37	0.64	1.23
*	*Vulpes vulpes*	2	Terr	D	138.8	33.3	13.7	19.5	26.5	67.6	0.49	0.71	1.26
	Vulpes zerda	1	Terr	D	74.5	18.5	7.2	11.4	14.4	37.8	0.51	0.64	1.28
	Ailuridae												
	Ailurus fulgens	1	Arb	P	116.1	26.1	8.4	17.2	21.4	36.0	0.31	0.49	1.22
	Ursidae												
	Ailuropoda melanoleuca	3	Scan	P	271.0	64.5	13.3	50.8	56.7	49.3	0.18	0.26	1.14
	Helarctos malayanus	3	Scan	P	251.8	53.8	11.4	41.1	48.3	47.7	0.19	0.28	1.11

Appendix 13.1 (*cont.*)

Taxon	N	Locomotion	Posture	FemLen	CalcLen	AntCalc	PostCalc	SustPos	MetaIII	Met/Fem (R1)	Ant/Post (R2)	Sust/Calc (R3)
Melursus ursinus	2	Scan	P	319.5	71.0	12.8	55.3	65.7	57.5	0.18	0.23	1.08
Tremarctos ornatus	1	Scan	P	286.0	56.0	10.8	42.4	47.1	53.5	0.19	0.25	1.19
* *Ursus americanus*	49	Scan	P	297.4	63.1	18.6	45.6	54.8	59.4	0.20	0.41	1.15
* *Ursus arctos*	4	Scan	P	443.8	95.3	19.4	75.9	84.3	95.1	0.22	0.27	1.13
Ursus maritimus	3	Terr	P	464.3	101.8	18.0	82.8	90.3	97.3	0.21	0.22	1.13
Ursus thibetanus	2	Scan	P	319.0	70.8	11.8	55.1	60.8	55.7	0.17	0.21	1.16
Procyonidae												
* *Bassaricyon gabbii*	42	Arb	P	69.7	17.9	6.0	10.9	14.3	27.8	0.40	0.55	1.25
* *Bassariscus astutus*	1	Terr	SD	68.6	16.3	5.2	10.1	12.9	24.4	0.35	0.52	1.27
* *Bassariscus sumichrasti*	1	Terr	SD	78.5	19.5	8.9	11.6	14.7	30.1	0.38	0.76	1.33
* *Nasua narica*	1	Scan	P	119.1	28.9	11.8	18.1	23.4	31.8	0.27	0.66	1.23
Nasua nasua	2	Scan	P	94.5	24.8	7.0	16.1	19.5	26.6	0.28	0.43	1.27
* *Potos flavus*	2	Arb	P	93.1	23.3	7.8	14.6	19.8	29.0	0.31	0.55	1.18
* *Procyon cancrivorus*	5	Scan	SD	145.4	37.8	15.6	24.2	31.4	52.0	0.36	0.65	1.21
* *Procyon lotor*	3	Scan	SD	119.2	29.3	11.9	17.9	23.6	36.9	0.31	0.69	1.24
Mustelidae												
Mustelinae												
* *Arctonyx collaris*	1	SemFos	SD	104.9	26.9	9.5	19.1	23.7	27.6	0.26	0.50	1.14

*	*Eira barbara*	1	Scan	P	103.1	27.0	11.2	16.8	22.2	33.1	0.32	0.67	1.22
	Galictis cuja	1	SemFos	P	63.6	15.5	5.9	11.0	14.1	22.4	0.35	0.54	1.11
*	*Galictis vittata*	1	SemFos	P	60.2	17.5	6.8	11.0	15.0	22.3	0.37	0.62	1.17
*	*Gulo gulo*	1	SemFos	SD	141.5	41.5	15.0	28.6	34.2	55.5	0.39	0.52	1.21
	Ictonyx libyca	2	Scan	P	32.6	8.5	3.2	5.2	7.4	12.0	0.37	0.60	1.14
	Ictonyx striatus	1	Scan	P	63.0	20.8	8.9	12.7	17.3	36.6	0.58	0.71	1.20
*	*Martes americana*	2	Scan	P	71.9	18.1	5.9	11.2	14.5	32.0	0.44	0.53	1.25
	Martes foina	2	Scan	SD	92.1	22.9	7.4	14.6	18.8	37.1	0.40	0.51	1.22
	Martes martes	1	Scan	SD	74.7	17.0	6.8	10.7	14.3	32.5	0.43	0.64	1.19
*	*Martes pennanti*	26	Scan	P	78.8	19.0	7.3	11.7	15.6	34.3	0.43	0.62	1.22
	Meles meles	2	SemFos	SD	92.3	28.7	7.3	20.1	24.0	28.9	0.31	0.36	1.19
	Mellivora capensis	2	SemFos	SD	102.4	27.3	8.9	17.3	21.1	28.7	0.28	0.51	1.29
	Melogale personata	2	SemFos	P	55.5	14.1	5.0	8.7	11.4	19.0	0.34	0.57	1.24
*	*Mustela erminea*	5	Terr	SD	28.0	6.1	2.5	4.2	5.3	11.1	0.40	0.62	1.15
	Mustela eversmanii	2	Terr	SD	59.5	15.3	6.2	10.0	12.7	23.6	0.40	0.62	1.20
*	*Mustela frenata*	1	Terr	SD	33.0	7.8	3.2	4.7	6.5	14.7	0.45	0.69	1.20
*	*Mustela nigripes*	22	Terr	SD	51.5	13.0	4.8	8.4	11.3	20.0	0.39	0.57	1.15
*	*Mustela nivalis*	2	Terr	SD	15.9	3.6	1.2	2.2	3.0	7.0	0.44	0.54	1.22
*	*Neovison vison*	2	Terr	SD	45.9	12.3	5.3	7.6	10.5	20.2	0.44	0.70	1.17
	Poecilogale albinucha	4	Terr	P	34.0	10.0	3.1	6.3	7.4	13.3	0.39	0.48	1.36
*	*Taxidea taxus*	1	SemFos	P	106.5	28.8	7.4	20.0	23.0	29.6	0.28	0.38	1.25

Appendix 13.1 (*cont.*)

	Taxon	N	Locomotion	Posture	FemLen	CalcLen	AntCalc	PostCalc	SustPos	MetaIII	Met/Fem (R1)	Ant/Post (R2)	Sust/Calc (R3)
	Lutrinae												
	Enhydra lutris	2	Nata	P	116.6	39.2	14.7	23.8	32.3	69.0	0.59	0.62	1.21
*	*Lontra canadensis*	1	Nata	SD	70.4	27.1	8.9	17.5	20.8	39.0	0.55	0.51	1.30
*	*Lontra longicaudis*	2	Nata	SD	68.5	23.0	8.2	16.6	18.9	33.5	0.49	0.50	1.22
	Mephitidae												
*	*Conepatus chinga*	2	Terr	SD	67.1	19.3	7.9	11.8	15.3	17.2	0.26	0.67	1.26
	Conepatus leuconotus	3	Terr	SD	77.1	21.6	9.0	13.8	18.4	20.1	0.26	0.67	1.18
*	*Conepatus semistriatus*	3	Terr	SD	69.2	19.8	8.4	11.9	17.0	20.2	0.30	0.71	1.16
*	*Mephitis macroura*	1	Terr	P	62.5	19.5	8.2	11.4	15.2	19.2	0.31	0.71	1.28
*	*Mephitis mephitis*	11	Terr	P	64.7	18.7	8.4	11.0	15.4	18.7	0.29	0.77	1.22
*	*Spilogale gracilis*	1	Terr	P	43.9	11.7	5.0	6.9	9.5	13.6	0.31	0.73	1.23
*	*Spilogale putorius*	1	Terr	P	47.1	12.1	4.5	7.5	9.9	14.6	0.31	0.60	1.22

14

Interpreting sabretooth cat (Carnivora; Felidae; Machairodontinae) postcranial morphology in light of scaling patterns in felids

MARGARET E. LEWIS AND MICHAEL R. LAGUE

Introduction

Reconstructing the behaviour and ecology of extinct felids, especially that of machairodontine felids, has been of great interest within the field of vertebrate paleontology. The anatomical design of these animals has been investigated with respect to dental function and prey acquisition behaviour, and, to a lesser degree, locomotion.

Few large felids exist today, and machairodontine felids were sometimes even larger than the largest extant felids, lions and tigers. This leads to the question of how much of the morphology observed in large machairodontines is simply an extension of size-related shape trends observed in modern felids. That is, to what extent are the morphological differences between machairodontines and smaller extant felids due to differences in size? Which extinct forms appear to be scaled-up versions of smaller felids, and which ones exhibit morphology indicative of functional differences?

This preliminary study investigates machairodontine postcranial morphology in light of scaling patterns in extant felids and examines how well trends in smaller extant felids predict the morphology of larger felids. We also look for any overall trends in machairodontine postcranial morphology that unite them as a group, much like the possession of machairodont dentition does.

Machairodontine variation

Felids traditionally have been split into two subfamilies: the sabretoothed cats (Machairodontinae) and the conical-toothed cats (Felinae). Machairodontine felids are typically placed into three tribes: Metailurini (e.g. *Adelphailurus,*

Carnivoran Evolution: New Views on Phylogeny, Form, and Function, ed. A. Goswami and A. Friscia. Published by Cambridge University Press. © Cambridge University Press 2010.

Dinofelis, Metailurus), Homotheriini (e.g. *Dinobastis, Machairodus, Homotherium*), and Smilodontini (e.g. *Megantereon, Smilodon*) (Turner and Antón, 1997), although Metailurini may be a wastebasket taxon (Werdelin *et al.*, in press).

In terms of morphology, machairodontines have been split into morphotypes that roughly follow tribal divisions (Martin, 1980). Scimitar-toothed cats (e.g. *Homotherium, Dinobastis*) have elongated limbs and short, coarsely serrated upper canines, while dirk-toothed cats have shortened limbs and long, finely serrated upper canines (e.g. *Smilodon*). Cranial shape also differs between more derived forms of the two morphotypes (Slater and Van Valkenburgh, 2008). As Martin (1980) noted, however, there are taxa that do not fit into this dichotomous scheme. Not all members of a tribe may share the same morphotype. For example, the recently described homotheriin *Xenosmilus* has scimitar-shaped canines with the short limbs of a dirk-tooth, leading to Martin and colleagues' (2000) suggestion that there is a third way to be a sabretooth cat. Likewise, the smilodontin *Megantereon* lacks serrated canines.

Metailurins are also often unclassifiable within this dichotomy. Their diverse morphology is reflected in the taxonomic and morphotypological limbo in which they reside. *Dinofelis*, for example, has been called a machairodontine (e.g. Beaumont, 1964, 1978; Berta and Galiano, 1983), a conical-toothed cat (e.g. Kretzoi, 1929; Hendey, 1974), or something intermediate between the two (e.g. Piveteau, 1961). Martin (1980) regarded *Dinofelis* as being more like scimitar-toothed cats despite the lack of serrations on the teeth. A later revision of the genus (Werdelin and Lewis, 2001) confirmed its taxonomic status as a machairodontine, and noted variation within the genus ranging from some *Panthera*-like taxa to more typically machairodontine taxa. While none of the species of *Dinofelis* have the elongated limbs of *Homotherium*, at least one taxon (*Dinofelis* sp. indet. D. from Olduvai Gorge, Tanzania, in Werdelin and Lewis, 2001) evolved relatively short limbs, if this is indeed *Dinofelis*. In short, a single machairodontine genus can encompass a larger degree of morphological variation than is typically noted in the literature, and thus can span traditional morphotypes.

Scaling relationships within the Felidae

Within a group of related taxa, such as the Felidae, one would expect to see morphological variation due to behavioural diversity, as well as variation due to differences in body size. For example, skull shape variation in *felines* is clearly driven by body size rather than by phylogeny or variation in function (Werdelin, 1983; Slater and Van Valkenburgh, 2008). In contrast, among *machairodontines*, canine length is a better predictor of skull shape than is body size (Slater and Van Valkenburgh, 2008). Once canines reach a certain length, accommodating their length becomes more important than body size in determining cranial shape. This difference between felines and machairodontines led Slater and Van Valkenburgh to conclude that sabretoothed taxa differed in their general predatory adaptations (e.g. killing prey faster in a high competition environment) from felines.

Within highly size-variant groups such as felids, body size can play an important role in driving interspecific variation in skeletal morphology. Peak functional stresses have been hypothesised to be similar in the skeleton, regardless of body mass, due to functional constraints that work to maintain uniform safety factors (Biewener, 1983; Biewener and Taylor, 1986; Bertram and Biewener, 1990). In other words, as a species increases in size, it will avoid greater stresses by changing some aspect of its design (e.g. morphology, posture) to avoid mechanical failure. Scaling relationships between body mass and measurements of postcranial elements have been explored in a wide range of vertebrate taxa (e.g. McMahon, 1975; Alexander, 1977; Alexander et al., 1979; Jungers, 1979, 1984a, 1984b, 1985; Prothero and Sereno, 1982; Anderson et al., 1985; Bou and Casinos, 1985; Preuschoft and Demes, 1985; Scott, 1985; Bou et al., 1987; Ruff, 1987, 2000; Gingerich, 1990; Scott, 1990; Alexander and Pond, 1992; Ruff and Runestad, 1992; Heinrich and Biknevicius, 1998; Christiansen, 1999a, 1999b, 1999c; Blob, 2000; Garcia and da Silva, 2004). Despite early suggestions that limb bone proportions of all terrestrial mammals scale isometrically with body mass (Alexander et al., 1979), different taxonomic groups, even within Mammalia, exhibit different scaling relationships.

Little attention has been paid to the impact of body size on felid morphology except when reconstructing body mass in extinct taxa (e.g. Van Valkenburgh, 1990; Anyonge, 1993; O'Regan and Turner, 2002; Andersson, 2004; Christiansen and Harris, 2005). While many of these studies shed light on the relationship (or non-relationship) between the size of craniodental structures and overall body mass, few investigate postcranial elements, despite this region of the body being the load-bearing portion of the animal. In addition, many scaling studies involving felids examine carnivorans in general, rather than felids in particular. This is important given the observation that different characters are better at predicting body mass within different carnivoran families (Anyonge, 1993). In other words, not only do different regions of the skeleton scale differently, different families exhibit different patterns of scaling.

Previous work involving felids has demonstrated that long bone length scales isometrically with body mass, particularly that of the tibia and ulna (Christiansen and Harris, 2005), and humerus and femur (Anyonge, 1993; Christiansen and Harris, 2005). Humeral circumference, on the other hand, appears to increase disproportionately with body mass (Anyonge, 1993;

Christiansen and Harris, 2005), suggesting an increase in humeral robusticity in larger-bodied felids. In contrast, the same studies found felid femoral circumference to scale isometrically. Christiansen and Harris (2005) also found size-related increases in robusticity of the ulna and tibia. In addition, Bertram and Biewener (1990) observed positive allometry of both anteroposterior and mediolateral midshaft diameters (relative to element length) for all four long bones they examined (humerus, radius, femur, and tibia).

Few studies have examined articular scaling in felids, although Anyonge (1993) observed positive allometry for distal femoral articular surface area, which was also found to be the best femur-based predictor of body size in felids. Positive allometry has also been demonstrated for distal humeral articular surface area by Andersson (2004), although this study included various non-felid carnivorans.

It is worth noting that Christiansen and Harris (2005) used total element lengths (maximum bone length) rather than functional element lengths (length from articular surface to articular surface), which are more typical in functional studies going back to Howell (1944). (See Bertram and Biewener, 1990, for a typical description of functional measurements.) Hence processes (e.g. greater tuberosity of the humerus, olecranon process of the ulna) were included in Christiansen and Harris' element lengths, despite not being part of the weight-bearing portion of the bones. These processes reflect whether a limb is designed to be moved quickly (relatively short process) or with great strength (relatively long process) for a given movement. As such, inclusion of processes extending beyond the articular surface not only inflates the measured length of an element, but inflates the element non-uniformly across taxa. For this reason, measurements of radial functional length provide a better proxy for forearm length than the ulna.

In sum, while various postcranial features, particularly cross-sectional properties and joint surface areas, have been deemed useful in predicting body mass in felids and other carnivorans, few researchers have looked at the underlying causation in the patterns they were studying. While identifying characters to predict body mass is important, it is equally important to consider the underlying causal factors in scaling relationships and to identify the reason for outliers. This last step is critical for understanding the functional morphology of extinct taxa, as one must determine whether an extinct taxon follows the general scaling pattern or falls with outliers.

Size-dependent scaling patterns

Not unexpectedly, the limb bones of carnivorans in general have been shown to increase in robusticity as body size increases (Bertram and Biewener, 1990), leading to greater resistance to bending. However, non-linear scaling of the limb bones across the order has been found with respect to the relationship between element length and AP diameter. Smaller taxa scale closer to isometry, while the larger taxa (>100 kg) scale with strong positive allometry (Bertram and Biewener, 1990). This size-related shift in scaling may be due to the fact that postural changes are only viable for forms below roughly 100–200 kg (Bertram and Biewener, 1990) or even 300 kg (Biewener, 1990) body mass. Hence, larger forms must maintain uniform skeletal stress through decreased locomotor performance and increased skeletal allometry (Biewener, 1990).

A recent study of scaling of felid limb posture (Day and Jayne, 2007) tested Bertram and Biewener's posture hypothesis. Despite including nine felid species ranging in mass from 3.3 to 192 kg, this study found that limb postures were uniform with respect to size. Even the relatively long-limbed cheetahs, lynxes, and servals were similar to other felids in limb posture. More scansorial taxa had a slightly more crouched posture on the ground, regardless of size. In addition, trends in limb posture (e.g. the scansorial trend) tended to occur across different speeds and gaits.

Since felids seem to be uniform in posture (Day and Jayne, 2007), the carnivoran pattern observed by Bertram and Biewener (1990) regarding large-bodied versus small-bodied scaling patterns may not hold for felids in particular. Among extant felids, only lions and tigers exceed Bertram and Biewener's cutoff of 100 kg. It is worth investigating, therefore, whether their morphology is well-predicted by scaling trends in smaller-bodied felids. Similarly, given the large body masses of many machairodontines, understanding scaling patterns within the extant Felidae becomes critical for understanding machairodontine shape variation.

Abbreviations

Institutions
AMNH, American Museum of Natural History, New York; ANSP, Academy of Natural Sciences, Philadelphia; BPI, Bernard Price Institute, Johannesburg; FMNH, Field Museum of Natural History, Chicago; ENM Ethiopia National Museum, Addis Ababa; SAM, South African Museum, Cape Town; KNM, Kenya National Museums, Nairobi; NMNH, National Museum of Natural History, Washington, DC; NHM, Natural History Museum, London; TMM, Texas Memorial Museum, Austin; UCMP, University of California Museum of Paleontology, Berkeley.

Other
AP, anteroposterior; ML, mediolateral; SI, superoinferior; CL, confidence limits.

Table 14.1 Measurements included in the study. Illustrations of some of these measurements can be found in Lewis (1995).

Humerus	Radius	Femur	Tibia
Functional length	Functional length	Functional length	Functional length
SI head diameter	Max head width	SI head diameter	Proximal ML width
ML head diameter	Min head width	ML head diameter	Medial condyle AP length
AP head diameter	ML midshaft diameter	AP head diameter	Lateral condyle AP length
ML midshaft diameter	AP midshaft diameter	ML midshaft diameter	ML midshaft diameter
AP midshaft diameter	Max distal articular width	AP midshaft diameter	AP midshaft diameter
Medial trochlear lip height		Bicondylar width	Distal ML width
Lateral trochlear lip height	Min distal articular width	Distal AP length (condyles)	Distal AP length
ML trochlear width			

Methods

Materials and measurements

Linear measurements (functional limb bone lengths, midshaft diameters, and articular dimensions; Table 14.1) were made in various extant and fossil carnivorans. Extant species ranged in size from the black-footed cat (*Felis nigripes*) to the Siberian tiger (*Panthera tigris altaica*; Table 14.2). Our taxonomy follows Johnson *et al.*, 2006. Data on extant taxa were collected at the AMNH, NHM, FMNH, SAM, and NMNH.

Body masses reported in Table 14.2 were taken from the literature (Sunquist and Sunquist, 2002). Species averages were weighted by the sample size listed for each geographic location, although we ignored body mass data from locations clearly not associated with our particular skeletal samples (i.e. we excluded mass data available for caracals and lions from outside Africa, as well as data for Sumatran tigers). For extant species in which sexes reportedly differ in average body mass by more than 10 kg (i.e. pumas, leopards, lions, and tigers), sexes were considered separately in all regression analyses; otherwise, conspecific sexes were pooled. Although male and female jaguars differ in average body

Table 14.2 Extant taxa arranged by average body mass.[1]

Species	Common name	Max n	Sex	Avg. body mass (kg)
Felis nigripes	Black-footed cat	2	B	1.6
Prionailurus bengalensis	Asian leopard cat	2	B	2.1
Leopardus geoffroyi	Geoffroy's cat	7	B	4.5
Felis lybica	African wild cat	1	B	4.6
Felis chaus	Jungle cat	3	B	7.3
Leopardus pardalis	Ocelot	4	B	8.7
Lynx rufus	Bobcat	4	B	10.3
Caracal caracal	Caracal	2	B	10.4
Caracal serval	Serval	4	B	11.2
Neofelis nebulosa	Clouded leopard	6	B	16.4
Panthera pardus	Leopard (Africa, Asia)	6	F	33.6
Puma concolor	Puma	3	F	39.5
Panthera uncia	Snow leopard	3	B	41.9
Acinonyx jubatus	Cheetah (Africa)	14	B	43.9
Panthera pardus	Leopard (Africa, Asia)	11	M	56.7
Puma concolor	Puma	1	M	58.3
Panthera onca	Jaguar	8	B	80.9
Panthera leo	Lion (Africa)	10	F	129.6
Panthera tigris tigris/altaica	Bengal and Siberian tigers	3	F	136.3
Panthera leo	Lion (Africa)	11	M	183.9
Panthera tigris tigris/altaica	Bengal and Siberian tigers	5	M	226.3

Note: [1] Body mass data based on Sunquist and Sunquist (2002). Taxonomy follows Johnson *et al.* (2006). B = both sexes, F = female, M = male.

mass by about 25 kg (based on weighted averages from Sunquist and Sunquist, 2002), the body masses reported in the literature are highly variable for both sexes. Since our jaguar specimens are of unknown sex and do not clearly fall into two size classes, we pooled our eight available individuals.

Given the hypothesised differences in how peak stresses are maintained in mammals above and below 100 kg (Bertram and Biewener, 1990), extant taxa were classified as either large-bodied (>100 kg) or small-bodied (<100 kg). 'Large-bodied' is equivalent to the carnivoran Size Class 4 of Lewis and Werdelin (2007), while 'small-bodied' includes Size Classes 1–3. (Size classes are based on a combination of body size, behaviour, and ecology.) Both sexes of extant lions and tigers are the only living Size Class 4 felids and are considered large-bodied, while all other extant taxa included are small-bodied for the purposes of this paper.

Fossil taxa included machairodontine taxa from various localities across Africa and North America (Table 14.3). Only complete humeri, radii, femora, and tibiae from these sites were included. As this is a preliminary study, future studies will include a broader range of skeletal elements and machairodontine taxa. Although *Dinobastis* was synonymised with *Homotherium* by Churcher (1966), we have followed later recommendations to revalidate the genus (Martin *et al.*, 1988; Werdelin *et al.*, in press), hence the use of the taxon name *Dinobastis serus* rather than *Homotherium serum*. Data on fossil specimens were collected from specimens housed at AMNH, ANSP, BPI, ENM, SAM, KNM, TMM, and UCMP.

Statistical analyses

Regressions: multivariate 'shape' vs. element size

Before exploring the scaling of individual linear dimensions (see next section), we wanted to capture overall size-related shape trends in extant felids. To consider allometric trends in a multivariate sense, we conducted a principal components analysis (PCA) of extant 'shape' data for each element (based on species means). 'Shape' data for a given element (for a given specimen) were derived by dividing each variable by a measure of overall element size (i.e. the geometric mean of all linear dimensions; cf. Darroch and Mosimann, 1985; Jungers *et al.*, 1995). For each element, we used regression to evaluate the relationship between element size (*x*-value) and the scores (*y*-value) along the first principal component (PC1); this procedure is similar to that used by O'Higgins and Jones (1998) to examine size-related shape variation in the mangabey facial skeleton, albeit with a different 'size' measurement (i.e. geometric mean rather than centroid size). In this way, we could gauge the extent to which the principal axis of shape variation among extant felids is related to size variation. We did not log-transform the PC1 scores, since they were negative for every element. Since we had no a-priori reason to believe that the relationships between element size and PC1 would be linear, we performed both linear and polynomial least squares regressions and assessed the resulting increase in the correlation coefficient when second- or third-order polynomial regression was used.

The intent of these PC1 regression analyses was to use size-related shape trends in extant taxa as a basis of comparison with which to evaluate the fossils; hence, fossil taxa were excluded from the PCA and regression calculations. In addition, to avoid the effect of outliers in defining extant shape trends, calculations were also done without several extant species that were clearly outliers in these regressions and/or subsequent linear regressions involving limb

Table 14.3 Extinct taxa included in the study. Dates taken from the literature (Graham, 1976; Hendey, 1981; Potts, 1988; Feibel et al., 1989; Cooke, 1991; Walter, 1994; Morgan and Hulbert, 1995; Ditchfield et al., 1999; Hulbert, 2001; Werdelin and Lewis, 2005; Klein et al., 2007).

Element	Species	Specimen #	Site	Horizon, Mb, or LMA	Age
Humerus	Homotherium cf. badarensis	AL 169-22	Hadar, Ethiopia	Denen Dora	3.22–3.18 Mya
	Homotherium sp.	OMO 230-73-2945	Omo, Ethiopia	Shungura G	2.33–1.88 Mya
	Homotherium sp.	KNM-ER 791 A	Koobi Fora, Kenya	Upper Burgi	2.0–1.88 Mya
	Dinofelis aronoki	KNM-ER 4419 A	Koobi Fora, Kenya	Upper Burgi	2.0–1.88 Mya
	Dinofelis sp. D	OLD 74-01	Olduvai Gorge, Tanzania	Bed I	1.87–1.7 Mya
	Dinofelis sp. D	OLD 74-54	Olduvai Gorge, Tanzania	Bed I	1.87–1.7 Mya
	Smilodon gracilis	F:AM 69227	Mcleod Pocket A (FL), USA	L. Irvingtonian	~700 Kya
Radius	Dinobastis serus	TMM 933-2421	Friesenhahn Cave (TX), USA	Rancholabrean	19–17 Kya
	Dinobastis serus	TMM 933-2565	Friesenhahn Cave (TX), USA	Rancholabrean	19–17 Kya
	Homotherium sp.	KNM-ER 3113	Koobi Fora, Kenya	Tulu Bor	3.36–2.68 Mya
	Dinofelis aronoki	KNM-ER 4419 C	Koobi Fora, Kenya	Upper Burgi	2.0–1.88 Mya
	Dinofelis cf. D. diastemata	PQ-L 20995	Langebaanweg, S. Africa	E Quarry	~5 Mya
Femur	Dinofelis piveteaui	KNM-KE 21 O	Kanam East, Kenya	—	1.0–0.9 Mya
	Smilodon gracilis	F:AM 69220	Mcleod Pocket A, FL, USA	L. Irvingtonian	~700 Kya
	Dinobastis serus	TMM 933-1570	Friesenhahn Cave, TX, USA	Rancholabrean	19–17 Kya
	Dinobastis serus	TMM 933-2668	Friesenhahn Cave, TX, USA	Rancholabrean	19–17 Kya
	Dinobastis serus	TMM 933-3498	Friesenhahn Cave, TX, USA	Rancholabrean	19–17 Kya
	Homotherium sp.	F 267-1	Omo, Ethiopia	Shungura G	2.33–1.88 Mya

Table 14.3 (*cont.*)

Element	Species	Specimen #	Site	Horizon, Mb, or LMA	Age
	Dinofelis barlowi	UCMP 69526	Bolt's Farm, S. Africa	—	~2.0 Mya
	Smilodon fatalis	TMM 933–373	Friesenhahn Cave, TX, USA	Rancholabrean	19–17 Kya
	Smilodon gracilis	F:AM 69229	Mcleod Pocket A, FL, USA	L. Irvingtonian	~700 Kya
	Smilodon gracilis	F:AM 69230	Mcleod Pocket A, FL, USA	L. Irvingtonian	~700 Kya
Tibia	*Megantereon whitei*	EFT 277	Elandsfontein, S. Africa	—	1–0.6 Mya
	Megantereon whitei	EFT 9486 C	Elandsfontein, S. Africa	—	1–0.6 Mya
	Smilodon fatalis	TMM 933–374	Friesenhahn Cave, TX, USA	Rancholabrean	19–17 Kya

length (described below); all six excluded taxa have been previously demonstrated to be unusually long-limbed (servals, cheetahs, African wildcats, lynxes) or short-limbed (clouded leopards, jaguars) (Sunquist and Sunquist, 2002; Christiansen and Harris, 2005; Day and Jayne, 2007). PCAs and PCi regressions were therefore based on a set of 15 extant taxa, ranging in size from approximately 1.6 to 226 kg.

For each element, fossil specimens were projected onto PCi to derive their corresponding PCi scores. Individual fossil specimens could then be assessed by their relationship to the extant-based regression line. Since all of our fossil specimens are within the upper size range of our extant taxa (i.e. close to lions and tigers), no extrapolation of established regression lines was necessary. (Extrapolation can be particularly problematic for polynomial regression.)

It is worth noting that our PCi regression lines are only descriptive of extant scaling trends and cannot be construed as lines of functional equivalence or functional competence among taxa (see Jungers, 1984b). That is, we have no reason to believe that those taxa which fall on a given regression line are doing so because of some shared underlying biomechanical factor associated with the demands of body size. Close proximity of lions and tigers to the regression line, for example, will have much to do with the simple fact that they represent the largest specimens included in the regression. This is particularly relevant given the observation (noted above) that large-bodied carnivorans exhibit different scaling patterns than small-bodied carnivorans (Bertram and Biewener, 1990). Nonetheless, the PCi regressions allow us to construct a comparative baseline with which to consider the overall morphology of the fossils.

The linear regression analyses described in the following section are designed to assess how well *all* large-bodied felids (both extant and extinct) fit with scaling trends defined by small-bodied (<100 kg) extant taxa.

Regressions: dimension vs. element size

Model II regression (RMA) of ln-transformed extant data was used to quantify scaling relationships and to test the null hypothesis of geometric similarity. (The choice of natural over base-10 logarithms is completely arbitrary.) Ideally, since we are interested in the effects of body mass on skeletal design, we would examine the scaling of each skeletal dimension relative to body mass. Unlike in extant taxa, however, body masses of extinct taxa can only be estimated from the preserved skeletal dimensions. Hence, a common approach when using fossils in scaling analyses is to perform regressions in which two linear dimensions are compared. While informative, this approach does not convey how each dimension relates to a measure of *overall size* (e.g. element size).

Hence, the results of such an approach can be potentially misleading, since taxa may exhibit similar deviations from the regression line, but for very different reasons. For example, taxa may fall below the regression line due to a relatively *large* X value, a relatively *small* Y value, or a combination of both; indeed, taxa with a relatively *large* Y value may still fall *below* the regression line due to a relatively large X value.

As an alternative to dimension vs. dimension scaling, we chose to regress each linear dimension of a given bone against a measure of overall bone size, whereby the size of the element was calculated as the geometric mean of all its linear dimensions (e.g. Darroch and Mosimann, 1985; Jungers *et al.*, 1995). The magnitude of a given measurement could thereby be assessed relative to the 'size' of the entire element, rather than relative to the magnitude of an additional (and perhaps arbitrary) skeletal dimension. This method is particularly useful in the case of fossils where an individual may be represented by a single bone rather than a skeleton. Obviously, a change in a single linear dimension of a specimen will also produce a change in its geometric mean, thereby potentially changing how one interprets the other dimensions relative to overall element size. Hence, the use of geometric means is not as ideal as the use of body mass. Nonetheless, for a given element in our data set, even if one artificially doubles the size of any linear dimension (without changing any other dimensions), the resulting change in our size measurement is relatively small and produces a minimal (and practically negligible) difference in the relationship of the unchanged variables to overall size.

The goal of our RMA regressions is to evaluate how well the morphology of machairodontines and extant large-bodied taxa (>100 kg) is predicted by scaling trends defined by small-bodied extant taxa. Hence, fossil taxa and large-bodied extant taxa (lions, tigers) were excluded from the RMA calculations. In addition, regressions involving limb bone *lengths* were run with and without expected outliers, particularly the unusually long-limbed or short-limbed taxa noted in the previous section. Hence, the majority of RMA regressions were based on a set of 17 taxa (ranging in size from approximately 1.6 to 80.9 kg), while additional limb length regressions were based on a reduced set of 11 taxa (ranging in size from approximately 1.6 to 58.3 kg).

Taxa not included in the RMA calculations (e.g. fossil specimens, tigers, lions) were later added to the RMA scatterplots to visually assess their position relative to the regression line. Although assessment of our larger specimens requires extrapolation of regression lines into size ranges beyond those for which the regressions were calculated, we do not consider this problematic given that these extrapolations are relatively small; indeed, the point of this research is to observe how well the morphology of larger-bodied felids is predicted by scaling trends in smaller-bodied felids. To quantify the relationship of a given taxon/fossil to a regression line, we calculated its 'Relative Deviation' (RD; Organ and Ward, 2006). In short, the RD is the RMA 'residual' standardised by the standard error and is expressed in standard error units:

$$RD = \sqrt{|residual|} \Big/ CSEE$$

RMA 'residuals' were calculated as the area of a triangle defined by the RMA line and the horizontal and vertical distances of the data point from the line (McCarthy, 2001). The CSEE is the corrected standard error of the estimate (SEE); equations for both SEE and CSEE are described by Zar (1984). Specimens that fall on the regression line would have an RD value of zero. Critical values from Student's t-distribution were used to determine whether the absolute value of RD for a given taxon/fossil was significantly high (i.e. whether the taxon/fossil fell outside the 95% confidence interval of the RMA line). For example, if $N = 10$, then a specimen with an $|RD|$ of 2.228 or less falls within the 95% confidence interval of the RMA line.

Rather than relying solely on standard regression plots, we examined the relationship of each taxon/fossil to the regression line via 'relative deviation plots'. On the RD plots, element size (ln-transformed) was plotted on the x-axis (as in the regression plots), but RD values were plotted on the y-axis. The RD plots provide clearer pictures of which taxa/fossils fall outside the 95% CL, which are easily included on the graphs.

For each element, in addition to using the linear dimensions themselves, we also calculated 'overall size' for each of three different regions: (1) proximal articular surface, (2) midshaft, and (3) distal articular surface. The 'size' of each of these regions was calculated as the geometric mean of its linear dimensions. Using RMA regression (as described above), we were thereby able to assess how each of these regions scales relative to overall element size.

For each regression, we tested the null hypothesis of geometric similarity. Significant positive ($k > 1$) or negative ($k < 1$) allometry was indicated when the confidence limits of the slope value (k) did not contain the expected value of the isometric slope ($k = 1$).

Phylogenetic inertia

Since species may share similarities in traits due to common ancestry (i.e. phylogenetic inertia), species used in a regression equation may not be independent examples of the relationship between traits (Clutton-Brock and Harvey, 1977; Harvey and Pagel, 1991; Smith, 1994). Hence, when each species is treated as independent, the degrees of freedom are inflated, resulting in underestimated standard errors and confidence intervals. To make valid statistical inferences based on more appropriate degrees of freedom, we calculated a smaller effective sample size (N_e) for each regression using the method described by Smith (1994). Under Smith's method, nested ANOVA is used to partition the total variation in the data set among the levels of a nested hierarchy defined by taxonomy. The number of groups at each level of the taxonomic hierarchy is weighted by the percentage of variation explained by that level of the nested ANOVA. An effective sample size (N_e) that takes phylogenetic constraint into account is calculated by summing these weighted values, whereby the maximum possible value of N_e equals the number of species. The effective sample size is calculated for every regression.

In our case, the 15 taxa used in our PC1 regressions were divisible into 6 genera, while the 17 taxa used in most of our RMA regressions were divisible into 8 genera (see Table 14.2 for taxonomic classifications). As noted above, RMA regressions involving limb length were based on a reduced set of 11 taxa; these were divided into 6 genera. In those cases where conspecific sexes were considered separately, each sex was counted as a separate 'species' within the genus. For each regression, N_e was rounded off to the nearest integer and used to determine the degrees of freedom for significance testing and confidence limits. For all regressions, we calculated an effective sample size well below the actual sample size (with N_e values ranging from 5 to 9), rendering our regression results relatively conservative.

Results

Humerus

The first PC based on mean 'shape' data accounts for the vast majority (98.7%) of shape variation among the 15 extant species. The correlations of each variable with the scores along PC1 are provided at the bottom of Table 14.4. The variable most strongly correlated with PC1 is clearly humeral length ($r = -1.000$). Other contributing variables include AP and SI head diameter, as well as AP midshaft diameter.

With respect to the relationship between PC1 and overall element size, a polynomial (quadratic) regression ($R^2 = 0.947$) is a significantly better fit ($P = 0.014$) than a linear regression ($R^2 = 0.887$). The R^2 value for the polynomial regression is significantly high ($P < 0.001$) even for the low effective sample size ($N_e = 9$) based on phylogenetic considerations. Given that PC1 (which is based on 'shape' data) is significantly correlated with element size,

Table 14.4 Relative deviation (RD) values and regression statistics from the RMA regressions of the **humerus**.[1]

Specimen ID	Taxon	Humeral head					Midshaft			Trochlea			
		Length	SI Diam	ML Diam	AP Diam	Overall Size	ML Diam	AP Diam	Overall size	Medial lip height	Lateral Lip Height	ML width	Overall size
KNM-ER 4419 A	*Dinofelis aronoki*	**-2.57**	-1.01	**-2.89**	-0.67	**-2.24**	**(-2.22)**	1.06	-0.43	2.93	-1.27	**2.55**	2.03
OLD 74-01	*Dinofelis* sp. D²	**-2.64**	0.93	0.59	**4.88**	2.53	0.82	0.96	1.07	1.59	**-3.29**	1.86	-1.65
OLD 74-54	*Dinofelis* sp. D	**-3.29**	-0.32	-1.16	**2.41**	0.06	-0.57	-0.54	-0.62	-0.79	0.98	0.01	1.05
OMO 230-73-2945	*Homotherium* sp.	**3.13**	1.56	1.54	1.15	**(2.18)**	0.48	**3.72**	**2.71**	**-4.04**	-0.90	-1.34	**-3.90**
KNM-ER 791 A	*Homotherium* sp.	**4.15**	1.39	**-4.15**	**-3.76**	**-2.39**	**(-2.11)**	**4.34**	1.78	**(-2.15)**	-0.14	0.30	-0.91
AL 169-22	*H.* cf. *badarensis*	1.53	-1.10	**-2.34**	**2.88**	-0.85	0.17	0.71	0.58	-1.65	0.06	0.35	-0.35
F:AM 69227	*Smilodon gracilis*	**-3.83**	**-2.45**	0.55	-1.37	-1.65	0.46	-1.03	-0.41	0.24	0.44	1.66	1.97
Large extant taxa	*Panthera leo* (F)	(-2.35)	-0.35	0.38	0.04	0.04	-1.69	-0.70	-1.30	0.35	0.18	0.93	1.31
	Panthera leo (M)	-1.26	-1.11	0.14	-0.05	-0.58	-1.19	-0.15	-0.69	0.05	-0.04	1.02	0.97
	Panthera tigris (F)	0.56	1.27	1.85	2.38	**2.55**	0.50	-1.14	-0.47	1.02	**-2.73**	1.77	-1.31

Table 14.4 (*cont.*)

Specimen ID		Humeral head					Midshaft			Trochlea			
	Taxon	Length	SI Diam	ML Diam	AP Diam	Overall Size	ML Diam	AP Diam	Overall size	Medial lip height	Lateral Lip Height	ML width	Overall size
	Panthera tigris (M)	−1.46	−0.13	0.08	−0.68	−0.21	−0.20	−0.59	−0.46	−1.34	0.13	1.45	0.63
Regression statistics	R	0.998	0.995	0.997	0.999	0.999	0.996	0.994	0.997	0.987	0.958	0.989	0.996
	N_e	7	10	10	9	10	10	10	10	10	10	9	10
	Slope (k)	(0.838)	(0.910)	1.015	(1.058)	0.993	1.060	(1.146)	(1.102)	0.9874	1.088	0.997	1.004
	95% L_1 (k)	0.784	0.838	0.957	1.019	0.953	0.985	1.047	1.028	0.865	0.866	0.875	0.934
	95% L_2 (k)	0.896	0.988	1.078	1.099	1.034	1.047	1.255	1.181	1.119	1.365	1.137	1.078
	y-intercept	2.695	0.514	0.327	0.212	0.355	−0.425	−0.459	−0.438	−1.041	−2.336	0.234	−0.995
Correlation with humeral PC$_1$		**−1.000**	**−0.814**	0.413	**0.802**	—	0.447	**0.876**	—	−0.164	0.162	0.497	—

Note: [1] The independent variable for each regression is 'overall size' of the humerus. All sample sizes equal 17, except for 'length' ($N = 11$). For RD values, bold indicates that the RD is outside the 95% confidence interval based on full df; parentheses indicate that the RD is not significantly high given the reduced df based on N_e. Correlations with PC$_1$ are bold if significantly high ($P < 0.05$). Slope values in parentheses indicate significant allometry (i.e. the 95% confidence limits of the slope do not include 1.00). All 95% confidence limits for slope values are based on reduced df.
[2] Designation from Werdelin and Lewis (2001).

and that it also accounts for virtually all of the shape variation observed among taxa, it appears that much of the variation in humeral shape among extant felids is associated with size variation. None of the 15 extant taxa used in the polynomial regression falls outside the 95% CL of the regression line.

The machairodontine specimens vary in their placement with respect to the extant polynomial regression line (Figure 14.1). One must remember that taxa can fall above or below the regression lines for different reasons since PC1 scores are based on multiple variables. Although the three specimens of *Dinofelis* are not particularly well-predicted, they all fall within the 95% CL. Based on full degrees of freedom (df), *Smilodon* falls above the upper 95% CL, and all three specimens of *Homotherium* (Omo, Koobi Fora, and Hadar)

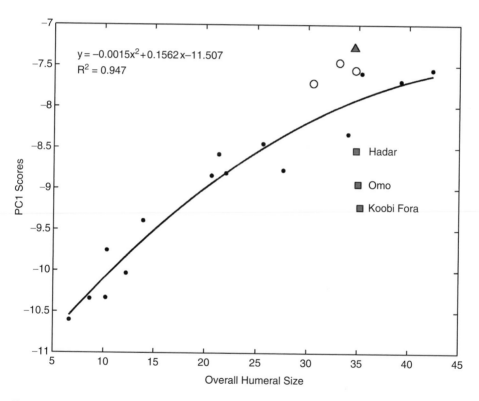

Figure 14.1 Polynomial regression of overall humeral 'shape' (i.e. PC1 scores based on mean 'shape' data for each species) vs. overall humeral size. The regression is based on species means for extant felids (dots) only. Unusually long-limbed or short-limbed taxa (identified in text) were not included. Shaded fossil specimens are those found to be outside the 95% confidence limits (based on full *N*). The machairodontines are variable in placement. Note that the Hadar specimen is the oldest of the three *Homotherium* humeri. Machairodontine taxa: ○ = *Dinofelis*, ■ = *Homotherium*, ▲ = *Smilodon*.

fall below the lower 95% CL. If the degrees of freedom are reduced (based on $N_e = 9$), *Smilodon* and one of the *Homotherium* specimens (Hadar) are no longer significantly deviant. None the less, it is clear that neither *Smilodon* nor *Homotherium* fall close to the shape trend established by the extant taxa.

Results of the RMA regressions between each linear dimension and overall humeral size are also presented in Table 14.4. All of the correlations are strong ($R > .95$). Several slope values (k) indicate significant allometry (i.e. the confidence limits of the slope do not include 1.00). As overall humeral size increases, humeral length decreases (relatively), while the AP diameters of both the humeral head and the midshaft increase. It would appear that as felids get larger, the humerus becomes more robust (larger diameter relative to length), particularly with respect to the AP dimension.

The large-bodied extant taxa (lions and tigers) are generally well-predicted by the RMA regressions based on smaller taxa. Only female tigers exhibit more than one significantly high RD value.

With the exception of *Homotherium* cf. *hadarensis* (AL 169–22), all of the machairodontines have RD values beyond the 95% CL with respect to humeral length (Figure 14.2). The pattern of RD values for humeral length reflects the pattern of residuals around the polynomial regression line. The two non-Hadar specimens of *Homotherium* (i.e. Omo and Koobi Fora) are relatively long and fall with servals, cheetahs, and bobcats beyond the upper 95% CL. Specimens assigned to *Dinofelis* and *Smilodon* are relatively short (below the lower 95% CL), and are similar to clouded leopards and jaguars.

The three specimens of *Homotherium* also exhibit a number of other significant RD values, although the results are not always consistent. Overall, as suggested by the polynomial regression results, the two younger specimens from Omo and Koobi Fora are more similar to one another than they are to the Hadar specimen, and are more 'extreme' for several dimensions (despite the fact that all three specimens are almost identical in overall size). For example, although all three specimens have positive RDs for humeral length, the humeral length of *H.* cf. *hadarensis* is not outside the 95% CL. Similar results are obtained for AP midshaft diameter (positive RDs) and medial trochlear lip height (negative RDs); all three *Homotherium* specimens share the same relative position (above or below the RMA line), but the Omo and Koobi Fora specimens are more extreme. Omo 230–73–2945 and KNM-ER 791A exhibit unusually large AP midshaft diameters, a positively allometric feature that is also significantly (and positively) correlated with PC1. In addition, these two specimens (and, to some extent, AL 169–22) have relatively small medial trochlear lip heights, although the RD value for KNM-ER 791A is not significantly high when reduced degrees of freedom are used.

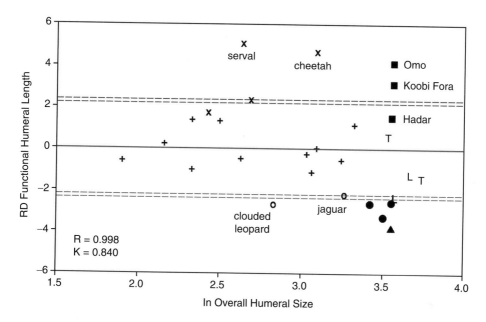

Figure 14.2 Relative deviations (RD) from the RMA regression of functional humeral length and overall humeral size. Only extant taxa indicated by plus signs (+) were included in the RMA regression. Dotted lines indicate 95% confidence limits based on full N (inner lines) and reduced N_e (outer lines). With the exception of *Homotherium* cf. *hadarensis*, *Homotherium* has significantly longer humeri relative to overall humeral size (like servals and cheetahs). *Dinofelis* and *Smilodon* have significantly shorter humeri relative to overall humeral size (like clouded leopards, jaguars, and female lions). Extant taxa not included in the regression: x = long-limbed taxa, ○ = short-limbed taxa, L = lions, T = tigers. Machairodontine taxa: ● = *Dinofelis*, ■ = *Homotherium*, ▲ = *Smilodon*.

As noted above, all three *Dinofelis* specimens show the opposite trend from *Homotherium* with respect to humeral length (i.e. they are extremely short relative to overall humeral size). In fact, this is the only feature for which all three specimens exhibit a significantly high RD. The two Olduvai specimens also appear to have relatively large AP humeral head diameters, which, along with unusually short lengths, would account for their position on the polynomial regression plot (since AP head diameter is positively correlated with PC1). The anteroposteriorly wide head is not shared, however, with KNM-ER 4419A (*D. aronoki*). In fact, *D. aronoki* is found to have a relatively small ML head diameter.

Based on our single specimen of *Smilodon gracilis*, this species differs from all of the other machairodontine taxa in having a relatively small SI humeral head diameter. The fact that both humeral length and SI head diameter are negatively correlated with PC1 would account for the high positive residual observed for *Smilodon* on the polynomial regression. Since we have included only one complete humeral specimen of *S. gracilis*, however, we cannot be confident that the relatively small SI head diameter is characteristic of this species in general.

Radius

As was the case with the humerus, the first PC based on radial 'shape' data accounts for the vast majority (99.3%) of shape variation among the 15 extant felids. The correlations between each variable and the PC1 scores are provided at the bottom of Table 14.5. As above, the variable most strongly correlated with PC1 is element length ($r = -1.000$); other contributing variables include AP midshaft diameter and maximum distal articular width, both of which are positively correlated with PC1.

With respect to the relationship between PC1 and overall radial size, the polynomial regression ($R^2 = 0.950$) is a significantly better fit ($P = 0.014$) than the linear regression ($R^2 = 0.872$). The R^2 value for the polynomial regression is significantly high ($P < 0.001$) even for the low effective sample size ($N_e = 9$). None of the extant taxa used in the polynomial regression falls outside the 95% CL of the regression line.

The machairodontine specimens vary in their placement with respect to the polynomial regression line (Figure 14.3). The *Smilodon* specimen and the three specimens of *Dinofelis* do not fall outside the 95% CL. Based on full df, *Homotherium* falls below the lower 95% CL, as do the two specimens of *Dinobastis serus*. On the other hand, if the degrees of freedom are reduced (based on $N_e = 9$), *Homotherium* is no longer significantly deviant, although it is clearly not well-predicted.

Results of the RMA regressions between each linear dimension and overall radial size are presented in Table 14.5. All of the correlations are strong ($R > .99$). As was the case with the humerus, the radius appears to increase anteroposterior bending strength as the overall size of the bone increases. That is, radial length is found to exhibit significant negative allometry, while AP midshaft diameter is found to exhibit significant positive allometry (while ML midshaft diameter shows an isometric trend).

The large-bodied extant taxa (lions and tigers) are particularly deviant with respect to both midshaft diameters. Interestingly, opposing patterns are observed for ML diameter and AP diameter within these taxa: significantly high *negative* RD values for ML diameter, and significantly high *positive* RD values for AP diameter. (Female lions are not beyond the 95% CL of

Table 14.5 Relative deviation (RD) values and regression statistics from the RMA regressions of the **radius**.[1]

Specimen	Taxon	Length	Radial head			Midshaft			Distal articular surface		
			Max Diam	Min Diam	Overall size	ML Diam	AP Diam	Overall size	Max width	Min width	Overall size
KNM-ER 4419 C	*Dinofelis aronoki*	-2.04	1.69	1.98	2.00	-0.77	-0.05	-0.55	-0.97	0.31	-0.45
PQ-L 20995	*Dinofelis* cf. *D. diastemata*	-0.69	1.19	**2.83**	2.07	0.15	0.83	0.75	**-2.57**	**(-2.11)**	-3.37
KNM-KE 21 O	*Dinofelis piveteaui*	-2.10	1.38	1.88	1.76	-0.21	-0.13	-0.19	0.56	-1.48	-0.51
KNM-ER 3113	*Homotherium* sp.	2.50	-2.05	0.72	-0.96	-0.50	0.99	0.41	**-2.70**	1.00	-1.36
TMM 933–2421	*Dinobastis serus*	2.08	-0.53	-0.65	-0.62	-1.30	**(2.13)**	0.61	-1.98	-0.38	-1.73
TMM 933–2565	*Dinobastis serus*	**(2.34)**	-0.05	1.52	0.69	-1.93	0.24	-1.21	-1.52	-0.36	-1.36
F:AM 69220	*Smilodon gracilis*	-1.83	0.47	**2.97**	1.67	-0.44	-0.48	-0.60	-2.07	**(2.16)**	-0.12
Large extant taxa	*Panthera leo* (F)	0.16	-0.66	**3.15**	1.03	**-2.95**	2.10	-0.68	0.36	-1.62	-0.75
	Panthera leo (M)	0.19	-0.73	0.55	-0.20	**-3.77**	**2.93**	-0.69	**(2.17)**	-1.15	0.99
	Panthera tigris (F)	-0.87	-0.83	-1.15	-1.04	**-4.74**	**3.28**	-1.19	**3.15**	2.05	**3.88**
	Panthera tigris (M)	-0.65	-0.59	-0.83	-0.74	**-3.99**	2.53	-1.14	(2.20)	1.98	3.10

Table 14.5 (*cont.*)

Specimen	Taxon	Radial head				Midshaft			Distal articular surface		
		Length	Max Diam	Min Diam	Overall size	ML Diam	AP Diam	Overall size	Max width	Min width	Overall size
Regression statistics	R	0.992	0.993	0.996	0.996	0.992	0.994	0.998	0.996	0.996	0.997
	N_e	7	10	10	10	10	10	10	10	10	10
	Slope (k)	(0.721)	1.018	1.030	1.023	1.092	(1.159)	1.012	1.076	0.953	1.121
	95% L_1 (k)	0.624	0.921	0.961		0.983	1.056		0.997	0.885	
	95% L_2 (k)	0.833	1.126	1.105		1.213	1.271		1.162	1.027	
	y-intercept	2.869	−0.146	−0.497	−0.319	−0.668	−1.328	−0.189	−0.158	−0.234	−0.986
Correlation with radial PC1		**−1.000**	−0.008	0.467	—	−0.303	**0.859**	—	**0.737**	−0.351	—

Note: [1] The independent variable for each regression is 'overall size' of the radius, except for 'length' ($N = 11$). All sample sizes equal 17, except for 'length' ($N = 11$). For RD values, bold indicates that the RD is outside the 95% confidence interval based on full df; parentheses indicate that the RD is not significantly high given the reduced df based on N_e. Correlations with PC1 are bold if significantly high ($P < 0.05$). Slope values in parentheses indicate significant allometry (i.e. the 95% confidence limits of the slope do not include 1.00). All 95% confidence limits for slope values are based on reduced df.

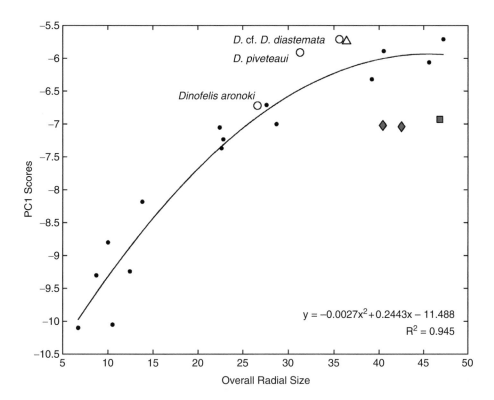

Figure 14.3 Polynomial regression of overall radial 'shape' (i.e. PC1 scores based on mean 'shape' data for each species) vs. overall radius size. The regression is based on species means for extant felids (dots) only. Unusually long-limbed or short-limbed taxa (identified in text) were not included. Shaded fossil specimens are those found to be outside the 95% confidence limits (based on full N). Once again, machairodontines are variable in placement. Machairodontine taxa: ◆=*Dinobastis*, ○=*Dinofelis*, ■=*Homotherium*, △=*Smilodon*.

the RMA line for the latter dimension, although they do have a high positive RD.) For felids of their size, lions and tigers have unusually large AP midshaft diameters combined with unusually small ML midshaft diameters. As a result of these opposing patterns, none of the lions or tigers are unusual with respect to overall size of the midshaft.

None of the machairodontine taxa exhibits the lion/tiger pattern of deviations. Indeed, the machairodontines in general exhibit few significantly high RD values. As noted above, *Dinobastis* and *Homotherium* were found to be the most deviant fossil taxa from the polynomial regression line, falling well below it. With respect to radial length (which is highly correlated with PC1), the specimens belonging to both of these taxa are relatively long; their RD values for radial length straddle the upper 95% confidence limit and fall close to the shortest of the long-limbed extant taxa (all of which are beyond the upper 95% CL). In contrast, *Dinofelis* and *Smilodon*, for the most part, fall near the lower 95% CL (though not beyond it), indicating relatively short radii (like clouded leopards and jaguars). *Dinofelis* cf. *D. diastemata* (PQ-L 20995), however, falls much closer to the RMA line and very near to extant leopards.

Based on the results for radial length, we combined the fossil and extant data into three groups: (1) long-limbed extant taxa, *Homotherium*, and *Dinobastis*, (2) short-limbed extant taxa, *Smilodon*, and *Dinofelis*, and (3) all other extant taxa, including lions and tigers. When separate RMA regressions are run for each group (radial length against overall radial size), three parallel scaling trends emerge (Figure 14.4): a long-limbed trend, a short-limbed trend, and an average length-limbed trend. The three slopes are very similar ($0.70 < k < 0.74$) and exhibit significant negative allometry, but the regression lines differ in the *y*-intercept, reflecting differences in relative radial length. The parallel regressions nicely demonstrate the three different groups of extant

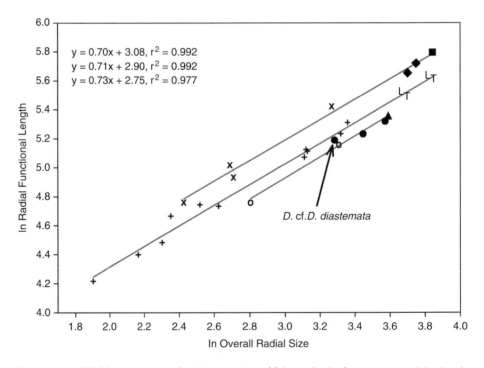

Figure 14.4 RMA regressions for three groups: (1) long-limbed extant taxa, *Dinobastis* (◆), and *Homotherium* (■), (2) short-limbed extant taxa, *Dinofelis* (●), and *Smilodon* (▲), and (3) all other extant taxa, including lions and tigers. All slopes indicate significant negative allometry. Note that these three lines are almost parallel.

and extinct large-bodied felids with respect to relative radial length. Lions and tigers fall along the average length-limbed trend, while most machairodontines fall along either the long-limbed trend (*Homotherium* and *Dinobastis*), or the short-limbed trend (*S. gracilis*, *Dinofelis aronoki*, and *Dinofelis piveteaui*). The oldest species of *Dinofelis* included (*D. cf. D. diastemata*) does not fall with the other *Dinofelis* specimens, but rather falls between the average length-limbed and short-limbed regression lines.

Another reason (in addition to long radii) why *Homotherium* and *Dinobastis* fall well below the polynomial regression line is their comparatively high negative RD values for maximum distal articular width (Figure 14.5). This variable is positively correlated with PC1 and scales isometrically with overall radial size. *Homotherium* (KNM-ER 3113) and the two specimens of *Dinobastis serus* have relatively small distal articular surfaces (at least in the widest dimension), resulting in negative RD values that either approach (*Dinobastis*) or surpass (*Homotherium*) the lower 95% CL. *Smilodon* also falls low with

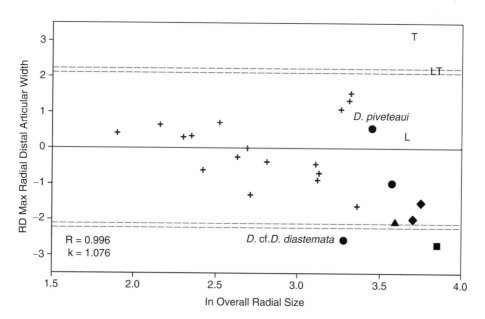

Figure 14.5 Relative deviations (RD) from the RMA regression of maximum radial distal articular width and overall radial size. All extant taxa (+) were included in the RMA regression except for lions (L) and tigers (T). Dotted lines indicate 95% confidence limits based on full N (inner lines) and reduced N_e (outer lines). Extinct and extant large-bodied felids generally exhibit opposite trends with respect to distal articular width, such that most machairodontines are comparatively narrow while most lions and tigers are relatively wide. Note the range of diversity in *Dinofelis*. Machairodontine taxa: ◆=*Dinobastis*, ●=*Dinofelis*, ■=*Homotherium*, ▲=*Smilodon*.

Dinobastis and *Homotherium*. In addition, *Dinofelis* cf. *D. diastemata* also falls below the 95% CL, and once again, differs from the other two *Dinofelis* specimens, which fall closer to the RMA line.

In addition to its small *maximum* distal articular width, *D.* cf. *D. diastemata* also has an unusually small *minimum* distal articular width, giving it a significantly small distal articular surface overall. On the other hand, *Smilodon* combines its unusually small maximum width with an unusually *large* minimum width (beyond the lower 95% CL). Hence, unlike *D.* cf. *D. diastemata*, its overall articular size is very close to the RMA line. The minimum dimension of the *Smilodon* radial head is also unusually large for overall radial size.

Tigers and lions are also unusual with respect to maximum width of the distal articular surface and, with the exception of female lions, unlike all other extant felids. All but female lions have significantly high RD values for this dimension, which is the opposite pattern to that observed among the machairodontine specimens noted above. (In fact, among machairodontines, only *D. piveteaui* falls above the regression line.) In short, with the exception of *D. piveteaui*, all included machairodontines have relatively narrow maximum distal articular widths relative to overall radial size. This condition is in contrast to the relatively wide width found in extant lions and tigers. A similar result has been found by other researchers (Meachen-Samuels, personal communication, 2008).

Femur

The first PC based on femoral 'shape' data accounts for the majority (97.8%) of shape variation among the 15 extant felids. Table 14.6 provides the correlation of each variable with the PC1 scores. As above, the variable most strongly correlated with PC1 is element length ($r = -1.000$); other contributing variables include AP diameter of the femoral head and bicondylar width, both of which have positive correlations with PC1.

With respect to the relationship between PC1 and overall femoral size, the polynomial regression ($R^2 = 0.740$) is not a significantly better fit ($P = 0.318$) than the linear regression ($R^2 = 0.713$; Figure 14.6). Indeed, the R^2 value for the linear least squares regression, although significantly high ($P < 0.001$; even for $N_e = 9$), indicates a much weaker relationship than demonstrated above for the two forelimb bones. As a result of the lower R^2 value and the wider confidence limits, none of the extant taxa are found to be significant outliers. The same holds true for the machairodontine specimens; none of them are beyond the 95% CL of the regression line. None the less, with the exception of the two *Smilodon gracilis*

Table 14.6 Relative deviation (RD) values and regression statistics from the RMA regressions of the **femur**.[1]

Specimen	Taxon	Femoral head					Midshaft			Distal articular surface		
		Length	AP Diam	SI Diam	ML Diam	Overall size	ML Diam	AP Diam	Overall size	Bicond. width	AP length	Overall size
UCMP 69526	*Dinofelis barlowi*	-1.02	-0.32	0.96	-0.74	-0.03	1.56	0.32	1.66	-1.84	0.00	-1.05
F 267–1	*Homotherium*	**-2.48**	1.82	**2.61**	0.77	1.98	2.26	**-2.40**	0.30	-1.21	-1.62	**(-2.17)**
TMM 933–1570	*Dinobastis serus*	-1.72	**2.77**	1.74	1.44	**(2.19)**	-0.01	-0.88	-0.57	**-3.22**	-0.70	**-2.49**
TMM 933–2658	*Dinobastis serus*	-1.33	1.98	-0.21	**2.77**	1.86	0.76	-1.48	-0.34	**-4.15**	-0.13	**-2.52**
TMM 933–3498	*Dinobastis serus*	-1.63	**3.10**	1.27	**2.56**	**2.65**	0.34	-2.09	-1.16	**-2.98**	-1.29	**-2.90**
TMM 933–373	*Smilodon fatalis*	-1.80	1.19	1.74	**(2.14)**	**(2.17)**	0.82	-1.42	-0.24	-1.34	**(-2.21)**	**-2.79**
F:AM 69229	*Smilodon gracilis*	-0.76	1.62	-0.20	**2.23**	1.49	-0.10	-0.65	-0.49	**(-2.12)**	-1.00	**(-2.13)**
F:AM 69230	*Smilodon gracilis*	-0.85	0.86	-0.76	**2.87**	1.41	-0.60	-0.41	-0.75	**-2.37**	-0.16	-1.50
Large extant taxa	*Panthera leo* (F)	-1.34	-0.42	0.43	1.04	0.65	0.66	-0.92	-0.03	-0.56	0.13	0.16
	Panthera leo (M)	-0.84	-0.35	-0.49	1.16	0.34	0.69	-1.97	-0.76	-0.59	1.19	0.80
	Panthera tigris (F)	0.36	0.57	0.00	-1.39	-0.56	0.08	-1.77	-1.16	-0.18	**(2.15)**	1.93

Table 14.6 (*cont.*)

	Femoral head					Midshaft			Distal articular surface		
Specimen / Taxon	Length	AP Diam	SI Diam	ML Diam	Overall size	ML Diam	AP Diam	Overall size	Bicond. width	AP length	Overall size
Panthera tigris (M)	−0.09	−0.18	−1.07	−0.85	−0.91	0.48	−1.31	−0.45	0.57	2.02	**2.26**
Regression statistics											
R	0.995	0.999	0.996	0.995	0.998	0.995	0.997	0.999	0.999	0.997	0.998
N_e	7	10	10	10	10	10	10	10	10	10	10
Slope (k)	0.905	1.035	1.008	0.997	1.012	0.967	1.047	1.040	(1.062)	1.020	1.004
95% L_1 (k)	0.807	0.996	0.941	0.915		0.890	0.982		1.020	0.957	
95% L_2 (k)	1.016	1.074	1.079	1.087		1.052	1.116		1.106	1.088	
y-intercept	2.303	−0.477	−0.467	−0.436	−0.457	−0.474	−0.791	0.107	0.033	0.172	−0.625
Correlation with femoral PC$_1$	**−1.000**	**0.697**	0.134	0.330	—	−0.081	−0.032	—	**0.637**	0.394	—

Note: [1] The independent variable for each regression is 'overall size' of the femur. All sample sizes equal 17, except for 'length' ($N = 11$). For RD values, bold indicates that the RD is outside the 95% confidence interval based on full df; parentheses indicate that the RD is not significantly high given the reduced df based on N_e. Correlations with PC1 are bold if significantly high ($P < 0.05$). Slope values in parentheses indicate significant allometry (i.e. the 95% confidence limits of the slope do not include 1.00). All 95% confidence limits for slope values are based on reduced df.

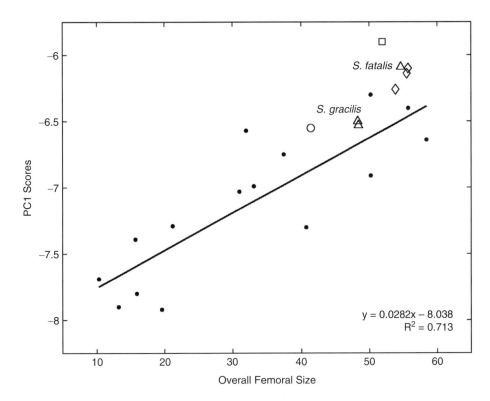

Figure 14.6 Linear regression of overall femoral 'shape' (i.e. PC1 scores based on mean 'shape' data for each species) vs. overall femur size. The regression is based on species means for extant felids (dots) only. Unusually long-limbed or short-limbed taxa (identified in text) were not included. None of the fossil specimens are outside the 95% confidence limits. The two species of *Smilodon* differ in placement due to differences in overall femoral size. Machairodontine taxa: $\Diamond = Dinobastis$, $\bigcirc = Dinofelis$, $\square = Homotherium$, $\triangle = Smilodon$.

specimens, the fossil femora are further above the regression line than the majority of extant taxa (especially *Homotherium*).

Results of the RMA regressions between each linear dimension and overall femoral size are presented in Table 14.6. All of the correlations are strong ($R > .99$). Unlike the lengths of the two forelimb bones, femoral length is not found to scale with significant allometry (although the slope is less than one), nor is any significant allometry detected in the femoral midshaft. In fact, the only significantly allometric dimension is the femoral bicondylar width, which scales with positive allometry ($k = 1.062$).

Lions and tigers are generally well-predicted by the RMA regressions. In addition, and in contrast to the forelimb results, all of the long-limbed and

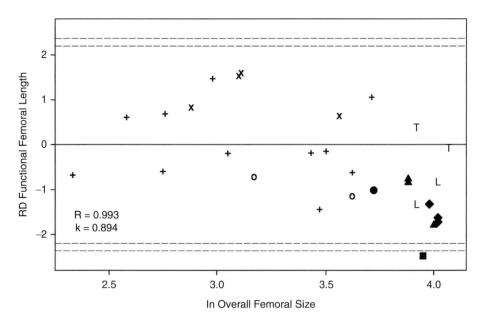

Figure 14.7 Relative deviations (RD) from the RMA regression of functional femoral length and overall femoral size. Only extant taxa indicated by plus signs (+) were included in the RMA regression. Dotted lines indicate 95% confidence limits based on full *N* (inner lines) and reduced N_e (outer lines). The machairodontines cluster together below the regression line, indicating relatively short femora, though only one specimen (*Homotherium*) has a significant RD value. Extant taxa not included in the regression: x = long-limbed taxa, o = short-limbed taxa, L = lions, T = tigers. Machairodontine taxa: ◆ = *Dinobastis*, ● = *Dinofelis*, ■ = *Homotherium*, ▲ = *Smilodon*.

short-limbed extant taxa have femoral lengths that are well-predicted (Figure 14.7), albeit with the same pattern of positive (long-limbed) and negative (short-limbed) deviations.

The comparatively deviant position of *Homotherium* (F 267–1) on the PC1 regression plot is likely related to the fact that it is the specimen with the most unusual femoral length, for which it has a significant negative RD. Although none of the other machairodontines falls outside the 95% CL with respect to femoral length, they similarly have negative RDs, with a number of specimens (apart from *Dinofelis* and the two *S. gracilis* specimens) falling at the lower extreme for RD values. This result is also reflected in the PC1 regression plot; in particular, the two *S. gracilis* fall closer to the PC1 regression line than any other fossil specimens (as noted above). Hence, all our machairodontine taxa, with the possible exception of *S. gracilis*, have relatively short femora relative to overall femoral size.

In general, the machairodontines do not appear remarkable with respect to femoral midshaft diameter. Indeed, with respect to overall midshaft size (i.e. geometric mean of AP and ML diameters), we observe very little variation among machairodontines, and all but *Dinofelis barlowi* clump together with deviations close to zero (Figure 14.8). The machairodontines appear conservative with respect to overall size of the midshaft, and are well-predicted by the slightly (but not significantly; $k = 1.04$) allometric trend observed among extant taxa. Hence, most of the machairodontine femoral specimens exhibit increased robusticity due to relatively short length (rather than relatively wide midshafts), at least when we consider linear dimensions relative to overall element size.

In contrast to the other fossil taxa, *Homotherium* has a significantly wider midshaft mediolaterally, as well as a significantly *smaller* midshaft anteroposteriorly. Its RD values for these two dimensions are similar in absolute value; in fact, as noted above, its overall midshaft size yields a relative deviation close to zero (0.30). Hence, based on one specimen, it would appear that *Homotherium* has an unusually shaped midshaft for a felid of its size (i.e. larger ML diameter

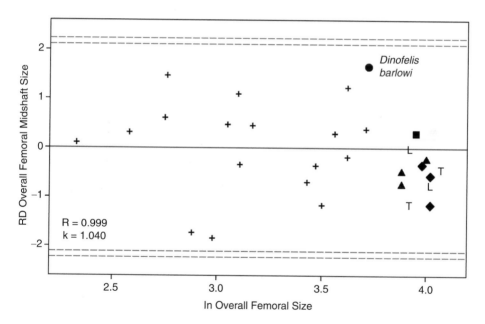

Figure 14.8 Relative deviations (RD) from the RMA regression of overall femoral midshaft size and overall femoral size. All extant taxa (+) were included in the RMA regression except for lions (L) and tigers (T). Dotted lines indicate 95% confidence limits based on full N (inner lines) and reduced N_e (outer lines). All of the large-bodied felids (both extinct and extant) cluster together with the exception of *Dinofelis barlowi*. Machairodontine taxa: ◆ = *Dinobastis*, ● = *Dinofelis*, ■ = *Homotherium*, ▲ = *Smilodon*.

relative to AP diameter), but that the overall size of the midshaft is not unusual. Although the other machairodontines have RD values close to zero with respect to ML midshaft diameter (unlike *Homotherium*), it is worth noting that all but *D. barlowi* have negative RD values with respect to AP midshaft diameter. Hence, the significantly narrow AP midshaft diameter of *Homotherium* appears to be an extreme case of a general machairodontine trend.

Aspects of femoral articular morphology effectively separate machairodontines from other felids. All of the machairodontines except *Dinofelis barlowi* have positive RD values for both AP and ML femoral head dimensions, as well as for overall size of the femoral head (i.e. the geometric mean of its three linear dimensions). Although not all fossil RD values for femoral head size are significantly high, they do distinguish most machairodontines from the extant taxa (Figure 14.9). In essence, the fossil femora are characterised by a large femoral head size relative to overall femur size. This enlarged femoral head is achieved largely by relative increases in the AP and ML dimensions, rather

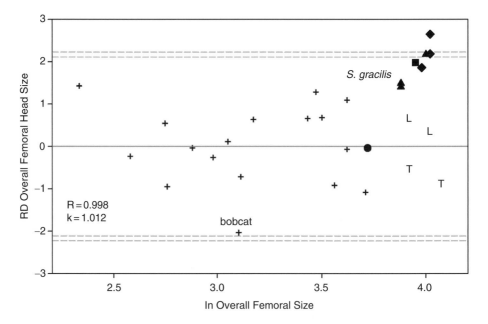

Figure 14.9 Relative deviations (RD) from the RMA regression of femoral head size and overall femoral size. All extant taxa (+) were included in the RMA regression except for lions (L) and tigers (T). Dotted lines indicate 95% confidence limits based on full N (inner lines) and reduced N_e (outer lines). While the overall femoral head size of *Dinofelis barlowi* is well-predicted, the other machairodontines tend to have large femoral heads relative to overall femoral size. Note the placement of lions and tigers. Machairodontine taxa: ◆ = *Dinobastis*, ● = *Dinofelis*, ■ = *Homotherium*, ▲ = *Smilodon*.

than the superoinferior dimension. Only *D. barlowi* (our only *Dinofelis* femoral specimen) falls along the regression line, despite having a femoral length close to that of the shortest specimen of *Dinobastis*. The relatively large AP head diameter of most machairodontines would also account for their respective positions above the PC1 regression line, since this dimension is positively correlated with this PC1.

The distal articular surface is also of interest. With respect to overall size of this region, most extant taxa fall well within the 95% CL (Figure 14.10). Male and female tigers straddle the upper 95% CL and are unusual among extant taxa in having a relatively large distal joint surface. In contrast, all of the machairodontines fall below the regression line; though not all of them have significant RD values, they all fall below most (if not all) extant taxa. All three specimens of *Dinobastis serus* have significant RDs, as well as two specimens of *Smilodon*, and our single specimen of *Homotherium*. In general, the unusually small articular surface of machairodontines is largely due to a

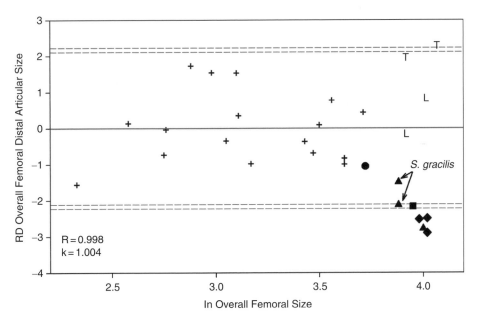

Figure 14.10 Relative deviations (RD) from the RMA regression of overall femoral distal articular size and overall femoral size. All extant taxa (+) were included in the RMA regression except for lions (L) and tigers (T). Dotted lines indicate 95% confidence limits based on full N (inner lines) and reduced N_e (outer lines). The distal articular size of most machairodontines is significantly small relative to overall femoral size. Note that lions and tigers differ from one another and from machairodontines. Machairodontine taxa: ◆ = *Dinobastis*, ● = *Dinofelis*, ■ = *Homotherium*, ▲ = *Smilodon*.

relatively narrow bicondylar width. The fossil RD values for distal AP length are also consistently negative (or 0.00 in the case of *D. barlowi*), although they are not typically as extreme as those for bicondylar width. As a result, the overall size of the distal articular surface is relatively small for all of the machairodontine femora. Hence, in general, most of the machairodontines exhibit an unusual combination of a large proximal articular surface, and a small distal articular surface.

Since bicondylar width is positively correlated with PC1, and machairodontines are relatively small in this dimension, it seems unusual that the fossil specimens would fall above the PC1 regression line. It is likely the case, however, that the combined influence of short femoral length (extreme negative correlation with PC1) and large AP femoral head diameter (positive correlation with PC1) overcomes the conflicting influence of bicondylar width.

Tibia

As consistent with the above results, the first PC based on tibial 'shape' data accounts for the large majority (99.2%) of shape variation among the 15 extant felids. As before, the variable most strongly correlated with PC1 is element length ($r = -1.000$); other contributing variables include AP length of the lateral condyle and AP midshaft diameter, both of which are positively correlated with PC1 (Table 14.7).

With respect to the relationship between PC1 and overall tibial size, the polynomial regression ($R^2 = 0.932$) is a somewhat better fit than the linear regression ($R^2 = 0.916$), although not quite significantly so ($P = 0.099$). The R^2 value for the polynomial regression is significantly high ($P < 0.001$) even for the low effective sample size ($N_e = 9$). None of the 15 extant taxa used in the polynomial regression falls outside the 95% CL of the regression line.

For the tibia, we have only three fossil specimens for comparison, since few complete tibiae of machairodontines have been found at the sites studied. Interestingly, all three specimens (two *Megantereon*, one *Smilodon*) fall well above the polynomial regression line, beyond the 95% CL (Figure 14.11) even using reduced degrees of freedom (based on $N_e = 9$).

Results of the RMA regressions between each linear dimension and overall tibial size are presented in Table 14.7. All of the correlations are strong ($R > 0.99$). As observed for the two forelimb bones, the results suggest that the tibia increases anteroposterior bending strength as element size increases. Tibial length exhibits significant negative allometry, while AP midshaft diameter exhibits significant positive allometry (and ML midshaft diameter shows an isometric trend). All other dimensions scale isometrically.

Table 14.7 Relative deviation (RD) values and regression statistics from the RMA regressions of the **tibia**.[1]

Specimen	Taxon	Proximal articular surface					Midshaft			Distal articular surface		
		Length	ML width	Medial condyle AP length	Lateral condyle AP length	Overall size	ML Diam	AP Diam	Overall Size	ML width	AP length	Overall size
EFT 277	*Megantereon whitei*	−5.11	−0.12	0.63	−0.19	0.32	1.61	2.61	2.79	0.30	−0.28	−0.06
EFT 9486 C	*Megantereon whitei*	−5.95	−2.41	2.06	0.67	1.22	0.12	2.59	1.53	0.52	0.79	0.86
TMM 933–374	*Smilodon fatalis*	−4.60	2.03	−0.10	2.70	3.15	−0.99	−1.50	−1.57	0.97	0.86	1.12
Large extant taxa	*Panthera leo* (F)	−2.60	−1.24	1.45	1.89	2.32	−0.12	−2.74	−1.52	−0.28	0.92	0.63
	Panthera leo (M)	−1.58	−1.04	1.28	0.65	1.22	−0.20	−2.07	−1.22	−0.44	1.17	0.75
	Panthera tigris (F)	−1.11	0.84	−0.36	−0.59	−0.32	−0.90	−0.63	−1.04	1.27	1.70	1.90
	Panthera tigris (M)	−1.25	0.34	−0.44	−0.71	−0.65	−0.65	−0.34	−0.66	0.62	2.03	1.89
Regression statistics	R	0.995	0.998	0.994	0.993	0.999	0.992	0.997	0.997	0.998	0.993	0.997
	N_e	7	10	10	10	10	10	10	10	10	10	10
	Slope (k)	(0.805)	1.025	1.024	1.082	1.040	1.009	(1.114)	1.059	1.018	0.975	0.995

Table 14.7 (*cont.*)

Specimen Taxon	Length	Proximal articular surface ML width	Medial condyle AP length	Lateral condyle AP length	Overall size	Midshaft ML Diam	AP Diam	Overall Size	Distal articular surface ML width	AP length	Overall size
95% L_1 (k)	0.718	0.979	0.939	0.981	0.999	0.911	1.048	0.998	0.968	0.885	0.937
95% L_2 (k)	0.902	1.074	1.117	1.193	1.082	1.118	1.184	1.123	1.071	1.074	1.057
Y-intercept	2.549	0.215	−0.286	−0.528	−0.187	−0.719	−0.939	−0.820	−0.090	−0.376	−0.228
Correlation with Tibial PC1	**−1.000**	0.276	0.434	**0.620**	—	−0.063	**0.682**	—	0.495	0.345	—

Note: [1] The independent variable for each regression is 'overall size' of the tibia. All sample sizes equal 17, except for 'length' ($N=11$). For RD values, bold indicates that the RD is outside the 95% confidence interval (using reduced df based on N_e). Correlations with PC1 are bold if significantly high ($P < 0.05$). Slope values in parentheses indicate significant allometry (i.e., the 95% confidence limits of the slope do not include 1.00). All 95% confidence limits for slope values are based on reduced df.

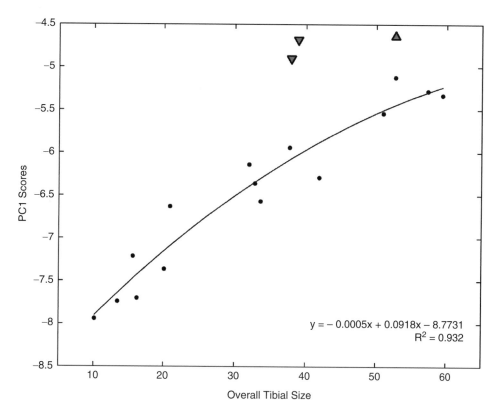

Figure 14.11 Polynomial regression of overall tibial 'shape' (i.e. PC1 scores based on mean 'shape' data for each species) vs. overall tibia size. The regression is based on species means for extant felids (dots) only. Unusually long-limbed or short-limbed taxa (identified in text) were not included. All three fossil specimens are outside the 95% confidence limits (based on N_e). Machairodontine taxa: ▲=*Smilodon*, ▼=*Megantereon*.

As observed on the RD graph for tibial length (Figure 14.12), all of the long-limbed felids have positive RD values, with servals and cheetahs falling above the upper 95% CL due to their relatively elongated tibiae. The two short-limbed taxa have negative RD values, although only jaguars fall below the 95% CL, indicating relatively shortened tibiae. All of the large-bodied extant felids also have negative RD values similar to *Neofelis*, although only female lions actually fall below the 95% CL. The few machairodontine specimens included (*Smilodon* and *Megantereon*) all fall far below the 95% CL, indicating extremely short tibiae relative to the overall size of the bone. It is likely that their extreme shortness largely accounts for the high position of these specimens on the polynomial regression plot, since tibia length has a strong negative correlation with PC1. As *Smilodon* and *Megantereon* are believed to be

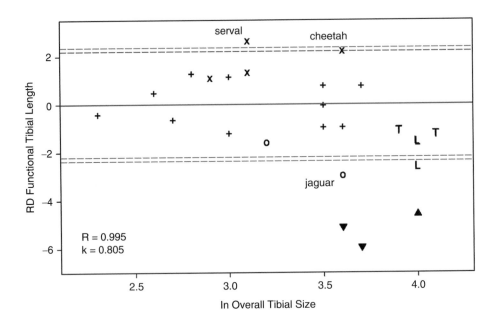

Figure 14.12 Relative deviations (RD) from the RMA regression of functional tibial length and overall tibial size. Only extant taxa indicated by plus signs (+) were included in the RMA regression. Dotted lines indicate 95% confidence limits based on full N (inner lines) and reduced N_e (outer lines). *Smilodon* and *Megantereon*, like jaguars and female lions, have significantly short tibiae relative to overall tibial size. Extant taxa not included in the regression: x = long-limbed taxa, o = short-limbed taxa, L = lions, T = tigers. Machairodontine taxa: ▲=*Smilodon*, ▼=*Megantereon*.

closely related (Kurtén, 1963; Berta and Galiano, 1983; Berta, 1987), only the inclusion of more genera in future studies will determine whether this is a machairodontine-wide pattern or found only within the Smilodontini.

Despite similarities in relative length, the two machairodontine taxa exhibit opposite extremes with respect to overall midshaft size. Among all extant and fossil specimens, the two *Megantereon* specimens have the greatest positive RDs (outside the 95% CL for EFT 277), and the *Smilodon* specimen has the most extreme negative RD, although it falls close to lions. The difference in overall midshaft size between the two fossil taxa is due to differences in both AP and ML midshaft diameters, though more so for the former than the latter. With respect to AP midshaft diameter (Figure 14.13), the two *Megantereon* specimens fall closely together beyond the upper 95% CL, while *Smilodon* falls close to (but not beyond) the lower 95% CL. Although lions and tigers have similar deviations for overall midshaft size, lions have extreme negative RDs for AP

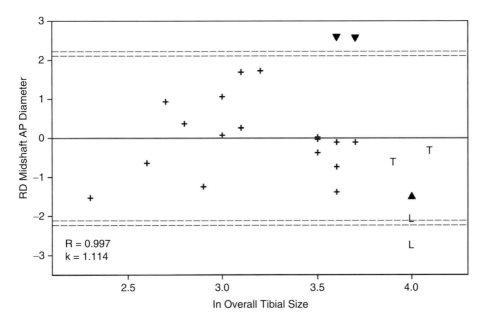

Figure 14.13 Relative deviations (RD) from the RMA regression of anteroposterior tibial midshaft diameter and overall tibial size. All extant taxa (+) were included in the RMA regression except for lions (L) and tigers (T). Dotted lines indicate 95% confidence limits based on full N (inner lines) and reduced N_e (outer lines). *Smilodon* and *Megantereon* differ in relative midshaft AP diameter, with both specimens of *Megantereon* being significantly large in this dimension. Machairodontine taxa: ▲=*Smilodon*, ▼=*Megantereon*.

midshaft diameter (beyond the 95% CL for females), whereas tigers fall close to zero. On balance, the tibia of *Smilodon* is similar to that of lions in being relatively short, with a relatively narrow AP midshaft diameter. In contrast, *Megantereon* has an extremely robust tibia, combining relative shortness with a relatively large AP midshaft.

The only other result of interest with respect to the fossil tibiae is the fact that *Smilodon* has an unusually large proximal articular surface (as do female lions) due mainly to a large lateral condyle (anteroposteriorly).

Discussion

Before launching into a larger discussion of scaling trends in felids, it is important to examine what has been learned about specific taxa of interest and their similarity to morphology predicted by the extant taxa. Each genus or subgroup will be summarised and discussed in turn.

Scaling trends in extant felids

As one would expect, some features of extant, average-limbed felid postcrania scale allometrically. For example, as overall element size increases, the lengths of the humerus, radius, and tibia decrease and anteroposterior (but not mediolateral) midshaft diameters increase, thereby increasing robusticity in these elements. Our findings with respect to scaling of element length appear contrary to those of previous felid scaling analyses that observed isometry for the humerus (Anyonge, 1993; Christiansen and Harris, 2005) and the tibia (Christiansen and Harris, 2005). It must be kept in mind, however, that our results are based on scaling of length with overall element size rather than with body mass; hence our scaling coefficients are not directly comparable to those of the aforementioned studies. (Furthermore, Christiansen and Harris used total lengths rather than functional lengths, as discussed in the Introduction.) In fact, our results mirror those of Bertram and Biewener (1990), who also found that AP midshaft diameter increases with positive allometry relative to length. Based on the scaling patterns observed in our study, the humerus, radius, and tibia all increase their ability to resist AP bending as body size increases. In contrast to the aforementioned elements, neither the length of the femur, nor its midshaft dimensions scale with significant allometry. Although Bertram and Biewener (1990) found significant positive allometry for femoral midshaft diameter relative to element length, it was far less pronounced than in the other long bones.

In general, the four long-limbed taxa (servals, cheetahs, African wildcats, lynxes) and two short-limbed taxa (clouded leopards, jaguars) tend to be outliers with respect to element length, although not all taxa fall consistently outside the 95% CL. Once again, the exception is the femur. In contrast to the results observed for the other limb bones, all of the short- and long-limbed taxa fall within the 95% CL for femoral length. Although the short-limbed taxa do have relatively short femora and the long-limbed taxa have relatively long femora, there are other taxa with relatively shorter femora (snow leopards and female lions) or relatively equally long femora (caracals, male pumas).

Contrary to biomechanical expectations, none of the elements exhibit allometric scaling of overall articular size (at least with respect to overall element size). None the less, size-related changes in articular shape are apparent in both the humerus (proximally) and the femur (distally). The positive allometry of the AP diameter of the humeral head is balanced by negative allometry of its SI dimension. The bicondylar width of the femur is also positively allometric, though the overall size of the distal femur scales with geometric similarity. Anyonge (1993) similarly found positive allometry of the distal femoral articular surface (relative to body mass) using an estimate of articular area (calculated as

the product of AP and ML condylar dimensions). Unfortunately, the distal
femur was the only articular surface he examined.

The lack of allometry for overall articular size among felids may be related to
the fact that limb joint size/shape is not strictly related to load-bearing, but
also reflects joint mobility. It is also possible that the nature of articular scaling
is dictated by the nature of loading through the joint, such that allometric
scaling is only necessary for those aspects of the articular surface that are most
directly involved in load transmission (Lague, 2000). This might account for
the size-related shape differences noted above, and potentially other size-
required shape differences not detected by our limited analysis of articular
morphology. Alternatively, the shape differences observed here may reflect
size-related differences in load distribution through the joint, such that animals
of different body mass are not loading their limb joints in the same way
(although Day and Jayne (2007) report no differences among size-variant felids
with respect to limb posture). Finally, it is important to bear in mind that
different results might emerge if we considered the scaling of articular size
relative to body mass, as opposed to overall element size.

Extant lions and tigers

Given the hypothesised scaling differences between large and small
carnivorans (Bertram and Biewener, 1990), we examined how well trends in the
smaller bodied felids (<100 kg) predict the morphology of the extant larger
felids (i.e. lions and tigers). In general, trends in the element proportions of
smaller felids (minus short- and long-limbed taxa) are only occasionally good
predictors of element proportions in lions and tigers.

The shape of the humerus and femur are generally well-predicted by trends
in average-limbed smaller felids. However, female tigers do differ in some
features of both elements. It is unclear whether this is an anomaly due to the
small sample size for female tigers ($n = 3$).

In contrast, the radius of lions and tigers is somewhat different from that
of other extant felids. For felids of their size, lions and tigers have large
AP midshaft diameters combined with small ML midshaft diameters. As a
result of these opposing patterns, lions and tigers are not unusual with respect
to the overall size of the midshaft. Hence, it would appear that while lions and
tigers follow the felid trend with respect to overall midshaft size, they require
an even larger AP diameter than expected for their body size. This difference
may relate to some type of change in relative load-bearing of the ulna and
radius not found in other extant or extinct felids. Their ability to capture much
larger prey relative to their body size than do smaller felids (Pienaar, 1969;

Schaller, 1972; Sunquist and Sunquist, 2002) may also be a factor. Flexing the elbow joint against resistance while holding onto relatively large prey may create greater AP forces in the radius than occurs in other extant taxa. However, machairodontine felids, who presumably also captured large prey relative to their body size, do not share this feature. The distal articular surface of the radius of lions and tigers is also unusually wide mediolaterally for its overall size. While one might be tempted to attribute this to a need to maintain peak functional stresses, the fact that machairodontines generally have relatively narrow radial articular surfaces mediolaterally suggests that this is not the case. Instead, lions and tigers may need this wider articular surface for greater agility during grappling with large prey. The wider surface allows greater abduction/adduction and rotatory movements at the radiocarpal joint. Radial morphology suggests a difference in prey procurement techniques between extant lions and tigers and machairodontine felids included here (see next section). Further analyses with increased sample sizes and taxa may resolve some of these issues.

Tibial measurements are relatively well-predicted with the exception of those belonging to female lions. Female lions have significantly shorter tibiae, significantly larger overall proximal size, and significantly smaller AP midshaft diameters. In all cases, this is an extreme of what is seen in the male lion. Tigers differ from lions in having tibial morphology that is closer to that predicted by smaller felids.

Why might male and female lions differ in some aspects of their morphology? Schaller (1972) stated that male lions are less agile and slower than females. He noted, however, that while males hunt less frequently than female lions, they are equally adept at dispatching prey. He suggested that males are not as adept at killing medium-sized, agile prey and may target larger, slow-moving prey more often than do female lions. Schaller hypothesised that this difference in prey procurement is due not only to sexual dimorphism in body size, but also possibly due to the difference in speed and agility mentioned above. Whether the anatomical differences observed here are reflecting the potential differences noted by Schaller will be examined elsewhere in more detail.

Lions and tigers, despite similarity in overall element size, do not always share the same proportions for any given element. This result is not surprising due to differences in their behaviour and ecology (see Sunquist and Sunquist, 2002). While both species can capture prey larger than themselves, lions tend to chase prey more often than the more ambush-oriented tigers. Lions also tend to inhabit more open and mixed habitats, while tigers are found primarily in more mixed and closed habitats. Unlike lions, tigers are excellent swimmers and have been known to swim in strong tidal currents (Sunquist and Sunquist, 2002). Lions also engage in cooperative group hunting, while tigers do not. Thus, the demands placed on the skeletons of lions and tigers are quite different.

Both the more cursorial nature of lions and their more open habitat suggests that lions should possess relatively longer limbs than tigers, although both species (with the exception of female lions) have limb lengths that are within the range predicted by trends in smaller felids. However, tigers have been noted to have hindlimbs that are slightly longer than their forelimbs, a feature which is believed to contribute to their great capacity for leaping (Sunquist and Sunquist, 2002). Postcranial morphology of lions may reflect a conflict between locomotor/habitat demands and the need to shift their centre of gravity posteriorly when contacting prey with their forelimbs while the hindlimb touches the ground. Tigers may have reduced the need to shift the centre of gravity by increasing their use of leaping at prey and using the effects of gravity on their own body weight to help bring the prey down. In short, behavioural studies with an eye towards biomechanical differences are needed to determine if there are subtle differences in prey procurement, grappling and manipulation. Overall, lions and tigers are clearly more similar to each other in morphology than either is to machairodontine felids.

Trends in machairodontine felids

The initial question asked at the beginning of this research was whether postcranial variation within machairodontines reflects size-related shape variation in extant felids. The answer to this question is that rarely is machairodontine morphology well-predicted by shape trends in extant felids, whether large or small-bodied. Even the machairodontine that is most like *Panthera*, *Dinofelis barlowi*, does not fall consistently within extant felid trends or with species of *Panthera*. In addition, there is a great deal of variation within machairodontines that is more than just size-related. As a result, there are very few trends that cut across all machairodontines.

As has been suggested in previous studies, machairodontines vary in relative limb length (e.g. Lewis, 1997; Turner and Antón, 1997). With respect to radial length, for example, *Dinofelis* and *Smilodon* fall with short-limbed extant taxa (clouded leopards and jaguars), lions and tigers fall with 'average-limbed' taxa, and *Homotherium* and *Dinobastis* fall with long-limbed taxa (cheetahs, servals, bobcats, and African wildcats). These three groups of felids form three parallel scaling trends. The same pattern of relative length is observed for extant taxa with respect to the tibia. Unfortunately, not enough complete fossil tibiae are included in this study to see whether the trend carries on through the machairodontines.

The great difference among large-bodied felids in distal articular size of the radius is perhaps one of the most interesting findings. Lions and tigers have an enlarged ML distal articular region in comparison to other extant taxa. Interestingly, snow leopards, which are known for having a large manus (Hemmer, 1972; Sunquist and Sunquist, 2002), have a slightly smaller than average maximum distal radial articular surface (RD = −0.892) relative to overall radial size, indicating that distal radial size does not translate into the size of the manus. In contrast to lions and tigers, most machairodontines have a relatively small ML distal articular size. This difference in the radiocarpal joint suggests that machairodontines were not grappling with prey in a manner similar to lions or tigers, nor were they behaving in a manner similar to smaller extant felids. Previous studies have noted the narrowed distal radius in *Homotherium* and interpreted it as an adaptation to cursorial behaviour that resulted in a loss or reduction in prey-grappling ability (Ballesio, 1963; Rawn-Schatzinger, 1992; Antón *et al.*, 2005). However, this is a feature found in other machairodontines with relatively shorter limb lengths (e.g. *Smilodon* and some *Dinofelis*), suggesting that reduction in radiocarpal rotatory ability is not specific to a cursorial adaptation.

Further evidence that machairodontines were not utilising their forelimbs in a manner similar to extant felids is the fact that machairodontines often have an enlarged first terminal phalanx coupled with a shortened, robust proximal phalanx in the first manual digit. While the enlarged dew claw has been interpreted as having allowed the presumably more cursorial *Homotherium* to capture prey (Antón *et al.*, 2005), this feature was common across machairodontines. Large dew claws have been noted in various species of *Dinofelis* (MEL, personal observation; Werdelin and Lewis, 2001), *Homotherium* (Ballesio, 1963; Antón *et al.*, 2005), *Lokotunjailurus* (Antón, 2003; Werdelin, 2003), *Machairodus* (Gaudry, 1862–1867), *Megantereon* (Christiansen and Adolfssen, 2007), and *Smilodon* (Méndez-Alzola, 1941; Cox and Jefferson, 1988), although the size of the dew claw relative to the other claws may differ from taxon to taxon. Clearly, machairodontines were utilising the manus in a manner distinct from any modern felid.

Machairodontines are more similar to each other in femoral morphology than in the morphology of the other elements considered in these analyses. All machairodontines have a relatively short femur relative to overall size. However, all taxa overlap with female lions and tigers with the exception of the extremely short femur (relatively) of *Homotherium*. Unfortunately, since only one complete *Homotherium* femur is included, it is difficult to know whether this would hold true for other members of the genus, particularly as the presumably closely related *Dinobastis* falls with the other machairodontines.

Overall femoral midshaft size in machairodontines is reasonably well-predicted by extant taxa. Thus, increased robusticity in the machairodontine femur has more to do with the decrease in length rather than an increase in relative shaft size. However, all large taxa (extant and machairodontine) tend to have a relatively large ML midshaft diameter and a relatively small AP midshaft diameter, resulting in generally small midshaft size overall (albeit not significantly).

With the exception of *Dinofelis barlowi*, all machairodontines have large femoral heads relative to overall femoral size, particularly in the AP and ML directions. This is not true for tigers and lions. In addition, compared to smaller felids, machairodontines have relatively narrow femoral bicondylar widths and slightly smaller distal AP lengths, leading to significantly smaller overall distal femora. Lions and tigers differ from machairodontines in that they have only slightly smaller bicondylar widths and larger (sometimes significantly so) distal AP lengths than predicted. Thus, large felids show two patterns: (1) the enlarged head and shrunken distal morphology of most machairodontines relative to overall femoral size, and (2) the average-sized femoral head and large distal AP length relative to overall femoral size in lions and tigers. It is unclear whether either one of these patterns is due to size-related shape requirements or if both are more affected by behavioural differences.

What could account for the unique femoral morphology in machairodontines? Clearly, if they had increased loads at the hip joint, one would expect increased loads at the knee joint, regardless of the position of the bone. To answer questions about the ability of the knee to withstand load-bearing, the relative size of individual condyles and the proportion of intercondylar space within the total mediolateral width of the distal femur must be examined in the future, as has been done in primates (e.g. Lague, 2000, 2009). It is hard to imagine a position, other than sitting, where the femoral head would be heavily loaded without the knee experiencing similar loads. Thus, it is more likely that the enlarged femoral head relates to an increase in rotatory ability.

In any case, since the AP diameter of the midshaft is relatively small for the overall size of the bone, either the femur was positioned in a way to minimise AP bending or smaller extant felids engage in behaviours that create a greater amount of AP bending than seen in lions, tigers, and most machairodontines.

While few included taxa were represented by complete tibia, a few things can be noted about *Smilodon* and *Megantereon*. Both have extremely short tibiae relative to overall size. However, in features of the articular surface, they often differ. More commentary on the tibia will be reserved until a larger sample size of machairodontines can be included.

Features of individual machairodontine genera

Dinobastis

No humerus or tibia of this taxon was included in the analysis. The radius of *Dinobastis* is unusually long with a mediolaterally narrow distal end. The femur is relatively short, although not significantly so. However, the overall size of the femoral head is significantly large relative to the overall size of the bone. While similar in some respects to African *Homotherium*, *Dinobastis* is often less extreme in its expression of postcranial characters, just as it is in its dentition.

Dinofelis

This genus includes a diverse group of taxa, not all of which are represented by all elements. While some species have been suggested to be similar to *Panthera*, this does not hold true for all species, nor does any single species completely mimic *Panthera* morphology.

The humerus of *Dinofelis* is generally short relative to its overall size. The Olduvai material, assuming that it is *Dinofelis*, is unusually short in length and has an unusually large AP head diameter in comparison to other members of the genus.

While the overall humeral size range of *Dinofelis* overlaps that of female lions and tigers (Size Class 4), the overall radial size overlaps medium-sized felids (Size Class 3; Lewis and Werdelin, 2007) such as leopards, jaguars, and pumas. The radius of *Dinofelis* is relatively short, as in extant clouded leopards and jaguars. The size of the distal articular surface is reasonably well predicted in comparison to that of tigers and lions or other machairodontines. However, the oldest and smallest radius of *Dinofelis* (partial skeleton with mandible, PQ-L 20995, *D.* cf. *D. diastemata*) shows a different pattern. This specimen, which is about the size of a leopard radius, is more similar in relative length to the radii of extant felids. It also has an extremely small distal articular surface in both ML and AP dimensions in comparison to other extants and machairodontines. While the small size may be primitive for *Dinofelis*, it is unclear what role the unique distal articular size played.

Dinofelis barlowi, the only member of this genus with a complete femur in this study, has a relatively short femur like other machairodontines, but is still within the range of extant taxa for its overall size. Interestingly, it has a relatively wide femoral midshaft, but is not significantly different from extant taxa. This is also the only machairodontine to have a femoral head size that is relatively well-predicted by extant taxa.

Homotherium

In general, *Homotherium* has a long humerus and a large midshaft in the AP direction in comparison to extant felids. However, it is not surprising that the much older humerus from Hadar is different from the younger Omo and Koobi Fora specimens. While they are all similar in size, the Hadar specimen is less extreme in some dimensions. Like *Dinobastis*, the radius of *Homotherium* is unusually long.

While all machairodontines have relatively short femora, *Homotherium* is the only taxon to have a relative femur length significantly different from extant taxa. In addition, the one specimen of this genus has a significantly wide femoral midshaft ML diameter. The addition of a significantly narrow AP midshaft diameter causes its overall midshaft size to fall within the range of modern taxa. Thus, this taxon increased resistance to mediolateral bending at the expense of resistance to both torsion and anteroposterior bending. This pattern is opposite to that seen in cheetahs, which have relatively narrow ML midshaft diameters and relatively thick AP midshaft diameters producing a midshaft shape that is more tubular than that of any other felid. Cheetah morphology resists bending in all directions and torsion, but is not designed to resist bending from any specific direction particularly well. *Homotherium* was clearly more concerned with side to side bending, suggesting movements beyond a cheetah-like form of cursoriality. *Homotherium* may have been more like spotted hyaenas, which are cursorial, but who have to resist ML bending in the femur based on carcass manipulation and carcass carrying by the head and neck. Given the canine morphology of *Homotherium*, it was unlikely to have carried large carcasses in the same manner as a spotted hyaena. In fact, taphonomic data have suggested that the close relative of *Homotherium*, *Dinobastis*, engaged in carcass disarticulation at the kill site before transporting meatier elements elsewhere (Marean and Ehrhardt, 1995). *Homotherium*, therefore, may have placed a large portion of its weight on the hindlimb during prey capture rather than during the transport of large carcasses. The longer forelimbs in combination with the shorter hindlimbs suggest a need to move the centre of gravity posteriorly. This similarity in posture to spotted hyaenas has been noted previously (Lewis, 2001; Antón *et al.*, 2005). While all lions, tigers and machairodontines, except *Dinofelis barlowi*, tend to reduce the AP diameter of their femoral midshaft, only *Homotherium* has taken the enlargement of the ML diameter to such an extreme.

Megantereon

Only two complete tibiae were included in this analysis, both from the last record of *Megantereon whitei* in Africa (Elandsfontein; Klein *et al.*, 2007). The tibia of this species is extremely short and robust, as in *Smilodon*. The overall midshaft size is relatively much larger than in any other taxon in contrast to the relatively smaller midshaft sizes of *Smilodon* and lions and tigers. Unlike *Smilodon*, *Megantereon* has a slightly larger ML diameter and significantly larger AP midshaft diameter.

Smilodon

Smilodon is characterised by short humeri and radii. The one humeral specimen had a relatively small SI head diameter, although this may not be characteristic of this taxon. While none of the machairodontine taxa are significantly different from extant taxa with respect to overall femoral size, *Smilodon gracilis* is the most similar. All machairodontines have relatively short femora, but *Smilodon* tends to overlap with the relative size of extant lions and tigers for its size. However, like most machairodontines, *Smilodon* has a relatively large femoral head, particularly with respect to the mediolateral diameter. The tibia is extremely short and robust, as in *Megantereon*. However, *Smilodon* has a significantly larger AP length of the lateral condyle leading to a significantly larger overall proximal size.

Implications

Size-related trends in extant taxa cannot fully predict machairodontine morphology due to the greater variability within this extinct subfamily than in extant felids. While some aspects of machairodontine morphology are well-predicted for certain taxa, a good deal of non-size-related shape variation exists among machairodontines that is likely due to differences in habitual limb use.

Hindlimb morphology (or at least femoral morphology) is more consistent within machairodontines than is forelimb morphology. However, the overall hindlimb pattern is not reflective of extant scaling trends. That is, machairodontines (and lions and tigers) are not simply enlarged versions of smaller felids. Presumably, there is some similar functional requirement among machairodontines in the use of the hindlimb, such as the need to stabilise the body in a similar way during prey capture.

Forelimb morphology, on the other hand, is much more variable among machairodontines, probably due to the greater impact of prey procurement technique. The forelimbs of *Megantereon* and *Smilodon* and even *Dinofelis* are much more clearly designed for grabbing or grappling with prey than the more elongated forelimbs of *Homotherium* and *Dinobastis*. Not surprisingly, functional demands related to prey procurement have also been suggested as driving the evolution of cranial shape diversity in machairodontines (Slater and Van Valkenburgh, 2008).

Members of the genus *Dinofelis* encompass a greater range of variability than is observed in other taxa. Members of this genus may at times be similar to species of *Panthera* (as has often been noted), or to *Smilodon*, or at times be simply unique. Although known from Eurasia and North America, *Dinofelis* persisted longer in Africa than other machairodontines (with the possible exception of *Megantereon* due to the range of dates at Elandsfontein; Klein *et al.*, 2007) despite upheavals going on within the carnivore guild during the Plio-Pleistocene (Werdelin and Lewis, 2001, 2005; Lewis and Werdelin, 2007). This variability within the genus may be an indication of a greater degree of adaptability that allowed them to survive when their less adaptable cousins became extinct. Even if Metailurini is a wastebasket taxon, the variability within *Dinofelis* suggests that further explorations of this nature should be made into metailurin morphology.

The fact that size-related shape trends in smaller-bodied extant felids are not always the best predictors of morphology in extinct larger felids is a particularly important point for future reconstructions of body mass in machairodontine felids. Machairodontines are not simply lions or tigers with enlarged canines. Future functional investigations of machairodontines must take into account the dual influence of behaviour and body size on morphology.

Likewise, the great amount of variation observed in machairodontine limb morphology means that a single species, genus, or tribe cannot be used as a stand-in for other machairodontine taxa in paleoecological, morphological, or other analyses. There truly is no 'typical' machairodontine felid. Changes in postcranial morphology may or may not coincide with craniodental change, even if they are being driven by the same selective pressures. Therefore, postcranial material, when available, must be included not only in future studies of machairodontine felid behaviour and ecology, but also in the study of machairodontine taxonomy and systematics.

Acknowledgements

We thank the conveners of the symposium 'Carnivora: Phylogeny, Form and Function' (SVP, 2007), A. Goswami and A. Friscia, for the invitation to participate in both the symposium and this volume. We would also like to thank the other participants and two anonymous reviewers for their comments. In addition, we thank M. Avery, J. Flynn, L. Gordon, P. Holroyd, F. C. Howell, J. Hooker, M. Leakey, E. Lundelius, R. MacPhee, E. Mbua, N. Mudida, N. Simmons, R. Thorington, and E. Westwig for access to specimens and the Governments of Ethiopia, Kenya, and Tanzania for permission to study fossils. Data collection was supported by NSF, Sigma Xi, and Stockton Distinguished Faculty Fellowships to MEL and a Leakey Foundation Grant to MEL and Kaye Reed. Travel to participate in the symposium was supported by a Stockton Career Development Grant to MEL.

REFERENCES

Alexander, R. M. (1977). Allometry of the limbs of antelopes (Bovidae). *Journal of Zoology (London)*, **183**, 125–46.

Alexander, R. M., Jayes, A. S., Maloiy, G. M. O. and Wathuta, E. M. (1979). Allometry of the limb bones of mammals from shrews (Sorex) to elephants (Loxodonta). *Journal of Zoology*, **189**, 305–14.

Alexander, R. M. and Pond, C. M. (1992). Locomotion and bone strength of the white rhinoceros (*Ceratotherium simum*). *Journal of Zoology (London)*, **227**, 63–69.

Anderson, J. F., Hall-Martin, A. and Russell, D. A. (1985). Long-bone circumference and weight in mammals, birds and dinosaurs. *Journal of Zoology, London*, **207**, 53–61.

Andersson, K. (2004). Predicting carnivoran body mass from a weight-bearing joint. *Journal of Zoology, London*, **262**, 161–72.

Antón, M. (2003). Notes on the reconstructions of fossil vertebrates from Lothagam. In *Lothagam: Dawn of Humanity in East Africa*, ed. M. G. Leakey and J. M. Harris. New York, NY: Columbia University Press, pp. 661–65.

Antón, M., Galobart, A. and Turner, A. (2005). Co-existence of scimitar-toothed cats, lions and hominins in the European Pleistocene. Implications of the post-cranial anatomy of Homotherium latidens (Owen) for comparative palaeoecology. *Quaternary Science Reviews*, **24**, 1287–301.

Anyonge, W. (1993). Body mass in large extant and extinct carnivores. *Journal of Zoology, London*, **231**, 339–50.

Ballesio, R. (1963). Monographie d'un Machairodus du Gisement villafranchien de Senèze: *Homotherium crenatidens* Fabrini. *Travaux du Laboratoire de Géologie de la Faculté des Sciences de Lyon*, **9**, 1–129.

Beaumont, G. de (1964). Remarques sur la classification des Felidae. *Ecologae Geologicae Helvetiae*, **57**, 837–45.

Beaumont, G. de (1978). Notes complémentaires sur quelques félidés (Carnivores). *Archives des Sciences, Genève*, **31**, 219–27.

Berta, A. (1987). The sabercat *Smilodon gracilis* from Florida and a discussion of its relationships (Mammalia, Felidae, Smilodontini). *Bulletin of the Florida State Museum, Biological Sciences*, **31**, 1–63.

Berta, A. and Galiano, H. (1983). *Megantereon hesperus* from the late Hemphillian of Florida with remarks on the phylogenetic relationships of machairodonts (Mammalia, Felidae, Machairodontinae). *Journal of Paleontology*, **57**, 892–99.

Bertram, B. C. R. and Biewener, A. A. (1990). Differential scaling of the long bones in the terrestrial Carnivora and other mammals. *Journal of Morphology*, **204**, 157–69.

Biewener, A. A. (1983). Allometry of quadrupedal locomotion: the scaling of duty factor, bone curvature and limb orientation to body size. *Journal of Experimental Biology*, **105**, 147–71.

Biewener, A. A. (1990). Biomechanics of terrestrial locomotion. *Science*, **250**, 1097–103.

Biewener, A. A. and Taylor, C. R. (1986). Bone strain: a determinant of gait and speed? *Journal of Experimental Biology*, **123**, 383–400.

Blob, R. W. (2000). Interspecific scaling of the hindlimb skeleton in lizards, crocodilians, felids and canids: does limb bone shape correlate with limb posture? *Journal of Zoology*, **250**, 507–31.

Bou, J. and Casinos, A. (1985). Scaling of bone mass to body mass in insectivores and rodents. In *Functional Morphology in Vertebrates*, ed. H. R. Duncker and G. Fleischer. Stuttgart: Gustav Fischer Verlag. pp. 61–64.

Bou, J., Casinos, A. and Ocana, J. (1987). Allometry of the limb long bones of insectivores and rodents. *Journal of Morphology*, **192**, 113–23.

Christiansen, P. (1999a). Long bone scaling and limb posture in non-avian theropods: evidence for differential allometry. *Journal of Vertebrate Paleontology*, **19**, 666–80.

Christiansen, P. (1999b). Scaling of mammalian long bones: small and large mammals compared. *Journal of Zoology (London)*, **247**, 333–48.

Christiansen, P. (1999c). Scaling of the limb long bones to body mass in terrestrial mammals. *Journal of Morphology*, **239**, 167–90.

Christiansen, P. and Adolfssen, J. S. (2007). Osteology and ecology of Megantereon cultridens SE311 (Mammalia; Felidae; Machairodontinae), a sabrecat from the Late Pliocene–Early Pleistocene of Senèze, France. *Zoological Journal of the Linnean Society*, **151**, 833–84.

Christiansen, P. and Harris, J. M. (2005). Body size of *Smilodon* (Mammalia: Felidae). *Journal of Morphology*, **266**, 369–84.

Churcher, C. S. (1966). The affinities of *Dinobastis serus* Cope 1893. *Quaternaria*, **8**, 263–75.

Clutton-Brock, T. H. and Harvey, P. H. (1977). Primate ecology and social organization. *Journal of Zoology, London*, **183**, 1–39.

Cooke, H. B. S. (1991). *Dinofelis barlowi* (Mammalia, Carnivora, Felidae) cranial material from Bolt's Farm, collected by the University of California African Expedition. *Palaeontologia Africana*, **28**, 9–21.

Cox, S. M. and Jefferson, G. T. (1988). The first individual skeleton of Smilodon from Rancho La Brea. *Current Research in the Pleistocene*, **5**, 66–67.

Darroch, J. N. and Mosimann, J. E. (1985). Canonical and principal components of shape. *Biometrika*, **72**, 241–52.

Day, L. M. and Jayne, B. C. (2007). Interspecific scaling of the morphology and posture of the limbs during the locomotion of cats (Felidae). *Journal of Experimental Biology*, **210**, 642–54.

Ditchfield, P., Hicks, J., Plummer, T. W., Bishop, L. C. and Potts, R. (1999). Current research on the Late Pliocene and Pleistocene deposits north of Homa Mountain southwestern Kenya. *Journal of Human Evolution*, **36**, 123–50.

Feibel, C. S., Brown, F. H. and McDougall, I. (1989). Stratigraphic context of fossil hominids from the Omo Group deposits: northern Turkana basin, Kenya and Ethiopia. *American Journal of Physical Anthropology*, **78**, 595–622.

Garcia, G. J. M. and da Silva, J. K. L. (2004). On the scaling of mammalian long bones. *Journal of Experimental Biology*, **207**, 1577–84.

Gaudry, A. (1862–1867). *Animaux fossiles et géologie de l'Attique*. Paris, France.

Gingerich, P. D. (1990). Prediction of body mass in mammalian species from long bone lengths and diameters. *Contributions from the Museum of Paleontology, University of Michigan*, **28**, 79–92.

Graham, R. W. (1976). *Pleistocene and Holocene Mammals, Taphonomy, and Paleoecology of the Friesenhahn Cave Local Fauna, Bexar County, Texas*. Unpublished PhD dissertation, University of Texas, Austin.

Harvey, P. and Pagel, M. (1991). *The Comparative Method in Evolutionary Biology*. Oxford: Oxford University Press.

Heinrich, R. E. and Biknevicius, A. R. (1998). Skeletal allometry and interlimb scaling patterns in mustelid carnivorans. *Journal of Morphology*, **235**, 121–34.

Hemmer, H. (1972). *Uncia uncia. Mammalian Species*, **20**, 1–5.

Hendey, Q. B. (1974). The late Cenozoic Carnivora of the south-western Cape Province. *Annals of the South African Museum*, **63**, 1–369.

Hendey, Q. B. (1981). Palaeoecology of the late Tertiary fossil occurrences in 'E' Quarry, Langebaanweg, South Africa, and a reinterpretation of their geological context. *Annals of the South African Museum*, **84**, 1–104.

Howell, A. B. (1944). *Speed in Animals, Their Specialization for Running and Leaping*. Chicago, IL: University of Chicago Press.

Hulbert, R. C., Jr. (2001). Florida's fossil vertebrates, an overview. In *The Fossil Vertebrates of Florida*, ed. R. C. Hulbert, Jr. Gainesville, FL: University Press of Florida, pp. 25–33.

Johnson, W. E., Eizirik, E., Pecon-Slattery, J., *et al.* (2006). The Late Miocene radiation of modern Felidae: a genetic assessment. *Science*, **311**, 73–77.

Jungers, W. L. (1979). Locomotion, limb proportions, and skeletal allometry in lemurs and lorises. *Folia Primatologia*, **32**, 8–28.

Jungers, W. L. (1984a). Aspects of size and scaling in primate biology with special reference to the locomotor skeleton. *Yearbook of Physical Anthropology*, **27**, 73–97.

Jungers, W. L. (1984b). Scaling of the hominoid locomotor skeleton with special reference to lesser apes. In *The Lesser Apes: Evolutionary and Behavioural Biology*. ed. H. Preuschoft, D. J. Chivers, W. Y. Brockelman, and N. Creel. Edinburgh: Edinburgh University Press, pp. 146–69.

Jungers, W. L. (1985). *Size and Scaling in Primate Biology*. New York, NY: Plenum Press.

Jungers, W. L., Falsetti, A. and Wall, C. (1995). Shape, relative size, and size-adjustments in morphometrics. *Yearbook of Physical Anthropology*, **38**, 137–61.

Klein, R. G., Avery, G., Cruz-Uribe, K. and Steele, T. E. (2007). The mammalian fauna associated with an archaic hominin skullcap and later Acheulean artifacts at Elandsfontein, Western Cape Province, South Africa. *Journal of Human Evolution*, **52**, 164–86.

Kretzoi, M. (1929). *Feliden-Studien. A Magyar Királyi Földtani Intézet Hazinyomdaja*, **24**, 1–22.

Kurtén, B. (1963). Notes on some Pleistocene mammal migrations from the Palaearctic to the Nearctic. *Eiszeitalter und Gegenwart*, **14**, 96–103.

Lague, M. R. (2000). *Patterns of sexual dimorphism in the joint surfaces of the elbow and knee of catarrhine primates*. PhD thesis, State University of New York at Stony Brook.

Lague, M. R. (2009). Patterns of knee joint shape dimorphism in guenons (*Cercopithecus*) reflect interspecific scaling trends among cercopithecoid monkeys. *American Journal of Physical Anthropology Supplement*, **48**, 172.

Lewis, M. E. (1995). *Plio-Pleistocene carnivoran guilds: implications for hominid paleoecology*. PhD thesis, State University of New York at Stony Brook.

Lewis, M. E. (1997). Carnivoran paleoguilds of Africa: implications for hominid food procurement strategies. *Journal of Human Evolution*, **32**, 257–88.

Lewis, M. E. (2001). Implications of interspecific variation in the postcranial skeleton of Homotherium (Felidae, Machairodontinae). *Journal of Vertebrate Paleontology*, **21**, 73A.

Lewis, M. E. and Werdelin, L. (2007). Patterns of change in the Plio-Pleistocene carnivorans of eastern Africa: implications for hominin evolution. In *Hominin Environments in the East African Pliocene: An Assessment of the Faunal Evidence*, ed. R. Bobe, Z. Alemseged and A. K. Behrensmeyer. The Netherlands: Springer-Verlag, pp. 77–105.

Marean, C. W. and Ehrhardt, C. L. (1995). Paleoanthropological and paleoecological implications of the taphonomy of a sabertooth's den. *Journal of Human Evolution*, **29**, 515–47.

Martin, L. D. (1980). Functional morphology and the evolution of cats. *Transactions of the Nebraska Academy of Sciences*, **8**, 141–54.

Martin, L. D., Schultz, C. B. and Schultz, M. R. (1988). Saber-toothed cats from the Plio-Pleistocene of Nebraska. *Transactions of the Nebraska Academy of Sciences*, **XVI**, 153–63.

Martin, L. D., Babiarz, J. P., Naples, V. L. and Hearst, J. (2000). Three ways to be a saber-toothed cat. *Naturwissenschaften*, **87**, 41–44.

McCarthy, R. C. (2001). Anthropoid cranial base architecture and scaling relationships. *Journal of Human Evolution*, **40**, 41–66.

McMahon, T. A. (1975). Allometry and biomechanics: limb bones of adult ungulates. *American Naturalist*, **107**, 547–63.

Méndez-Alzola, R. (1941). El *Smilodon bonaërensis* (Muñiz). Estudio osteológico y osteométrico del gran tigre de La Pampa comparado con otros félidos actuales y fósiles. *Annales del Museo Argentino de Ciencias Naturales*, **40**, 135–252.

Morgan, G. S. and Hulbert, R. C., Jr. (1995). Overview of the geology and vertebrate biochronology of the Leisey Shell Pit local fauna, Hillsborough County, Florida. *Bulletin of the Florida Museum of Natural History*, **37**, 1–92.

O'Higgins, P. and Jones, N. (1998). Facial growth in *Cercocebus torquatus*: an application of three-dimensional geometric morphometric techniques to the study of morphological variation. *Journal of Anatomy*, **198**, 251–72.

O'Regan, H. J. and Turner, A. (2002). The assessment of size in fossil Felidae. *Estudios geológicos*, **58**, 45–54.

Organ, J. M. and Ward, C. V. (2006). Contours of the hominoid lateral tibial condyle with implications for Australopithecus. *Journal of Human Evolution*, **51**, 113–27.

Pienaar, U. de V. (1969). Predator–prey relationships amongst the larger mammals of the Kruger National Park. *Koedoe*, **12**, 108–76.

Piveteau, J. (1961). *Les Carnivores. Traité de Paléontologie, Tome VI, Vol. 1*, ed. J. Piveteau. Paris: Masson et Cie, pp. 641–820.

Potts, R. (1988). *Early Hominid Activities at Olduvai*. New York, NY: Aldine de Gruyter Press.

Preuschoft, H. and Demes, B. (1985). Influence of size and proportions on the biomechanics of brachiation. In *Size and Scaling in Primate Biology*, ed. W. L. Jungers. New York, NY: Plenum Press, pp. 383–99.

Prothero, D. R. and Sereno, P. C. (1982). Allometry and ecology of middle Miocene dwarf rhinoceroses from the Texas Gulf coastal plain. *Paleobiology*, **8**, 16–30.

Rawn-Schatzinger, V. (1992). The scimitar cat *Homotherium serum* Cope: osteology, functional morphology, and predatory behavior. *Illinois State Museum Reports of Investigations*, **47**, 1–80.

Ruff, C. B. (1987). Structural allometry of the femur and tibia in Hominoidea in Primates. *Folia Primatologia*, **48**, 9–49.

Ruff, C. B. (2000). Body size, body shape and long bone strength in modern humans. *Journal of Human Evolution*, **38**, 269–90.

Ruff, C. B. and Runestad, J. A. (1992). Primate limb bone structural adaptations. *Annual Review of Anthropology*, **21**, 407–33.

Schaller, G. B. (1972). *The Serengeti Lion*. Chicago, IL: University of Chicago Press.

Scott, K. M. (1985). Allometric trends and locomotor adaptations in the Bovidae. *Bulletin of the American Museum of Natural History*, **179**, 197–288.

Scott, K. M. (1990). Postcranial dimensions as predictors of body mass. *Body Size in Mammalian Paleobiology: Estimation and Biological Implications*, ed. J. D. Damuth and B. J. MacFadden. Cambridge: Cambridge University Press, pp. 301–55.

Slater, G. J. and Van Valkenburgh, B. (2008). Long in the tooth: evolution of sabertooth cat cranial shape. *Paleobiology*, **34**, 403–19.

Smith, R. J. (1994). Degrees of freedom in interspecific allometry: an adjustment for the effects of phylogenetic constraint. *American Journal of Physical Anthropology*, **93**, 95–107.

Sunquist, M. E. and Sunquist, F. C. (2002). *Wild Cats of the World*. Chicago, IL: University of Chicago Press.

Turner, A. and Antón, M. (1997). *The Big Cats and their Fossil Relatives: An Illustrated Guide to their Evolution and Natural History*. New York, NY: Columbia University Press.

Van Valkenburgh, B. (1990). Skeletal and dental predictors of body mass in carnivores. In *Body Size in Mammalian Paleobiology: Estimation and Biological Implications*, ed. J. D. Damuth, and B. J. MacFadden. Cambridge: Cambridge University Press, pp. 181–205.

Walter, R. (1994). The age of Lucy and the First Family: single crystal 40Ar/39Ar dating of the Denen Dora and lower Kada Hadar Members of the Hadar Formation, Ethiopia. *Geology*, **22**, 6–10.

Werdelin, L. (1983). Morphological patterns in the skull of cats. *Biological Journal of the Linnean Society*, **19**, 375–91.

Werdelin, L. (2003). Mio-Pliocene Carnivora from Lothagam, Kenya. In *Lothagam: The Dawn of Humanity in Eastern Africa*, ed. M. G. Leakey and J. M. Harris. New-York, NY: Columbia University Press, pp. 261–678.

Werdelin, L. and Lewis, M. E. (2001). A revision of the genus *Dinofelis* (Mammalia, Felidae). *Zoological Journal of the Linnean Society*, **132**, 147–258.

Werdelin, L. and Lewis, M. E. (2005). Plio-Pleistocene Carnivora of eastern Africa: species richness and turnover patterns. *Zoological Journal of the Linnean Society*, **144**, 121–44.

Werdelin, L., Yamaguchi, N., Johnson, W. E. and O'Brien, S. J. (in press). Felid phylogeny and evolution. In *The Biology and Conservation of Wild Felids*, ed. D. Macdonald and A. Loveridge. Oxford: Oxford University Press.

Zar, J. H. (1984). *Biostatistical Analysis*, 2nd ed. Englewood Cliffs, NJ: Prentice-Hall.

15

Cranial mechanics of mammalian carnivores: recent advances using a finite element approach

STEPHEN WROE

Introduction

Analyses of functional morphology typically apply phenomenological methodologies, where investigators seek to identify links between morphological and ecological data, or mechanistic approaches, where engineering principles are applied to explain how and why particular morphologies are associated with specific behaviours and ecologies. Often these two are combined within a comparative context.

Although there is certainly debate over basic assumptions and the degree to which structure might be expected to predict function (Gould, 2002), for the majority of biologists, determining relationships between form and function is fundamental to understanding the evolution of behaviours and ecologies, the nature of morphological convergence, and the prediction of habitus in living and extinct taxa.

With respect to the mammalian carnivore skull, numerous studies invoking a range of phenomenological and mechanistic approaches have identified some correspondence between these variables (Savage, 1977; Buckland-Wright, 1978; Radinsky, 1981a, 1981b; Werdelin, 1986; Van Valkenburgh, 1989; Therrien, 2005a; Wroe *et al.*, 2005; Christiansen and Wroe, 2007; Wroe and Milne, 2007). However, the degree to which any skull might be optimised for feeding is not well understood. The vertebrate skull is not simply a food-processing mechanism, it also houses major sensory and neural apparatuses (Dumont *et al.*, 2005), and, in mammalian carnivores, it may be subject to considerable external stresses generated in the subjugation and killing of prey (Preuschoft and Witzel, 2004). Consequently, skull morphology may represent compromise between various competing influences (Hylander *et al.*, 1991;

Carnivoran Evolution: New Views on Phylogeny, Form, and Function, ed. A. Goswami and A. Friscia. Published by Cambridge University Press. © Cambridge University Press 2010.

Preuschoft and Witzel, 2004). Identifying such compromise is difficult to achieve using traditional methods.

Finite element analysis (FEA) represents a relatively new mechanistic approach to questions surrounding form and function in biology with the potential to provide robust and detailed answers. In FE-based studies of biological structures, two- or three-dimensional finite element models (FEMs) are typically assembled from computerised tomography (serial X-ray data) and converted into simulations that comprise discrete numbers of elements. These elements are connected by nodes and assigned material properties such as Young's modulus (a measure of stiffness or the resistance of an elastic material to deformation by an applied force) and Poisson's ratio (ratio of transverse contracting strain to elongation strain when an object is stretched by forces applied at its ends parallel to its axis). The nodes act as information conduits, providing points at which models can be constrained and to which forces (typically measured in Newtons [N]) can be applied. The result is a digital model from which displacements can be computed at the nodes providing the data required to further predict the distribution of stress (force per unit area [σ]) and strain (deformation that results from a given load [ϵ]).

Benefits offered by FEA include its ability to simulate very complex, precise, and repeatable loading conditions, and to reconstruct or predict mechanical behaviour in extremely fine three-dimensional (3D) detail. It has the further advantage of being non-destructive. In theory, FEA can allow the identification of strict performance limits on biological structure and the determination of which particular structures might be best or least well-adapted to perform any hypothesised function.

As with any current methodology, care must be taken in the interpretation of results, e.g. demonstration that a structure is capable of, or even well-adapted to perform a particular role does not mean that this is necessarily its function, but, at the very least FEA has the potential to convincingly rule out mechanically unviable interpretations. Well-validated FEMs, backed up by correspondence with robust experimental data for all or part of the model, have the further potential to identify mechanical failure and other strict performance limits.

Primary disadvantages associated with FEA in biology have included the time consuming nature of model generation (with single models typically taking weeks or months to assemble), a need for computational resources that often exceeds those offered by standard desktop computers, and difficulties in capturing and assigning the multiple properties of bone, or simulating musculature and other soft tissues.

Despite these drawbacks to the modelling of biological structures, the power and insight afforded by FEA has attracted growing interest from outside the engineering fraternity, beginning with researchers pursuing questions in the field of orthopaedic biomechanics (Rybicki *et al.*, 1972) and more recently still from students of zoology (Guillet *et al.*, 1985) and palaeontology (Daniel *et al.*, 1997). FE models of varying complexity have now been used to examine the mechanical behaviour of crania for at least some representatives of a wide range of taxa, including gorgonopsid and therocephalian synapsids (Jenkins *et al.*, 2002), primates (Kupczik *et al.*, 2007; Strait *et al.*, 2007; Wroe *et al.*, 2007b), bats (Dumont *et al.*, 2005), living reptiles and extinct dinosaurs (Rayfield *et al.*, 2001; Rayfield, 2005; McHenry *et al.*, 2006; Moreno *et al.*, 2008) and fish (Wroe *et al.*, 2008b).

Until very recently, the application of FEA to the skulls of mammalian carnivores has been extremely limited. In an experiment designed to address biomedical rather than broader form–function questions, a study by Verrue *et al.* (2001) represents perhaps the first example using 3D FEA to examine a mammalian carnivore cranium. Using what would now be considered a very low-resolution FEM comprising ~3000 elements, Verrue *et al.* (2001) simulated and validated bone displacements in the cranium of a domestic dog under orthopaedic loads. None the less, the modelling achieved results that were in close to reasonable agreement with in-vivo measurements taken using strain gauges. In a number of respects this was a sophisticated simulation, with multiple material properties assigned for cancellous and cortical bone, as well as enamel and sutures. On the other hand, it exemplifies the sometimes laborious nature of FEM construction in biology, with the FEM built on the basis of manual tracings from 2D tomographic data and material properties also assigned to individual elements by hand.

Here I briefly review recent developments in the application of the FE method to the understanding of relationships between form and function in the mammalian carnivore skull and the prediction of behaviour in extinct species (McHenry *et al.*, 2007; Wroe *et al.*, 2007a; Bourke *et al.*, 2008; Clausen *et al.*, 2008; Wroe, 2008). The findings I will discuss are largely the product of collaborative efforts by members of the Computational Biomechanics Research Group (University of New South Wales and University of Newcastle, Australia). Protocols and approaches developed and applied by this team offer the following advantages: model assembly times have been considerably reduced, with high-resolution solid FE models of up to 3,000,000 'brick' elements generally completed within a day; multiple properties can be assigned semi-automatically; crania and mandibles can be treated together in correct anatomical positions; the action of muscles can be quickly and more realistically simulated; and, importantly regarding access and affordability, both pre- and post-processing can be completed using standard desktop or laptop computers.

Table 15.1 Values used to derive Young's modulus (E) for element property types in FE models.

Material property type	Mean HU	Heterogeneous density (kg m^{-3})	E (GPa)	Homogeneous density (kg m^{-3})	E (GPa)	Hypothetical density (kg m^{-3})	E (GPa)
1	−768	251	1.53	2060	18	132	0.65
2	−256	292	1.87	2060	18	644	5.34
3	256	333	2.22	2060	18	978	9.29
4	768	1094	10.79	2060	18	1234	12.65
5	1279	1856	21.73	2060	18	1490	16.24
6	1791	2191	27.08	2060	18	1746	20.04
7	2303	2526	32.70	2060	18	2002	24.02
8	2815	2861	38.58	2060	18	2258	28.18

Before discussing the results of individual analyses, I will outline procedures and features common to the assembly of the FEMs considered in these treatments so as to avoid repetition. Each FEM was constructed on the basis of CT scan data in dicom format acquired by a Toshiba Aquilon 16 Scanner (Mater Hospital, Newcastle). Segmentation was performed using the image processing software Mimics (Version 9.1.1; Materialise Inc., Leuven, Belgium). Solid FEMs comprising four-noded tetrahedral elements (tet4s) were created in the finite element programme Strand7 (vers. 2.3; Strand7 Pty. Ltd., Sydney, Australia), and for heterogeneous models eight material properties were assigned on the basis of CT attenuation data (see Table 15.1 and McHenry *et al.*, 2007). This approach does not attempt to model actual trabecular struts and spaces within cancellous bone (which would require enormous computational resources, high-resolution microCT data, and thousands of computer hours to solve – see Verhulp, 2006), but represents an approximation at macroscopic scale taken from X-ray attenuation in individual voxels. In each model it was further assumed that bone reacts uniformly irrespective of the direction of loading (isotropic). In reality, bone does not behave uniformly (anisotropic). This remains a challenging area in FE modelling of biological structure, and simulations incorporating anisotropic material properties have rarely been attempted (but see Strait *et al.*, 2005). However, it should be noted that, to date, models assuming isotropy have produced results that achieve relatively high levels of realism as confirmed by validation against experimental strain gauge data (Kupczik *et al.*, 2008).

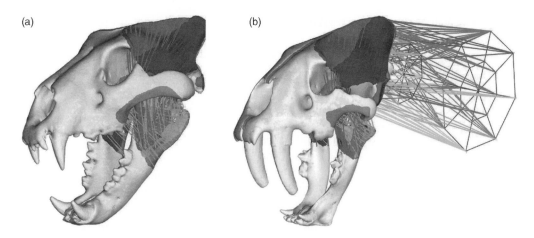

Figure 15.1 For colour version, see Plate 14. Construction of models. a, *Panthera leo* model showing the Temporalis (blue) and Masseter (red) systems, muscle 'beams' and attachment areas. b, *Smilodon fatalis* model showing the neck assembly. Different coloured beams on the neck correspond to neck muscle groups in felids. From McHenry *et al.* (2007).

Crania and mandibles were connected using beams released to enable free rotation. The maximum force that can be generated by a muscle is proportional to its maximum cross-sectional area lying within a range between ~15 and 39 N cm^{-2} (Rayfield *et al.*, 2001). Forces for jaw closing muscles were determined by applying an intermediate value of 30 N cm^{-2} (Weijs and Hillen, 1985), equivalent to 300 kilopascals (kPa), to maximal cross-sectional areas for primary jaw adducting muscles estimated following the 'dry-skull' method (Thomason, 1991; Wroe *et al.*, 2005). The muscles themselves were simulated using multiple pre-tensioned trusses (beams that carry axial loads only). Using this approach muscle vectors are determined automatically and forces are spread more evenly and realistically (Figure 15.1). Refinement of this process may allow still closer approximation by simulation that accounts for curvature of muscle around bone, as recently demonstrated by Grosse *et al.* (2007).

Although in theory it is possible to achieve perfect equilibrium between loads applied to an FEM, in practice this is all but impossible (Richmond *et al.*, 2005). Consequently, FE models must be constrained in order to prevent rigid movement once external forces are applied. In the models discussed below, each was constrained at varying bite points on the teeth, either bilaterally or unilaterally, dependent on the nature of the behaviour being simulated. A further constraint was applied at the occipital condyle to approximate the skull's attachment to the neck. In two models the crania were further constrained by simulated cervical musculature (McHenry *et al.*, 2007). Application of constraints to single nodes can result in artefactual point loadings (Dumont *et al.*, 2005). In order to more evenly spread forces, a network of fine metal beams was introduced at all constraints.

In most instances these FEMs were subject to both 'intrinsic' and 'extrinsic' loadings, where intrinsic loads are those generated solely by the jaw adductors and extrinsic loads simulated forces applied either by the predator's own postcranial musculature, or struggling prey. While there are no data currently available with respect to the nature of actual forces involved in the subjugation of relatively large prey by mammalian carnivores, there can be little doubt that these can be considerable and that the need for a capacity to resist them has played some role in mammalian carnivore evolution. Indeed, even the estimates of loadings developed at the occipital condyle of a large dog vigorously shaking rabbit-sized prey approach those of the forces developed by the animal's jaw muscles (Preuschoft and Witzel, 2004). In the examples discussed below, we did not attempt to estimate actual extrinsic loadings encountered by different species, but only to scale forces in order to permit assessment on a comparative basis. Thus, extrinsic load magnitudes were calculated as proportions of the predators' body masses and include simulations of a lateral 'shake', a dorsoventral head depression, a pull, and an axial twist. All analyses alluded to here were linear-static and all stresses were Von Mises.

Reconstructing predatory behaviour in the dirk-toothed sabrecat, *Smilodon fatalis*

The Pleistocene American sabrecat *Smilodon fatalis* represents among the most specialised of mammalian carnivores. Just how and to what purpose the sabrecat's greatly hypertrophied upper canine teeth were deployed remains the focus of debate (Warren, 1853; Akersten, 1985; Bicknevicus and Van Valkenburgh, 1996; Therrien, 2005b; Christiansen, 2007; Wroe *et al.*, 2008a). Long-standing questions have centred on whether the sabrecat's canines were inserted using a head-swinging stabbing motion or a variant of the clamp and hold technique commonly used by extant conical-toothed cats; whether its bite was relatively powerful or weak; the respective contributions of jaw adductor versus cervical musculature in its application; and whether the bite was typically applied to the preys' throat or belly.

In the first application of FEA to a fossil mammalian carnivore skull, McHenry *et al.* (2007) examined these questions using a comparative approach to FEMs of *S. fatalis* and *Panthera leo* (African lion), with both subject to equivalent intrinsic and extrinsic loadings. Estimates for body mass were 229 and 267 kg, respectively. In addition to loads calculated on the basis of muscle cross-sectional areas, a further loading case was performed for the sabrecat in which it was assumed that *S. fatalis* generated bite forces comparable to those of a 'typical' 229 kg felid using data for the regression of bite force on body mass (Wroe *et al.*, 2005). These models were further refined through the addition of simulated cervical musculature to allow the modelling of neck-driven as well as jaw-muscle-driven bites.

Results of this study showed that under loadings simulating the influence of struggling prey, the sabrecat's skull showed much higher stress than in the *P. leo* model for lateral and torsional forces. This strongly suggested that *S. fatalis* was not well-adapted to bite into unrestrained prey animals. Regarding bite force, the application of two-dimensional methods had previously suggested a relatively weak bite for the sabrecat (Wroe *et al.*, 2005). Finite element modelling of intrinsic loadings showed that when 3D muscle geometry was considered, the sabrecat's jaw adductor-driven bite force was weaker still, around 33% of that of the lion's. This indicates that differences in in-lever geometry, as well as lesser muscle cross-sectional area, contributed to low jaw adductor-driven bite force in the sabrecat. It was further demonstrated that if *S. fatalis* could develop the jaw muscle forces that would be expected for a typical felid of its size, its cranium and mandible would exhibit far higher stresses than did the lion FEM. Consistent with these findings, another FE study, concentrating on the mechanics of carnivoran teeth, showed that the strength of sabrecat's canines was considerably lower than would be expected in a same-sized conical-toothed cat (Freeman and Lemen, 2007). However, results of modelling by McHenry *et al.* (2007), wherein jaw adductors were augmented by cervical musculature in the dorsoventral plane, showed that the skull of *S. fatalis* was well-adapted to deploy these muscle groups simultaneously (Figure 15.2).

Considered together, and in the context of data demonstrating great postcranial strength and stability in *S. fatalis* (Anyonge, 1996; Wroe *et al.*, 2008a), these findings suggest an intriguing mix of power and precision in the sabrecat's kill technique. The most parsimonious interpretation of the FE data is that application of the bite was most likely constrained to a dorsoventral motion. Furthermore, these data are consistent with the argument that the 'bite' was augmented by activation of the head-depressing musculature. In order to achieve such a bite, prey would have to have been robustly secured so as to avoid damage from laterally directed loadings to the sabrecat's own skull and teeth. McHenry *et al.* (2007) concluded that the bite was most probably directed at the prey's throat, because it is easier to restrain struggling prey by the head and neck than by the abdomen.

(a)

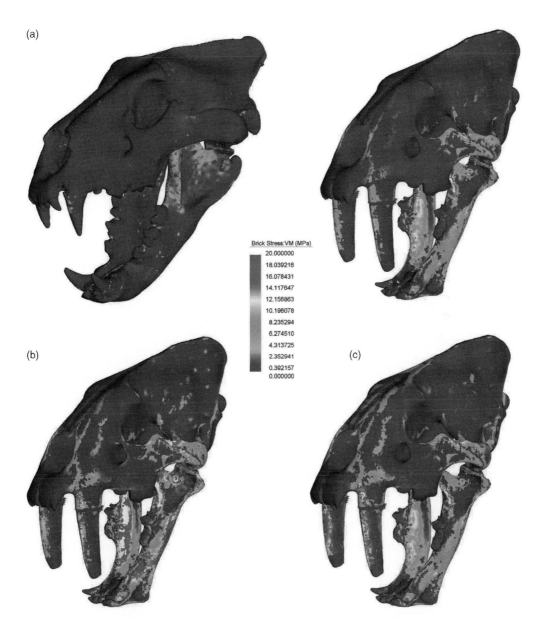

(b)

(c)

Brick Stress:VM (MPa)
20.000000
18.039216
16.078431
14.117647
12.156863
10.196078
8.235294
6.274510
4.313725
2.352941
0.392157
0.000000

Figure 15.2 For colour version, see Plate 15. Von Mises stress under intrinsic loads (bilateral canine bites). a, Bite force predicted by 3D dry skull method, adjusted to account for pennation; shown are lion biting at 3388 N (*Left*) and *S. fatalis* biting at 1104 N (*Right*). b and c, *S. fatalis* biting at the forces calculated using regression of bite force on body mass for a 229-kg felid (2110 N), powered by jaw adductors only (b) and by neck + jaw muscles (c). From McHenry *et al.* (2007).

A comparison of cranial mechanics in marsupial and placental lions

Few species have roused greater controversy or more disparate interpretation of feeding behaviour and ecology than has Australia's largest mammalian carnivore, the extinct Pleistocene marsupial lion, *Thylacoleo carnifex*. Although first described as an active predator (Owen, 1859), subsequent construal suggested habits ranging from scavenging to oophagy, and even specialisation on native cucumbers (Flower, 1868; De Vis, 1883; Anderson, 1929). However, the cheek dentition of *T. carnifex* has none of the adaptations that would be expected in either a specialist osteophage, oophage, or herbivore (Wroe *et al.*, 1998; Wroe, 2002).

Although serious debate over whether *T. carnifex* was carnivorous was effectively silenced following publication of a seminal work by Wells *et al.* (1982), its most unusual craniodental morphology left many questions open. *T. carnifex* possessed the hallmark of a terrestrial mammalian hypercarnivore, i.e. a highly carnassialised cheek dentition (Wroe *et al.*, 1998). Indeed, in relative terms its carnassial teeth are the largest of any mammal (Savage, 1977). Moreover, using 2D methodology, Wroe *et al.* (2005) determined that the species was also able to generate extraordinarily high bite forces, and both this and a latter study (Christiansen and Wroe, 2007) have shown that high bite forces, when normalised for body mass allometry, correlate with predation on relatively large prey. However, *T. carnifex*, like all thylacoleonids, lacked the large canine teeth that are so conspicuous in other mammalian carnivores. Its anterior toothrow was instead dominated by large, tusk-like incisors. This absence of well-developed canines has underpinned historical scepticism over the animal's prowess as a predator, but it has remained difficult to reconcile a capacity for extreme bite force with a lack of dental specialisation toward bone-cracking and an inability to kill large prey. If *T. carnifex* was not well-adapted to cracking bones, killing large prey, or processing plant material, then to what end were its massive jaw muscles applied?

Unlike other mammalian carnivores, the rostrum of *T. carnifex* narrows markedly and abruptly anterior to the carnassial teeth. Wroe *et al.* (2005) suggested that this, in combination with high bite forces, may have allowed the marsupial lion to apply a unique killing technique, wherein its greatly hypertrophied carnassials were used to scissor through soft tissue and effect the rapid dispatch of large prey, rather than being used only in the butchery of carcasses as in other mammalian carnivores. To further examine questions surrounding bite force and kill technique in *T. carnifex*, Wroe (2008) assembled a 3D FE model of a marsupial lion skull and compared its mechanical behaviour to that of an African lion (*Panthera leo*), under four intrinsic and four extrinsic loadings.

3D analysis confirmed the proposition that *T. carnifex* was able to produce high bite forces for an animal of its size. Although the small marsupial lion used in this study was around only 31% of the African lion's body mass, it generated a bite reaction force that was 80% of the *P. leo* model, suggesting high mechanical efficiency as well as high muscular force.

Analysis of stress distributions under intrinsic loadings showed that the marsupial lion's cranium was well-optimised to resist high forces during unilateral biting at the carnassial teeth, but less well-adapted than *P. leo* to resist forces generated in biting at the anterior, caniniform teeth (Figure 15.3). Under extrinsic loadings posterior regions of the marsupial lion's cranium received comparable or considerably lower stress than did that of *P. leo* (Figure 15.4). Although these FE-derived results did not prove the proposition that *T. carnifex* actively dispatched large prey through purposeful employment of its carnassial teeth, they are consistent with it.

Computer simulation of feeding behaviour in the thylacine and dingo

Convergence is a key theme in the study of evolution (Winter and Oxnard, 2001). Among mammalian carnivores, the thylacine, or marsupial wolf (*Thylacinus cynocephalus*) and its proposed placental counterpart, the grey wolf (*Canis lupus*), are a commonly cited supporting example. However, the degree to which apparent morphological similarity might reflect functional parallels in these two taxa remains an open question, and different approaches have yielded conflicting results (Werdelin, 1986; Johnson and Wroe, 2003; Jones, 2003; Wroe and Milne, 2007). Some analyses of skull morphology (Werdelin, 1986) and canine tooth breakage rates (Jones, 2003) have suggested that the thylacine specialised on small prey; geometric morphometric analysis (Wroe and Milne, 2007) has suggested specialisation on medium-sized prey; while estimates of bite force (Wroe *et al.*, 2005) and canine shape (Jones, 2003) have indicated that the thylacine was capable of taking relatively large animals.

In addition to shedding light on the question of convergence, gauging the degree of functional similarity between these two species can help to clarify the role of competitive exclusion in the extinction of mainland thylacines (Johnson and Wroe, 2003). This is because a subspecies of grey wolf (*Canis lupus dingo*) has been widely implicated in the disappearance of mainland thylacines around 3000 years ago. Wroe *et al.* (2007a) assembled and compared FEMs of both *T. cynocephalus* and *C. l. dingo* to address these questions.

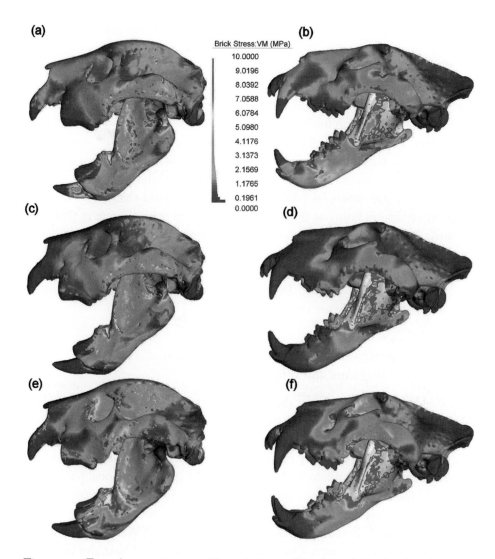

Figure 15.3 For colour version, see Plate 16. Stress (Von Mises) distributions in lateral views for intrinsic loading cases in *Thylacoleo carnifex* (a, c, e) and *Panthera leo* (b, d, f): (a, b) bilateral bite at canines; (c, d) bilateral bite at carnassials; and (e, f) unilateral bite at left carnassial. MPa = mega pascals. From Wroe (2008).

The results of this study revealed many broad similarities with respect to the general distribution of stresses in crania of both taxa under intrinsic and extrinsic loadings (Figure 15.5). For example, in both models, under intrinsic loadings, mean brick stresses were higher for canine bites than carnassial bites, lowest in the rostrum for both bilateral and unilateral bites, and lower on the balancing than working sides for unilateral loadings. FE analyses simulating

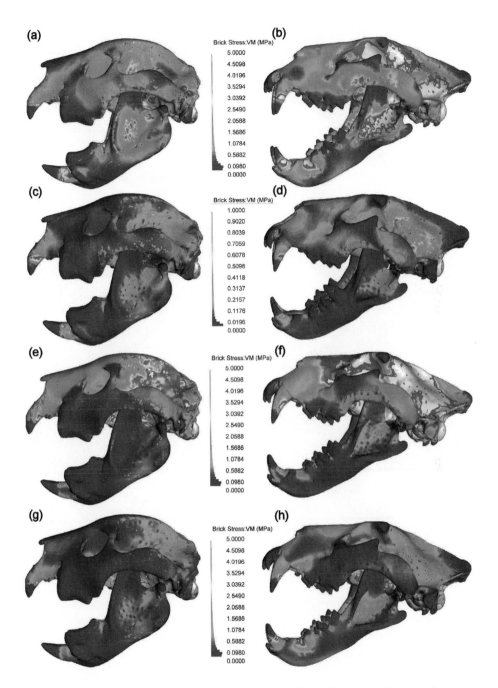

Figure 15.4 For colour version, see Plate 17. Stress (Von Mises) distributions in lateral views for extrinsic loading cases in *Thylacoleo carnifex* (a, c, e, g) and *Panthera leo* (b, d, f, h): (a, b) lateral shake; (c, d) axial twist; (e, f) dorsoventral head depression; and (g, h) pull-back. MPa = mega pascals. From Wroe (2008).

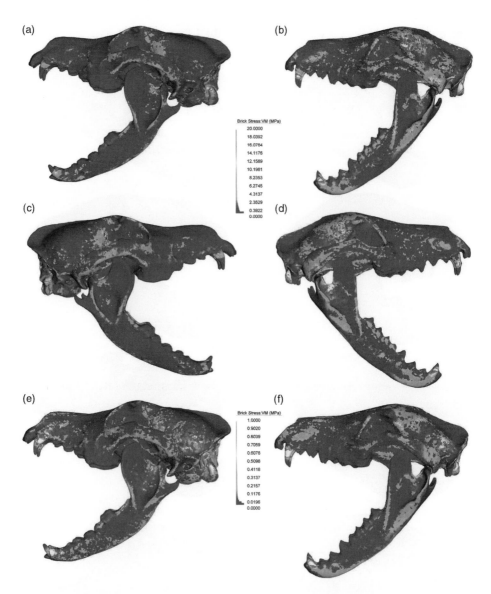

Figure 15.5 For colour version, see Plate 18. Stress (Von Mises) distributions in lateral views of FE models of *Canis lupus dingo* (a–c) and *Thylacinus cynocephalus* (e–g) under two load cases: (a, d) lateral shake (left); (b, e) lateral shake (right); and (c–f) axial twist. MPa = mega pascals. From Wroe *et al.* (2007).

the influences of struggling prey showed that for both thylacine and dingo the posterior of the cranium exhibited the greatest stress, while the highest mean stresses were under dorsoventral loadings, and the lowest stresses were under 'pull-back' loadings.

There were, however, notable differences in the mechanical behaviours of both models. 3D modelling further supported the contention that the thylacine had a powerful bite, but also that mean brick stress was higher in the marsupial under most intrinsic loadings, particularly in the mandible. Under extrinsic loads, mean brick stresses were consistently higher in the posterior cranium of the thylacine, with the exception of axial twisting.

Although mean brick stresses were higher in the thylacine under most intrinsic loadings, all else being equal, this might be expected given that the marsupial generated considerably greater bite force. Safety factors might be lower in *T. cynocephalus*, or it may be that the relationship between muscle cross-sectional area and muscle force differs between marsupials and placentals. However, results for extrinsic loadings also suggest that the thylacine was less well-adapted to take large prey, which is likely to involve the imposition of considerable, and often erratic forces on the predator's skull. The likelihood that *C. l. dingo* was better equipped to kill relatively large animals is further compounded by its capacity for social hunting, which was probably limited in *T. cynocephalus* (Corbett, 1995; Paddle, 2000).

Another possibility is that the thylacine may have required more bite force to kill equivalent-sized prey because it used less precise or less sophisticated predation techniques, relying on repeated, less-well directed bites to immobilise prey, as has been reported for some extant carnivorous marsupials (Ewer, 1969).

Regarding competition with the dingo as a cause for mainland extinction of *T. cynocephalus*, these results suggest that, once differences in size between the two predators are considered, there was considerable niche overlap, increasing the likelihood that the dingo played a role in the thylacine's demise.

Effects of gape and tooth position on bite force and skull stress

Theoretical modelling suggests strong correlations among bite force, gape angle, and tooth position (Herring and Herring, 1974; Pruim *et al.*, 1980; Greaves, 1982; Lindauer *et al.*, 1993). The closer the tooth is to the fulcrum (temporomandibular joint) and the shallower the gape angle, the greater the resultant bite force. However, few empirical studies have examined relationships between these factors in non-human species (Fields *et al.*, 1986), and none has considered their influence on the distribution of stress. It has been argued that the geometry of the mammalian carnivore cranium and its jaw closing musculature may be optimised to generate higher bite forces at wider gape angles than in herbivorous or omnivorous taxa (Dumont and Herrel, 2003).

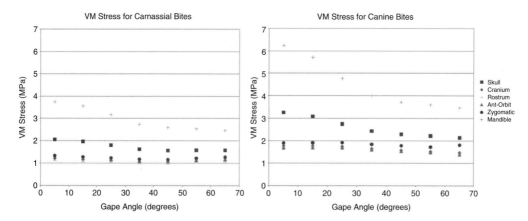

Figure 15.6 For colour version, see Plate 19. Comparison of mean 'brick' stress (Von Mises) distributions in selected cranial regions of *C. l. dingo* in bites at seven different gape angles. From Bourke *et al.* (2008).

To examine relationships between bite force, bite point, gape angle, and cranial stress, Bourke *et al.* (2008) analysed an FE model of the cranium and mandible of *Canis lupus dingo* at two different bite points (canines and carnassials) and seven gape angles, ranging from 5 to 65 degrees.

Bourke *et al.* (2008) found that maximal bite force was highest at the smallest gape angle (5°) as expected; however, they also found that the greatest increase in bite force occurred at 25° for a canine bite and 35° for a carnassial bite (Figure 15.6). Moreover, biting at shallower angles than these corresponded with marked increases in mean brick stress levels, suggesting that the dingo's masticatory apparatus may be optimised to bite at these gapes. It will be interesting to conduct such analyses on a variety of mammalian carnivores. If the skulls of different species are optimised to bite at different gape angles, this may offer a useful predictor of feeding behaviour and diet.

Future directions

To date, FE-based investigations into form–function relationships in the mammalian carnivore skull have applied a comparative approach without validation. This permits relative comparisons of performance, but models constructed so far cannot be used to predict absolute performance limits. So, at present, while we might reasonably deduce that the skull of one taxon is better or less well-adapted to perform a particular function than another, we cannot predict at what point material failure might occur, or estimate safety factors. Preliminary results of collaboration between the Computational

Biomechanics Research Group (CBRG) and P. O'Higgins and colleagues (Hull York Medical School) suggest close correlation for strain distributions and magnitudes between experimentally derived values and those obtained using FE modelling of a primate cranium assembled using the protocols described above. However, in-vitro and in-vivo studies of mammalian carnivore skulls are much needed. Further investigations will include material property testing, better correlation of relationships between muscle cross-sectional area and muscle force, as well as validation of strain magnitudes and distributions.

The reduction of model assembly times will allow broader reaching investigations into form and function. The CBRG has now constructed heterogeneous FE solids for eight extant and one extinct felid, with assembly of another six anticipated in coming months. This will enable the first FE based study of differences in mechanical behaviour across a large sample of species within a carnivoran family. The ability to rapidly generate FEMs will not only facilitate more detailed analyses of relationships between skull shape and function across taxa, but also better understanding of differences in mechanical behaviour through ontogeny, differences between sexes, and the influence of allometry.

Acknowledgements

Work was funded by Australian Research Council Discovery Grants (DP0666374, DP0987985), an Australia Pacific Science Foundation Grant and a University of New South Wales Internal Strategic Initiatives Grant (to SW). Special thanks to C. McHenry and P. Clausen; and to E. Cunningham and the Mater Hospital (Newcastle) for the provision of CT data.

REFERENCES

Akersten, W. (1985). Canine function in *Smilodon* (Mammalia, Felidae, Machairodontinae). *Los Angeles County Museum Contributions in Science*, **356**, 1–22.

Anderson, C. (1929). Palaeontological note no. 1. *Macropus titan* Owen and *Thylacoleo carnifex* Owen. *Records of the Australian Museum*, **17**, 35–45.

Anyonge, W. (1996). Locomotor behaviour in Plio-Pleistocene sabre-tooth cats: a biomechanical analysis. *Journal of Zoology (London)*, **238**, 395–413.

Bicknevicus, A. R. and Van Valkenburgh, B. (1996). Design for killing: craniodental adaptations of predators. In *Carnivore Behavior, Ecology and Evolution*, ed. J. L. Gittleman. Ithaca, NY: Cornell University Press, pp. 393–428.

Bourke, J., Wroe, S., Moreno, K., McHenry, C. and Clausen, P. (2008). Effects of gape and tooth position on bite force and skull stress in the dingo (*Canis lupus dingo*) using a 3-dimensional finite element approach. *PLoS ONE*, **3** (e2200).

Buckland-Wright, J. C. (1978). Bone structure and the patterns of force transmission in the cat skull (*Felis catus*). *Journal of Morphology*, **155**, 35–61.

Christiansen, P. (2007). Comparative bite forces and canine bending strength in feline and sabretooth felids: implications for predatory ecology. *Zoological Journal of the Linnean Society*, **151**, 423–37.

Christiansen, P. and Wroe, S. (2007). Bite forces and evolutionary adaptations to feeding ecology in carnivores. *Ecology*, **88**, 347–58.

Clausen, P., Wroe, S., McHenry, C., Moreno, K. and Bourke, J. (2008). The vector of jaw muscle force as determined by computer-generated three dimensional simulation: a test of Greaves' model. *Journal of Biomechanics*, **41**, 3184–88.

Corbett, L. (1995). *The Dingo in Australia and Asia*. Sydney: University of New South Wales Press.

Daniel, T. L., Helmuth, B. S., Saunders, W. B. and Ward, P. D. (1997). Septal complexity in ammonoid cephalopods increased mechanical risk and limited depth. *Paleobiology*, **24**, 470–81.

De Vis, C. W. (1883). On tooth-marked bones of extinct marsupials. *Proceedings of the Linnean Society of New South Wales*, **8**, 187–90.

Dumont, E. R. and Herrel, A. (2003). The effect of gape angle and bite point on bite force in bats. *The Journal of Experimental Biology*, **206**, 2117–23.

Dumont, E. R., Piccirillo, J. and Grosse, I. R. (2005). Finite-element analysis of biting behavior and bone stress in the facial skeletons of bats. *The Anatomical Record Part A*, **283A**, 319–30.

Ewer, R. F. (1969). Some observations on the killing and eating of prey by two dasyurid marsupials: the mulgara, *Dasycercus cristicauda*, and the Tasmanian devil, *Sarcophilus harrisii*. *Zeitschrift fur Tierpsychologie*, **26**, 23–38.

Fields, H. W., Proffitt, W. R., Case, J. C. and Vig, K. W. L. (1986). Variables affecting measurements of vertical occlusal force. *Journal of Dental Research*, **62**, 135–38.

Flower, W. H. (1868). On the affinities and probable habits of of the extinct Australian marsupial *Thylacoleo carnifex* Owen. *Quarterly Journal of the Geological Society*, **24**, 307–19.

Freeman, P. W. and Lemen, C. (2007). An experimental approach to modeling the strength of canine teeth. *Journal of Zoology*, **271**, 162–69.

Gould, S. J. (2002). *The Structure of Evolutionary Theory*. Cambridge, MA: Belknap Press of Harvard University Press.

Greaves, W. S. (1982). A mechanical limitation on the position of the jaw muscles in mammals: the one third-rule. *Journal of Mammalogy*, **63**, 261–66.

Grosse, I. R., Dumont, E. R., Coletta, C. and Tolleson, A. (2007). Techniques for modeling muscle-induced forces in finite element models of skeletal structures. *The Anatomical Record*, **290**, 1069–88.

Guillet, A., Doyle, W. S. and Ruther, H. (1985). The combination of photogrammetry and finite elements for a fine grained functional analysis of anatomical structures. *Zoomorphology*, **105**, 51–59.

Herring, S. W. and Herring, S. E. (1974). The superficial masseter and gape in mammals. *American Naturalist*, **108**, 561–76.

Hylander, W. L., Picq, P. G. and Johnson, K. R. (1991). Masticatory-stress hypotheses and the supraorbital region of primates. *American Journal of Physical Anthropology*, **86**, 1–36.

Jenkins, I., Thomason, J. J. and Norman, D. B. (2002). Primates and engineering principles: applications to craniodental mechanisms in ancient terrestrial predators. *Senckenbergiana Lethaea*, **82**, 223–40.

Johnson, C. N. and Wroe, S. (2003). Causes of extinction of vertebrates during the Holocene of mainland Australia: arrival of the dingo, or human impact? *The Holocene*, **13**, 941.

Jones, M. E. (2003). Convergence in ecomorphology and guild structure among marsupial and placental carnivores. In *Predators with Pouches: The Biology of Carnivorous Marsupials*, ed. M. E. Jones, C. Dickman and M. Archer. Collingwood: CSIRO Publishing, pp. 265–89.

Kupczik, K., Dobson, C. A., Fagan, M. J., Crompton, R. H., Oxnard, C. E. and O'Higgins, P. (2007). Assessing mechanical function of the zygomatic region in macaques:validation and sensitivity testing of finite element models. *Journal of Anatomy*, **210**, 41–53.

Kupczik, K., Dobson, C. A., Crompton, R. H., *et al.* (2008). Masticatory loading and bone adaptation in the supraorbital torus of developing macaques. *American Journal of Physical Anthropology*, **9999**, NA.

Lindauer, S. J., Gay, T. and Rendell, J. (1993). Effect of jaw opening on masticatory muscle EMG-force characteristics. *Journal of Dental Research*, **72**, 51–55.

McHenry, C. R., Clausen, P. D., Daniel, W. J. T., Meers, M. B. and Pendharkar, A. (2006). The biomechanics of the rostrum in crocodilians: a comparative analysis using finite element modelling. *Anatomical Record, Part A*, **288**, 827–49.

McHenry, C. R., Wroe, S., Clausen, P. D., Moreno, K. and Cunningham, E. (2007). Supermodeled sabercat, predatory behavior in *Smilodon fatalis* revealed by high-resolution 3D computer simulation. *Proceedings of the National Academy of Sciences (USA)*, **104**, 16010–15.

Moreno, K., Wroe, S., Clausen, P., *et al.* (2008). Cranial performance in the Komodo dragon (*Varanus komodoensis*) as revealed by high-resolution 3-D finite element analysis. *Journal of Anatomy*, **212**, 736–46.

Owen, R. (1859). On the fossil mammals of Australia. Part II. Description of an almost entire skull of the *Thylacoleo carnifex*, Owen, from a freshwater deposit, Darling Downs Queensland. *Philosophical Transactions of the Royal Society*, **149**, 309–22.

Paddle, R. (2000). *The Last Tasmanian Tiger: The History and Extinction of the Thylacine*. Cambridge: Cambridge University Press.

Preuschoft, H. and Witzel, U. (2004). A biomechanical approach to craniofacial shape in primates, using FESA. *Annals of Anatomy*, **186**, 397–404.

Pruim, G. J., De Jongh, H. J. and Ten Bosch, J. J. (1980). Forces acting on the mandible during bilateral static biting at different bite force levels. *Journal of Biomechanics*, **13**, 755–63.

Radinsky, L. B. (1981a). Evolution of skull shape in carnivores. II. Additional modern carnivores. *Biological Journal of the Linnean Society*, **16**, 337–55.

Radinsky, L. B. (1981b). Evolution of skull shape in carnivores. I. Representative modern carnivores. *Biological Journal of the Linnean Society*, **15**, 369–88.

Rayfield, E. J. (2005). Aspects of comparative cranial mechanics in the theropod dinosaurs *Coelophysis*, *Allosaurus* and *Tyrannosaurus*. *Zoological Journal of the Linnean Society*, **144**, 309–16.

Rayfield, E. J., Norman, D. B., Horner, C. C., *et al.* (2001). Cranial design and function in a large theropod dinosaur. *Nature*, **409**, 1033–37.

Richmond, B. G., Wright, B. W., Grosse, I., *et al.* (2005). Finite element analysis in functional morphology. *The Anatomical Record, Part A*, **283A**, 259–74.

Rybicki, E. F., Simonen, F. A. and Weis, E. B. (1972). On the mathematical analysis of stress in the human femur. *Journal of Biomechanics*, **5**, 203–15.

Savage, R. J. G. (1977). Evolution in carnivorous mammals. *Palaeontology*, **20**, 237–71.

Strait, D. S., Richmond, B. G., Spencer, M. A., Ross, C. F., Dechow, P. C. and Wood, B. A. (2007). Masticatory biomechanics and its relevance to early hominid phylogeny: an examination of palatal thickness using finite-element analysis. *Journal of Human Evolution*, **52**, 585–99.

Strait, D. S., Wang, Q., Dechow, P. C., *et al.* (2005). Modeling elastic properties in finite-element analysis: how much precision is needed to produce an accurate model? *The Anatomical Record, Part A*, **283A**, 275–87.

Therrien, F. (2005a). Mandibular force profiles of extant carnivorans and implications for the feeding behaviour of extinct predators. *Journal of Zoology*, **267**, 249–70.

Therrien, F. (2005b). Feeding behaviour and bite force of sabretoothed predators. *Zoological Journal of the Linnean Society*, **145**, 393–426.

Thomason, J. J. (1991). Cranial strength in relation to estimated biting forces in some mammals. *Canadian Journal of Zoology*, **69**, 2326–33.

Van Valkenburgh, B. (1989). Carnivore dental adaptations and diet: a study of trophic diversity within guilds. In *Carnivore Behaviour, Ecology and Evolution*, ed. J. L. Gittleman. New York, NY: Cornell University Press, pp. 410–33.

Verhulp, E. (2006). *Analyses of Trabecular Bone Failure*, vol. PhD. Eindhoven: University of Technology.

Verrue, V., Dermaut, L. and Verhegghe, B. (2001). Three-dimensional finite element modelling of a dog skull for the simulation of initial orthopaedic displacements. *European Journal of Orthodontics*, **23**, 517–27.

Warren, J. C. (1853). Remarks on *Felis smylodon*. *Proceedings of the Boston Society of Natural History*, **4**, 256–58.

Weijs, W. A. and Hillen, B. (1985). Cross-sectional areas and estimated intrinsic strength of the human jaw muscles. *Acta Morphology Neerlandiae-Scandinaviae*, **23**, 267–74.

Wells, R. T., Horton, D. R. and Rogers, P. (1982). *Thylacoleo carnifex* Owen (Thylacoleonidae): marsupial carnivore? In *Carnivorous Marsupials*, ed. M. Archer. Sydney: Royal Zoological Society of New South Wales, pp. 573–86.

Werdelin, L. (1986). Comparison of skull shape in marsupial and placental carnivores. *Australian Journal of Zoology*, **34**, 109–17.

Winter, W. D. and Oxnard, C. E. (2001). Evolutionary radiations and convergences in the structural organization of mammalian brains. *Nature*, **409**, 710–14.

Wroe, S. (2002). A review of terrestrial mammalian and reptilian carnivore ecology in Australian fossil faunas and factors influencing their diversity: the myth of reptilian domination and its broader ramifications. *Australian Journal of Zoology*, **49**, 603–14.

Wroe, S. (2008). Cranial mechanics compared in extinct marsupial and extant African lions using a finite-element approach. *Journal of Zoology (London)*, **274**, 332–39.

Wroe, S. and Milne, N. (2007). Convergence and remarkable constraint in the evolution of mammalian carnivore skull shape. *Evolution*, **61**, 1251–60.

Wroe, S., Brammall, J. and Cooke, B. (1998). The skull of *Ekaltadeta ima* (Marsupialia, Hypsiprymnodontidae) an analysis of some marsupial cranial features and a reinvestigation of propleopine phylogeny, with notes on the inference of carnivory in mammals. *Journal of Paleontology*, **72**, 735–51.

Wroe, S., McHenry, C. and Thomason, J. (2005). Bite club: comparative bite force in big biting mammals and the prediction of predatory behaviour in fossil taxa. *Proceedings of the Royal Society of London, Series B*, **272**, 619–25.

Wroe, S., Clausen, P., McHenry, C., Moreno, K. and Cunningham, E. (2007a). Computer simulation of feeding behaviour in the thylacine and dingo as a novel test for convergence and niche overlap. *Proceedings of the Royal Society of London, Series B*, **274**, 2819–28.

Wroe, S., Moreno, K., Clausen, P., McHenry, C. and Curnoe, D. (2007b). High resolution three-dimensional computer simulation of hominid cranial mechanics. *The Anatomical Record, Part A*, **290**, 1248–55.

Wroe, S., Lowry, M. B. and Anton, M. (2008a). How to build a mammalian super-predator. *Zoology*, **111**, 196–203.

Wroe, S., Huber, D., Lowry, M., *et al.* (2008b). Three-dimensional computer analysis of white shark jaw mechanics: how hard can a great white bite? *Journal of Zoology (London)*, **276**, 336–42.

Index

Page numbers in bold type refers to figures.

Acinonyx jubatus, 226, 304
Actiocyon, 115, 125
active replacement, 311
Adelphailurus, 411
Ailuridae, 11, 30, 92–126, 231, 304
   diagnosis, 116
Ailurinae, 92, 116–20
Ailuropoda, 15, 20, 31, 94, 104, 107, 226
Ailurus, 117, 126
Ailurus fulgens, **3**, **30**, **93**, 126
alisphenoid canal, 82, 107, 110
allometry, 43, 165, 168
   multiphasic, 172–6, 178–85
   postcranial, 411–59
Alopecocyon, 112, 113, 115, 122, 125
Alopex lagopus, 392
American lion, *see Panthera leo* cf. *atrox*
Amphictis, 30, 92, 109, 112, 115, 116,
      123–124
Amphicynodon, 109
Amphicynopsis, 297
Amphicyon major, 299
Amphicyonidae, 17, 34, 41, 142, 193, 295
Ancient DNA, 25
Andrewsarchus, 304
Aonyx, 236
aquatic species, 4, *see* Pinnipedia
ArcGIS, 379
Arctictis, 104, 261, 303
Arctictis binturong, **2**, 8, 79, 226
Arctocephalus australis, 345, 363
Arctocephalus gazella, **347**
arctocyonids, 270, 271, 273
Arctodus, 15
Arctogalidia, 261
Arctogalidia trivirgata, 75, 79

Arctoidea, 12, 27, 30, 31, 40, 94, 104
Artiodactyla, 312
Atilix paludinosis, 9

badgers, *see* Mustelidae
bamboo, 12, 94
Barbourofelinae, 10, 34, 295, 301
basicranium, 37, 65, 107, 155, 157, 158, 159
Bassaricyon, 11
Bassaricyon lasius, 377
Bassariscus, 231, 238
Bassariscus astutus, 388
Bathygale, 32
bear-dogs, *see* Amphicyonidae
behaviour, 411, 454
Bergmann's Rule, 396
biogeography, 225–39, 247–65, 361
biomechanics
   cranial, 466–81
   postcranial, 450–9
bite force, 466–81
body size, 39–43, 226, 248, 249, 269,
      270, 301, 314, 325, 330, 412, 413
bone cracking, 8, 16, 19, 289, 304
brain size, 39, 43–7
   lions, 168
Buxolestes, 271

C_4 grasslands, 237
Callorhinus ursinus, 345, 352, 363
Canidae, 16, 94, 192–203, 229, 230,
      233, 234, 235, 294, 314, 315,
      392, 393
   body size evolution, 41–3
   Borophaginae, 16, 41, 233
   brain size evolution, 46–7

Caninae, 16, 203, 233
  Hesperocyoninae, 16, 41, 203, 233
Caniformia, 1, 10, 27, 31, 94
  body size evolution, 40–1
  brain size evolution, 44
Canis, 468
Canis (Lycaon) pictus, 226, 231
Canis aureus, 295
Canis dirus, 177
Canis latrans, 392
Canis lupus, **30**, 233
carnassials, 4, 17, 20, 192, 196, 230
Carnivoraformes, 26, 36–8, 47
Carnivoramorpha, 4, 26–36
carnivoran
  definition of, 1
carnivore
  definition of, 1
Cephalogale, 109
character correlation, 37, 141
  effects on phylogenetic analyses,
    146–7, 158
Chasmaporthetes, 8, 203
cheetah, *see Acinonynx*
Chriacus, 303
Chrotogale owstoni, 68, 79
civets, *see* Viverridae
Civettictis, 264
Civettictis civetta, 65, 82, 257, 258
climate change, 375
competition, 194, 197, 311–35
competitive replacement, 311
condylarths, 303
constraint, 190, 192–4, 197, 361
  developmental, 213
  functional, 213
convergence, 19, 359–61, 475
Cope's Rule, 315, 332 *see* Canidae, body
    size evolution
correspondence analysis, 226, 230
Creodonta, 15, 17, 27, 271, 301, 311–35
  competition with Carnivora, 17
  Hyaenodontidae, 17, 270, 273, 275,
    295, 297, 301, 304, 314
  Limnocyoninae, 275
  Machairoidinae, 275

Oxyaenidae, 17, 270, 273, 289, 314
  Proviverrinae, 275
Crocuta crocuta, **2**, **30**
Cryptoprocta ferox, **2**, 9, 66, 67, 203, 204,
    226, 231, 233, 234
Cynogale bennettii, 78
Cynogale lowei, 78
Cynoidea, *see* Canidae
Cystophora cristata, 352, 353, 356, 359,
    360, 362, 363

development, 142, 143, 145, 172, 174, 176,
    190, 213, 362, 363
Diatryma, 304
Didymictis, 47, 49
Didymictis protenus, **49**
didymoconids, 270, 271
diet, 226, 269, 270, 271, 314, 342, 466
dimorphism, 345, 352, 356, 361, 363–4
dingo, 475–9
Dinictis, 329
Dinobastis, 412, 418, 433, 434, 435, 453,
    455, 456, 459
Dinobastis serus, 443
Dinofelis, 412, 427, 429, 430, 434, 453,
    454, 456, 458, 459
Dinofelis barlowi, 441, 442, 453
Diplogale hosei, 79
directional selection, 197, 212
dirk-toothed cats, 412
disparity, 225, 227, 233–4, 238, 239,
    246, 247–65, 316
dog-bears, *see* Ursidae, Hemicyoninae
Dollo's Law, 190

ecological replacement, 334
ecomorphology, 18, 192, 225, 247–65,
    269, 299, 311–35, 359, 362,
    374–97, 453
ecoregions, 382
elevation, 381
Enaliarctos, 13
Enaliarctos mealsi, 344
encephalisation, *see* brain size
Enhydra lutris, 231, 236, 385
Enhydrocyon, 202, 203

Erignathus barbatus, 14, 342, 353, 359, 360, 362
Eupleres goudotii, 9, **30**, 66, 68, 204,
        231, 233, 234
Eupleridae, 9, 31, 67–72, 229, 231, 233, 234
    Euplerinae, 70
    Galidiinae, 70
evolutionarily stable systems, 190
evolutionary ratchet, 212
evolutionary rates, 195, 197, 205–15
exaptation, 114

false sabre-toothed cats, *see* Nimravidae
false-thumb, 114
Felidae, 7, 194, 203, 229, 230, 231, 233,
        234, 235, 248, 294, 297, 393
    Felinae, 411
    Machairodontinae, 7, 411–59
        Homotheriini, 412
        Metailurini, 411
        Smilodontini, 412
Feliformia, 1, 6, 27, 64, 247
Feloidea, 27
Finite Element Analysis, 466–81
Fissipedia, 34
Fossa fossana, 66, 68, 204
fossil record, 4
    ancestral state reconstruction, 39, 198
    phylogenetic analyses, 25
fur seals, *see* Otariidae

Galerella, 256, 261
Galerella pulverulenta, 258
Galerella sanguinea, 258
Galictis, 294, 301
Galicitis cuja, 295
Galidia elegans, 69
Galidictis fasciata, 69
Galidictis grandidieri, 69
Genetta, 78, 80, 254, 258, 261, 301
Genetta abyssinica, 81, 257, 260
Genetta angolensis, 81
Genetta bourloni, 81
Genetta cristata, 81
Genetta felina, 81
Genetta genetta, 65, 81, 295
Genetta johnstoni, 81

Genetta maculata, 81
Genetta pardina, 81
Genetta piscivora, 80, 81
Genetta poensis, 81
Genetta servalina, 81
Genetta tigrina, 81, 255
Genetta victoriae, 81
Genetta thierryi, 81
geometric morphometrics, 147, 345
Giant panda, *see Ailuropoda*
Great American Biotic Interchange,
        7, 10, 12, 16
guild structure, 269–305
Gulo diaphorus, 121
Gulo gulo, **30**

habitat selectivity, 265
Halichoerus grypus, 353, 354, 359, 360, 362
Harpagolestes, 304
Helogale, 294
Hemicyon, 299
Hemigalus, 261
Hemigalus derbyanus, 79
Herpestes, 256, 261, 264, 294
Herpestes ichneumon, 256, 258
Herpestes vitticollis, 255
Herpestidae, 8, 31, 64, 70, 203, 229,
        234, 238, 247–65, 294, 315
    Herpestinae, 70
    Mungotinae, 70
Herpestides, 77
Herpestoidea, 27
Homotherium, 25, 412, 418, 427, 428, 430,
        433, 434, 435, 440, 441, 443, 453,
        454, 457, 459
Hyaenictis, 203
Hyaenidae, 8, 31, 193, 203, 229, 231, 233,
        234, 289, 294
Hyaenodon, 325, 329, 331, 332
Hyaenodon gigas, 304
Hyainailouros, 304
Hydrurga leptonyx, 14, 342, 352–3, 354, 359
hypercarnivory, 19, 20, 192–215, 230, 254,
        258, 260, 271, 315
hypocarnivory, 230, 254, 255, 271, 304
hypsodonty, 396

Ichneugale, 122
Ictonyx, 260
independent contrasts, 351
insectivory, 271, 315
Ischyrictis, 297, 299, 301

Laurasiatheria, 5
Leptonychotes weddelli, 354
Leptoplesictis, 299
Limnocyon, 330
Limnocyonidae, 314
linsang, *see* Prionodontidae
Lobodon carcinophagus, 14, 342, 354
locomotion, 4, 19, 39, 269, 270, 271,
        302, 374–97, 411 *see* postcranial
        morphology
    digitigrade, 376, 379
    plantigrade, 376, 379
    semidigitigrade, 379
Lokotunjailurus, 454
Lontra canadensis, **3**, **30**
Lophocyoninae, 303
Lutrinae, 235
Lycyaena, 202, 203
Lynx canadensis, 388
Lynx rufus, **30**, 388

Machaeroides, 297, 301, 325
Machairodus, 412, 454
Macrogalidia musschenbroekii, 75
Madagascar, 9
Magerictis, 116, 119, 125
Magerictis imperialis, 112, 115
Malagasy carnivorans, *see* Eupleridae
Malagasy mongooses, 69, *see* Eupleridae
marsupial carnivores, 18
    Borhyaenoidea, 18
Martes, 235, 297
Martes americana, 388
Megantereon, 412, 444, 447, 448, 454,
        458, 459
Megaviverra, 264
Megistotherium, 304, 333
Megistotherium osteothlastes, 17
Meles, 104, 235
Melursus, 231

Mephitidae, 12, 30, 229, 231, 233
Mephitis mephitis, **3**, **30**, 388
Mesonychia, 270, 271, 273, 289
Mesonyx, 304
Metailurus, 412
Miacidae, 297, 301, 314 *see* Miacoidea
Miacis, 38, 47, 330
Miacis uintensis, **49**
Miacoidea, 5, 34–6, 37 *see* Carnivoraformes
Mimocyon, 325
Miocyon, 329, 330, 332
Miracinonyx, 25
Mirounga leonina, 40, 356, 359, 360,
        362, 363
modularity, 37, 141, 142–6, 159
molecular phylogenetics, 4, 26, 31, 33, 66,
        93–100
mongooses, *see* Herpestidae
Monte Carlo simulations, 150
morphological bias, 189–215
morphological diversity, 194
morphological integration, 141, 142–6, 150,
        156, 190, 213
morphospace occupation, 195, 227, 232,
        234–8, 316, 353, 358, 361
Mungos mungo, **2**, **30**
Mungotictis decemlineata, 69
Mustela, 202, 203, 235
Mustela frenata, 30
Mustela nivalis, 40, 226
Mustelictis, 109
Mustelidae, 10, 28, 30, 32, 104, 193, 203,
        229, 231, 235, 247, 294, 297, 301,
        315, 334, 393
    relationships, 10, 32 *see* Musteloidea
Musteloidea, 10–12, 27, 30, 95, 115
Mydaus, 104, 231

Nandinia binotata, **30**, 66–7
Nandiniidae, 9, 30, 231
Nasua, 11
Nasua narica, **3**
Neofelis, 447
Nimravidae, 9, 15, 34, 142, 194, 203
non-metric multidimensional scaling
        (NMDS), 324

Odobenidae, 13, 30, 342
    Dusignathinae, 13
    Odobeninae, 13
    suction feeding, 13
Odobenus rosmarus, **3**, 342, 353
Ommatophoca rossi, 354
ontogeny, 345, 352, 354, 362
Oödectes, 38, 330
Oödectes herpestoides, **49**
Otaria byronia, 345, 353, 356, 360, 361, 362, 363
Otariidae, 13, 14, 30, 342
    Arctocephalinae, 14
    Otariinae, 14
Otocyon megalotis, 231, 234
otters, see Mustelidae, Lutrinae
Oxyaena, 330

Pachycrocuta, 235
Paguma, 261
Paguma larvata, 79
Palaearctonyx, 297
palaeobiogeography, 5–17, 344
palaeoclimate, 269
palaeoenvironmental reconstruction, 269,
    374–97
Palaeonictis, 329
Paleomustelidae, 32
pangolins, see Pholidota
Panthera, 165, 459
Panthera leo, **2**, 294, 451–3, **470**, 471–2,
    473, 474–5, **476**
Panthera leo cf. atrox, 165–85, **470**
Panthera onca, 295
Panthera pardus, 295
Panthera spelea, 165
Panthera tigris, 451–3
Pantherinae, 291
pantolestids, 270, 271, 273, 304
Paracynictis, 261
Paradoxurus, 79, 261
Paradoxurus hermaphroditus, 79
Paradoxurus jerdoni, 79
Paradoxurus zeylonensis, 80
Parailurus, 92, 111, 112, 116, 117–19
Paralutra, 295, 299
paroxyclaenids, 303

passive replacement, 312
Patriofelis, 304, 329, 331
Percrocuta, 299
Percrocutidae, 289
perineal gland, 65, 75, 78
Perissodactyla, 312
Philotrox, 203
Phoca vitulina, **30**
Phocidae, 13, 14, 30, 342
    Monachinae, 14
    Phocinae, 14
Pholidota, 5, 27
phylogenetic inertia, 423
Pinnipedia, 12, 27, 30, 41, 342–65
    relationships, 12, 14, 343
Pinnipedimorpha, 27
Plesictis, 32
Pliocrocuta, 235
Poecilogale, 260
Poiana, 65, 75, 80, 258, 261
Poisson's ratio, 467
postcranial morphology, 47–51, 415,
    see locomotion
Potos, 11, 104, 303
Potos flavus, **30**
precipitation, 381
prehensile tail, 4, 11
prey acquisition, 411, 454
Primates, 312
principal component analysis, 269
principal components analysis, 195, 272,
    349, 418
principal coordinates analysis, 153, 158
Prionodon linsang, 66
Prionodon pardicolor, 66
Prionodontidae, 31, 68, 72–5, 231
Pristinailurus, 112, 116, 119, *126*
Procynodictis, 38, 47
Procyon, 104
Procyon gloveralleni, 377
Procyonidae, 11, 31, 94, 95, 104, 112, 229,
    231, 235, 238, 294, 303, 315, 393
Proputorius, 297, 299
Proteles cristata, 8, 226, 231, 234
Protictitherium, 299
Protursus, 113, 121, 125

Pseudaelurus, 297, 299
Puijila darwini, 13

Quercitherium, 304
Quercygale, 38

raccoons, *see* Procyonidae
radial sesamoid, *see* false-thumb
Rangifer tarandus, 391
Red pandas, *see* Ailuridae
relationships, *see* fossil record, phylogenetic
    analyses, molecular phylogenetics
  Carnivora, 25
  placental superorders, 4
  total evidence, 26, 31
reproductive strategies, 342
Rodentia, 312

sabre-toothery, 7, 19, 413
Salanoia concolor, 69
Sansanosmilus, 297, 301
Sarkastodon, 304
Scimitar-toothed cats, 412
sea lions, *see* Otariidae
seals, *see* Phocidae
Semigenetta, 77, 297
short-faced bear, *see* Arctodus
Simocyon, 112, 113, 116, 121, 125
Simocyon batalleri, 114
Simocyoninae, 11, 30, 113–16, 120–3
Sinopa rapax, 272
skunks, *see* Mephitidae
Smilodon, 25, 141, 412, 427, 430, 434, 443,
    444, 447, 449, 453, 454, 458, 459
Smilodon fatalis, **2**, 177, **470**, 471–2, **473**
Smilodon gracilis, 429, 436
specialisation, 190–2, 362
speciation, 246
species richness, 387
spectacled bear, *see* Tremarctos
Speothos, 295
Spilogale pygmaea, 377
stabilising selection, 190, 194, 197, 212
stem carnivorans, *see* Carnivoramorpha
stink badgers, *see* Mephitidae
strain, 467

stress, 467
Sunkahetanka, 203
Suricata suricatta, 9

taphonomy, 302
Tapocyon, 38, 47
Tapocyon robustus, **49**
Taxidea, 235
taxon-free analysis, 247, 269
temperature, 375, 381
Teratodon, 304
Thylacinus cynocephalus, 475–9, *see* marsupial
    carnivores
Thylacoleo carnifex, 474–5, **476**, *see* marsupial
    carnivores
Tremarctos, 15
Tremarctos ornatus, 236
Tritemnodon, 297, 301, 330
Trocharion, 295
Trochictis, 299
Trochotherium, 295

Uintacyon, 38, 301
Uintacyon major, 297
Urocyon cinereoargenteus, 388
Ursidae, 12, 15, 30, 31, 41, 94, 95, 193,
    229, 231, 295, 393
  Amphicynodontinae, 15
  Hemicyoninae, 15
  polar bear, 15, 226
  sloth bear, 15
  Ursinae, 15
Ursus americanus, **30**
Ursus arctos, **3**, 237, 304
Ursus spelaeus, 25

vegetation cover, 382
Viretius, 122
Viverra, 78, 254, 261, 264
Viverra bakeri, 264
Viverra civettina, 82
Viverra durandi, 264
Viverra howelli, 264
Viverra leakeyi, 264
Viverra leptorhyncha, 122
Viverra megaspila, 82

Viverra tainguensis, 83
Viverra tangalunga, 82
Viverra zibetha, **30**, 82
Viverravidae, 5, 34–6, 38, 47, 275, 295, 313
    locomotion, 49–51
Viverravus, 49
Viverravus acutus, 6, **49**
Viverravus gracilis, 37
Viverricula, 261, 264
Viverricula indica, 67, 82
Viverridae, 7, 28, 31, 104, 193, 229, 231,
        233, 234, 238, 247–65, 294, 301,
        303, 315, 334
    Genettinae, 75, 78, 80
    Hemigalinae, 66, 75, 77, 78

Paradoxurinae, 66, 75, 77, 79
    relationships, 64–83
    Viverrinae, 66, 75, 77, 81
Vulpavus, 38, 47, **49**, 297
Vulpes vulpes, **3**, 392
Vulpes zerda, 233

walrus, *see* Odobenidae
weasels, *see* Mustelidae

Xenosmilus, 412

Young's modulus, 467

Zalophus californianus, **3**, **30**